Lecture Notes in Computer Science 2701

Edited by G. Goos, J. Hartmanis, and J. van Leeuwen

Springer
Berlin
Heidelberg
New York
Barcelona
Hong Kong
London
Milan
Paris
Tokyo

Martin Hofmann (Ed.)

Typed Lambda Calculi and Applications

6th International Conference, TLCA 2003
Valencia, Spain, June 10-12, 2003
Proceedings

Springer

Volume Editor

Martin Hofmann
Universität München
Institut für Informatik
Oettingenstr. 67, 80538 München, Germany
E-mail: mhofmann@informatik.uni-muenchen.de

Cataloging-in-Publication Data applied for

A catalog record for this book is available from the Library of Congress.

Bibliographic information published by Die Deutsche Bibliothek
Die Deutsche Bibliothek lists this publication in the Deutsche Nationalbibliografie; detailed bibliographic data is available in the Internet at <http://dnb.ddb.de>.

CR Subject Classification (1998): F.4.1, F.3, D.1.1, D.3

ISSN 0302-9743
ISBN 3-540-40332-9 Springer-Verlag Berlin Heidelberg New York

Springer-Verlag Berlin Heidelberg New York
a member of BertelsmannSpringer Science+Business Media GmbH

http://www.springer.de

Printed in Germany

Typesetting: Camera-ready by author, data conversion by PTP-Berlin GmbH
Printed on acid-free paper SPIN: 10928660 06/3142 5 4 3 2 1 0

Preface

This volume represents the proceedings of the 6th International Conference on Typed Lambda Calculi and Applications, TLCA 2003, held in Valencia, Spain on 10–12 June 2003 in conjunction with CADE and RTA. It contains 21 contributions which were selected from 40 submissions. Three invited talks by D. McAllester, G. Gonthier, and R. Loader are not included in this volume.

The editor wishes to thank the members of the Program Committee and the referees for their help in putting together a very attractive program.

April 2003 Martin Hofmann

Program Committee

A. Asperti
(University of Bologna)
T. Coquand
(Chalmers University, Göteborg)
V. Danos
(University Paris VII)
M. Hofmann
(Chair, University of Munich)
P.-A. Mellies (CNRS, Paris)
J. Palsberg
(Purdue University)
H. Schwichtenberg
(University of Munich)
N. Shankar
(SRI International, Menlo Park)
P. Urzyczyn
(Warsaw University)

Steering Committee

S. Abramsky
(Chair, Oxford University)
M. Dezani-Ciancaglini
(University of Torino)
H. Barendregt
(University of Nijmegen)
R. Hindley
(University of Swansea)

Referees

A. Abel
T. Altenkirch
A. Asperti
P. Baillot
G. Barthe
M. Benke
N. Benton
U. Berger
J. Chrzaszcz
A. Compagnoni
T. Coquand
R. di Cosmo
P.-L. Curien
V. Danos
R. Dyckhoff
T. Ehrhard
C. Faggian
C. Fuehrmann
H. Geuvers
J. Goubault
P. de Groote
R. Hasegawa
H. Herbelin
M. Hofmann
G. Hutton
B. Jay
F. Joachimski
J.B. Joinet
A.J. Kfoury
U. Dal Lago
F. Lamarche
H. Leiss
T. Loew
J. Longley
P. Manoury
S. Martini
R. Matthes
M. Mauny
J. McKinna
P.A. Mellies
S. Owre
J. Palsberg
R. Pollack
S. Prasad
L. Regnier
E. Robinson
P. Rosolini
H. Ruess
C. Sacerdoti Coen
A. Schubert
H. Schwichtenberg
P. Scott
P. Selinger
N. Shankar
R. Staerk
T. Streicher
K. Terui
T. Uribe
P. Urzyczyn
T. Weber
M. Zaionc

Table of Contents

Termination and Productivity Checking with Continuous Types

Andreas Abel*

Department of Computer Science, University of Munich
Oettingenstr. 67, 80538 München, Germany
abel@informatik.uni-muenchen.de

Abstract. We analyze the interpretation of inductive and coinductive types as sets of strongly normalizing terms and isolate classes of types with certain continuity properties. Our result enables us to relax some side conditions on the shape of recursive definitions which are accepted by the type-based termination calculus of Barthe, Frade, Giménez, Pinto and Uustalu, thus enlarging its expressivity.

1 Introduction and Related Work

Interactive theorem provers like Coq [13], LEGO [20] and Twelf [18] support proofs by induction on finite-depth (inductive) structures (like natural numbers, lists, infinitely branching trees) and infinite-depth (coinductive) structures (like streams, processes, trees with infinite paths) in the form of recursive programs. However, these programs constitute valid proofs only if they denote *total* functions. In the last decade, considerable effort has been put on the development of means to define total functions in the type theories of the abovementioned theorem provers.

The first solution was to restrict programs to specific recursion schemes like iteration or primitive recursion (like `natrec`). Since these schemes come with complicated reduction behavior, they were inconvenient to use and so the search started how to incorporate the simpler *general recursion* (fix resp. letrec) known from functional programming. Programs using fix are in general not *terminating* (resp. *productive*, for the case of infinite structures), hence could denote partial or undefined functions. Thus, the use of fix has to be restricted in some manner.

Static analysis of the program. The first approach is to check the *code* of a recursive program to make sure some argument is decreasing in each recursive call (cf. Abel and Altenkirch [1,4], Pientka [19], Lee, Jones and Ben-Amram [14]). For programs defining infinite objects like streams, the recursive calls need to be *guarded*. Guardedness checks have been devised e.g. by Coquand [8], Telford and Turner [22] and Giménez [9].

* Research supported by the *Graduiertenkolleg Logik in der Informatik* (PhD Program Logic in Computer Science) of the Deutsche Forschungsgemeinschaft (DFG). The author thanks Martin Hofmann and Ralph Matthes for helpful discussions.

M. Hofmann (Ed.): TLCA 2003, LNCS 2701, pp. 1–15, 2003.

$$
\begin{array}{l}
\mathsf{pivot} : \mathsf{int} \to \forall Y \approx \mathsf{int\ list}.\, Y \to Y \times Y \\[4pt]
\mathsf{pivot}\ a\ [\,]^{Y1} \qquad\qquad = ([\,]^{Y1},\ [\,]^{Y1}) \\
\mathsf{pivot}\ a\ (x :: \mathsf{xs}^Y)^{Y1} \ \ = \mathsf{let}\ (l^Y,\ r^Y) = \mathsf{pivot}\ a\ \mathsf{xs}\ \ \mathsf{in} \\
\qquad\qquad\qquad\qquad\qquad \mathsf{if}\ a > x\ \ \mathsf{then}\ ((x :: l)^{Y1},\ r^{Y \leq Y1})\ \mathsf{else}\ (l^{Y \leq Y1},\ (x :: r)^{Y1}) \\[8pt]
\mathsf{qsapp} : \forall Y \approx \mathsf{int\ list}.\, Y \to \mathsf{int\ list} \to \mathsf{int\ list} \\[4pt]
\mathsf{qsapp}\ [\,]^{Y1}\ \ \mathsf{ys} \qquad\qquad = \mathsf{ys} \\
\mathsf{qsapp}\ (x :: \mathsf{xs}^Y)^{Y1}\ \ \mathsf{ys} = \mathsf{let}\ (l^Y,\ r^Y) = \mathsf{pivot}\ x\ \mathsf{xs}\ \ \mathsf{in}\ \ \mathsf{qsapp}\ l^Y\ \ (x :: \mathsf{qsapp}\ r^Y\ \ \mathsf{ys}) \\[8pt]
\mathsf{quicksort} : \mathsf{int\ list} \to \mathsf{int\ list} \\
\mathsf{quicksort}\ l = \mathsf{qsapp}\ l\ [\,]
\end{array}
$$

Fig. 1. Example: Quick-sort.

For programs like quicksort (cf. Fig. 1, ignore type annotations for now), input arguments are transformed *via another function* (here: pivot), before fed into the recursive call. Hence, analyses need to infer the size relation between input and output of a function, or at least detect non-size increasing functions like pivot. Well-designed static analyses [19,14,15,22] support some size checking for first-order programs, but fail for higher-order programs. These analyses usually have some other drawbacks: they are quite sophisticated and thus hard to comprehend and to trust, and they are sensitive to little changes in the program code: a program which passed a termination check might not pass any longer if a redex is introduced (cf. discussion in Barthe *et al.* [7]). Furthermore, they all fail for non-strictly positive datatypes.

Type-based analysis tries to address these problems by assigning a specific *type* to a total program instead of analyzing the code: Types are preserved under reduction and scale to higher-order functions and non-strictly positive data. Type theory has a long tradition in the programming language and theorem proving community and people are trained to understand typing rules. So a type-based approach might be easier to comprehend and easier trusted.

The first type-based approach to termination is attributed to Mendler [16]. He used universally quantified type variables to restrict recursive calls. His ideas have been further developed by many [6,21]; we will only consider approaches here which also detect non-size increasing functions.

Type-based termination in a nutshell. Inductive structures can be viewed as trees and thus classified by their *height*. First-order data structures like lists and binary trees have a finite height $< \omega$ which can be denoted by a natural number. Proof theory deals also with higher-order data structures like infinitely branching trees whose height can only be denoted by an infinite ordinal $\alpha \geq \omega$ below the least uncountable ordinal Ω. The dual concept to height is *definedness* or *observability* upto depth α and applies to coinductive structures.

In type-based termination, total functions are defined by induction on the height/definedness of structures. Let $\nabla X\sigma$ define a inductive ($\nabla \equiv \mu$) or coinductive ($\nabla \equiv \nu$) datatype. Then $\rho \approx \nabla X\sigma$ denotes a subtype or *approximation* ρ of $\nabla X\sigma$ which contains elements below a certain height/definedness α. Following Barthe *et al.* [7] we will refer to such elements as elements at *stage* α. Next, ρ^n denotes the type of elements at stage $\alpha + n$. Types involving bounded quantification $\forall Y{\approx}\nabla X\sigma.\,\tau(Y)$ can be instantiated at any approximation $\rho \approx \nabla X\sigma$ of a datatype to obtain the type $\tau(\rho)$. Recursive functions of result type τ involving $\nabla X\sigma$ are introduced by the rule

$$\frac{Y{\approx}\nabla X\sigma,\ g{:}\tau(Y) \vdash M : \tau(Y^1)}{\mathsf{fix}^{\nabla} g.M : \forall Y{\approx}\nabla X\sigma.\,\tau(Y)}.$$

The premise incarnates the step case of induction, taking the function from stage α to stage $\alpha+1$. The base case $\alpha = 0$ is handled by side conditions on the type τ which ensures that τ at stage 0 is interpreted as the whole universe of programs. This holds trivially for the class

$$\begin{array}{ll} \tau(Y) \equiv Y \to \tau'(Y) & \text{if } \nabla = \mu, \\ \tau(Y) \equiv \tau'(Y) \to Y & \text{if } \nabla = \nu. \end{array}$$

The fact that τ' may depend on Y enables the type system to tract stage dependencies between input and output of a function like $\mathsf{pivot} : \mathsf{int} \to \forall Y \approx \mathsf{int\ list}.\, Y \to Y \times Y$. The type of pivot guarantees that the length of both output lists is bounded by the length of the input list. This feature is very valuable for defining Euclidian division, merge sort, the stream of Fibonacci numbers (see below) etc.

For higher-order datatypes $\nabla X\sigma$ we have to continue induction transfinitely and also handle the limit case $\alpha = \lambda$. More precisely, we have to show that if the recursive program $\mathsf{fix}^{\nabla} g.M$ inhabits τ at stages $\beta < \lambda$ then it inhabits τ also at the limit stage λ. For types in the special class above this is always the case if $\tau'(Y)$ is monotone in Y. But this is not a necessary condition and there *are* interesting types which do not fall into the above class. E.g., let $\mathsf{Stream} = \nu X.\,\mathsf{Nat}{\times}X$ denote streams of natural numbers, $\mathsf{fold} : \forall Y{\approx}\mathsf{Stream}.\,\mathsf{Nat}{\times}Y \to Y^1$ the injection into Stream and

$$\begin{array}{l} \mathsf{sum} : \forall Y{\approx}\mathsf{Stream}.\,Y \to Y \to Y \\ \mathsf{sum}\ (x :: xs)\,(y :: ys) = x + y :: \mathsf{sum}\,xs\,ys \\ \mathsf{fib}' \ : \forall Y{\approx}\mathsf{Stream}.\,\mathsf{Nat} \times Y \\ \mathsf{fib}' \ = (0,\ 1 :: \mathsf{sum}\,(\mathsf{fold}\,\mathsf{fib}')\,(\mathsf{snd}\,\mathsf{fib}')) \end{array}$$

from which we can obtain the stream of Fibonacci numbers $\mathsf{fib} = \mathsf{fold}\,\mathsf{fib}'$. In the type system of Barthe *el al.* [7], the type of fib' is not accepted and it is not clear how to define the Fibonacci stream in a similarly elegant way. As we will see in Sect. 4, both definitions are perfectly legal in our system.

Table 1. Recent work on type-based termination.

Approach	Flavor	Polym.	Norm.	Datatypes	Stages	Arithmetic	Continuity
Xi [23]	Church	yes	closed	FO-μ	$< \omega$	Presb., $*/$	—
Hughes/Pareto+ [12,11,17]	Curry	yes	closed	FO-$\mu\nu$	$\leq \omega$	Presb.	yes
Barthe/Giménez+ [7]	Curry	no	SN	HO-$\mu\nu$	$< \Omega$	$\_+1$ only	no
this	Curry	no	SN	HO-$\mu\nu$	$< \Omega$	$\_+1$ only	yes

Contribution. This work addresses the question: Which types *are* legal as the result type τ of a recursive definition? We cannot give a final answer yet, but we have identified a class of types exceeding Barthe *el al.* [7] which follow a regular pattern. We analyzed the interpretation of types by sets of strongly normalizing terms and found that τ needs to be *continuous* for corecursive and something which we called *paracontinuous* for recursive definitions.

Hughes' *et al.* [12] system accepts types τ which are ω*-undershooting* and pass a *bottom-check*, but he worked in a domain theoretic denotational semantics with only first-order datatypes and it is not clear yet whether all of his results carry over when considering *strong* normalization and higher-order datatypes.

Table 1 compares some recent approaches to type-based termination. Xi [23] aims at functional programming with a cbv-reduction semantics for closed terms and first-order inductive datatypes, hence its sufficient for him to consider only finite stages and he does not need to address the problem of limit stages and continuity. His stage expressions are most expressive and support full arithmetic, but this makes type-checking semi-decidable if multiplication of stages is involved. Hughes and Pareto aim at embedded functional programming with streams and processes, they have treated the problem of continuity for ω-instantiation. Barthe and Giménez aim at theorem proving and support higher-order datatypes needed for proof theory. They show strong normalization, but do not investigate continuity, and here this works steps in.

Our calculus $\lambda^{\square\times\mu\nu}$ might serve as the core of a total functional programming language or an interactive proof assistant. Since it also handles non-strictly positive types, it could help to investigate programs with continuation types or algorithms extracted from classical proofs via double-negation translation.

This article is organized as follows. In Sect. 2 we will present the untyped term language with reduction semantics, and in Sect. 3 we will define types, subtyping and type interpretation. Sect. 4 is devoted to continuous types and Sect. 5 to co- and paracontinuous types. In Sect. 6 we will introduce the typing rules and prove soundness.

2 Untyped Calculus

Figure 2 presents terms, evaluation contexts and reduction relations of $\lambda^{\square\times\mu\nu}$, a core functional programming language with iso-recursive categorical datatypes, (fold, unfold), recursion (fix^μ) and corecursion (fix^ν). "fold" is the general constructor, sometimes called in; "unfold" the general destructor, also called out.

Terms $M, N \in \mathsf{TM}$.

$$M, N ::= x \mid \lambda x.M \mid M\,N \mid \mathsf{inl}\,M \mid \mathsf{inr}\,M \mid (\mathsf{case}\,M \text{ of } \mathsf{inl}\,x_1 \Rightarrow M_1 \mid \mathsf{inr}\,x_2 \Rightarrow M_2)$$
$$\mid (M_1, M_2) \mid \mathsf{fst}\,M \mid \mathsf{snd}\,M \mid \mathsf{fold}\,M \mid \mathsf{unfold}\,M \mid \blacksquare x^{\mu} g.M \mid \blacksquare x^{\nu} g.M$$

Evaluation contexts.

$$E[\bullet] ::= \bullet \mid E[\bullet]\,M \mid \mathsf{unfold}\,E[\bullet] \mid (\mathsf{case}\,E[\bullet] \text{ of } \mathsf{inl}\,x_1 \Rightarrow M_1 \mid \mathsf{inr}\,x_2 \Rightarrow M_2)$$
$$\mid \mathsf{fst}\,E[\bullet] \mid \mathsf{snd}\,E[\bullet] \mid (\blacksquare x^{\mu} g.M)\,E[\bullet]$$

β-Reduction axioms: Lambda calculus with sums and products +

$$\mathsf{unfold}\,(\mathsf{fold}\,M) \longrightarrow_\beta M$$
$$(\blacksquare x^{\mu} g.M)\,(\mathsf{fold}\,N) \longrightarrow_\beta ([\blacksquare x^{\mu} g.M/g]M)(\mathsf{fold}\,N)$$
$$\mathsf{unfold}\,E[\blacksquare x^{\nu} g.M] \longrightarrow_\beta \mathsf{unfold}\,E[[\blacksquare x^{\nu} g.M/g]M]$$

Reduction relations.

$\longrightarrow_\beta$	β-reduction
$\longrightarrow$	one-step reduction: closure of $\longrightarrow_\beta$ under all term constructors
$\longrightarrow^+$	transitive closure of $\longrightarrow$
$\longrightarrow^*$	reflexive-transitive closure of $\longrightarrow$

Fig. 2. $\lambda^{\blacksquare\chi\mu\nu}$: Terms, Evaluation Contexts and Reduction

Reduction axioms are standard, solely of interest are the rules for the fixed-points: Recursive functions fix^μ are only unfolded when applied to an argument guarded by a fold and infinite objects fix^ν only under a unfold-destructor. These restrictions are essential for establishing strong normalization. The reduction relation $\longrightarrow$, which results from closing $\longrightarrow_\beta$ under all term constructors, is confluent, which can be shown with the parallel-reduction method by Tait and Martin-Löf.

The set of *strongly normalizing terms* SN is defined as the wellfounded part of the set TM of terms wrt. the reduction relation $\longrightarrow$. All variables x and all subterms of strongly normalizing terms are strongly normalizing. Furthermore, the set SN is closed under reduction.

Proofs of strong normalization typically involve some standardization argument. In our case we use *weak head reduction*; it is a reduction of the form $E[M_1] \longrightarrow E[M_2]$ where $M_1 \longrightarrow_\beta M_2$. Terms of the form $E[x]$ are called *neutral*. A set of terms $P \subseteq \mathsf{TM}$ is called *saturated*, written $P \in \mathcal{SAT}$, if it contains only strongly normalizing terms, all strongly normalizing neutral terms, and if it is closed under weak head expansion. The *saturation* of a set P is defined as the closure under the following rules:

$$\frac{M \in P}{M \in P^*} \qquad \frac{E[x] \in \mathsf{SN}}{E[x] \in P^*} \qquad \frac{M \in \mathsf{SN} \quad M \longrightarrow_\beta M' \quad E[M'] \in P^*}{E[M] \in P^*}$$

The set SN is saturated, and so is the *function space* $P \to Q := \{M \mid M\,N \in Q \text{ for all } N \in P\}$, provided $P, Q \in \mathcal{SAT}$.

Types ($\nabla \in \{\mu, \nu\}$, $n \in \mathbb{N}$).

$$\rho, \sigma, \tau ::= X \mid \sigma \to \tau \mid \sigma + \tau \mid \sigma \times \tau \mid \nabla X \sigma \mid Y^n \mid \forall Y \approx \nabla X \sigma.\, \tau$$

Contexts.

$$\Gamma ::= \cdot \mid \Gamma, x{:}\tau \mid \Gamma, X \mid \Gamma, Y \approx \nabla X \sigma$$

Judgments.

Γ cxt	Γ is a wellformed context.
$\Gamma \vdash X$ pos σ	Variable X appears positively in σ
$\Gamma \vdash X$ neg σ	Variable X appears negatively in σ
$\Gamma \vdash X$ only pos in σ	All occurrences of X in σ are positive.
$\Gamma \vdash \tau$: type	τ is a wellformed type.
$\Gamma \vdash \rho \approx \nabla X \sigma$	ρ is an approximation of $\nabla X \sigma$.
$\Gamma \vdash \rho \leq \sigma$	ρ is a subtype of σ.

Approximations.

$$\frac{(Y \approx \nabla X \sigma) \in \Gamma}{\Gamma \vdash Y^n \approx \nabla X \sigma} \approx Y \qquad \frac{}{\Gamma \vdash \nabla X \sigma \approx \nabla X \sigma} \approx \nabla$$

Subtyping.

$$\frac{\rho \in \{\mu X \sigma, \nu X \sigma\}}{\Gamma \vdash \rho \leq \rho} \leq \textsc{Base} \qquad \frac{(Y \approx \mu X \sigma) \in \Gamma}{\Gamma \vdash Y^n \leq \mu X \sigma} \leq Y_1^\mu \qquad \frac{(Y \approx \mu X \sigma) \in \Gamma \qquad n \leq m}{\Gamma \vdash Y^n \leq Y^m} \leq Y_2^\mu$$

$$\frac{(Y \approx \nu X \sigma) \in \Gamma}{\Gamma \vdash \nu X \sigma \leq Y^n} \leq Y_1^\nu \qquad \frac{(Y \approx \nu X \sigma) \in \Gamma \qquad n \leq m}{\Gamma \vdash Y^m \leq Y^n} \leq Y_2^\nu$$

$$\frac{\Gamma \vdash \sigma_1 \leq \rho_1 \qquad \Gamma \vdash \rho_2 \leq \sigma_2}{\Gamma \vdash \rho_1 \to \rho_2 \leq \sigma_1 \to \sigma_2} \leq\to \qquad \frac{\Gamma \vdash \rho_1 \leq \sigma_1 \qquad \Gamma \vdash \rho_2 \leq \sigma_2 \qquad \star \in \{+, \times\}}{\Gamma \vdash \rho_1 \star \rho_2 \leq \sigma_1 \star \sigma_2} \leq\star$$

$$\frac{\Gamma \vdash \rho \approx \nabla X \sigma \qquad \Gamma \vdash [\rho/Y]\tau \leq \tau'}{\Gamma \vdash (\forall Y \approx \nabla X \sigma.\, \tau) \leq \tau'} \leq\forall L \qquad \frac{\Gamma, Y \approx \nabla X \sigma \vdash \rho \leq \tau}{\Gamma \vdash \rho \leq \forall Y \approx \nabla X \sigma.\, \tau} \leq\forall R$$

Fig. 3. $\lambda^{\square\times\mu\nu}$: Types and Subtyping.

3 Types

Figure 3 presents types and type related judgments. Most of these are standard, like wellformedness of contexts and positive and negative occurrences of variables. A type τ is wellformed ($\Gamma \vdash \tau$: type) if it follows the grammar with the restriction that if $\tau \equiv \nabla X \sigma$ is a (co)inductive type, X may appear only positively in σ and σ may not contain quantifiers. Approximation types Y^n and bounded quantification $\forall Y \approx \nabla X \sigma.\, \tau$ have been explained in the introduction already. Note that unit and empty type can be defined: $1 = \nu X.X$, $0 = \mu X.X$.

We regard all variables bound in a context as distinct. (Capture-avoiding) substitution for types is defined as usual and written $[\rho/X]\sigma$. Sometimes we exhibit a free variable in a type $\sigma(X)$ and write $\sigma(\rho)$ for the result of substitution. For approximations we define a special substitution given by the axioms $[X^m/Y]Y^n = X^{n+m}$ and $[\nabla X \sigma/Y]Y^n = \nabla X \sigma$ plus congruence rules for all type constructors. The second axiom is justified by the fact that $\nabla X \sigma$ denotes a fixed-point which is unaffected by n more iterations.

Subtyping is induced by approximation types. Since the semantics $[\![Y]\!]$ of some bounded type variable $Y \approx \mu X \sigma$ denotes some subset of $[\![\mu X \sigma]\!]$, we introduce the subtyping rule $Y \leq \mu X\sigma$. Furthermore, since the approximations of an inductive type form an ascending chain bounded by the fixed-point of the approximation operator, we can safely add subtyping rules $Y^i \leq Y^j$ and $Y^i \leq \mu X\sigma$ for all indices i and j with $i \leq j$. Note that two different type variables $X \neq Y$ are incomparable. Subtyping rules for approximations of coinductive types are obtained by dualization.

To increase expressivity of subtyping, we have incorporated a reflexivity rule for base types, congruence rules for sum, product and function types (contravariant on the left!) and rules for quantification. The subtyping relation $\leq$ is reflexive and transitive. Since most rules are syntax-directed, it is not hard to come up with a decision algorithm for subtyping. The only critical rules are the ones for quantification: rule $\leq\forall R$ always has to be applied before $\leq\forall L$. When the rule $\leq\forall L$ fires, an approximation ρ has to be guessed. This can be implemented using existential variables and unification.

Type interpretation and soundness. In the following, we give an interpretation of types as sets of terms which we will later show to be saturated. Approximations will be interpreted as iterations Φ^α of monotonic operators $\Phi : \mathcal{P}(\mathsf{TM}) \to \mathcal{P}(\mathsf{TM})$. To distinguish between approximation from above and below, we define iterates as $\Phi^{\nabla,\alpha}$ with a tag $\nabla \in \{\mu, \nu\}$.

$$\begin{aligned} \Phi^{\mu,0} &= \emptyset^* & \Phi^{\nu,0} &= \mathsf{SN} \\ \Phi^{\mu,\alpha+1} &= \Phi(\Phi^{\mu,\alpha}) & \Phi^{\nu,\alpha+1} &= \Phi(\Phi^{\nu,\alpha}) \\ \Phi^{\mu,\lambda} &= \bigcup_{\alpha<\lambda} \Phi^{\mu,\alpha} & \Phi^{\nu,\lambda} &= \bigcap_{\alpha<\lambda} \Phi^{\nu,\alpha} \end{aligned}$$

When clear from the context, we omit the tags μ, ν. To interpret context as sets of substitutions, let P, Q from here denote sets of terms. Raw substitutions are given by the grammar $\theta ::= \cdot \mid \theta, x \mapsto M \mid \theta, X \mapsto P \mid \theta, Y \mapsto (\Phi, \nabla, \alpha)$. Note that approximation variables Y are mapped to a triple consisting of a monotonic operator Φ, a flag $\nabla \in \{\mu, \nu\}$ denoting whether Y approximates a least fixed-point from below or a greatest fixed-point from above, plus an ordinal α denoting the current iteration. Γ-substitutions $\theta : \Gamma$ are those with $\mathsf{dom}(\theta) = \mathsf{dom}(\Gamma)$.

Let $\Gamma \vdash \tau : \mathsf{type}$ and $\theta : \Gamma$. We define the type interpretation $[\![\tau]\!]\theta \subseteq \mathsf{TM}$ by recursion on τ.

$$\begin{aligned} [\![\sigma + \tau]\!]\theta &= \{\mathsf{inl}\, M \mid M \in [\![\sigma]\!]\theta\}^{\square} \cup \{\mathsf{inr}\, M \mid M \in [\![\tau]\!]\theta\}^{\square} \\ [\![\sigma \times \tau]\!]\theta &= \{M \mid \mathsf{fst}\, M \in [\![\sigma]\!]\theta, \mathsf{snd}\, M \in [\![\tau]\!]\theta\} \\ [\![\sigma \to \tau]\!]\theta &= [\![\sigma]\!]\theta \to [\![\tau]\!]\theta & [\![X]\!]\theta &= \theta(X) \\ [\![\mu X\sigma]\!]\theta &= \Phi^{\Omega}_{\mu X\sigma,\theta} & \Phi_{\mu X\sigma,\theta}(Q) &= \{\mathsf{fold}\, M \mid M \in [\![\sigma]\!](\theta, X \mapsto Q)\}^{\square} \\ [\![\nu X\sigma]\!]\theta &= \Phi^{\Omega}_{\nu X\sigma,\theta} & \Phi_{\nu X\sigma,\theta}(Q) &= \{M \mid \mathsf{unfold}\, M \in [\![\sigma]\!](\theta, X \mapsto Q)\} \\ [\![Y^n]\!]\theta &= \Phi^{\alpha+n} & &\text{where } \theta(Y) = (\Phi, \nabla, \alpha) \\ [\![\forall Y \approx \nabla X\sigma.\, \tau]\!]\theta &= \square_{\alpha} [\![\tau]\!](\theta, Y \mapsto (\Phi_{\square X\sigma,\theta}, \nabla, \alpha)) \end{aligned}$$

The "introduction-based" semantics of $+$ and μ has to be saturated explicitely (note the *), which is not necessary for "elimination-based" constructions ($\to$, $\times$, and ν). The semantics of (co)inductive types $\nabla X\sigma$ are defined as iterates of operators at stage Ω, which is a (greatest) least fixed-point iff the semantics $[\![\tau]\!]$ is monotonic for each substitution θ, i.e., monotonic in every variable that occurs positively and antitonic in every variable that occurs negatively. To make this observation precise, we define *inclusion for substitutions*. Let $\Gamma \vdash \tau : \mathsf{type}$. Then $\Gamma \vdash_\tau \theta_1 \subseteq \theta_2$ is defined to hold iff θ_1 and θ_2 are Γ-substitutions and

- $\theta_1(X) \subseteq \theta_2(X)$ for all X with $\Gamma \vdash X \ \mathsf{pos}\ \tau$,
- $\theta_1(X) \supseteq \theta_2(X)$ for all X with $\Gamma \vdash X \ \mathsf{neg}\ \tau$, and
- $\theta_1(Y) = \theta_2(Y)$ for all $(Y \approx \nabla X \sigma) \in \Gamma$.

Lemma 1 (Soundness of Subtyping). *If $\Gamma \vdash \rho \leq \sigma$ and $\Gamma \vdash_\rho \theta_1 \subseteq \theta_2$ then $[\![\rho]\!]\theta_1 \subseteq [\![\sigma]\!]\theta_2$.*

Proof. By induction on $\Gamma \vdash \rho \leq \sigma$.

By reflexivity of subtyping, this lemma entails monotonicity of all types which, again, entails soundness of our fixed-point construction. By simple induction proofs we also establish that substitution for types and approximations is sound. Finally, we show is that types are interpreted as saturated sets. To this end, we define the semantical version $\theta \in [\![\Gamma]\!]$ of $\theta : \Gamma$.

$$\frac{}{\cdot \in [\![\cdot]\!]} \qquad \frac{M \in [\![\tau]\!]\theta \qquad \theta \in [\![\Gamma]\!]}{(\theta, x \mapsto M) \in [\![\Gamma, x{:}\tau]\!]} \qquad \frac{P \in \mathcal{SAT} \qquad \theta \in [\![\Gamma]\!]}{(\theta, X \mapsto P) \in [\![\Gamma, X]\!]}$$

$$\frac{\alpha \text{ ordinal} \qquad \theta \in [\![\Gamma]\!]}{(\theta, Y \mapsto (\Phi_{\nabla X \sigma, \theta}, \nabla, \alpha)) \in [\![\Gamma, Y \approx \nabla X \sigma]\!]}$$

Lemma 2 (Saturatedness). *If Γ cxt then for all $\theta \in [\![\Gamma]\!]$ and all X the set $\theta(X)$ is saturated. If $\Gamma \vdash \tau : \mathsf{type}$ then for all $\theta \in [\![\Gamma]\!]$ it holds that $[\![\tau]\!]\theta \in \mathcal{SAT}$.*

4 Continuous Types

In this section, we will identify a set of legal result types $\tau(Y)$ for corecursion (see Sect. 1). The key requirement on τ is continuity. An operator Φ is *continuous* if for all families P^α $(\alpha \in I)$ of term sets it holds that

$$\bigcap_{\alpha \in I} \Phi(P^\alpha) \subseteq \Phi\left(\bigcap_{\alpha \in I} P^\alpha\right).$$

In the following we will motivate why continuity is a necessary condition on τ. Consider the typing rule for corecursion specialized to streams.

$$\frac{Y \approx \mathsf{Stream},\ g{:}\tau(Y) \vdash M : \tau(Y^1)}{\mathsf{fix}^\nu g.M : \forall Y \approx \mathsf{Stream}.\, \tau(Y)}$$

Let $S(Q) = \{M \in \mathsf{TM} \mid \mathsf{hd}\, M \in [\![\mathsf{Nat}]\!] \text{ and } \mathsf{tl}\, M \in Q\}$ be the semantical operator to construct the set of streams. Then $[\![\mathsf{Stream}]\!] = \bigcap_{\alpha<\omega} S^\alpha$. Hence, the corecursion rule for streams can be proven sound by transfinite induction upto ω. In the limit case ω, we can use the induction hypothesis for all smaller stages, i.e., we can assume

$$\mathsf{fix}^\nu g.M \in \bigcap_{\alpha<\omega} [\![\tau]\!](S^\alpha) \quad \text{to show} \quad \mathsf{fix}^\nu g.M \in [\![\tau]\!]\Big(\bigcap_{\alpha<\omega} S^\alpha\Big) = [\![\tau]\!]([\![\mathsf{Stream}]\!]).$$

Obviously, we require $[\![\tau]\!]$ to be continuous.

A grammar for continuous types. We introduce a new judgment $\Gamma \vdash \boldsymbol{X} \text{ cont } \tau$, meaning that τ is continuous in the variables $\boldsymbol{X}$.

$$\frac{}{\Gamma \vdash \boldsymbol{X} \text{ cont } X_i} \qquad \frac{\boldsymbol{X} \notin \mathsf{FV}(\tau)}{\Gamma \vdash \boldsymbol{X} \text{ cont } \tau} \qquad \frac{\Gamma \vdash \boldsymbol{X} \text{ only pos in } \sigma \qquad \Gamma \vdash \boldsymbol{X} \text{ cont } \tau}{\Gamma \vdash \boldsymbol{X} \text{ cont } \sigma \to \tau}$$

$$\frac{\Gamma \vdash \boldsymbol{X} \text{ cont } \sigma, \tau}{\Gamma \vdash \boldsymbol{X} \text{ cont } \sigma + \tau} \qquad \frac{\Gamma \vdash \boldsymbol{X} \text{ cont } \sigma, \tau}{\Gamma \vdash \boldsymbol{X} \text{ cont } \sigma \times \tau} \qquad \frac{\Gamma, X \vdash \boldsymbol{X}, X \text{ cont } \sigma}{\Gamma \vdash \boldsymbol{X} \text{ cont } \nabla X \sigma}$$

$$\frac{(Y \approx \nabla X \sigma) \in \Gamma \qquad \Gamma \vdash \boldsymbol{X} \text{ cont } \nabla X \sigma}{\Gamma \vdash \boldsymbol{X} \text{ cont } Y^n} \qquad \frac{\Gamma, Y \approx \nabla X \sigma \vdash \boldsymbol{X} \text{ cont } \tau}{\Gamma \vdash \boldsymbol{X} \text{ cont } \forall Y \approx \nabla X \sigma.\, \tau}$$

For function types to be continuous, the domain just needs to be monotonic and the range continuous. Fixed-points are continuous if the respective operator is continuous and (of course) monotonic. Surprising is the fact that *least* fixed-points are continuous, which requires that union preserves continuity. Let Φ be the operator of an inductive type. Then

$$\bigcap_\alpha \bigcup_{\beta<\lambda} \Phi^\beta(P^\alpha) \subseteq \bigcup_{\beta<\lambda} \Phi^\beta\Big(\bigcap_\alpha P^\alpha\Big)$$

does not hold simply by set-theoretic means, even if Φ is continuous. In our semantics however, an intersection *can* be pulled into a union. Informally this may be explained as follows:

Consider polymorphic lists $\mathsf{List}(X)$ with operator $L(A)(Q) = \{\mathsf{nil}\} \cup \{a :: l \mid a \in A, l \in Q\}$. Note that for some fixed set A, $L(A)^n$ denotes the nth approximant to $[\![\mathsf{List}]\!](A)$. Recall that S was defined as the operator for streams. We say a stream s has goodness n if $s \in S^n$. Let M be some list of streams with the following property: At the same time, let M be a list of length m_0 of streams of goodness 0, a list of length m_1 of streams of goodness 1, ... a list of length m_n of streams of goodness n, ... Formally, $M \in L(S^i)^{m_i}$ for all i. Since a list can only have one length, all m_i must be equal. Hence $m_i = m_0$ for all i, and

$$\begin{aligned} &\textstyle\bigcap_n [\![\mathsf{List}]\!](S^n) = \bigcap_n (\bigcup_m (L(S^n))^m) = \bigcap_n (L(S^n))^{m_0} \\ &\subseteq (L(\textstyle\bigcap_n S^n))^{m_0} = (L([\![\mathsf{Stream}]\!]))^{m_0} \subseteq [\![\mathsf{List}]\!]([\![\mathsf{Stream}]\!]) \end{aligned}$$

This example shows that $\mathsf{List}(A)$ is continuous in A. To generalize this result for all inductive types, we make the following general observation that new inductive data is only generated by successor iterates:

Lemma 3 (Successor Iterates are Sufficient). *If* $\mathsf{fold}\, M \in \Phi^{\alpha}_{\mu X\sigma,\theta}$ *then* $\mathsf{fold}\, M \in \Phi^{\beta+1}_{\mu X\sigma,\theta}$ *for some* $\beta < \alpha$.

Now we give a formulation of the above explained independence result for inductive types $\mu Y \sigma$ with a free variable X.

Lemma 4 (Independence). *Let* $[\![\sigma(X, Y)]\!]$ *be continuous in the variables* X *and* Y *and* $\Phi(P)(Q) = \{\mathsf{fold}\, M \mid M \in [\![\sigma]\!](X \mapsto P, Y \mapsto Q)\}^*$. *Furthermore, let* (β_α) *a sequence of ordinals and* P^α *a sequence of sets of terms. Then*

$$\bigcap_\alpha (\Phi(P^\alpha))^{\beta_\alpha} \subseteq \Phi(\bigcap_\alpha P^\alpha)^{\beta_0}$$

This lemma is proven by induction on β_0 and can be generalized to continuous types $\sigma(\boldsymbol{X}, Y)$ by defining intersection on substitutions θ. It then yields soundness of syntactic continuity.

Theorem 1 (Soundness of cont). *Let* $\Gamma \vdash \tau : \mathsf{type}$. *If* $\Gamma \vdash \boldsymbol{X}\ \mathsf{cont}\ \tau$, *then* $[\![\tau]\!]$ *is continuous in* $\boldsymbol{X}$.

As a side product of our considerations, we notice that all *strictly positive* coinductive types are continuous. Thus, their fixed-point is reached at iteration ω.

Types for infinite objects. Can all continuous types be used for fix^ν-definitions? Given our reduction semantics, the answer is no. Sums, inductive and coinductive types may be continuous, but they do not provide the necessary guardedness. For example, let

$$\mathsf{map}_0 \;=\; \mathsf{map}\,(\lambda x.0 :: x) \;:\; \forall Y{\approx}\mathsf{Stream}.\mathsf{List}(Y) \to \mathsf{List}(Y^1)$$

Then $M = \mathsf{fix}^\nu g.\, \mathsf{map}_0\, g$ could be ascribed type $\forall Y{\approx}\mathsf{Stream}.\, \mathsf{List}(Y)$. But unfolding of M diverges: $\mathsf{unfold}\, M \longrightarrow_\beta \mathsf{unfold}(\mathsf{map}_0\, M) \longrightarrow_\beta \mathsf{unfold}(\mathsf{map}_0\,(\mathsf{map}_0\, M)) \longrightarrow_\beta \dots$ Similar examples can be found for types with $+$ and ν. What remains as legal types for fix^ν, are function space and product. By a judgment $Y\ \mathsf{legal}^\nu\ \tau$, we give a grammar for these types.

$$\frac{}{Y\ \mathsf{legal}^\nu\ Y} \qquad \frac{Y\ \mathsf{only\ pos\ in}\ \sigma \qquad Y\ \mathsf{legal}^\nu\ \tau}{Y\ \mathsf{legal}^\nu\ \sigma \to \tau} \qquad \frac{Y\ \mathsf{legal}^\nu\ \sigma \qquad Y\ \mathsf{legal}^\nu\ \tau}{Y\ \mathsf{legal}^\nu\ \sigma \times \tau}$$

Examples for legal types τ according to this grammar which are *not* of the form $\tau(Y) \to Y$ (τ monotone), like $Y \to Y \to Y$ and $\mathsf{Nat} \times Y$, have been given in the introduction (Fibonacci streams).

5 Co- and Paracontinuous Types

As we have answered for fix^ν, we can ask for the recursive function constructor fix^μ: Which types $\tau(Y)$ would be legal in the rule

$$\frac{Y{\approx}\mu X\sigma,\ g{:}\tau(Y) \vdash M : \tau(Y)}{\mathsf{fix}^\mu g.M : \forall Y{\approx}\mu X\sigma.\, \tau(Y)}$$

This rule would be proven sound by transfinite induction on the approximations Φ^α of the inductive type $\mu X\sigma$. For limit steps λ, we may assume

$$\mathsf{fix}^\mu g.M \in \bigcap_{\alpha<\lambda} [\![\tau]\!](\Phi^\alpha) \qquad \text{to show} \qquad \mathsf{fix}^\mu g.M \in [\![\tau]\!](\bigcup_{\alpha<\lambda} \Phi^\alpha).$$

We call such types $\tau(Y)$, which permit this inference, *paracontinuous*. Obviously paracontinuous are types of the form $Y \to \tau(Y)$ for Y only pos in τ. However, this does not include the most precise type for, e.g., the maximum function $\mathsf{max} : \forall Y \approx \mathsf{Nat}.\, Y \to (Y \to Y)$. Here Y occurs negatively in $Y \to Y$, but the type is still paracontinuous. On our journey to identify paracontinuous types syntactically, we will first consider cocontinuity, the concept dual to continuity.

Cocontinuity. An operator $\Phi : \mathcal{P}(\mathsf{TM}) \to \mathcal{P}(\mathsf{TM})$ is called *cocontinuous*, if for all chains P (with $P^\alpha \subseteq P^\beta$ for $\alpha < \beta$) it holds that

$$\Phi(\bigcup_\alpha P^\alpha) \subseteq \bigcup_\alpha \Phi(P^\alpha)$$

In the following we will identify the types τ for which the semantics $[\![\tau]\!]$ is cocontinuous. Let us first consider products.

Lemma 5 (Products preserve Cocontinuity). *Let $\Gamma \vdash \sigma(X), \tau(X) : \mathsf{type}$. If $[\![\sigma]\!]$ and $[\![\tau]\!]$ are cocontinuous and monotonic in X, so is $[\![\sigma \times \tau]\!]$.*

Proof. Assume $M \in [\![\sigma \times \tau]\!](\bigcup_\alpha P^\alpha)$. Then by definition, $\mathsf{fst}\, M \in [\![\sigma]\!](\bigcup_\alpha P^\alpha)$ and $\mathsf{snd}\, M \in [\![\tau]\!](\bigcup_\alpha P^\alpha)$. Since σ and τ are cocontinuous by assumption, there are ordinals α, β such that $\mathsf{fst}\, M \in [\![\sigma]\!](P^\alpha)$ and $\mathsf{snd}\, M \in [\![\tau]\!](P^\beta)$. Now let $\gamma = \max(\alpha, \beta)$. Since P is a chain and $[\![\sigma]\!]$ and $[\![\tau]\!]$ are monotonic, $\mathsf{fst}\, M \in [\![\sigma]\!](P^\gamma)$ and $\mathsf{snd}\, M \in [\![\tau]\!](P^\gamma)$. We conclude $M \in [\![\sigma \times \tau]\!](P^\gamma)$.

Thus, products are cocontinuous because the binary maximum always exists. Since function types and coinductive types can be viewed as infinite products, they are not cocontinuous in general. For example, $\lambda x.x \in \mathsf{List} \to \bigcup_{\alpha<\omega} L^\alpha$, but it does not inhabit $\mathsf{List} \to L^\alpha$ for any $\alpha < \omega$. We expect that the same holds for quantification. Sums and inductive types are trivially cocontinuous, provided they are monotonic. Hence, we can give a grammar for cocontinuous types via a judgment $\Gamma \vdash \boldsymbol{X}\ \mathsf{cocont}\ \tau$.

$$\frac{}{\Gamma \vdash \boldsymbol{X}\ \mathsf{cocont}\ X_i} \qquad \frac{X \notin \mathsf{FV}(\tau)}{\Gamma \vdash X\ \mathsf{cocont}\ \tau} \qquad \frac{\Gamma \vdash \boldsymbol{X}\ \mathsf{cocont}\ \sigma, \tau}{\Gamma \vdash \boldsymbol{X}\ \mathsf{cocont}\ \sigma + \tau, \sigma \times \tau}$$

$$\frac{\Gamma, X \vdash \boldsymbol{X}, X\ \mathsf{cocont}\ \sigma}{\Gamma \vdash \boldsymbol{X}\ \mathsf{cocont}\ \mu X\sigma} \qquad \frac{(Y \approx \mu X\sigma) \in \Gamma \qquad \Gamma \vdash \boldsymbol{X}\ \mathsf{cocont}\ \mu X\sigma}{\Gamma \vdash \boldsymbol{X}\ \mathsf{cocont}\ Y^n}$$

Note that $\Gamma \vdash \boldsymbol{X}\ \mathsf{cocont}\ \tau$ implies $\Gamma \vdash \boldsymbol{X}$ only pos in τ. The grammar describes polynomial interleaved (nested) inductive types, which includes, e.g., natural numbers, lists and finitely branching trees, i.e., almost all inductive types used in practical programming. Furthermore all of these types are continuous and close at ω.

Lemma 6 (Soundness). *If* $\Gamma \vdash \boldsymbol{X}$ cocont τ, *then* $[\![\tau]\!]$ *is cocontinuous and monotonic in* $\boldsymbol{X}$.

Proof. By induction on $\Gamma \vdash \boldsymbol{X}$ cocont τ, using Lemma 5.

Paracontinuity. The considerations at the start of Sect. 5 suggested a definition of paracontinuity immediately. However, to make it work, we have to strengthen it a little.

An operator Φ is *paracontinuous* if for all chains P it holds that

$$\text{for all } \alpha_0 : \bigcap_{\alpha_0 \leq \alpha} \Phi(P^\alpha) \subseteq \Phi(\bigcup_\alpha P^\alpha).$$

Obviously all monotonic operators are paracontinuous. For function types, the domain has to be *cocontinuous.*

Lemma 7 (Paracontinuous Function Types). *If* $[\![\tau]\!]$ *is paracontinuous in* X *and* $[\![\sigma]\!]$ *is cocontinuous and monotonic in* X, *then* $[\![\sigma \to \tau]\!]$ *is paracontinuous in* X.

Proof. Fix some α_0 and assume $M \in \bigcap_{\alpha_0 \leq \alpha} [\![\sigma \to \tau]\!](P^\alpha)$ and $N \in [\![\sigma]\!](\bigcup_\alpha P^\alpha)$. Since σ is cocontinuous, $N \in [\![\sigma]\!](P^\beta)$ for some β. Let $\gamma = \max(\alpha_0, \beta)$. Now, since P^α is a ascending chain, we can exploit the monotonicity of $[\![\sigma]\!]$ to infer $N \in [\![\sigma]\!](P^\alpha)$ for all $\alpha \geq \gamma$. By assumption, $M\,N \in [\![\tau]\!](P^\alpha)$ for all $\alpha \geq \gamma$. Finally, since τ is paracontinuous, $M\,N \in [\![\tau]\!](\bigcup_\alpha P^\alpha)$.

For types $\sigma(X, Y)$ which are *continuous* in Y, fixed-point formation $\nabla Y \sigma$ preserves paracontinuity in X. The case of inductive types is especially interesting, since it makes again use of the Independence Lemma.

Lemma 8 (Paracontinuous Inductive Types). *If* $[\![\sigma(X, Y)]\!]$ *is paracontinuous in* X *and continuous and monotonic in* Y, *then* $[\![\mu Y \sigma]\!]$ *is paracontinuous in* X.

Proof. By transfinite induction on the stages of $[\![\mu Y \sigma]\!]$, using Lemma 4.

For coinductive types $\nu Y \sigma$ the proof is quite similar, but can be done without Lemma 4. The proven lemmata can be generalized to paracontinuity in several variables $\boldsymbol{X}$.

We give a grammar for paracontinuous types by the judgment $\Gamma \vdash \boldsymbol{X}$ para τ.

$$\frac{}{\Gamma \vdash \boldsymbol{X} \text{ para } X_i} \qquad \frac{\boldsymbol{X} \notin \mathsf{FV}(\tau)}{\Gamma \vdash \boldsymbol{X} \text{ para } \tau} \qquad \frac{\Gamma \vdash \boldsymbol{X} \text{ cocont } \sigma \qquad \Gamma \vdash \boldsymbol{X} \text{ para } \tau}{\Gamma \vdash \boldsymbol{X} \text{ para } \sigma \to \tau}$$

$$\frac{\Gamma \vdash \boldsymbol{X} \text{ para } \sigma, \tau}{\Gamma \vdash \boldsymbol{X} \text{ para } \sigma + \tau, \sigma \times \tau} \qquad \frac{\Gamma, X \vdash X \text{ cont } \sigma \qquad \Gamma, X \vdash \boldsymbol{X}, X \text{ para } \sigma}{\Gamma \vdash \boldsymbol{X} \text{ para } \nabla X \sigma}$$

$$\frac{(Y \approx \nabla X \sigma) \in \Gamma \qquad \Gamma \vdash \boldsymbol{X} \text{ para } \nabla X \sigma}{\Gamma \vdash \boldsymbol{X} \text{ para } Y^n} \qquad \frac{\Gamma, Y \approx \nabla X \sigma \vdash \boldsymbol{X} \text{ para } \tau}{\Gamma \vdash \boldsymbol{X} \text{ para } \forall Y \approx \nabla X \sigma . \tau}$$

Theorem 2 (Soundness). *Let* $\Gamma \vdash \tau : \mathsf{type}$. *If* $\Gamma \vdash \boldsymbol{X}\ \mathsf{para}\ \tau$ *then* $[\![\tau]\!]$ *is paracontinuous in* $\boldsymbol{X}$.

Proof. By induction on $\Gamma \vdash \boldsymbol{X}\ \mathsf{para}\ \tau$, using Lemmata 7 and 8.

Note that $\Gamma \vdash \boldsymbol{X}\ \mathsf{cont}\ \tau$ implies $\Gamma \vdash \boldsymbol{X}\ \mathsf{para}\ \tau$. A *non*-paracontinuous type is $\tau(Y) = \mathsf{Stream}(Y) \to \mathsf{Bool}$ (acknowledgment: John Hughes). A function "double" which checks whether a stream has duplicate elements would be total for streams of numbers bounded by some $\alpha < \omega$ (returning always true) but not for streams of arbitrary natural numbers $\alpha = \omega$.

As for continuous types we raise the question: Are all paracontinuous types legal result types for the recursion constructor fix^μ? Again, the answer is no. Our reduction semantics only permits types $Y \to \tau(Y)$ for $Y\ \mathsf{para}\ \tau$, which is still a great improvement over Y only pos in τ.

Typing (τ and all types in Γ closed):

$$\Gamma \vdash M : \tau$$

(Lambda-calculus with products and sums +) Folding.

$$\frac{\Gamma \vdash M : [\nabla X\sigma/X]\sigma}{\Gamma \vdash \mathsf{fold}\, M : \nabla X\sigma} \qquad \frac{(Y \approx \nabla X\sigma) \in \Gamma \qquad \Gamma \vdash M : [Y^n/X]\sigma}{\Gamma \vdash \mathsf{fold}\, M : Y^{n+1}}$$

Unfolding.

$$\frac{\Gamma \vdash M : \nabla X\sigma}{\Gamma \vdash \mathsf{unfold}\, M : [\nabla X\sigma/X]\sigma} \qquad \frac{(Y \approx \nabla X\sigma) \in \Gamma \qquad \Gamma \vdash M : Y^{n+1}}{\Gamma \vdash \mathsf{unfold}\, M : [Y^n/X]\sigma}$$

Fixed-Points.

$$\frac{\Gamma,\ Y \approx \mu X\sigma,\ g : Y \to \tau(Y) \vdash M : Y^1 \to \tau(Y^1) \qquad \Gamma \vdash Y\ \mathsf{para}\ \tau(Y)}{\Gamma \vdash \mathsf{fix}^\mu g.M : \forall Y \approx \mu X\sigma.\, Y \to \tau(Y)}$$

$$\frac{\Gamma,\ Y \approx \nu X\sigma,\ g : \tau(Y) \vdash M : \tau(Y^1) \qquad \Gamma \vdash Y\ \mathsf{legal}^\nu\ \tau(Y)}{\Gamma \vdash \mathsf{fix}^\nu g.M : \forall Y \approx \nu X\sigma.\, \tau(Y)}$$

Subsumption.

$$\frac{\Gamma \vdash M : \sigma \qquad \Gamma \vdash \sigma \le \tau}{\Gamma \vdash M : \tau} \qquad \frac{\Gamma, Y \approx \nabla X\sigma \vdash M : \tau}{\Gamma \vdash M : \forall Y \approx \nabla X\sigma.\, \tau}$$

Fig. 4. $\lambda^{\mathsf{fix}\mu\nu}$: Typing rules.

6 Typing and Soundness

Figure 4 displays the typing rules of $\lambda^{\mathsf{fix}\mu\nu}$. We excluded polymorphism, hence terms are only assigned *closed* types. By this we mean that τ contains no free type variables X, whereas approximations Y^n of fixed-point types are permitted.

The rules for lambda-calculus, sums and products are omitted since they are standard. For wrapping (fold) and unwrapping (unfold) (co)inductive types there are two rules each, one rule for the fixed-point and one dealing with approximations Y^{n+1}, keeping track of the stage. The rules for (co)recursion carry the side conditions developed in the previous sections. Furthermore, we have rules for subsumption (which includes $\forall$-instantiation) and $\forall$-introduction independent of a fix^∇.

Reduction is type preserving, I have shown this for a slightly simpler system [2]. The proof of soundness of typing (implying strong normalization) is similar to Barthe/Giménez [7], although I developed it independently following Abel/Altenkirch [3]. Due to lack of space I can only give a sketch.

Let the application $M\theta$ of a substitution θ to a term M be defined as expected.

Lemma 9 (Soundness for fix^μ). *Let $\Gamma \vdash Y$ para τ, $\theta \in [\![\Gamma]\!]$ a substitution and $\Phi = \Phi_{\mu X\sigma,\theta}$. Assume $M\theta' \in [\![Y^1 \to \tau(Y^1)]\!]\theta'$ for all $\theta' \in [\![\Gamma, Y \approx \mu X\sigma, g : Y \to \tau(Y)]\!]$. Then*

$$\mathsf{fix}^\mu g.M\theta \in \bigcap_{\alpha<\Omega} \Phi^\alpha \to [\![\tau]\!](\theta, Y \mapsto (\Phi, \mu, \alpha)).$$

Proof. By transfinite induction on α. The base case holds since Φ^0 contains no canonical terms. The step case makes use of saturatedness, the ind. hyp. and the assumption on M. The limit case follows by paracontinuity.

The soundness of fix^ν requires a little more work. First we give a judgment $E[\bullet] \in [\![\tau \multimap \rho]\!]\theta$ that types some evaluation contexts.

$$\frac{}{\bullet \in [\![\rho \multimap \rho]\!]\theta} \qquad \frac{E[\bullet] \in [\![\tau \multimap \rho]\!]\theta \qquad N \in [\![\sigma]\!]\theta}{E[\bullet\, N] \in [\![(\sigma \to \tau) \multimap \rho]\!]}$$

$$\frac{E[\bullet] \in [\![\tau_1 \multimap \rho]\!]}{E[\mathsf{fst}\,\bullet] \in [\![\tau_1 \times \tau_2 \multimap \rho]\!]} \qquad \frac{E[\bullet] \in [\![\tau_2 \multimap \rho]\!]}{E[\mathsf{snd}\,\bullet] \in [\![\tau_1 \times \tau_2 \multimap \rho]\!]}$$

The linear arrow $\multimap$ is not a first-class type constructor, rather some notation. It is motivated by the fact that evaluation contexts E have exactly one hole $\bullet$.

Lemma 10. *Let $\Gamma \vdash Y$ legal$^\nu$ τ and some $\theta \in [\![\Gamma]\!]$. Assume that for all $E[\bullet] \in [\![\tau \multimap Y]\!]\theta$, $E[M] \in [\![Y]\!]\theta$. Then $M \in [\![\tau]\!]\theta$.*

Proof. By induction on $\Gamma \vdash Y$ legal$^\nu$ τ.

Lemma 11 (Soundness of fix^ν). *Let $\Gamma \vdash Y$ legal$^\nu$ τ, substitution $\theta \in [\![\Gamma]\!]$ and $\Phi = \Phi_{\nu X\sigma,\theta}$. Assume $M\theta' \in [\![\tau(Y^1)]\!]\theta'$ for all $\theta' \in [\![\Gamma, Y \approx \nu X\sigma, g : \tau(Y)]\!]$. Then*

$$\mathsf{fix}^\nu g.M\theta \in \bigcap_{\alpha<\Omega} [\![\tau]\!](\theta, Y \mapsto (\Phi, \nu, \alpha))$$

Proof. By transfinite induction on α. Note that the assumption implies $G \equiv \mathsf{fix}^\nu g.M\theta \in \mathsf{SN}$. The base case follows by Lemma 10 since $\Phi^0 = \mathsf{SN}$, and the limit case is a consequence of continuity of $[\![\tau(Y)]\!]$. The step case makes crucial use of Lemma 10, since we consider $\mathsf{unfold}\, E[G] \longrightarrow_\beta \mathsf{unfold}\, E[M\theta']$ for an appropriate extension θ' of θ and all $E[\bullet] \in [\![\tau \multimap Y]\!](\theta, Y \mapsto \Phi^{\alpha+1})$. The r.h.s. can be shown to be in the semantics by assumption, so the l.h.s. as well by saturation, which entails the goal. □

Theorem 3 (Soundness of Typing). $\Gamma \vdash M : \tau$, $\theta \in [\![\Gamma]\!]$ *imply* $M\theta \in [\![\tau]\!]$.

Soundness, proved by induction on $\Gamma \vdash M : \tau$, now directly entails strong normalization (choose some θ which keeps all term variables fixed).

References

1. Andreas Abel. Specification and verification of a formal system for structurally recursive functions. In *TYPES '99*, vol. 1956 of *LNCS*, pages 1–20. Springer, 2000.
2. Andreas Abel. Termination checking with types. Technical Report 0201, Institut für Informatik, Ludwigs-Maximilians-Universität München, 2002.
3. A. Abel and T. Altenkirch. A predicative strong norm. proof for a λ-calculus with interleaving inductive types. In *TYPES '99*, vol. 1956 of *LNCS*. Springer, 2000.
4. Andreas Abel and Thorsten Altenkirch. A predicative analysis of structural recursion. *Journal of Functional Programming*, 12(1):1–41, January 2002.
5. Thorsten Altenkirch. *Constructions, Inductive Types and Strong Normalization.* PhD thesis, University of Edinburgh, November 1993.
6. Roberto M. Amadio and Solange Coupet-Grimal. Analysis of a guard condition in type theory. In *FoSSaCS '98*, volume 1378 of *LNCS*. Springer, 1998.
7. G. Barthe, M. J. Frade, E. Giménez, L. Pinto, and T. Uustalu. Type-based termination of recursive definitions. *Math. Struct. in Comp. Sci.*, 2002. To appear.
8. Thierry Coquand. Infinite objects in type theory. In *TYPES '93*, volume 806 of *LNCS*, pages 62–78. Springer, 1993.
9. Eduardo Giménez. Codifying guarded definitions with recursive schemes. In *TYPES´94*, volume 996 of *LNCS*, pages 39–59. Springer, 1995.
10. Martin Hofmann. Non strictly positive datatypes for breadth first search. TYPES mailing list, 1993.
11. John Hughes and Lars Pareto. Recursion and dynamic data-structures in bounded space: Towards embedded ML programming. In *ICFP'99*, pages 70–81, 1999.
12. John Hughes, Lars Pareto, and Amr Sabry. Proving the correctness of reactive systems using sized types. In *POPL'96*, pages 410–423. ACM Press, 1996.
13. INRIA. *The Coq Proof Assistant Reference Manual*, version 7.0 edition, April 2001.
14. Chin Soon Lee, Neil D. Jones, and Amir M. Ben-Amram. The size-change principle for program termination. In *POPL'01*. ACM Press, 2001.
15. David McAllester and Kostas Arkoudas. Walther Recursion. In *CADE-13*, volume 1104 of *LNCS*. Springer, 1996.
16. Nax Paul Mendler. Inductive types and type constraints in the second-order lambda calculus. *Annals of Pure and Applied Logic*, 51(1–2):159–172, 1991.
17. Lars Pareto. *Types for Crash Prevention.* PhD thesis, Chalmers University of Technology, 2000.
18. Frank Pfenning and Carsten Schürmann. Twelf – a meta-logical framework for deductive systems. In *CADE-16*, volume 1632 of *LNAI*. Springer, 1999.
19. Brigitte Pientka. Termination and reduction checking for higher-order logic programs. In *IJCAR 2001*, volume 2083 of *LNAI*, pages 401–415. Springer, 2001.
20. Randy Pollack. *The Theory of LEGO*. PhD thesis, University of Edinburgh, 1994.
21. Christophe Raffalli. Data types, infinity and equality in System AF2. In *CSL '93*, volume 832 of *LNCS*, pages 280–294. Springer, 1994.
22. Alastair J. Telford and David A. Turner. Ensuring streams flow. In *AMAST '97*, volume 1349 of *LNCS*, pages 509–523. Springer, 1997.
23. Hongwei Xi. Dependent types for program termination verification. *Journal of Higher-Order and Symbolic Computation*, 15:91–131, 2002.

Derivatives of Containers

Michael Abbott[1], Thorsten Altenkirch[2], Neil Ghani[1], and Conor McBride[3]

[1] Department of Mathematics and Computer Science, University of Leicester
michael@araneidae.co.uk, ng13@mcs.le.ac.uk

[2] School of Computer Science and Information Technology, Nottingham University
txa@cs.nott.ac.uk

[3] Department of Computer Science, University of Durham
c.t.mcbride@durham.ac.uk

Abstract. We are investigating McBride's idea that the type of one-hole contexts are the formal derivative of a functor from a categorical perspective. Exploiting our recent work on containers we are able to characterise derivatives by a universal property and show that the laws of calculus including a rule for initial algebras as presented by McBride hold — hence the differentiable containers include those generated by polynomials and least fixpoints. Finally, we discuss abstract containers (i.e. quotients of containers) — this includes a container which plays the role of e^x in calculus by being its own derivative.

1 Introduction

In his classic functional pearl Huet (1997) shows how to represent a tree with one of its subtrees 'in focus' by a pair of the subtree and the one-hole context (or 'zipper') in which it sits. The unpublished article McBride (2001) gives a 'generic program' for computing the type of one-hole contexts for any regular inductive datatype: remarkably, the key step is to *differentiate* the functor which generates the datatype by the rules we learned from Leibniz (1684). It was an observation in search of an explanation. In this paper, we find such an explanation from a categorical perspective.

Our categorical presentation of *containers* in Abbott et al. (2003) provides the key to the mystery. We now specify differentiation by a universal property in the category of containers, more precisely $\partial F \cong \mathrm{Id} \multimap F$ where $H \multimap -$ is the right adjoint of $- \times H$ in the category $\widehat{\mathcal{G}}$ of *cartesian* morphisms between containers. We thus uncover the linear notion of *tangent* which McBride's programs mechanise.

Given this specification by universal property, we verify Leibniz's laws, McBride's extension for initial algebras (via the chain rule) and further conjecture an extension to terminal coalgebras. We note also that there are also containers which are not differentiable, such as $(\mathbb{N} \Rightarrow \mathbb{N}) \Rightarrow -$.

Our present work rediscovers the notion of a derivative of an analytic functor, Joyal (1986), from a computational perspective. We also generalise his approach, considering a more general class of functors and categories. Inspired by Joyal, we introduce the notion of an *abstract container*, which arises when closing containers under coequalisers. In particular we seek out the container which corresponds to the exponential function: just like e^x, it is its own derivative.

M. Hofmann (Ed.): TLCA 2003, LNCS 2701, pp. 16–30, 2003.

2   Background

McBride's article gives the computational intuition to differentiating analytic functors. He presents *syntactically* a class of functors F, closed under polynomial operations and (first-order) least fixed point. He then shows that (for a general X), $(\partial F)X$ represents 'an FX with one hole for an X', and gives the corresponding *linear application* ('plugging-in') operator, $@_F$ in the diagram below:

$$(\partial F)X \quad d \quad @_F \qquad = \qquad X \quad FX \quad d$$

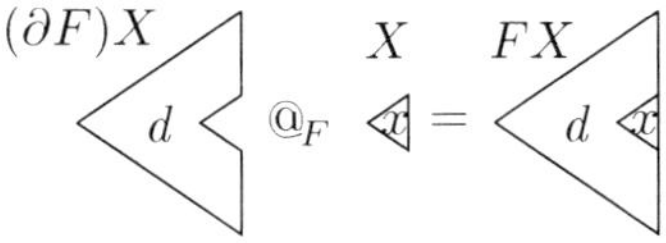

Curried, $@_F$ embeds $(\partial F)X$ into the function space $X \to FX$, of which it is the linear fragment. One immediate application is to make Huet's notion of 'zipper' generic over inductive datatypes μF wherever ∂F can be defined. The single constructor $\mathrm{in}_F \;:\; F(\mu F) \to \mu F$ making an element of μF from a 'container' of its immediate subelements, is an isomorphism. Hence, $(\partial F)(\mu F)$ gives 'elements of μF with a hole for one immediate sub-μF'. Just as the 'subterm' relation is the reflexive-transitive closure of the 'immediate subterm' relation, we can represent 'elements of μF with a hole for one sub-μF' by *lists*, $\mathrm{List}(\partial F)(\mu F))$, where each element is a step on the path from hole to root, recording the surrounding data. Here, lists grow 'on the right', echoing Huet's account of 'zippers' as *stacks*. The corresponding application operation iterates $@_F$:

$$@_F^\mu \;:\; \mathrm{List}((\partial F)(\mu F)) \times \mu F \to \mu F$$

$$\left([] :: d_n \cdots :: d_1 \right) @_F^\mu u \;=\; \mathrm{in}_F \left\{ d_n @_F \cdots \mathrm{in}_F \left\{ d_1 @_F u \right. \right.$$

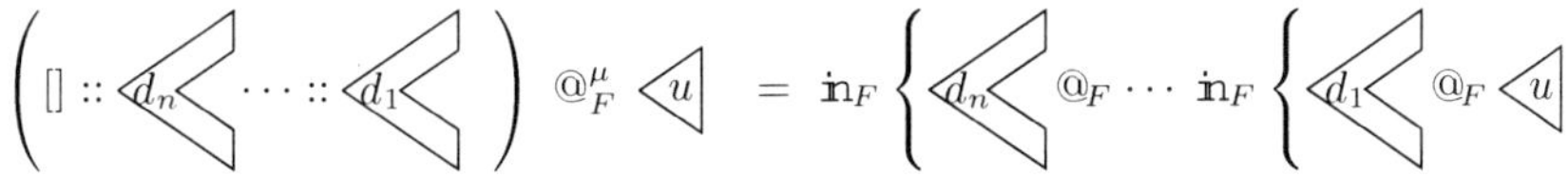

The zipper data structure has many uses: Huet's example is 'structure editing', with $\mathrm{List}((\partial F)(\mu F)) \times \mu F$ exactly representing μF with a 'cursor' at one node. Many tree-rewriting operations can thus be expressed in terms of more primitive navigating (moving the cursor) and editing (replacing the term at the cursor) steps. Context-sensitive operations need merely inspect the list.

Moreover, we gain a language with which we can express general properties of datatypes more easily, and we gain tools with which to prove them. For example, 'inductive datatypes have no cycles' becomes

$$ds \;@_F^\mu\; u = u \;\Rightarrow\; ds = [] \ .$$

Generically, structural induction on μF amounts to showing $P(\mathrm{in}_F\, t)$ given hypotheses $P(u)$ for each d, u such that $t = d \;@_F\; u$.

2.1   What Is ∂?

How can we explain ∂F in terms of F? Intuitively, $d \;@_F\; x$ computes an element of FX, using x exactly once, and without inspecting x. That is, as Huet (2003) suggests, ∂F corresponds to $X \multimap FX$, the *linear* fragment of $X \to FX$, in a sense which we make precise in Definition 4.2. We can think of this linear approximation, $X \multimap FX$, as a kind of *tangent* to F at X. Similarly, Ehrhard and Regnier (2001) use differentiation to compute linear approximants to λ-terms. Hence it is not so remarkable that ∂ behaves like differentiation—let us check some familiar laws informally. Consider an element of $(F \times G)$, containing a particular chosen element of X.

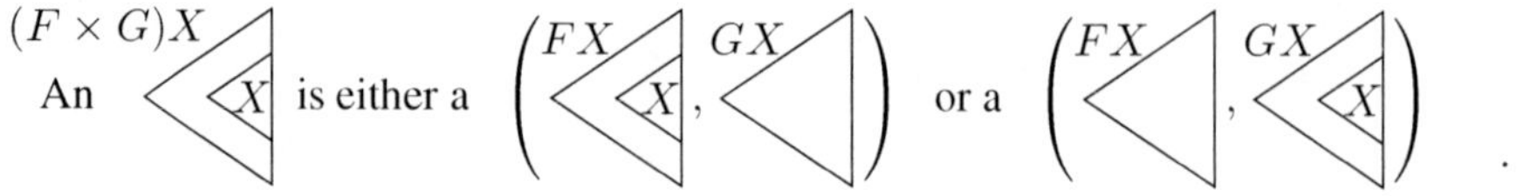

$$(\partial(F \times G))X \quad \cong \quad (\partial F)X \quad \times \quad GX \quad + \quad FX \quad \times \quad (\partial G)X$$

The X is either in the FX part or the GX part. Now, $(\partial(F \times G))X$ must record the possible contexts for our chosen X — the 'chevrons' which remain when the X is cut out. If the X lies within the FX, we must record a $(\partial F)X$ and the whole of the GX, and correspondingly in the other case. Hence the law holds.

Next, choosing an X within an $(F \circ G)X$ amounts to choosing a GX within an $F(GX)$, then an X within the GX.

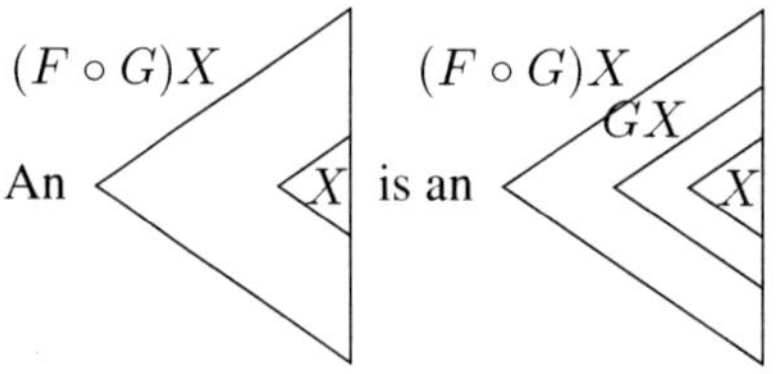

$$(\partial(F \circ G))X \cong (\partial F)(GX) \times (\partial G)X$$

The X's context comes in two parts, corresponding with Leibniz's 'chain rule'.

2.2 Containers and Derivatives

Containers are a formalisation of the idea that many important datatypes consist of templates where data is stored. For example, any element of the type of lists List X can be uniquely written as a natural number n, the length of the list, together with a function $\{1, \ldots, n\} \to X$ which labels each position in the list with an element from X:

$$n : \mathbb{N} \ , \qquad \sigma : \{1 \,..\, n\} \to X \ .$$

Similarly, any binary tree is given by its underlying shape (obtained by deleting the data stored at the leaves) and a mapping from positions in this shape to the data thus:

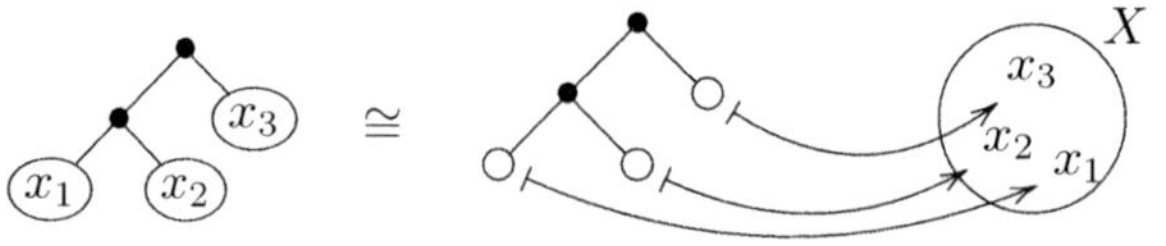

Thus we are led to consider datatypes which are given by a set of shapes S and, for each $s \in S$, a family of positions $(P_s)_{s \in S}$. This presentation of the datatype defines an endofunctor $X \mapsto \coprod_{s \in S} X^{P_s}$ on **Set**. More generally, Abbott et al. (2003) formalises these intuitions by considering families of objects in a locally cartesian closed category

$\mathbb{C}$, where the family $s{:}S \vdash P(s)$ is represented by an object $P \in \mathbb{C}/S$, and the associated functor or *container* $T_{S \rhd P} : \mathbb{C} \to \mathbb{C}$ is defined by

$$T_{S \rhd P} X \equiv \Sigma s : S.\, (P(s) \Rightarrow X) \enspace .$$

In effect, this gives a generalised 'power series' for functors representing containers of X. We might hope that their derivatives might obey a law like

$$\partial T_{S \rhd P} X \cong \Sigma s : S.\, (P(s) \times ((P(s) - 1) \Rightarrow X))$$

and in fact they do, provided we can explain what we mean by $(P(s) - 1)$. Deleting an X within some $(s, f) : T_{S \rhd P} X$ amounts to choosing the $p : P(s)$ which points to it. In doing so, we effectively select the shape of the context for the X; then such a context contains an X for each position in $P(s)$ *except* p. The analogue of $P(s) - 1$ is thus

$$\Sigma p' : P(s).\, \neg \mathrm{Eq}(p, p')$$

where p is the position of the missing X. For this to behave properly, $P(s)$ must possess a *decidable equality*, so that we can tell if any p' is the hole position, p:

$$\mathrm{Eq}(p, p') + \neg \mathrm{Eq}(p, p') \cong 1$$

so, indeed, for each $p : P(s)$:

$$P(s) \cong \Sigma p' : P(s).\, \neg \mathrm{Eq}(p, p') + 1 \enspace .$$

2.3 Notation and Conventions

We use the same general semantic domain as in Abbott et al. (2003) where the category $\mathbb{C}$ is extensive[1], locally cartesian closed and locally finitely presentable (Adámek and Rosický, 1994).

We write $a : A \vdash B(a)$ or just $A \vdash B$ for $B \in \mathbb{C}/A$, and similarly $\Sigma a : A.\, B(a)$ and $\Sigma_A B$ for the domain of B regarded as an object. We'll write $a{:}A, b{:}B(a) \vdash C(a, b)$ as a shorthand for $(a, b) : \Sigma_A B \vdash C(a, b)$. We omit variables and weakenings when the meaning is clear from the context. For example, given $A \vdash B$ we can interchangeably write this as $a : A, c : C \vdash B(a)$ or $A \times C \vdash \pi^* B$ (for $\pi : A \times C \to A$).

For $A \vdash B$ write $\pi_B : \Sigma_A B \to A$ for the map projecting out the first component and write $\pi'_B : 1 \to \pi_B^* B$ for the morphism in $\mathbb{C}/\Sigma_A B$ corresponding to the second variable $a{:}A,\ b{:}B(a) \vdash b{:}B(a)$. Here we follow Hofmann (1994). When A and B are objects of $\mathbb{C}$ we may interchangeably write $A^* B$ or $\pi_A^* B$ for the object $\pi : A \times B \to A$ of $\mathbb{C}/A$.

Given $A \vdash B$ and $u : C \to A$ write $u_B : \Sigma_C u^* B \to \Sigma_A B$ for the map making the following square a pullback

$$\begin{array}{ccc} \Sigma_C u^* B & \xrightarrow{\;u_B\;} & \Sigma_A B \\ {\scriptstyle \pi_{u^* B}} \downarrow & \lrcorner & \downarrow {\scriptstyle \pi_B} \\ C & \xrightarrow[\;u\;]{} & A \end{array}$$

[1] In general a category $\mathbb{C}$ with coproducts is *extensive* iff coproducts are preserved by pullbacks, coprojections into coproducts are monomorphisms, and the pullback of distinct coprojections $\kappa_i : A_i \to \coprod_{i \in I} A_i$ into a coproduct is always the initial object 0. In a cartesian closed category all conditions except the last automatically hold.

We'll write $\neg B \equiv 0^B$. We'll interchangeably write $B \Rightarrow X$ and X^B for the local exponent. Given an object $B \in \mathbb{C}$ the equality type $B \times B \vdash \mathrm{Eq}_B$ (or Eq when the type B is obvious from context) is given by the morphism $\delta : B \to B \times B$.

We denote the *global coproduct* of $A \vdash B$ and $C \vdash D$ by $A + C \vdash B \mathbin{\dot{+}} D$. We denote the fibred coproduct of objects $A \vdash B$ and $A \vdash D$ over a common base by $A \vdash B + D$.

3 The Category of Containers

In Abbott et al. (2003), we began the investigation of functors which arise from containers by defining

$$T_{S \rhd P} X \equiv \Sigma s : S.\, (P(s) \Rightarrow X)$$

where S is the object of shapes and P is the object of positions over S. We proved that such functors include constant functors, the identity and projections and are closed under fixed exponents, sums, products, least fixed points and greatest fixed points. These results are based upon two ideas:

1. To prove closure of container functors under least and greatest fixed points we generalised from containers and endofunctors (as described above) to containers in several parameters and functors $F : \mathbb{C}^n \to \mathbb{C}$.
2. We defined container morphisms and thereby obtain a category $\mathcal{G}$ of *containers*. T then becomes a functor $T : \mathcal{G} \to [\mathbb{C}, \mathbb{C}]$ which is full and faithful (when $[\mathbb{C}, \mathbb{C}]$ is understood as the category of *fibred* endo-functors and natural transformations) and which preserves limits, coproducts and filtered colimits of *cartesian diagrams*.

In the rest of this subsection we briefly describe these results so as to lay the groundwork for defining the derivatives of containers. The reader should consult Abbott et al. (2003) for further details and proofs.

To understand containers in several parameters, consider the bifunctor $1 + X \times Y$ and, in particular, how this bifunctor can be understood in terms of shapes and positions. Firstly, there are two shapes where data may be stored corresponding to the different summands of the coproduct. Above the first shape there are no positions whereas above the second shape there is one position for storing elements of X and one position for storing elements of Y. Crucially, where elements of X are stored is independent from where elements from Y are stored. This independence of how data is stored at one position from how data is stored at another position is a defining hallmark of containers. Consequently, a container in n variables in a locally cartesian closed category $\mathbb{C}$ is an object $A \in \mathbb{C}$ together with n objects over A, i.e. $A \vdash (B_i)_{i \in 1..n}$. The functor generated by such a container sends $\vec{X}$ to $\Sigma a : A.\, \prod_{i \in 1..n} (B_i(a) \Rightarrow X_i)$. For example, the bifunctor $1 + X \times Y$ arises with $n = 2$, $A = \{s_1, s_2\}$, $B_1(s_1) = B_2(s_1) = 0$ and $B_1(s_2) = B_2(s_2) = 1$.

The second element of our analysis is the definition of container morphisms and the properties of the functor T from containers to functors.

Definition 3.1. *Given an index set I define the* category of containers $\mathcal{G}_I$ *as follows:*

- *Objects are pairs* $(A \in \mathbb{C},\ B \in (\mathbb{C}/A)^I)$*; write this as* $(A \rhd B) \in \mathcal{G}_I$
- *A morphism* $(A \rhd B) \to (C \rhd D)$ *is a pair* (u, f) *for* $u : A \to C$ *in* $\mathbb{C}$ *and* $f : (u^*)^I D \to B$ *in* $(\mathbb{C}/A)^I$.

As an intuition for container morphisms, consider the the map $\mathsf{tail} : \mathsf{List}\, X \to 1 + \mathsf{List}\, X$ taking the empty list to 1 and otherwise yielding the tail of the given list:

tail of list

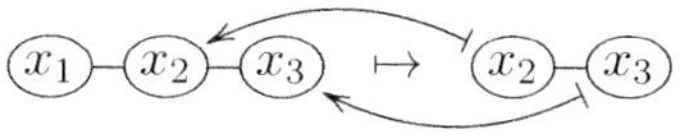

This map is defined by i) for each shape s in $\mathsf{List}\, X$ the choice of a shape $u(s)$ in $1 + \mathsf{List}\, X$; and ii) for each position above a shape $u(s)$, the choice of a position above s. This second component allows us to transform a labelling of the positions above a specific shape s to a labelling of the positions above $u(s)$. Notice the essential contravariance of this second component which results in a non-trivial mathematical theory. The first key result is that such container morphisms do indeed define natural transformations. More generally we can define:

Definition 3.2. *Define the functor* $T : \mathcal{G}_I \to [\mathbb{C}^I, \mathbb{C}]$ *taking each container* $(A \rhd B) \in \mathcal{G}_I$ *to its "extension", the* container functor $T_{A \rhd B}$, *as follows. For each* $X \in \mathbb{C}^I$ *define*

$$T_{A \rhd B} X \equiv \Sigma a : A. \prod\nolimits_{i \in I} (B_i(a) \Rightarrow X_i) \ ,$$

and for $(u, f) : (A \rhd B) \to (C \rhd D)$ *define* $T_{u,f} : T_{A \rhd B} \to T_{C \rhd D}$ *to be the natural transformation* $T_{u,f} X : T_{A \rhd B} X \to T_{C \rhd D} X$ *thus:*

$$(a, g) : T_{A \rhd B} X \ \vdash \ T_{u,f} X(a, g) \equiv (u(a), (g_i \cdot f_i)_{i \in I}) \ .$$

A typical functor which is not in the image of T is $\neg\neg(X) = (X \to 0) \to 0$. The key properties of containers and the functor T were proved in Abbott et al. (2003) and are summarised below.

Theorem 3.3. *The following are true*

- *For each container* $F \in \mathcal{G}_I$ *and each container morphism* $\alpha : F \to G$ *the functor* T_F *and natural transformation* T_α *are fibred over* $\mathbb{C}$.
- *The following containers define the identity, projection and constant functors*

$$\mathrm{Id} \equiv (1 \rhd 1) \qquad \mathrm{Id}_j \equiv (1 \rhd (\mathrm{Eq}(i, j))_{i \in I}) \qquad K_C \equiv (C \rhd 0)$$

  where $\mathrm{Eq}(i, j)$ *is* 1 *if* $i = j$ *and* 0 *otherwise.*
- *Containers are closed under sums, products and substitution as given by the formulae*

$$(A \rhd B) \times (C \rhd D) \cong (A \times C \rhd \pi^{\square} B + \pi^{\square\square} D) = (a : A,\ c : C \rhd B(a) + D(c))$$
$$(A \rhd B) + (C \rhd D) \cong (A + C \rhd B \overset{\square}{+} D)$$
$$(A \rhd B, E)[(C \rhd D)] \cong \big(a : A, f : C^{E(a)} \rhd B(a) + \Sigma e : E(a).\, D(fe)\big)$$

- *The functor* $T : \mathcal{G}_I \to [\mathbb{C}^I, \mathbb{C}]$ *is full and faithful and preserves limits and coproducts.*

Let us consider, for example, closure of containers under coproducts. Intuitively, if F and G are container functors generated by the containers $(A \triangleright B)$ and $(C \triangleright D)$, then an element of $F + G$ must be either an element from F or from G. Thus the shapes of $F + G$ must be $A + C$. In addition, the positions above a shape $\kappa(a)$ should just be $B(a)$ and similarly the positions above $\kappa'(c)$ should be $D(c)$. Thus the object of positions must be the global coproduct $B \mathbin{\text{⨥}} D$. A formal proof uses the fact that the formula defines the coproduct in $\mathcal{G}_I$ and that T preserves coproducts. A similar argument shows that containers are closed under products. This construction can be also found in Dybjer (1996) where it is used as an invariant in the proof that strictly positive types can be represented by W-types.

The substitution formula is used to construct chains (cochains) of containers which underpin their closure under least (greatest) fixed points. In particular, given a container in $\mathcal{G}_{I+1}$, we wish to fix the first I parameters and use substitution to build the relevant chain (cochain) whose colimit (limit) defines the least (greatest) fixed point. In effect, substitution is a functor $-[-]:\mathcal{G}_{I+1} \times \mathcal{G}_I \to \mathcal{G}_I$. When we substitute the $I+1$-parameter container $(A \triangleright B, E)$ (where B is the I-indexed object of positions and E is the $I+1$'th object of positions) with the I-parameter container $(C \triangleright D)$, we leave the B positions unaffected and replace each E-position with a shape in C above which data may be stored. Thus, the shapes of the resulting container will be a shape $a : A$ and a function $f : E(a) \Rightarrow C$. Above such a shape we will then have all the old B positions together with the D positions above fe for any $e : E(a)$.

Once we have substitution in place, we can defined fixed points by chains and cochains. Although $\mathcal{G}_I$ has colimits they are not in general preserved by T. This also applies to filtered colimits which underpin the construction of initial algebras. Fortunately, there exists a class of filtered colimits which is both sufficient for the construction of initial algebras and which *are* preserved by T. These diagrams are called *cartesian diagrams* as they are those diagrams which are built from *cartesian* morphisms defined as follows:

Definition 3.4. *A morphism $(u, f) : (A \triangleright B) \to (C \triangleright D)$ of containers is* cartesian *iff f is iso. The category of containers and cartesian morphisms is written $\widehat{\mathcal{G}}$.*

A natural transformation $\alpha : F \to G$ between two functors is cartesian *iff each naturality square of α is a pullback. The category of functors and cartesian natural transformations is written $\widehat{\mathcal{F}}$.*

In a cartesian morphism the map on positions merely rearranges them, and so morphisms in $\widehat{\mathcal{G}}$ cannot forget or copy data, hence they are in effect *linear* morphisms.

The following tells us that T restricts to a full and faithful functor $\widehat{T} : \widehat{\mathcal{G}} \to \widehat{\mathcal{F}}$.

Proposition 3.5. *A natural transformation $\alpha : T_F \to T_G$ between container functors is cartesian iff the corresponding container morphism (given by T full and faithful) is cartesian.* □

Further, taking products restricts to a monoidal operation on $\widehat{\mathcal{G}}$ and $\widehat{\mathcal{F}}$ (as the projection $\pi : F \times G \to F$ is not cartesian, the monoidal operation is not the cartesian product).

Finally, we have all the elements in place to define least and greatest fixed points of containers. Given a container $(A \triangleright B, E)$ in $\mathcal{G}_{I+1}$, substitution gives a functor $(A \triangleright B, E)[-] : \mathcal{G}_I \to \mathcal{G}_I$ and hence we can define the least fixed point $\mu(A \triangleright B, E)$ as the colimit of the chain

$$0 \longrightarrow (A \triangleright B, E)[0] \longrightarrow (A \triangleright B, E)^2[0] \longrightarrow \cdots$$

Crucially, this chain is a filtered cartesian diagram and hence its colimit is preserved by T. This has two important consequences i) our construction is correct:

$$T_{\bigvee (A \triangleright B, E)^i[0]} \cong \bigvee T_{(A \triangleright B, E)^i[0]}$$

and ii) the colimiting container $\bigvee (A \triangleright B, E)^i[0]$ has a concrete description as

$$\bigvee (A \triangleright B, E)^i[0] \cong (\bigvee A_i \triangleright \bigvee B_i)$$

where A_i is the object of shapes of $(A \triangleright B, E)^i[0]$ and B_i is the object of positions of $(A \triangleright B, E)^i[0]$.

Greatest fixed points are easier. One uses substitution to define the appropriate cochain. Since $\mathcal{G}_I$ has all limits, $(A \triangleright B, E)[-]$ preserves connected limits and T preserves all limits, containers are closed under greatest fixed points.

4 Derivatives of Containers

We now make the notion of derivative of a container explicit.

Definition 4.1. *If for containers F and H there exists a universal arrow in $\widehat{\mathcal{G}}$ from the functor $- \times H$ to F say that the* linear exponential of F by H *exists. Write the linear exponential as $H \multimap F$.*

In this situation we have a bijection of morphisms in $\widehat{\mathcal{G}}$

$$\frac{G \times H \longrightarrow F}{\overline{\overline{G \longrightarrow H \multimap F}}} \; .$$

In this paper we investigate the special case when $H = \mathrm{Id}$.

Note that $\times$ here refers to the cartesian product in $\mathcal{G}$, it is just a tensor product in $\widehat{\mathcal{G}}$. Now, $\multimap$ gives us a convenient way to introduce the type of *one-hole contexts*:

Definition 4.2. *Say that a container F is* differentiable *iff the linear exponential* $\mathrm{Id} \multimap F$ *exists. Call this the* derivative *of F, written $\partial F \equiv \mathrm{Id} \multimap F$.*

Similarly for a container F in I arguments say that F is differentiable at $i \in I$ *iff* $\mathrm{Id}_i \multimap F$ *exists and write this as $\partial_i F$.*

The notion of *decidability* turns out to be central to the construction of derivatives.

Definition 4.3. *Say that an object $A \vdash B$ is* decidable *iff for each $a : A$ the equality relation on $B(a)$ is decidable, in other words $b, b' : B \vdash \mathrm{Eq}(b, b') + \neg\mathrm{Eq}(b, b') \cong 1$.*

Say that a container $F \equiv (A \triangleright B)$ is decidable *iff B is decidable. Say that a container $F \equiv (A \triangleright (B_i)_{i \in I})$ in I parameters is* decidable at j *iff B_j is decidable, and that F* is decidable *iff it is decidable at each $j \in I$.*

The following key theorem will be proved later on in section 4.2.

Theorem 4.4. *Every decidable container $F \equiv (A \triangleright B)$ is differentiable with derivative*

$$\partial(A \triangleright B) \cong (\Sigma_A B \triangleright B') \ .$$

Similarly, every container $F \equiv (A \triangleright (B_i)_{i \in I})$ decidable at j is differentiable with respect to j with derivative

$$\partial_j(A \triangleright (B_i)_{i \in I}) \cong (\Sigma_A B_j \triangleright (B_i'^j)_{i \in I}) \ ,$$

where $B_i'^j \equiv \pi^*_{B_j} B_i$ when $i \neq j$, and $B_i'^i \equiv B'$.

In the special case of a two parameter container $(A \triangleright B, E)$ this is

$$\partial_1(A \triangleright B, E) \cong (\Sigma_A B \triangleright B', \pi^*_B E) \qquad \partial_2(A \triangleright B, E) \cong (\Sigma_A E \triangleright \pi^*_E B, E') \ .$$

As we will show in section 4.2, the following facts about follow from theorem 4.4.

Proposition 4.5. *Derivatives of decidable containers satisfy the following*

$$\partial(F+G) \cong \partial F + \partial G \qquad \partial_i \mathrm{Id}_j \cong \mathrm{Eq}(i,j)$$
$$\partial(F \times G) \cong \partial F \times G + F \times \partial G \qquad \partial K \cong 0$$

and for F a container in 2 parameters

$$\partial(F[G]) \cong \partial_1 F[G] + \partial_2 F[G] \times \partial G$$
$$\partial \mu F \cong \mu(\partial_1 F[\mu F] + \partial_2 F[\mu F] \times \mathrm{Id}) \ .$$

We also conjecture that the derivative of the terminal coalgebra is given by

$$\partial \nu F \cong \mu(\partial_1 F[\nu F] + \partial_2 F[\nu F] \times \mathrm{Id}) \ .$$

Note that the derivative of a coalgebra is a *least* fixed point — this corresponds to the fact that a "hole" in the datatype must be accessible.

The use of containers in definition 4.2 is essential, we could not have derived the laws of calculus by using cartesian maps between functors instead. Indeed, if we assume that $\partial \mathrm{Id} = \mathrm{Id} \multimap \mathrm{Id} \cong 1$ for functors we can construct a cartesian morphism $\neg\neg \to 1$ which would imply (theorem 8.1, Abbott et al., 2003) that $\neg\neg$ is a container.

4.1 Properties of Decidable Objects

First we talk about decidable objects and their general properties. Decidability on B allows us to construct an object which acts as a kind of "complement" to elements of B.

Proposition 4.6. *If $A \vdash B$ is decidable then there exists an object $\Sigma_A B \vdash B'$, called the* complement of B, *satisfying the isomorphism $\pi^*_B B \cong 1 + B'$.*

Proof. Define $B' \equiv \Sigma_B \neg \mathrm{Eq}$ and calculate

$$A, b : B \vdash B \cong \Sigma b' : B.\, 1 \cong \Sigma b' : B.\, (\mathrm{Eq}(b,b') + \neg\mathrm{Eq}(b,b'))$$
$$\cong \Sigma b' : B.\, \mathrm{Eq}(b,b') + \Sigma b' : B.\, \neg\mathrm{Eq}(b,b') \cong 1 + B'(b) \ . \quad \square$$

The following lemma follows from the result that in an extensive category there is a cancellation rule $1 + A \cong 1 + B \implies A \cong B$.

Lemma 4.7. *If $A \vdash B$ is decidable and $\pi_B^* B \cong 1 + D$ then $B' \cong D$.* □

Some important properties of decidable objects follow.

Lemma 4.8. *If $A \vdash B$ is decidable then its complement $\Sigma_A B \vdash B'$ is decidable.*

Proof. Fixing $b : B$ then $a, c : B'(b)$ is equivalent to $a, c : B$ satisfying $a \neq b$ and $c \neq b$, and clearly decidability in B can then be used to compute decidability in B'. □

Lemma 4.9. *If $A \vdash B$ is decidable then for each $u : C \to A$ the pullback $u^* B$ is decidable with $(u^* B)' \cong u_B^* B'$.*

Proof. Observing that $b : u^* B$ can be written as $c : C \vdash b : B(uc)$ it is clear that decidability of B is inherited by $u^* B$. To verify the equation for $(u^* B)'$, calculate $\pi_{u^* B}^* u^* B \cong u_D^* \pi_B^* B \cong u_D^*(1 + B') \cong 1 + u_D^* B'$. □

Lemma 4.10. *Complements are closed under products and coproducts, and satisfy the following equations.*

$$A + B \vdash (A + B)' \cong (A' + \pi_A^* B) \mathbin{⯒} (\pi_B^* A + B')$$
$$A \times B \vdash (A \times B)' \cong \pi^* A' + \pi'^* B' + \pi^* A' \times \pi'^* B'$$
$$0 \vdash 0' \cong 0 \qquad 1 \vdash 1' \cong 0 \ .$$

Proof. The results for $0'$ and $1'$ are immediate.

To show that $A + B$ is decidable, note first that since $\mathbb{C}$ is extensive we can analyse a binding $x : A + B$ into cases: $a : A \vdash x = \kappa a$ or $b : B \vdash y = \kappa' b$, and we know that $\kappa a \neq \kappa' b$ universally holds. Thus given $x, y : A + B$ decidability of equality reduces to decidability of equality separately in A and B.

To verify the complement of $A + B$ calculate:

$$\begin{aligned} \pi_{A+B}^*(A + B) &\cong \pi_{A+B}^* A + \pi_{A+B}^* B \cong \pi_A^*(A + B) \mathbin{⯒} \pi_B^*(A + B) \\ &\cong (\pi_A^* A + \pi_A^* B) \mathbin{⯒} (\pi_B^* A + \pi_B^* B) \\ &\cong (1 + A' + \pi_A^* B) \mathbin{⯒} (\pi_B^* A + 1 + B') \\ &\cong 1 + ((A' + \pi_A^* B) \mathbin{⯒} (\pi_B^* A + B')) \ . \end{aligned}$$

$A \times B$ is immediately decidable. The calculation of its complement is not needed in this paper and so is omitted. □

The corresponding results for Σ and $\mathbin{⯒}$ require a little more attention to the base.

Lemma 4.11. *If $A \vdash B$ and $C \vdash D$ are both decidable then $A + C \vdash B \mathbin{⯒} D$ is decidable with complement*

$$(\Sigma_{A+C}(B \mathbin{⯒} D) \vdash (B \mathbin{⯒} D)') \cong (\Sigma_A B + \Sigma_C D \vdash B' \mathbin{⯒} D') \ .$$

If A is also decidable then so is $\Sigma_A B$ with complement

$$\Sigma_A B \vdash (\Sigma_A B)' \cong B' + \Sigma_{A'} B \ . \ \square$$

4.2 Theorems about Derivatives

We restate theorem 4.4 and prove it here by establishing the universal property.

Theorem 4.12. *If F is decidable (definition 4.3) then its derivative ∂F exists and is given by the formula*

$$\partial(A \triangleright B) \cong (\Sigma_A B \triangleright B') \ .$$

Proof. Let $F \equiv (A \triangleright B)$ be a decidable container and write $\phi : \pi_B^* B \cong 1 + B'$ for the isomorphism derived from the decidability of B. Write $dF \equiv (\Sigma_A B \triangleright B')$ and observe that $dF \times \mathrm{Id} \cong (\Sigma_A B \triangleright 1 + B')$. Now define the container morphism $@_F$ thus (here ϕ acts as a map of container positions, and is therefore contravariant to the shape map π_B):

$$@_F \equiv (\pi_B, \phi) : dF \times \mathrm{Id} \longrightarrow F \ ;$$

this map is obviously cartesian and is a map in $\widehat{\mathcal{G}}$. We will now show that $@_F$ is a universal arrow, and that therefore $dF \cong \partial F$ as required by the theorem.

Given a cartesian morphism $G \times \mathrm{Id} \to F$ where $G \equiv (C \triangleright D)$ by unrolling the definitions of morphisms in $\mathcal{G}$ this can understood as a pair of maps $(u : C \to A,\ f : 1 + D \cong u^* B)$, and the universal property of $@_F$ corresponds to a bijection

$$\frac{u : C \longrightarrow A \qquad f : 1 + D \cong u^* B}{v : C \longrightarrow \Sigma_A B \qquad g : D \cong v^* B'} \quad \text{such that} \quad \begin{array}{l} u = \pi_B \cdot v \\ f = v^* \phi \cdot (1 + g) \end{array} .$$

So given (u, f) for $c : C$ note that from the isomorphism $f_c : 1 + D(c) \cong B(uc)$ we can choose $b : B(uc)$ such that $D(c) \cong B(uc)'(b)$. This assignment $c \mapsto b$ together with the associated isomorphism gives us morphisms $v : C \to \Sigma_A C$ and $g : D \cong v^* B'$ as required, and it is clear from the construction that this assignment is unique. □

We can now use the representation of derivatives established by this theorem to directly compute a number of important results about the derivatives of decidable containers.

To begin with, the following proposition tells us that if ∂F exists then so does $\partial^n F$ for every natural number n.

Proposition 4.13. *If a container is decidable then so is its derivative.*

Proof. This follows from lemma 4.8 that B' is decidable over B. □

The elementary type operations interact with differentiation as one might expect.

Proposition 4.14. *Derivatives of decidable containers satisfy the following*

$$\partial(F + G) \cong \partial F + \partial G \qquad \partial(F \times G) \cong \partial F \times G + F \times \partial G$$
$$\partial K \cong 0 \qquad \partial_i \mathrm{Id}_j \cong \mathrm{Eq}(i, j)$$

The equations for $F + G$, $F \times G$ and K are unchanged if ∂ is replaced by ∂_i.

Proof. Each of these isomorphisms follows pretty directly from the results already proved for decidable objects. Let $F \equiv (A \triangleright B)$, $G \equiv (C \triangleright D)$, then:

$$\begin{aligned}\partial(F+G) &\cong \partial((A \triangleright B) + (C \triangleright D)) \cong \partial(A + C \triangleright B \mathbin{\overset{\circ}{+}} D)\\ &\cong (\Sigma_{A+C}(B \mathbin{\overset{\circ}{+}} D) \triangleright (B \mathbin{\overset{\circ}{+}} D)') \cong (\Sigma_A B + \Sigma_C D \triangleright B' \mathbin{\overset{\circ}{+}} D')\\ &\cong (\Sigma_A B \triangleright B') + (\Sigma_C D \triangleright D') \cong \partial F + \partial G \ .\end{aligned}$$

The following derivation of $\partial(F \times G)$ omits both variables and explicit weakening, but these can be inferred unambiguously:

$$\begin{aligned}\partial(F \times G) &\cong \partial((A \triangleright B) \times (C \triangleright D)) \cong \partial(A \times C \triangleright B + D)\\ &\cong (\Sigma_{A\times C}(B+D) \triangleright (B+D)^{\square}) \cong (\Sigma_{A\times C}(B+D) \triangleright (B^{\square} + D) \mathbin{\overset{\square}{+}} (B + D^{\square}))\\ &\cong (\Sigma_{A\times C} B \triangleright B^{\square} + D) + (\Sigma_{A\times C} D \triangleright B + D^{\square})\\ &\cong (\Sigma_A B \triangleright B^{\square}) \times (C \triangleright D) + (A \triangleright B) \times (\Sigma_C D \triangleright D^{\square}) \cong \partial F \times G + F \times \partial G \ .\end{aligned}$$

A constant type $K \equiv (K \triangleright 0)$ has derivative $\partial K = (\Sigma_K 0 \triangleright 0') \cong (0 \triangleright 0) \cong 0$. Similarly $\partial_i \mathrm{Id}_j = (\mathrm{Eq}(i,j) \triangleright \mathrm{Eq}(i,j)') \cong (\mathrm{Eq}(i,j) \triangleright 0) \cong \mathrm{Eq}(i,j)$. □

An analogous form of the chain rule also holds for derivatives.

Proposition 4.15. *For decidable F and G the chain rule holds:*

$$\partial(F[G]) \cong \partial_1 F[G] + \partial_2 F[G] \times \partial G \ .$$

Proof. First note that in general

$$\begin{aligned}\partial(A \triangleright B + C) &\cong (\Sigma_A(B+C) \triangleright (B+C)')\\ &\cong (\Sigma_A(B+C) \triangleright (B' + \pi_B^* C) \mathbin{\overset{\circ}{+}} (\pi_C^* B + C'))\\ &\cong (\Sigma_A B \triangleright B' + \pi_B^* C) + (\Sigma_A C \triangleright \pi_C^* B + C') \ .\end{aligned}$$

Now let $F \equiv (A \triangleright B, E)$ and $G \equiv (C \triangleright D)$ and then (omitting explicit variables and weakening where possible)

$$\begin{aligned}&\partial(F[G]) = \partial(\Sigma_A f : C^E \triangleright B + \Sigma e : E.\, D(fe)) \cong (1) + (2)\\ &\text{where} \quad (1) \equiv (\Sigma_A \Sigma f : C^E.\, B \triangleright B' + \Sigma e : E.\, D(fe))\\ &\qquad\qquad (2) \equiv (\Sigma_A \Sigma f : C^E.\, \Sigma e : E.\, D(fe) \triangleright B + (\Sigma e' : E.\, D(fe'))') \ .\end{aligned}$$

The first part, $(1) \cong \partial_1 F[G]$ follows pretty immediately:

$$\begin{aligned}(1) &\cong (\Sigma_A \Sigma_B f : C^E \triangleright B' + \Sigma e : E.\, D(fe))\\ &\cong (\Sigma_A B \triangleright B', E)[(C \triangleright D)] \cong \partial_1 F[G] \ .\end{aligned}$$

To reduce (2) to $\partial_2 F[G] \times \partial G$ it is necessary to be a little more explicit about the variables and context. First note in context A, $f : C^E$, $e : E$, $d : D(fe)$ that

$$(\Sigma e' : E.\, D(fe'))' \cong D(fe)'(d) + \Sigma e' : E'.\, D(fe') \ .$$

Next observe that we can rearrange the context to bring $e : E$ before $f : C^E$ and can then write $C^E \cong C^{E'+1} \cong C^{E'} \times C$ giving us new bindings $f' : C^{E'}$ and $c : C$. The function value fe must then be replaced by c, and so finally in context A, $e : E$, $f : C^{E'}$, $c : C$, $d : D(c)$ we can write

$$D(c)'(d) + \Sigma e' : E' . D(fe')$$

which is isomorphic (modulo the change of base just described) to $(\Sigma e' : E . D(fe'))'$. This now allows us to write

$$\begin{aligned}(2) &\cong (\Sigma_A \Sigma_E \Sigma f : C^{E'} . \Sigma_C D \triangleright B + D' + \Sigma e' : E' . D(fe')) \\ &\cong (\Sigma_A E \triangleright B, E')[(C \triangleright D)] \times (\Sigma_C D \triangleright D') \cong \partial_2 F[G] \times \partial G \ . \quad \Box\end{aligned}$$

The datatype μF arises as the iteration of the container substitution operation $F[-]$. If F has derivatives in both parameters (the fixed and the iterated parameter) then the derivative of μF can also be computed.

First we need the following technical lemma which will allow us to lift the construction of the μ type to the construction of derivatives.

Lemma 4.16. *The complement of a cartesian filtered colimit of decidable containers is the filtered colimit of their complements, hence* $\partial(\bigvee_i F_i) \cong \bigvee_i \partial F_i$.

Proof. Let each $F_i \equiv (A_i \triangleright B_i)$ and first observe that $\bigvee F \cong (\bigvee A \triangleright \bigvee B)$. Then for each $i \to j$ the following diagram can be constructed in $\mathbb{C}$:

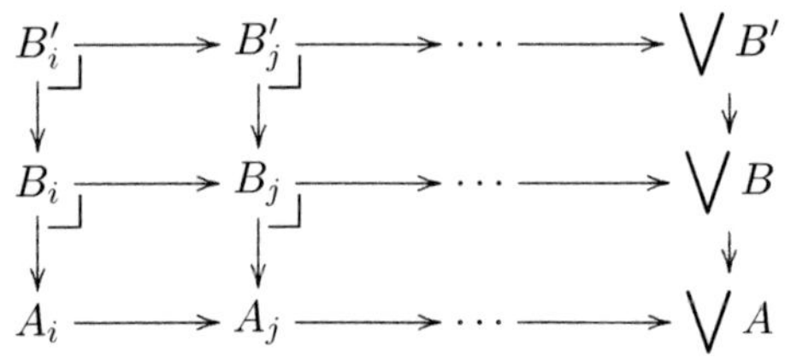

Since the bottom row consists of pullback squares then so does the top row (since the pullback of a complement is a complement), and thus $\bigvee \partial F$ is the top right hand arrow.

The object $\bigvee B'$ is the required complement of $\bigvee B$. $\Box$

Proposition 4.17. *If F is a decidable container in two parameters then the derivative of μF exists and is given by the equation*

$$\partial \mu F \cong \mu(\partial_1 F[\mu F] + \partial_2 F[\mu F] \times -) \ .$$

Proof. Define $G[-] \equiv \partial_1 F[\mu F] + \partial_2 F[\mu F] \times -$ and use the chain rule to observe

$$\partial \mu F \cong \partial(F[\mu F]) \cong \partial_1 F[\mu F] + \partial_2 F[\mu F] \times \partial \mu F = G[\partial \mu F] \ ;$$

this gives us a morphism $\theta : \mu G \to \partial \mu F$.

Conversely note that $\mu F \cong \bigvee_n (F^n[0])$ and so $\partial \mu F \cong \partial \bigvee_n (F^n[0]) \cong \bigvee_n \partial(F^n[0])$. We can therefore construct morphisms $\alpha_n : \partial(F^n[0]) \to G^n[0]$ inductively by setting

$$\alpha_{n+1} \equiv \partial_1 F[\kappa_n] + \partial_2 F[\kappa_n] \times \alpha_n$$

where $\kappa_n : F^n[0] \to \mu F$ is the colimiting cone of F. From the morphisms α we obtain a morphism $\phi : \partial \mu F \to \mu G$. It remains only to show that θ and ϕ are inverses. $\Box$

5 Abstract Containers

In future work we plan to include additional material on *abstract containers* and their derivatives, which we sketch below. Abstract containers arise when generalising the concept of a container such that includes structures where the order of elements does not matter, such as bags (or multisets). A bag is a list where sequences which can be obtained via an isomorphism on positions are identified. We follow the definition of analytic functors, as given in Hasegawa (2002), and allow any subgroup G of the automorphism group Aut_P of isomorphisms on P. Note however, that we work in a more general setting here as far as the ambient category is concerned and also that the type of positions is not necessarily finitely representable.

We introduce the notion of an *abstract container*[2] by the following data: an object of shapes S, a family $s : S \vdash P(s)$ and a family of subgroups of the automorphism groups $G(s) \subseteq \mathrm{Aut}_{P(s)}$ over $s : S$. We denote this abstract container by $(S \triangleright P/G)$. The associated functor $T_{S \triangleright P/G} : \mathbb{C} \to \mathbb{C}$ is defined by

$$T_{S \triangleright P/G} X \equiv \Sigma s : S.\, (P(s) \Rightarrow X / \sim_{G(s)}) \ .$$

where $f \sim_{G(s)} g \iff \exists \phi \in G(s). g = f \cdot \phi$. Categorically this can represented by a coequaliser. Morphisms between abstract containers $(A \triangleright B/G)$ and $(C \triangleright D/H)$ are given by an underlying container morphism $u{:}A \to C, a{:}A \vdash f{:}D(u(a)) \to B(a)$ such that $\forall \phi \in G(a).\, \exists \psi \in H(a).\, \phi{\cdot}f = f{\cdot}\psi$. We can extend T to a functor by noting that the extension of the underlying container morphism gives rise to a natural transformation between the abstract containers. We omit the straightforward generalisation to I-ary abstract containers. There is a full and faithful embedding from the category of containers to the category of abstract containers obtained by setting $G(s) = \{\mathrm{Id}_{P(s)}\}$. We believe that it is possible to extend all the properties shown for containers in Abbott et al. (2003) to abstract containers and additionally that abstract containers are closed under coequaliser of cartesian morphisms.

The notion of derivative as given before extends naturally to abstract containers, the derivative of an abstract container $F \equiv (A \triangleright B/G)$ is given by the derivative of the underlying container and by modifying G:

$$\partial F = \big(a : A,\, b : B(a) \triangleright \Sigma c : B(a).\, \neg \mathrm{Eq}(c, b)/G'(a, b)\big)$$

where $\phi \in G'(a, b) \iff \phi \in G(a) \wedge \phi(b) = b$. The characterisation of derivatives by $\partial F = \mathrm{Id} \multimap F$ extends to abstract containers.

The bag functor, as mentioned before, is represented by the abstract container

$$\mathrm{Bag} = (n : \mathbb{N} \triangleright n/\mathrm{Aut}_n)$$

Just by applying the rules of derivatives we obtain $\partial\mathrm{Bag} \cong \mathrm{Bag}$ and hence Bag plays the rule of e^x in calculus. This is maybe less surprising when noting that the extension of Bag

$$T_{\mathrm{Bag}}(X) = \Sigma n : \mathbb{N}.\, X^n/\mathrm{Aut}_n$$

corresponds to Euler's series defining e^x, since $|\mathrm{Aut}_n| = n!$.

[2] In analogy to abstract datatypes which are quotients of concrete datatypes.

Note, however, that one of the standard examples for abstract datatypes — finite sets — turns out not to be an abstract container. The reason for this is that the cardinality of positions depends on the equality of the argument type.

6 Conclusions

In the present paper, we have given an abstract characterisation of the derivative of a container, formalising the tangent of a container using a notion of linear function space. This should be useful in further understanding the relation between ordinary calculus and the calculus of containers developed here. We have shown that the usual laws of derivatives and McBride's law for least fixpoints can be derived from this notion using our previous work on containers (Abbott et al., 2003) and we have discussed abstract containers. We will make this precise and extend it to greatest fixpoints in further work.

Acknowledgements. Our thanks to the anonymous referees who provided helpful insights, and to Peter Hancock for many interesting discussions.

References

M. Abbott, T. Altenkirch, and N. Ghani. Categories of containers. In *Proceedings of Foundations of Software Science and Computation Structures*, 2003.

J. Adámek and J. Rosický. *Locally Presentable and Accessible Categories*. Number 189 in LMS Lecture Note Series. CUP, 1994.

P. Dybjer. Representing inductively defined sets by wellorderings in Martin-Löf's type theory. *Theoretical Computer Science*, 170(1–2):245–276, 1996.

T. Ehrhard and L. Regnier. The differential lambda-calculus. Available electronically, 2001. URL `http://iml.univ-mrs.fr/ftp/regnier/Articles.html`.

R. Hasegawa. Two applications of analytic functors. *Theoretical Comput. Sci.*, 272(1-2):112–175, 2002.

M. Hofmann. On the interpretation of type theory in locally cartesian closed catetories. In *Computer Science Logic*, volume 933 of *LNCS*, 1994.

G. Huet. The zipper. *Journal of Functional Programming*, 7(5):549–554, 1997.

G. Huet. Linear contexts and the sharing functor: Techniques for symbolic computation. In *Workshop on Thirty Five Years of Automath*, 2003.

A. Joyal. Foncteurs analytiques et espéces de structures. In *Combinatoire énumérative*, number 1234 in LNM, pages 126 – 159. 1986.

G. Leibniz. Nova methodus pro maximis et minimis, itemque tangentibus, qua nec irrationals quantitates moratur. *Acta eruditorum*, 1684.

C. McBride. The derivative of a regular type is its type of one-hole contexts. Available electronically, 2001. URL `http://www.dur.ac.uk/c.t.mcbride/`.

Max-Plus Quasi-interpretations

Roberto M. Amadio

Laboratoire d'Informatique Fondamentale, Marseille
amadio@cmi.univ-mrs.fr

Abstract. Quasi-interpretations are a tool to bound the size of the values computed by a first-order functional program (or a term rewriting system) and thus a mean to extract bounds on its computational complexity. We study the synthesis of quasi-interpretations selected in the space of polynomials over the *max-plus* algebra determined by the non-negative rationals extended with $-\infty$ and equipped with binary operations for the maximum and the addition. We prove that in this case the synthesis problem is NP-hard, and in NP for the particular case of *multi-linear* quasi-interpretations when programs are specified by rules of bounded size. The relevance of multi-linear quasi-interpretations is discussed by comparison to certain syntactic and type theoretic conditions proposed in the literature to control time and space complexity.

Keywords: Functional languages and term rewriting. Function algebras and implicit computational complexity. Static analysis. Polynomial interpretations and max-plus algebras.

1 Introduction

The extraction of complexity bounds from first-order functional programs has a long history and a variety of motivations. In a fundamental perspective one is interested in providing *functional algebras* characterizations of small complexity classes (see [Clo95] for a survey). In a more applied perspective, one is interested in estimating the resources required by a program for its execution. This is particularly interesting in the framework of mobile and/or embedded code (see, *e.g.*, [San01] for an elaboration of this point).

Cobham's characterization of polynomial time functions [Cob65] is based on definitions by primitive recursion on binary notation where the size of the result of the defined function is explicitly bounded by a polynomial. From a programming point of view, the annoying aspect of bounded recursion on notation is that the programmer has *to find* the bound while defining functions by *primitive recursion*. In other terms, bounded recursion on notation does not offer much support for automatically finding this bound and moreover imposes the use of primitive recursion.

Some years ago, Bellantoni-Cook [BC92] and Leivant [Lei94] have introduced a notion of *ramification*. In [BC92] this is expressed as a distinction between *normal* and *safe* arguments and a restriction on the way primitive recursion can be applied and functions composed ([Lei94] introduces a related notion of *tier*). By complying to this programming discipline the programmer is relieved from the problem of explicitly providing a bound. This bound is implicit in the constraints imposed by ramification and when needed can be explicitly computed.

M. Hofmann (Ed.): TLCA 2003, LNCS 2701, pp. 31–45, 2003.

It has been observed [Cas97] that this programming discipline rules out many natural algorithms. To improve the situation, one approach, followed by Marion *et al.* [Mar00, MM00,BMM01] has been to extend the notion of ramification to various types of *recursive path orders* (a family of simplification ordering such as lexicographic, multi-set,...). In a quite different direction, Jones [Jon97] and Hofmann [Hof02] have proposed to consider *general* recursive programs for which a bound can be found on the size of the values handled during the computation. Jones obtains this bound by imposing a (quite severe) syntactic restriction on the programs, while Hofmann introduces an 'affine' type system with a *resource* type that guarantees that values obtained in the course of the computation are *non-size increasing*.

Quasi-interpretations have been proposed by Marion *et al.* in the context of the work on ramified recursive path orders we mentioned above. More precisely, quasi-interpretations are extracted from a termination proof in the *ramified* path order and they are then used as a tool to bound the *space* required to compute the result of the program. Alternatively, quasi-interpretations have been considered in combination with recursive path orderings (see section 6.3).

Quasi-interpretations are obviously inspired by *polynomial –simplification– interpretations* which are one of the traditional tools used in proving the *termination* of *term rewriting systems* (TRS), see, *e.g.*, [BN98]. The limit of (quasi-)interpretations is that first, one has to *synthesize* one, and second, one has to *verify* that the interpretation fits a given TRS. For polynomial interpretations even the verification problem has a high complexity.

In this paper we address the problem of the automatic *synthesis* of quasi-interpretations for general recursive programs (and not just those admitting a termination proof by ramified path order). We share our goal if not the technical approach with a recent work by Hofmann and Jost [HJ03] (see also [HPS96,Par00] for earlier work on the verification problem). In particular, quasi-interpretations offer a bound on the 'heap' memory needed to evaluate a program (cf. theorem 1).

For the efficiency reasons mentioned above, we propose to restrict our attention to *multi-variate* polynomials over the so called max-plus algebra [BCOQ92]. We anticipate that polynomials over the max-plus algebra have a growth rate that is *linear* in the size of the argument. This growth rate is indeed a very severe restriction if we think in terms of traditional interpretations, *i.e.*, of the –time– taken by the computation to terminate. However, as pointed out above we are interested in *quasi*-interpretations as a mean to bound the –space– needed to compute a function. In particular, following Hofmann's work on type systems ensuring non-size increasing computations [Hof00], one can still accommodate within this framework all functions computable in exponential time whose output's size is bounded by the input's size; this is a respectable class of functions including for instance all decision problems decidable in time $2^{O(n)}$.

We propose to study the synthesis of quasi-interpretations selected in the space of polynomials over the *max-plus* algebra determined by the non-negative rationals extended with $-\infty$ and equipped with binary operations for the maximum and the addition. We prove that in this case the synthesis problem is NP-hard and in NP for the particular case of *multi-linear* quasi-interpretations when the size of the rules is bound by a constant (section 5).

We also relate multi-linear quasi-interpretations to certain syntactic and type-theoretic restrictions that have been proposed in the literature [Jon97,Hof02] in order to limit the computational complexity of the functions represented by a first-order functional language (section 6).

Most proofs are omitted and can be found in [Amadio02].

2 A First-Order Functional Language

We consider a first-order, simply typed functional language operating over inductively defined data types according to a call-by-value evaluation strategy. A *program* in this context is given by a collection of data types declarations, a collection of mutually recursive function definitions' relying on pattern matching, and a function symbol which is designated as initial. Following Marion *et al.*, an alternative framework for this study could be term-rewriting rules with a distinction between constructors and function symbols.[1]

Types The collection of types is the least set such that

$$\mu t.(\mathsf{c}_1 : \tau_{1,1}, \cdots, \tau_{1,n_1} \to t, \ldots, \mathsf{c}_m : \tau_{m,1}, \cdots, \tau_{m,n_m} \to t)$$

is a type provided, $\tau_{i,j}$ is either a type or the bound type variable t, and all constants c_i are distinct and do not occur in previously defined types $\tau_{i,j}$. Traditionally, the symbols $\mathsf{c}_1, \ldots, \mathsf{c}_n$ are the *constructors* of the data type. We assume that for a given program, constructors names are chosen so that no confusion may arise on what type a constructor belongs to. These types allow for the definition of basic data structures. For instance, we can regard $bool \equiv \mu t.(\mathsf{tt} : t, \mathsf{ff} : t)$ as the type of booleans, $tnat \equiv \mu t.(\mathsf{0} : t, \mathsf{s} : t \to t)$ as the type of tally natural numbers, and $tnatlist \equiv \mu t.(\mathsf{nil} : t, \mathsf{cons} : tnat, t \to t)$ as the type of lists of tally natural numbers.

Expressions We reserve: $\mathsf{c}, \mathsf{c}', \ldots$ for constructor symbols, $f, f', \ldots$ for function symbols, and $x, x' \ldots$ for first-order variables. Moreover, we introduce the syntactic categories of *values*, *patterns*, and *expressions* as follows:

$$\begin{array}{ll} v ::= \mathsf{c}(v, \ldots, v) & \text{(values)} \\ p ::= x \mid \mathsf{c}(p, \ldots, p) & \text{(patterns)} \\ e ::= x \mid \mathsf{c}(e, \ldots, e) \mid f(e, \ldots, e) & \text{(expressions).} \end{array}$$

We denote with $Var(e)$ the collection of variables occurring in the expression e. Note that values are closed patterns, *i.e.*, $Var(v) = \emptyset$, and patterns are expressions without function symbols.

We denote with $[\tau/t]\tau'$ and $[e/x]e'$ the substitution in types and expressions, respectively. A *signature* Σ attributes to every function symbol f a functional type $\Sigma(f) \equiv \tau_1, \ldots, \tau_n \to \tau$. As usual if u is either a constructor or a function symbol

[1] In this perspective, note that we work with orthogonal TRS and that for these TRS termination is equivalent to innermost termination [Gra96].

we denote with $arity(u)$ the number of expected arguments as specified by its type. A *context* Γ is a finite list $x_1 : \tau_1, \ldots, x_n : \tau$ where $x_i \neq x_j$ if $i \neq j$. We use the judgment $\Gamma \vdash_\Sigma e : \tau$ to state that the expression e has type τ with respect to the signature Σ and the context Γ. Provable typing judgment are defined by the following inference system:

$$\frac{x : \tau \in \Gamma}{\Gamma \vdash_\Sigma x : \tau} \quad \frac{\tau \equiv \mu t.(\ldots, \mathsf{c} : \tau_1, \ldots, \tau_n \to t, \ldots) \;\; \Gamma \vdash_\Sigma e_i : [\tau/t]\tau_i \;\; i = 1, \ldots, n}{\Gamma \vdash_\Sigma \mathsf{c}(e_1, \ldots, e_n) : \tau}$$

$$\frac{\Sigma(f) = \tau_1, \ldots, \tau_n \to \tau \;\; \Gamma \vdash_\Sigma e_i : \tau_i \;\; i = 1, \ldots, n}{\Gamma \vdash_\Sigma f(e_1, \ldots, e_n) : \tau} .$$

Functions' definitions Function symbols are defined by a finite system of mutually recursive equations so that each function symbol is defined by exactly one equation. If $\Sigma(f) = \tau_1, \ldots, \tau_n \to \tau$ then the equation defining f has the shape:

$$\begin{array}{l} f(x_1, \ldots, x_n) = \\ x_1 = p_{1,1}, \ldots, x_n = p_{1,n} \quad \Rightarrow e_1 \\ \ldots \\ x_1 = p_{m,1}, \ldots, x_n = p_{m,n} \Rightarrow e_m \end{array}$$

where the formal parameters $x_1, \ldots, x_n$ are distinct and

(1) Patterns are *linear*, *i.e.* in $p_{i,j}$ no variable occurs more than once and $Var(p_{i,j}) \cap Var(p_{i,j'}) = \emptyset$ if $j \neq j'$. We assume that if $Var(p_{i,j}) = \emptyset$ then $p_{i,j}$ is a constant constructor.[2] In practice, one may the freedom of omitting trivial patterns of the shape $x_i = x_i$.

(2) Patterns do not superpose, *i.e.*, if $i \neq j$ then the equations $p_{i,1} = p_{j,1}, \ldots, p_{i,n} = p_{j,n}$ are not unifiable. In particular, this entails that the programs we consider are *deterministic*.

(3) $Var(e_k) \subseteq \bigcup_{j=1,\ldots,n} p_{k,j}$, that is expressions' variables are contained in patterns' variables.

(4) Patterns and expressions are well typed, *i.e.*, for $i = 1, \ldots, m$ there are contexts Γ_i such that:

$$\Gamma_i \vdash_\Sigma p_{i,j} : \tau_j \text{ for } j = 1, \ldots, n \text{ and } \Gamma_i \vdash_\Sigma e_i : \tau .$$

We call *rule* a clause of the shape $x_1 = p_1, \ldots, x_n = p_n \Rightarrow e$.

Evaluation A program is a finite collection of inductive types and a finite system of functions' definitions with a selected *main* function. Expression evaluation follows a call-by-value strategy which is specified as follows:

$$\frac{e_j \mapsto v_j \;\; j = 1, \ldots, n}{\mathsf{c}(e_1, \ldots, e_n) \mapsto \mathsf{c}(v_1 \ldots, v_n)} \qquad \frac{e'_j \mapsto v_j, \;\; \sigma p_{i,j} = v_j, j = 1, \ldots, n \;\; \sigma(e_i) \mapsto v}{f(e'_1, \ldots, e'_n) \mapsto v}$$

assuming the function f is defined as in section 1 and σ denotes a pattern-matching substitution.

[2] This is a technical condition that does not impair the expressivity of the language as general patterns without free variables can be simulated by introducing auxiliary function symbols.

3 Quasi-interpretations

In this section we introduce the notion of quasi-interpretation and its basic properties.

Definition 1 (size). *The size of a value v is defined by* [3]

$$|\mathsf{c}(v_1, \ldots, v_n)| = \begin{cases} 0 & \textit{if } n = 0 \\ 1 + \Sigma_{i=1,\ldots,n}|v_i| & \textit{if } n > 0 \,. \end{cases}$$

Definition 2 (assignment). *Given a program, an assignment associates:*

(1) *To every constructor* c *with* k *arguments a function* $q_{\mathsf{c}} : (\mathbf{Q}^+)^k \to \mathbf{Q}^+$ *such that:*

(1.1) $q_{\mathsf{c}} = 0$ *if* c *has arity* 0 *and*

(1.2) $q_{\mathsf{c}} = d + \Sigma_{i=1,\ldots,n} x_i$ *for some* $d \geq 1$*, otherwise.*

(2) *To every function symbol* f *with* k *arguments a function* $q_f : (\mathbf{Q}^+)^k \to \mathbf{Q}^+$ *such that:*

(2.1) $q_f(n_1, \ldots, n_k) \geq n_i$ *for* $i = 1, \ldots, k$ *and*

(2.2) $q_f(n_1, \ldots, n_k) \geq q_f(m_1, \ldots, m_k)$ *if* $n_i \geq m_i$ *for* $i = 1, \ldots, k$*.*

Definition 3 (extension of the assignment). *Given an assignment and an expression* e *with* $Var(e) = \{x_1, \ldots, x_k\}$ *we can define a function* $q_e : (\mathbf{Q}^+)^k \to \mathbf{Q}^+$ *by induction on* e *as follows:*

$$q_x = x, \quad q_{\mathsf{c}(e_1,\ldots,e_n)} = q_{\mathsf{c}}(q_{e_1}, \ldots, q_{e_n}), \quad q_{f(e_1,\ldots,e_n)} = q_f(q_{e_1}, \ldots, q_{e_n}) \,.$$

Definition 4 (quasi-interpretation). *Given a program, the related assignment* q *is a quasi interpretation if for every function definition of the shape presented in section 1 the following condition holds for* $i = 1, \ldots, m$ *(where functions are ordered pointwise):*

$$q_f(q_{p_{i,1}}, \ldots, q_{p_{i,n}}) \geq q_{e_i} \,. \tag{1}$$

The notion of assignment we consider is obviously inspired by the *–simplification–interpretation method* used in termination proofs of TRS. The specific conditions on constructors correspond to the notion of *kind* 0 quasi-interpretation defined in [BMM01]. However, we work over the non-negative rationals rather than over the natural numbers and we force the interpretation 0 for constants. This last condition allows to simplify some interpretations by neglecting the space needed to store constant values. The following proposition summarizes the basic properties of quasi-interpretations.

Proposition 1. *Suppose* q *is a quasi-interpretation for a given program. Then:*

(1) *There is a constant* d *such that for any value* v*,* $|v| \leq q_v \leq d|v|$*.*

(2) *If* $e \mapsto v$ *then* $q_e \geq q_v \geq |v|$*. In particular, if* $f(v_1, \ldots, v_n) \mapsto v$ *then* $|v| \leq q_f(d|v_1|, \ldots, d|v_n|)$*.*

[3] The alternative definition of size where we assign a positive size to constants is equivalent to the present one within a constant multiplicative factor.

A point that seems to have gone unnoticed is that quasi-interpretations –by themselves– already provide a bound on the complexity of the program.

Theorem 1. *Suppose q is a quasi-interpretation for a given program. Then there is an evaluation strategy that given a function symbol f with arguments $v_1, \ldots, v_n$ returns the value v iff $f(v_1, \ldots, v_n) \mapsto v$ and a special symbol $\bot$ otherwise. The procedure runs in time $2^{O(q_{f(v_1,\ldots,v_n)})}$.*

Proof hint. Let $B = q_{f(v_1,\ldots,v_n)}$. The size of the values computed during the computation of $f(v_1, \ldots, v_n)$ is in $O(B)$. It follows that the number of values to which a function can be applied in the course of the computation is in $2^{O(B)}$.

Let E be a 'call-by-value' evaluation context E defined as follows:

$$E ::= [\,] \mid \mathsf{c}(v_1, \ldots, v_{i-1}, E, e_{i+1}, \ldots, e_n) \mid g(v_1, \ldots, v_{i-1}, E, e_{i+1}, \ldots, e_n)\ .$$

Any closed expression e which is not a value admits a unique decomposition in an evaluation context and a function application. We define an evaluation function *Eval* that performs an innermost leftmost evaluation of an expression.

$$\begin{array}{l} \mathit{Eval}(e) = \mathsf{case} \\ e \text{ value } : e \\ e \equiv E[f(v_1, \ldots, v_n)] \text{ and } \exists \sigma, i\ (\sigma(p_{i,j}) = v_j, j = 1, \ldots, n) : \mathsf{let}\ v' = \mathit{Eval}(\sigma(e_i))\ \mathsf{in} \\ \qquad\qquad\qquad\qquad\qquad\qquad\qquad\qquad\qquad\qquad \mathit{Eval}(E[v']) \\ \mathsf{else} : \mathit{Return}\ \bot \end{array}$$

where we assume that the function f is defined as in section 1 and that invoking $\mathit{Return}\bot$ stops the computation returning $\bot$ as result. Let k be the maximum number of function symbols that occur in an expression on the right hand side of $\Rightarrow$ in a function definition. We note that the evaluation function initially applied to the expression $f(v_1, \ldots, v_n)$ maintains the invariant that the number of function symbols in an argument e is bound by $max(k, 1)$. It follows that also the size of the expressions involved in the evaluation has size in $O(B)$. Then a stack frame for *Eval* has size in $O(B)$ and *Eval* can be implemented on an *auxiliary deterministic pushdown automata* with auxiliary memory which is in $O(B)$. Finally, by a well-known result by S. Cook [Coo71], the function can also be implemented to run on a Turing Machine in time $2^{O(B)}$ using a 'table' of size in $2^{O(B)}$ to store intermediate results (a so called *memoization* technique). □

4 Max-Plus Polynomials and Synthesis Problem

We consider the set $\mathbf{Q}^+ \cup \{-\infty\}$ equipped with two internal composition laws *max* and *plus* (denoted $+$) where it is understood that:

$$max(-\infty, x) = max(x, -\infty) = x \quad -\infty + x = x + (-\infty) = -\infty\ .$$

We briefly refer to this structure as $\mathbf{Q}^+_{max}$. We note that $\mathbf{Q}^+_{max}$ is a commutative and idempotent monoid for max with neutral element $-\infty$ and a commutative monoid for *plus* with neutral element 0. Moreover, *plus* distributes over *max*: $x + max(y, z) =$

$max(x+y, x+z)$. In the max-plus literature one regards *max* as an addition and *plus* as a multiplication. Consequently, for α natural number, αx can be interpreted in the max-plus algebra as x raised to the exponent α.

A *monomial* with coefficient in $a \in \mathbf{Q}^+_{max}$ and indeterminates $x_1, \ldots, x_n$ can be written as

$$\alpha_1 x_1 + \cdots + \alpha_n x_n + a \tag{2}$$

where $\alpha_i \in \mathbf{N}$. We say that a monomial has *degree* d if $\alpha_i \leq d$ for $i = 1, \ldots, n$.[4] A *polynomial* is now written as $max_{i \in I} m_i$ where m_i are monomials of the type specified above. We say that a polynomial has degree d if all monomials m_i have degree d. A polynomial of degree d with n indeterminates can be represented as

$$max_{I:\{1,\ldots,n\} \rightarrow \{0,\ldots,d\}}(I(1)x_1 + \cdots + I(n)x_n + a_I) \tag{3}$$

and it is therefore specified by the $(d+1)^n$ coefficients $\{a_I \mid I : \{1, \ldots, n\} \rightarrow \{0, \ldots, d\}\}$.

An *assignment* (cf. definition 2) of max-plus polynomials of degree d is determined as follows:

(1) For every constructor c with positive arity a coefficient a^{c} subject to the constraint $a^{\mathsf{c}} \geq 1$.

(2) For every function symbol f with $n \geq 1$ arguments a set of coefficients $\{a^f_I \mid I : \{1, \ldots, n\} \rightarrow \{0, \ldots, k\}\}$ subject to the constraints for $i = 1, \ldots, n$

$$max\{a^f_I \mid I(i) \geq 1\} \geq 0\,. \tag{4}$$

This last constraint is necessary and sufficient for the condition (2.1) $q_f(n_1, \ldots, n_k) \geq n_i$ of definition 2 to hold. We note that the following monotonicity condition (2.2) is always satisfied.

Definition 5 (synthesis problem). *Given a program, the synthesis problem amounts to determine whether there is a polynomial max-plus quasi-interpretation.*

If the program includes l constructors of positive arity and m functional symbols of arity at most n an assignment of polynomials of degree at most d is determined by at most $l + m(d+1)^n$ coefficients. The assignment is a quasi-interpretation iff it satisfies the constraints above and those induced by the condition (1) $q_f(q_{p_{i,1}}, \ldots, q_{p_{i,n}}) \geq q_{e_i}$.

Some simple instances of max-plus polynomial quasi-interpretations are given in [Amadio02]. Of course, the rule of the game is to get quasi-interpretations as small as possible and in this respect the *max* operator is quite useful. Moreover, in many examples where a variable occurs several times on the right-hand side, it is simply not possible to find a (max-plus) quasi-interpretation that does not rely on *max*. We note that in general the existence of a polynomial max-plus interpretation of a *given degree* can be reduced to the validity of an $\exists\forall$ formula in Pressburger arithmetic over $\mathbf{Q}^+_{max}$. This problem can be attacked with tools such as the OMEGA test [Pug92]. We can also consider a restricted

[4] This is a slightly improper terminology; we should say that the monomial restricted to any of its indeterminates has degree at most d.

problem where one looks for polynomials over $\mathbf{N}_{max} = \mathbf{N} \cup \{-\infty\}$ (with coefficients in $\mathbf{N}_{max}$). In this case, the satisfaction of a $\exists\forall$ formulae can also be attacked with automata theoretic tools.

Theorem 2. *The synthesis problem is* NP*-hard and it remains so if any combination of the following restrictions is considered:*

(1) *Rules of bounded size (for a small bound).*

(2) *Max-plus polynomials of bounded degree* $d \geq 1$.

(3) *Uniform choice of the coefficients of the constructors:* $a^{\mathsf{c}} = a^{\mathsf{c}'}$ *for all constructors* c, c' *of positive arity.*

PROOF HINT. This is based on a rather technical reduction from 3-SAT. As a first step, we produce rules forcing the quasi-interpretation of a function symbol f with n arguments to be $q_f = max(x_1, \ldots, x_n)$. If a non-uniform choice of constructors is allowed then we code literals as coefficients of unary constructors. Otherwise, some more work is needed to show that for a binary function symbol g one can force the quasi-interpretation $q_g = max(x_1 + a_1, x_2 + a_2)$. Then we use the coefficients a_1, a_2 to represent a boolean variable. □

5 Synthesis of Multi-linear Quasi-interpretations

We consider the synthesis problem when the degree is 1. We start by pointing out some specific properties of this case (section 5.1), then we give effective methods to compute and compare quasi-interpretations (section 5.2), and finally we show how the generated conditions can be reduced to linear programming (section 5.3).

5.1 Multi-linear Assignments

Following a rather standard terminology, we will refer to monomials (polynomials) of degree 1 as *multi-linear* monomials (polynomials). We note that a multi-linear polynomial in n indeterminates is specified by 2^n coefficients $\{a_I \mid I \subseteq \{1, \ldots, n\}\}$ and can be written as follows:

$$max_{I \subseteq \{1,\ldots,n\}}(\Sigma_{i \in I} x_i + a_I) . \tag{5}$$

Equivalently, if the multi-linear polynomial depends on the variables $x_1, \ldots, x_n$ we will also write:

$$max_{V \subseteq \{x_1,\ldots,x_n\}}(\Sigma_{v \in V} v + a_V) . \tag{6}$$

The following is easily checked.

Proposition 2 (normal form). *For every multi-linear polynomial* $P(x_1, \ldots, x_n)$ *there is an equivalent multi-linear polynomial* $P'(x_1, \ldots, x_n)$ *with coefficients* $\{a'_I \mid I \subseteq \{1, \ldots, n\}\}$ *satisfying the condition*

$$J \subseteq K \subseteq \{1, \ldots, n\} \;\Rightarrow a'_J \geq a'_K . \tag{7}$$

In the following we assume that a program has been fixed and that $\mathsf{c}_1, \ldots, \mathsf{c}_l$ are the constructors of positive arity occurring in the program. We say that a multi-linear polynomial is in *normal form* if its coefficients satisfy condition (7). For such polynomials condition (4) on assignments can be reformulated as $a^f_{\{i\}} \geq 0$ for every function f with $arity(f) = n$. Given a multi-linear assignment we will show that $q_{f(p_1,\ldots,p_n)}$ is always a multi-linear polynomial; a property that may fail for a general expression e.

Proposition 3. *Let P_1 be a multi-linear polynomial and P_2 be a polynomial. If $P_1 \geq P_2$ then P_2 must be multi-linear.*

5.2 Computing Multi-linear Quasi-interpretations

We explicitly compute the shape of the different polynomials arising from a multi-linear assignment. The proofs require some involved notation but just rely on elementary arithmetic considerations.

Proposition 4 (left-hand-side). (1) *Suppose q is a multi-linear assignment and p is a pattern in a function definition then q_p is a multi-linear polynomial of the shape*

$$q_p = \Sigma_{v \in Var(p)} v + \Sigma_{j=1,\ldots,l} \alpha_j a^{\mathsf{c}_j} \tag{8}$$

for some $\alpha_j \in \mathbf{N}$.

(2) *Suppose q is a multi-linear assignment and the function f contains the rule $x_1 = p_1, \ldots, x_n = p_n \Rightarrow e$. Then $q_{f(p_1,\ldots,p_n)}$ is always a multi-linear function and assuming $q_{p_i} = \Sigma_{v \in Var(p_i)} v + \Sigma_{j=1,\ldots,l} \alpha_{i,j} a^{\mathsf{c}_j}$ then the coefficient b_V for $V \subseteq \bigcup_{i=1,\ldots,n} Var(p_i)$ is given by*

$$b_V = a^f_{K_V} + \Sigma_{j=1,\ldots,l} (\Sigma_{k \in K_V} \alpha_{k,j}) a^{\mathsf{c}_j} \tag{9}$$

where $K_V = \{k \in \{1, \ldots, n\} \mid V \cap Var(p_k) \neq \emptyset\}$.

We now turn to the polynomial q_e. This polynomial is obtained by arbitrary composition of multi-linear polynomials and may fail to be multi-linear. However, in this case we know from propositions 3 and 4(2) that the inequality $q_{f(p_1,\ldots,p_n)} \geq q_e$ cannot hold. So our next task is to generate constraints that are necessary and sufficient to guarantee that q_e is multi-linear. To this end, we introduce in figure 1 a little formal system with judgments of the shape (e, C) where e is an expression and C is a set of constraints on the coefficients of the functions occurring in e. As usual we introduce a special constraint $\perp$ with the hypothesis that no assignment can satisfy it.

Proposition 5 (right-hand-side). *Suppose q is a multi-linear assignment in normal form.*

(1) *If $q_{e_i} = max_{U_i \subseteq V_i}(\Sigma_{v \in U_i} v + a^i_{U_i})$, $i = 1, \ldots, n$, are multi-linear polynomials where $V_i = Var(e_i)$ and $V = \bigcup_{i=1,\ldots,n} V_i$. Then:*

(1.1) *$q_{\mathsf{c}(e_1,\ldots,e_n)}$ is a multi-linear polynomial iff $i \neq j$ implies $V_i \cap V_j = \emptyset$, and in this case the coefficients b_U for $U \subseteq V$ are determined by:*

$$b_U = \Sigma_{i=1,\ldots,n} a^i_{U \cap V_i} + a^{\mathsf{c}} \tag{10}$$

$$\overline{(x, \emptyset)}$$

$$\frac{(e_i, C_i),\ i = 1, \dots, n \quad Var(e_i) \cap Var(e_j) = \emptyset \ \text{ for all } i \neq j}{(\mathsf{c}(e_1, \dots, e_n), \bigcup_{i=1,\dots,n} C_i)}$$

$$\frac{(e_i, C_i),\ i = 1, \dots, n \quad Var(e_i) \cap Var(e_j) \neq \emptyset \ \text{ for some } i \neq j}{(\mathsf{c}(e_1, \dots, e_n), \bigcup_{i=1,\dots,n} C_i \cup \{\bot\})}$$

$$\frac{(e_i, C_i),\ i = 1, \dots, n}{(f(e_1, \dots, e_n), \{a^f_{i,j} = -\infty \mid i \neq j,\ Var(e_i) \cap Var(e_j) \neq \emptyset\} \cup \bigcup_{i=1,\dots,n} C_i)}$$

Fig. 1. Constraints enforcing multi-linearity of q_e

(1.2) *Whenever* $q_{f(e_1,\dots,e_n)}$ *is a multi-linear polynomial the coefficients* b_U *for* $U \subseteq V$ *are determined by:*

$$b_U = max_{I \subseteq \{1,\dots,n\}, \downarrow I, U \subseteq \bigcup_{i \in I} V_i} (\Sigma_{i \in I} a^i_{U \cap V_i} + a^f_I) \tag{11}$$

where by definition $\downarrow I$ *if* $i, j \in I$ *and* $i \neq j$ *implies* $V_i \cap V_j = \emptyset$.

(2) *If* $\vdash (e, C)$. *Then* q_e *is multi-linear iff* q *satisfies* C.

Thus given a rule $x_1 = p_1, \dots, x_n = p_n \Rightarrow e$ and a generic multi-linear assignment q we determine the conditions under which q_e is multi-linear and then formally compute its coefficients. Next, we have to find necessary and sufficient conditions on the coefficients to compare multi-linear polynomials.

Proposition 6 (comparison). *Suppose* P_1 *and* P_2 *are multi-linear polynomials with* n *indeterminates and coefficients* $\{a_I \mid I \subseteq \{1, \dots, n\}\}$ *and* $\{b_I \mid I \subseteq \{1, \dots, n\}\}$, *respectively. Then*

(1) $P_1 \geq P_2$ *iff the following condition holds:*

$$max\{a_K \mid K \supseteq J\} \geq b_J \quad \textit{for all } J \subseteq \{1, \dots, n\}\,. \tag{12}$$

(2) *If moreover,* P_1 *is in normal form then the condition (12) is equivalent to* $a_J \geq b_J$ *for all* $J \subseteq \{1, \dots, n\}$.

To summarize, we have shown how to generate a system $\mathcal{S}$ of inequality constraints on the coefficients of multi-linear polynomials so that the constraints can be satisfied in $\mathbf{Q}^+_{max}$ iff the corresponding polynomials determine a multi-linear assignment and a quasi-interpretation.

5.3 Reduction to Linear Programming

For programs with rules of *bounded* size we show that the synthesis problem can be solved in non-deterministic polynomial time thus matching the lower bound given by theorem 2. The criteria (12) introduces inequalities of the shape $max(A_1,\ldots,A_m) \geq B$, where A_i are coefficients of the type specified by proposition 4(2) *not* containing the *max* operation. We remove the *max* by non-deterministically *guessing* the maximum A_i among $A_1, \ldots, A_m$ and transforming the inequality into $A_i \geq B$. We show next that the resulting system can be solved in deterministic polynomial time. Thus the quest of synthesis problems with (deterministic) *polynomial* time complexity seems to depend crucially on the possibility of removing the *max* operation on the left-hand side of the inequalities generated by the comparison criteria. Some interesting cases where this is actually possible are discussed in propositions 10 and 12.

We reserve $x_1, \ldots, x_n$ for the variables corresponding to the coefficients a^f_I or for auxiliary variables, and $y_1, \ldots, y_l$ for the variables corresponding to the coefficients a^{c_j} which are all subject to the constraint $y \geq 1$. Let $\mathcal{S}(\boldsymbol{x}, \boldsymbol{y})$ be the system of inequalities over $\mathbf{Q}^+_{max}$ that we have derived for the synthesis problem over multi-linear polynomials after elimination of the *max* on the left-hand side.

Proposition 7 (right *max*-elimination). *The system $\mathcal{S}(\boldsymbol{x}, \boldsymbol{y})$ can be transformed in polynomial time into a system $\mathcal{S}_1(\boldsymbol{x}, \boldsymbol{x}', \boldsymbol{y})$ over $\mathbf{Q}^+_{max}$ with additional auxiliary variables $\boldsymbol{x}'$ such that:*

(1) *The inequalities in $\mathcal{S}_1$ have one of the following 3 shapes assuming $\boldsymbol{x}, \boldsymbol{x}' \equiv x_1, \ldots, x_n$ and $\boldsymbol{y} \equiv y_1, \ldots, y_l$.*

$$(a)\ x = -\infty \ \text{ provided } x \in \{\boldsymbol{x}\} \qquad (b)\ y \geq 1 \ \text{ for all } y \in \{\boldsymbol{y}\}$$
$$(c)\ x + \Sigma_{j=1,\ldots,l}\alpha_j y_j \geq \Sigma_{j=1,\ldots,n}\beta_j x_j + \Sigma_{j=1,\ldots,l}\gamma_j y_j$$
$$\text{where: } \alpha_j, \beta_j, \gamma_j \in \mathbf{N}, x \in \{\boldsymbol{x}, \boldsymbol{x}'\}\ .$$

(2) *An assignment ρ satisfies $\mathcal{S}$ iff for some $\boldsymbol{w}$, $\rho[\boldsymbol{w}/\boldsymbol{x}']$ satisfies $\mathcal{S}_1$.*

PROOF HINT. An inequality $A \geq max_{i\in I}(\Sigma_{j\in J_i} B_{i,j} + C)$ can be transformed into

$$A \geq x' \quad x' \geq \Sigma_{j\in J_i} x'_{i,j} + C \text{ for } i \in I \quad x'_{i,j} \geq B_{i,j} \text{ for } i \in I, j \in J_i$$

where $x', x'_{i,j}$ are fresh variables. It can be easily verified that the derived system is satisfiable iff the initial one is. If $B_{i,j}$ contains again the max operator then we apply recursively the transformation to the inequality $x'_{i,j} \geq B_{i,j}$. □

Proposition 8 ($-\infty$-elimination). *The system $\mathcal{S}_1(\boldsymbol{x}, \boldsymbol{x}', \boldsymbol{y})$ over $\mathbf{Q}^+_{max}$ obtained in proposition 7 can be transformed in polynomial time into a system $\mathcal{S}_2(\boldsymbol{x}'', \boldsymbol{y})$ over $\mathbf{Q}^+$ where (i) $\{\boldsymbol{x}''\} \subset \{\boldsymbol{x}, \boldsymbol{x}'\}$, (ii) the constraints have the shape (b) and (c) in proposition 7, and assuming $\{\boldsymbol{z}\} = \{\boldsymbol{x}, \boldsymbol{x}'\} \backslash \{\boldsymbol{x}''\}$ an assignment ρ satisfies $\mathcal{S}_1$ iff $\rho[-\infty/\boldsymbol{z}]$ satisfies $\mathcal{S}_2$.*

PROOF HINT. We describe the proof strategy in a simplified case. Consider the conjunction of boolean formulae of the shape $\bigvee_{j\in J} x_j$ or $x \Rightarrow \bigvee_{j\in J} x_j$. Its satisfiability can be decided by applying the following rules:

$$\begin{array}{ll}
S, x, (x \Rightarrow \bigvee_{j\in J} x_j) & \to S, x, \bigvee_{j\in J} x_j \\
S, x, x \vee \bigvee_{j\in J} x_j & \to S, x \quad \text{if } J \neq \emptyset \\
S, (x \Rightarrow \bot), x \vee \bigvee_{j\in J} x_j & \to S, (x \Rightarrow \bot), \bigvee_{j\in J} x_j \quad \text{if } J \neq \emptyset \\
S, x', (x \Rightarrow (x' \vee \bigvee_{j\in J} x_j)) & \to S, x' \\
S, (x' \Rightarrow \bot), (x \Rightarrow (x' \vee \bigvee_{j\in J} x_j)) & \to S, (x' \Rightarrow \bot), (x \Rightarrow \bigvee_{j\in J} x_j)
\end{array}$$

where as usual $\bot$ stands for the empty disjunction, disjunction is treated as an associative and commutative operator, and ',' stands for conjunction. These simplification rules obviously terminate in a system S' that is satisfiable iff the original one is. Moreover, if $\bot \notin S'$ then the boolean variables X can be *partitioned* in three sets X_1, X_0, X_2 where $X_1 = \{x \mid s \in S'\}$ and $X_0 = \{x \mid (x \Rightarrow \bot) \in S'\}$. Then a satisfying assignment is obtained by taking $\rho(x) = 1$ if $x \in X_1 \cup X_2$ and $\rho(x) = 0$ if $x \in X_0$. This proof strategy is repeated for the system over $\mathbf{Q}^+_{max}$ where the constraint $x = -\infty$ corresponds to x and the constraint $x \geq 0$ to $(x \Rightarrow \bot)$. □

Our method also applies to multi-linear polynomial over $\mathbf{N}_{max}$: if the system of inequalities admits a solution over $\mathbf{Q}^+$ then by multiplying by the least common denominator we obtain a solution in $\mathbf{N}$. Also, since we reduce to a linear programming problem, it is possible to look for *minimal* solutions with respect to a given linear cost function. We summarize below our analysis for programs whose rules have *bounded* size. In general, the method is exponential in the size of the rules since the number of coefficients we have to determine is exponential in the number of the variables in a rule. However, in practice one expects the size of the rules to be a slow growing function in the size of the program.

Theorem 3. *The synthesis problem over multi-linear polynomials for programs with rules of bounded size is* NP*-complete.*

6 Expressivity

We present within our framework a 'no cons' syntactic restriction and a 'type system for in-place update' that have been proposed in the literature to control the time and space complexity. We also mention some upper and lower bounds on the complexity of functions representable by programs admitting a max-plus quasi-interpretation.

6.1 No cons Syntactic Condition

Jones' syntactic condition [Jon97] concerns first-order functional programs defined over the type of booleans $bool \equiv \mu t.(\mathsf{tt} : t, \mathsf{ff} : t)$ and the type of lists of booleans $blist \equiv \mu t.(\mathsf{nil} : t, \mathsf{cons} : bool, t \rightarrow t)$. The syntactic restriction requires that in a function definition of the shape presented in section 1 the cons constructor does not appear in the expressions e_i on the right-hand side of pattern matching. The following can be easily checked.

Proposition 9. *A program conforming to Jones' restriction admits the following multi-linear quasi-interpretation (assuming $arity(\mathsf{c}) = arity(f) = n \geq 1$):*

$$q_{\mathsf{c}} = 1 + \Sigma_{i=1,\ldots,n} x_i \quad q_f = max(x_1, \ldots, x_n) \; .$$

We consider a restricted class of multi-linear quasi-interpretations where:

$$q_{\mathsf{c}} = a + \Sigma_{i=1,\ldots,n} x_i, \; a \geq 1 \quad q_f = max(x_1 + a^f, \ldots, x_n + a^f), \; a^f \geq 0 \; . \quad (13)$$

We note that (i) all constructors have the *same* coefficient a, (ii) every function is determined by exactly *one* coefficient a^f, and (iii) the interpretation in proposition 9 falls in this family. We refer to this class of quasi-interpretations as *max*-multi-linear. The following is proven by noting that under the conditions (13) the *max* operation is not needed on the left-hand side of an inequality.

Proposition 10. *The synthesis problem over max-multi-linear interpretations can be solved in polynomial time.*

6.2 Type System for In-Place Update

Hofmann [Hof00] proposes a first-order functional language that can be compiled into code not requiring dynamic heap memory allocation. This is achieved by means of an –empty– 'resource type' $\diamond$ and 'affine' typing rules. Elements of resource type have to be understood as memory cells. Constructors of inductive types require an argument of resource type. Also functions may take as arguments elements of resource type. We look at a little fragment of this type system[5] composed of programs over the types:

$$\begin{array}{ll} \diamond \equiv \mu t.() & \text{(resource type)} \\ W \equiv \mu t.(\epsilon : t, \mathsf{0} : \diamond, t \to t, \mathsf{1} : \diamond, t \to t) & \text{(binary words)}. \end{array}$$

For every function f we assume $\Sigma(f)$ has the shape $(\diamond, \ldots, \diamond, W, \ldots, W) \to W$ and let $r(f) < arity(f)$ be the number of arguments of resource type. As usual patterns and expressions in functions' definitions have to be well typed (cf. section 1). This means that assuming $\Sigma(f) = (\tau_1, \ldots, \tau_n) \to W$ for every rule in the definition of f there is a context Γ such that:

$$\Gamma \vdash_{\Sigma} p_{i,j} : \tau_j \text{ for } j = 1, \ldots, n \quad \Gamma \vdash_{\Sigma} e_i : W . \tag{14}$$

Without loss of generality, we may assume that Γ contains only the variables occurring in the patterns $p_{i,j}$. Now we say that the typing is *affine* if in the typing of e_i the hypotheses in the context Γ are used at most once. Note that the typing of the patterns is always affine since we deal with linear patterns.

Resource arguments can be regarded as annotations for the compiler but no real computation is performed on them. Indeed, it is not even possible to create (closed) values of resource type. However, there is an obvious way to *erase* resource arguments and obtain the 'intended' program. In our simple case, the resulting program will operate over the type $w \equiv \mu t.(\epsilon : t, \mathsf{0} : t \to t, \mathsf{1} : t \to t)$. The erasure function *er* is defined as follows over expressions:

$$\begin{array}{c} er(x) = x \quad er(\epsilon) = \epsilon \quad er(\mathsf{0}(x, e)) = \mathsf{0}(er(e)) \quad er(\mathsf{1}(x, e)) = \mathsf{1}(er(e)) \\ er(f(e_1, \ldots, e_n)) = f(er(e_{r(f)+1}), \ldots, er(e_n)) . \end{array}$$

Proposition 11. *If a program has an affine typing then its erasure admits the following multi-linear quasi-interpretation:*

$$q_0 = q_1 = x + 1, \quad q_f = \Sigma_{i=1,\ldots,n} x_i + r(f) .$$

[5] In particular, we neglect enumerated, product, and higher-order types.

We consider a restricted class of multi-linear quasi-interpretations where:

$$q_f = a^f + \Sigma_{i=1,\ldots,n} x_i \quad a^f \geq 0 \, . \tag{15}$$

We note that (i) constructors are subject to the general conditions of assignments, (ii) every function is determined by exactly *one* coefficient a^f, and (iii) the interpretation in proposition 11 falls in this family. We refer to this class of quasi-interpretations as *sum*-multi-linear. The proof strategy for the following proposition is the same as for proposition 10.

Proposition 12. *The synthesis problem over sum-multi-linear interpretations can be solved in polynomial time.*

6.3 Lower and Upper Bounds on Complexity

If a program admits a polynomial max-plus interpretation of degree k, then $q_{f(x_1,\ldots,x_n)} \leq k(\Sigma_{i=1,\ldots,n} x_i) + c$ for some constant c. Hence $q_f(v_1, \ldots, v_n)$ is in $O(\Sigma_{i=1,\ldots,n}|v_i|)$. Then it follows from theorem 1 that a program admitting a polynomial max-plus interpretation can be evaluated in time $2^{O(n)}$ on data of size n.

For a lower bound, we refer to [Hof00] where it is shown that 'non-size increasing' recursive programs can simulate Turing machines (TM) running in time $2^{O(n)}$ (this is inspired by the simulation of an exponential time Turing Machine by a linearly bounded auxiliary pushdown automaton in Cook's theorem [Coo71]).

The exponential upper bound derived in theorem 1 might be too loose in practice. To improve the situation, quasi-interpretations can be combined with various methods enforcing *program termination*. In particular, in [BMM01] it is shown that a program terminating by *lexicographic path-order* (lpo[6]) and admitting a polynomially bounded quasi-interpretation (polynomial in the usual sense) can be evaluated in PSPACE. For a lower bound, one can program a checker for the validity of quantified boolean formulas that terminates by lpo and admits a multi-linear max-plus quasi-interpretation. By imposing further conditions on the termination method (product path-order) it is also possible to characterize PTIME [Mar00]. To prove these results, one can still rely on the basic evaluation strategies presented in the proof of theorem 1.

7 Conclusion

Cobham's celebrated result [Cob65] is about primitive recursive functions with polynomial bounds on the size of the results. Here we have looked at general recursive functions with max-plus polynomial bounds on the size of the results. We have shown that the synthesis problem in the multi-linear case is NP-complete (when rules have bounded size). This case appears to be a reasonable compromise between complexity of the decision procedure and power of the analysis. The synthesis problem for max-plus polynomials of degree higher than 1 remains to be analyzed.

[6] A popular termination method that can be synthesized in non-deterministic polynomial time; see, *e.g.*, [BN98].

Acknowledgement. This work was done while on leave at the *Ludwig-Maximilian Universität München* and was partially supported by the *IST-Global Computing Mobile Resource Guarantee Project* and the *Action Spécifique Méthodes formelles pour la mobilité*. The author benefitted from a number of discussions with Martin HOFMANN.

References

[Amadio02] R. Amadio. Max-plus quasi-interpretations. Research Report *Laboratoire d'Informatique Fondamentale de Marseille* 10–2002, December 2002.

[BC92] S. Bellantoni and S. Cook. A new recursion-theoretic characterization of the polytime functions. *Computational Complexity*, 2:97–110, 1992.

[BCOQ92] F. Baccelli, G. Cohen, G. Olsder, and J.-P. Quadrat. *Synchronization and linearity*. Wiley, 1992.

[BMM01] G. Bonfante, J.-Y. Marion, and J.-Y. Moyen. On termination methods with space bound certifications. In *Ershov 4th Int. Conf.*, Springer LNCS, 2001.

[BN98] F. Baader and T. Nipkow. *Term rewriting and all that*. Cambridge University Press, 1998.

[Cas97] V. Caseiro. *Equations for defining polytime functions*. PhD thesis, University of Oslo, 1997.

[Clo95] P. Clote. Computation models and function algebras. In *Proc. Logic and computational complexity, Springer Lecture Notes in Comp. Sci. 960*, 1995.

[Cob65] A. Cobham. The intrinsic computational difficulty of functions. In *Proc. Logic, Methodology, and Philosophy of Science II, North Holland*, 1965.

[Coo71] S. Cook. Characterizations of pushdown machines in terms of time-bounded computers. *Journal of the ACM*, 18(1):4–18, 1971.

[Gra96] B. Gramlich. On proving termination by innermost termination. In *Proc. RTA*, Springer Lecture Notes in Comp. Sci. 1103, 1996.

[HJ03] M. Hofmann and S. Jost. Static prediction of heap space usage for first-order functional programs. In *Proc. ACM POPL*, 2003.

[Hof00] M. Hofmann. A type system for bounded space and functional in-place update. *Nordic Journal of Computing*, 7(4):258–289, 2000.

[Hof02] M. Hofmann. The strength of non size-increasing computation. In *Proc. ACM POPL*, 2002.

[HPS96] R. Hughes, L. Pareto, and A. Sabry. Proving the correctness of reactive systems using sized types. In *Proc. ACM POPL*, 1996.

[Jon97] N. Jones. *Computability and complexity, from a programming perspective*. MIT-Press, 1997.

[Lei94] D. Leivant. Predicative recurrence and computational complexity i: word recurrence and poly-time. *Feasible mathematics II, Clote and Remmel (eds.)*, Birkhäuser: 320–343, 1994.

[Mar00] J.-Y. Marion. *Complexité implicite des calculs, de la théorie à la pratique*. PhD thesis, Universitè Nancy, 2000. Habilitation à diriger des recherches.

[MM00] J.-Y. Marion and J.-Y. Moyen. Efficient first order functional program interpreter with time bound certifications. In *Proc. LPAR*, Springer Lecture Notes in Comp. Sci. 1955, 2000.

[Par00] L. Pareto. *Types for crash prevention*. PhD thesis, Chalmers University, 2000.

[Pug92] W. Pugh. The omega test: a fast and practical integer programming algorithm for dependence analysis. In *Communications of the ACM, 102–114*, 1992.

[San01] D. Sannella. Mobile resource guarantee. Ist-global computing research proposal, U. Edinburgh, 2001. http://www.dcs.ed.ac.uk/home/mrg/.

Inductive Types in the Calculus of Algebraic Constructions

Frédéric Blanqui

Laboratoire d'Informatique de l'École Polytechnique
91128 Palaiseau Cedex, France
blanqui@lix.polytechnique.fr

Abstract. In a previous work, we proved that almost all of the Calculus of Inductive Constructions (CIC), the basis of the proof assistant Coq, can be seen as a Calculus of Algebraic Constructions (CAC), an extension of the Calculus of Constructions with functions and predicates defined by higher-order rewrite rules. In this paper, we prove that CIC as a whole can be seen as a CAC, and that it can be extended with non-strictly positive types and inductive-recursive types together with non-free constructors and pattern-matching on defined symbols.

1 Introduction

There has been different proposals for defining inductive types and functions in typed systems. In Girard's polymorphic λ-calculus or in the Calculus of Constructions (CC) [9], data types and functions can be formalized by using impredicative encodings, difficult to use in practice, and computations are done by β-reduction only. In Martin-Löf's type theory or in the Calculus of Inductive Constructions (CIC) [10], inductive types and their induction principles are first-class objects, functions can be defined by induction and computations are done by ι-reduction. For instance, for the type nat of natural numbers, the recursor $rec : (P : nat \Rightarrow \star)(u : P0)(v : (n : nat)Pn \Rightarrow P(sn))(n : nat)Pn$ is defined by the following ι-rules:

$$\begin{array}{lcl} rec\, P\; u\; v\; 0 & \rightarrow_\iota & u \\ rec\, P\; u\; v\; (s\; n) & \rightarrow_\iota & v\; n\; (rec\; P\; u\; v\; n) \end{array}$$

Finally, in the algebraic setting [11], functions are defined by using rewrite rules and computations are done by applying these rules. Since both β-reduction and ι-reduction are particular cases of higher-order rewriting [16], proposals soon appeared for integrating all these approaches. Starting with [15,2], this objective culminated with [4,5,6] in which almost all of CIC can be seen as a Calculus of Algebraic Constructions (CAC), an extension of CC with functions and predicates defined by higher-order rewrite rules. In this paper, we go one step further in this direction and capture all previous proposals, and much more.

Let us see the two examples of recursors that are allowed in CIC but not in CAC [20]. The first example is a third-order definition of finite sets of natural numbers (represented as predicates over nat):

M. Hofmann (Ed.): TLCA 2003, LNCS 2701, pp. 46–59, 2003.

$$\begin{aligned}
\mathit{fin} &: (\mathit{nat} \Rightarrow \star) \Rightarrow \star\\
\mathit{femp} &: \mathit{fin}\,\emptyset\\
\mathit{fadd} &: (x : \mathit{nat})(p : \mathit{nat} \Rightarrow \star)\mathit{fin}\,p \Rightarrow \mathit{fin}(\mathrm{add}\ x\ p)\\
\mathit{rec} &: (Q : (\mathit{nat} \Rightarrow \star) \Rightarrow \star)Q\emptyset\\
&\quad \Rightarrow ((x : \mathit{nat})(p : \mathit{nat} \Rightarrow \star)\mathit{fin}\,p \Rightarrow Qp \Rightarrow Q(\mathrm{add}\ x\ p))\\
&\quad \Rightarrow (p : \mathit{nat} \Rightarrow \star)\mathit{fin}\,p \Rightarrow Qp
\end{aligned}$$

where $\emptyset = [y : \mathit{nat}]\bot$ represents the empty set, add $x\ p = [y : \mathit{nat}]y = x \vee (p\ y)$ represents the set $\{x\} \cup p$, and the weak recursor *rec* (recursor for defining objects) is defined by the rules:

$$\begin{aligned}
\mathit{rec}\ Q\ u\ v\ p'\ \mathit{femp} &\rightarrow u\\
\mathit{rec}\ Q\ u\ v\ p'\ (\mathit{fadd}\ x\ p\ h) &\rightarrow v\ x\ p\ h\ (\mathit{rec}\ Q\ u\ v\ p\ h)
\end{aligned}$$

The problem comes from the fact that, in *fin*(add x p), the output type of *fadd*, the predicate p is not a parameter of *fin*.[1] This can be generalized to any big/impredicative dependent type, that is, to any type having a constructor with a predicate argument which is not a parameter. Formally, if $C : (\boldsymbol{z} : \boldsymbol{V})\star$ is a type and $c : (\boldsymbol{x} : \boldsymbol{T})C\boldsymbol{v}$ is a constructor of C then, for all predicate variable x occurring in some T_j, there must be some argument $v_{\iota_x} = x$, a condition called (I6) in [5].

The second example is John Major's equality which is intended to equal terms of different types [18]:

$$\begin{aligned}
\mathit{JMeq} &: (A : \star)A \Rightarrow (B : \star)B \Rightarrow \star\\
\mathit{refl} &: (A : \star)(x : A)(\mathit{JMeq}\ A\ x\ A\ x)\\
\mathit{rec} &: (A : \star)(x : A)(P : (B : \star)B \Rightarrow \star)(P\ A\ x)\\
&\quad \Rightarrow (B : \star)(y : B)(\mathit{JMeq}\ A\ x\ B\ y) \Rightarrow (P\ B\ y)
\end{aligned}$$

where *rec* is defined by the rule:

$$\mathit{rec}\ C\ x\ P\ h\ C\ x\ (\mathit{refl}\ C\ x) \rightarrow h$$

Here, the problem comes from the fact that the argument for B is equal to the argument for A. This can be generalized to any polymorphic type having a constructor with two equal type parameters. From a rewriting point of view, this is like having pattern-matching or non-linearities on predicate arguments, which is known to create inconsistencies in some cases [14]. Formally, a rule $f\boldsymbol{l} \rightarrow r$ with $f : (\boldsymbol{x} : \boldsymbol{T})U$ is *safe* if, for all predicate argument x_i, l_i is a variable and, if x_i and x_j are two distinct predicate arguments, then $l_i \neq l_j$. An inductive type is *safe* if the corresponding ι-rules are safe.

By using what is called in Matthes' terminology [17] an *elimination-based* interpretation instead of the *introduction-based* interpretation that we used in [5], we prove that recursors for types like *fin* or *JMeq* can be accepted, hence that CAC essentially subsumes CIC. In addition, we prove that it can be extended to non-strictly positive types (Section 7) and to inductive-recursive types [12] (Section 8).

[1] This is also the reason why the corresponding strong recursor, that is, the recursor for defining types or predicates, is not allowed in CIC (p could be "bigger" than *fin*).

2 The Calculus of Algebraic Constructions (CAC)

We assume the reader familiar with typed λ-calculi [3] and rewriting [11]. The Calculus of Algebraic Constructions (CAC) [5] simply extends CC by considering a set $\mathcal{F}$ of *symbols*, equipped with a total quasi-ordering $\geq$ (precedence) whose strict part is well-founded, and a set $\mathcal{R}$ of *rewrite rules*. The terms of CAC are:

$$t ::= s \mid x \mid f \mid [x:t]u \mid tu \mid (x:t)u$$

where $s \in \mathcal{S} = \{\star, \square\}$ is a *sort*, $x \in \mathcal{X}$ a *variable*, $f \in \mathcal{F}$, $[x:t]u$ an *abstraction*, tu an *application*, and $(x:t)u$ a *dependent product*, written $t \Rightarrow u$ if x does not freely occur in u. We denote by $\mathrm{FV}(t)$ the set of free variables of t, by $\mathrm{Pos}(t)$ the set of Dewey's positions of t, and by $\mathrm{dom}(\theta)$ the *domain* of a substitution θ.

The sort $\star$ denotes the universe of types and propositions, and the sort $\square$ denotes the universe of predicate types (also called *kinds*). For instance, the type *nat* of natural numbers is of type $\star$, $\star$ itself is of type $\square$ and $nat \Rightarrow \star$, the type of predicates over *nat*, is of type $\square$.

Every symbol f is equipped with a sort s_f, an *arity* α_f and a type τ_f which may be any closed term of the form $(\boldsymbol{x}:\boldsymbol{T})U$ with $|\boldsymbol{x}| = \alpha_f$ ($|\boldsymbol{x}|$ is the length of $\boldsymbol{x}$). We denote by Γ_f the environment $\boldsymbol{x}:\boldsymbol{T}$. The terms only built from variables and applications of the form $f\boldsymbol{t}$ with $|\boldsymbol{t}| = \alpha_f$ are called *algebraic*.

A rule for typing symbols is added to the typing rules of CC:

$$(\mathrm{symb}) \qquad \frac{\vdash \tau_f : s_f}{\vdash f : \tau_f}$$

A *rewrite rule* is a pair $l \rightarrow r$ such that (1) l is algebraic, (2) l is not a variable, and (3) $\mathrm{FV}(r) \subseteq \mathrm{FV}(l)$. A symbol f with no rule of the form $f\boldsymbol{l} \rightarrow r$ is *constant*, otherwise it is (partially) *defined*. We also assume that, in every rule $f\boldsymbol{l} \rightarrow r$, the symbols occurring in r are smaller than or equivalent to f.

Finally, in CAC, $\beta\mathcal{R}$-equivalent types are identified. More precisely, in the type conversion rule of CC, $\downarrow_\beta$ is replaced by $\downarrow_{\beta\mathcal{R}}$:

$$(\mathrm{conv}) \qquad \frac{\Gamma \vdash t : T \quad T \downarrow_{\beta\mathcal{R}} T' \quad \Gamma \vdash T' : s}{\Gamma \vdash t : T'}$$

where $u \downarrow_{\beta\mathcal{R}} v$ iff there exists a term w such that $u \rightarrow^*_{\beta\mathcal{R}} w$ and $v \rightarrow^*_{\beta\mathcal{R}} w$, $\rightarrow^*_{\beta\mathcal{R}}$ being the reflexive and transitive closure of $\rightarrow = \rightarrow_\beta \cup \rightarrow_\mathcal{R}$. This rule means that any term t of type T in the environment Γ is also of type T' if T and T' have a common reduct (and T' is of type some sort s). For instance, if t is a proof of $P(2+2)$ then t is also a proof of $P(4)$ if $\mathcal{R}$ contains the following rules:

$$\begin{array}{rcl} x + 0 & \rightarrow & x \\ x + (s\ y) & \rightarrow & s\ (x + y) \end{array}$$

This allows to decrease the size of proofs by an important factor, and to increase the automation as well. **All over the paper, we assume that $\rightarrow$ is confluent.**

A substitution θ *preserves typing from* Γ *to* Δ, written $\theta : \Gamma \leadsto \Delta$, if, for all $x \in \mathrm{dom}(\Gamma)$, $\Delta \vdash x\theta : x\Gamma\theta$, where $x\Gamma$ is the type associated to x in Γ. Type-preserving substitutions enjoy the following important property: if $\Gamma \vdash t : T$ and $\theta : \Gamma \leadsto \Delta$ then $\Delta \vdash t\theta : T\theta$.

For ensuring the *subject reduction* property (preservation of typing under reduction), every rule $f\boldsymbol{l} \to r$ is equipped with an environment Γ and a substitution ρ such that, if $f : (\boldsymbol{x} : \boldsymbol{T})U$ and $\gamma = \{\boldsymbol{x} \mapsto \boldsymbol{l}\}$, then $\Gamma \vdash f\boldsymbol{l}\rho : U\gamma\rho$ and $\Gamma \vdash r : U\gamma\rho$. The substitution ρ allows to eliminate non-linearities due to typing. For instance, the concatenation on polymorphic lists (type $list : \star \Rightarrow \star$ with constructors $nil : (A : \star)listA$ and $cons : (A : \star)A \Rightarrow listA \Rightarrow listA$) of type $(A : \star)listA \Rightarrow listA \Rightarrow listA$ can be defined by:

$$\begin{array}{rcl} app\ A\ (nil\ A')\ l' & \to & l' \\ app\ A\ (cons\ A'\ x\ l)\ l' & \to & cons\ A\ x\ (app\ A\ x\ l\ l') \\ app\ A\ (app\ A'\ l\ l')\ l'' & \to & app\ A\ l\ (app\ A\ l'\ l'') \end{array}$$

with $\Gamma = A : \star, x : A, l : listA, l' : listA$ and $\rho = \{A' \mapsto A\}$. For instance, $app\ A\ (nil\ A')$ is not typable in Γ (since $A' \notin \mathrm{dom}(\Gamma)$) but becomes typable if we apply ρ. This does not matter since, if an instance $app\ A\sigma\ (nil\ A'\sigma)$ is typable then $A\sigma$ is convertible to $A'\sigma$. Eliminating non-linearities makes rewriting more efficient and the proof of confluence easier.

3 Strong Normalization

Typed λ-calculi are generally proved strongly normalizing by using Tait and Girard's technique of *reducibility candidates* [13]. The idea of Tait, later extended by Girard to the polymorphic λ-calculus, is to strengthen the induction hypothesis. Instead of proving that every term is strongly normalizable (set $\mathcal{SN}$), one associates to every type T a set $[\![T]\!] \subseteq \mathcal{SN}$, the *interpretation* of T, and proves that every term t of type T is *computable*, *i.e.* belongs to $[\![T]\!]$. Hereafter, we follow the proof given in [7] which greatly simplifies the one given in [5].

Definition 1 (Reducibility candidates.) *A term t is* neutral *if it is not an abstraction, not of the form $c\boldsymbol{t}$ with $c : (\boldsymbol{y} : \boldsymbol{U})C\boldsymbol{v}$ and C constant, nor of the form $f\boldsymbol{t}$ with f defined and $|\boldsymbol{t}| < \alpha_f$. We inductively define the complete lattice $\mathcal{R}_t$ of the interpretations for the terms of type t, the ordering $\leq_t$ on $\mathcal{R}_t$, and the greatest element $\top_t \in \mathcal{R}_t$ as follows.*

- *$\mathcal{R}_t = \{\emptyset\}$, $\leq_t = \subseteq$ and $\top_t = \emptyset$ if $t \neq \Box$ and $\Gamma \not\vdash t : \Box$.*
- *$\mathcal{R}_s$ is the set of subsets $R \subseteq \mathcal{T}$ such that:*
  (R1) *$R \subseteq \mathcal{SN}$ (strong normalization).*
  (R2) *If $t \in R$ then $\to\!(t) = \{t' \mid t \to t'\} \subseteq R$ (stability by reduction).*
  (R3) *If t is neutral and $\to\!(t) \subseteq R$ then $t \in R$ (neutral terms).*
  Furthermore, $\leq_s = \subseteq$ and $\top_s = \mathcal{SN}$.

– $\mathcal{R}_{(x:U)K}$ *is the set of functions* R *from* $\mathcal{T} \times \mathcal{R}_U$ *to* $\mathcal{R}_K$ *such that* $R(u,S) = R(u',S)$ *whenever* $u \rightarrow u'$, $R \leq_{(x:U)K} R'$ *iff, for all* $(u,S) \in \mathcal{T} \times \mathcal{R}_U$, $R(u,S) \leq_K R'(u,S)$, *and* $\top_{(x:U)K}(u,S) = \top_K$.

Note that $\mathcal{R}_t = \mathcal{R}_{t'}$ whenever $t \rightarrow t'$ and that, for all $R \in \mathcal{R}_s$, $\mathcal{X} \subseteq R$.

Definition 2 (Interpretation schema.) *A* candidate assignment *is a function* ξ *from* $\mathcal{X}$ *to* $\bigcup\{\mathcal{R}_t \mid t \in \mathcal{T}\}$. *An assignment* ξ validates *an environment* Γ, *written* $\xi \models \Gamma$, *if, for all* $x \in \mathrm{dom}(\Gamma)$, $x\xi \in \mathcal{R}_{x\Gamma}$. *An* interpretation *for a symbol* f *is an element of* $\mathcal{R}_{\tau_f}$. *An* interpretation *for a set* $\mathcal{G}$ *of symbols is a function which, to each symbol* $g \in \mathcal{G}$, *associates an interpretation for* g.

The interpretation *of* t *w.r.t. a candidate assignment* ξ, *an interpretation* I *for* $\mathcal{F}$ *and a substitution* θ, *is defined by induction on* t *as follows.*

- $[\![t]\!]^I_{\xi,\theta} = \top_t$ *if* t *is an object or a sort,*
- $[\![x]\!]^I_{\xi,\theta} = x\xi$,
- $[\![f]\!]^I_{\xi,\theta} = I_f$,
- $[\![(x:U)V]\!]^I_{\xi,\theta} = \{t \in \mathcal{T} \mid \forall u \in [\![U]\!]^I_{\xi,\theta}, \forall S \in \mathcal{R}_U, tu \in [\![V]\!]^I_{\xi^S_x,\theta^u_x}\}$,
- $[\![[x:U]v]\!]^I_{\xi,\theta}(u,S) = [\![v]\!]^I_{\xi^S_x,\theta^u_x}$,
- $[\![tu]\!]^I_{\xi,\theta} = [\![t]\!]^I_{\xi,\theta}(u\theta, [\![u]\!]^I_{\xi,\theta})$,

where $\xi^S_x = \xi \cup \{x \mapsto S\}$ *and* $\theta^u_x = \theta \cup \{x \mapsto u\}$. *A substitution* θ *is* adapted *to a* Γ*-assignment* ξ *if* $\mathrm{dom}(\theta) \subseteq \mathrm{dom}(\Gamma)$ *and, for all* $x \in \mathrm{dom}(\theta)$, $x\theta \in [\![x\Gamma]\!]^I_{\xi,\theta}$. *A pair* (ξ,θ) *is* Γ-valid, *written* $\xi,\theta \models \Gamma$, *if* $\xi \models \Gamma$ *and* θ *is adapted to* ξ.

Note that $[\![t]\!]^I_{\xi,\theta} = [\![t]\!]^{I'}_{\xi',\theta'}$ whenever ξ and ξ' agree on the predicate variables free in t, θ and θ' agree on the variables free in t, and I and I' agree on the symbols occurring in t. The difficult point is then to define an interpretation for predicate symbols and to prove that every symbol f is computable (*i.e.* $f \in [\![\tau_f]\!]$).

Following previous works on inductive types [19,23], the interpretation of a constant predicate symbol C is defined as the least fixpoint of a monotone function $I \mapsto \varphi^I_C$ on the complete lattice $\mathcal{R}_{\tau_C}$. Following Matthes [17], there is essentially two possible definitions that we illustrate by the case of *nat*. The *introduction-based* definition:

$$\varphi^I_{nat} = \{t \in \mathcal{SN} \mid t \rightarrow^* su \Rightarrow u \in I\}$$

and the *elimination-based* definition:

$$\varphi^I_{nat} = \{t \in \mathcal{T} \mid \forall (\xi,\theta)\ \Gamma\text{-valid},\ rec\ P\theta\ u\theta\ v\theta\ t \in [\![Pn]\!]^I_{\xi,\theta^t_n}\}$$

where $\Gamma = P : nat \Rightarrow \star, u : P0, v : (n:nat)Pn \Rightarrow P(sn)$. In both cases, the monotonicity of φ_{nat} is ensured by the fact that *nat* occurs only positively[2] in

[2] X occurs positively in $Y \Rightarrow X$ and negatively in $X \Rightarrow Y$. In Section 8, we give an extended definition of positivity for dealing with inductive-recursive types [12].

the types of the arguments of its constructors, a common condition for inductive types.[3].

In [5], we used the introduction-based approach since this allows us to have non-free constructors and pattern-matching on defined symbols, which is forbidden in CIC and does not seem possible with the elimination-based approach. Indeed, in CAC, it is possible to formalize the type *int* of integers by taking the symbols $0 : int$, $s : int \Rightarrow int$ and $p : int \Rightarrow int$, together with the rules:

$$\begin{array}{rcl} s\ (p\ x) & \rightarrow & x \\ p\ (s\ x) & \rightarrow & x \end{array}$$

It is also possible to have the following rule on natural numbers:

$$x \times (y + z) \quad \rightarrow \quad (x \times y) + (x \times z)$$

To this end, we extended the notion of constructor by considering as *constructor* any symbol c whose output type is a constant predicate symbol C (perhaps applied to some arguments). Then, the arguments of c that can be used to define the result of a function are restricted to the arguments, called *accessible*, in the type of which C occurs only positively. We denote by $\mathrm{Acc}(c)$ the set of accessible arguments of c. For instance, x is accessible in sx since *nat* occurs only positively in the type of x. But, we also have x and y accessible in $x + y$ since *nat* occurs only positively in the types of x and y. So, $+$ can be seen as a constructor too.

With this approach, we can safely take:

$$\varphi^I_{nat} = \{t \in \mathcal{SN} \mid \forall f, t \rightarrow^* f\boldsymbol{u} \Rightarrow \forall j \in \mathrm{Acc}(f), u_j \in [\![U_j]\!]^I_{\xi,\theta}\}$$

where $f : (\boldsymbol{y} : \boldsymbol{U})C\boldsymbol{v}$ and $\theta = \{\boldsymbol{y} \mapsto \boldsymbol{u}\}$, whenever an appropriate assignment ξ for the predicate variables of U_j can be defined, which is possible only if the condition (I6) is satisfied (see the type *fin* in Section 1).

4 Extended Recursors

As we introduced an extended notion of constructor for dealing with the introduction-based method, we now introduce an extended notion of recursor for dealing with the elimination-based method.

Definition 1 (Extended recursors). *A* pre-recursor *for a constant predicate symbol* $C : (\boldsymbol{z} : \boldsymbol{V})\star$ *is any symbol* f *such that:*

- *the type of* f *is of the form*[4] $(\boldsymbol{z} : \boldsymbol{V})(z : C\boldsymbol{z})W$,

[3] Mendler proved that recursors for negative types are not normalizing [19]. Take for instance an inductive type C with a constructor $c : (C \rightarrow nat) \rightarrow C$. Assume now that we have $p : C \rightarrow (C \rightarrow nat)$ defined by the rule $p(cx) \rightarrow x$ (case analysis). Then, by taking $\omega = [x : C](px)x$, we get $\omega(c\omega) \rightarrow_\beta p(c\omega)(c\omega) \rightarrow \omega(c\omega) \rightarrow_\beta \ldots$

[4] Our examples may not always fit in this form but since, in an environment, two types that do not depend on each other can be permuted, this does not matter.

- *every rule defining f is of the form $f\boldsymbol{z}\boldsymbol{t}\boldsymbol{u} \to r$ with* $\mathrm{FV}(r) \cap \{\boldsymbol{z}\} = \emptyset$,
- *$f\boldsymbol{v}t\boldsymbol{u}$ is head-reducible only if t is constructor-headed.*

A pre-recursor f is a recursor *if it satisfies the following* positivity conditions:[5]

- *no constant predicate $D > C$ or defined predicate F occurs in W,*
- *every constant predicate $D \simeq C$ occurs only positively in W.*

A recursor of sort $\star$ (resp. $\square$) is weak *(resp.* strong*). Finally, we assume that every type C has a set $\mathcal{R}ec(C)$ (possibly empty) of recursors.*

For the types C whose set of recursors $\mathcal{R}ec(C)$ is not empty, we define the interpretation of C with the elimination-based method as follows. For the other types, we keep the introduction-based method.

Definition 2 (Interpretation of inductive types). *If every t_i has a normal form t_i^* then $\varphi_C^I(\boldsymbol{t}, \boldsymbol{S})$ is the set of terms t such that, for all $f \in \mathcal{R}ec(C)$ of type $(\boldsymbol{z} : \boldsymbol{V})(z : C\boldsymbol{z})(\boldsymbol{y} : \boldsymbol{U})V$, $\boldsymbol{y}\xi$ and $\boldsymbol{y}\theta$, if $\xi_{\boldsymbol{z}}^{\boldsymbol{S}}, \theta_{\boldsymbol{z} z}^{\boldsymbol{t}\, t} \models \boldsymbol{y} : \boldsymbol{U}$ then $f\boldsymbol{t}^* t \boldsymbol{y}\theta \in [\![V]\!]^I_{\xi_{\boldsymbol{z}}^{\boldsymbol{S}}, \theta_{\boldsymbol{z} z}^{\boldsymbol{t}\, t}}$. Otherwise, $\varphi_C^I(\boldsymbol{t}, \boldsymbol{S}) = \mathcal{SN}$.*

The fact that φ is monotone, hence has a least fixpoint, follows from the positivity conditions. One can easily check that φ_C^I is stable by reduction: if $\boldsymbol{t} \to \boldsymbol{t}'$ then $\varphi_C^I(\boldsymbol{t}, \boldsymbol{S}) = \varphi_C^I(\boldsymbol{t}', \boldsymbol{S})$. We now prove that $\varphi_C^I(\boldsymbol{t}, \boldsymbol{S})$ is a candidate.

Lemma 3. *$\varphi_C^I(\boldsymbol{t}, \boldsymbol{S})$ is a candidate.*

Proof. (R1) Let $t \in R$. We must prove that $t \in \mathcal{SN}$. Since $\mathcal{R}ec(C) \neq \emptyset$, there is at least one recursor f. Take $y_i\theta = y_i$ and $y_i\xi = \top_{U_i}$. We clearly have $\xi_{\boldsymbol{z}}^{\boldsymbol{S}}, \theta_{\boldsymbol{z} z}^{\boldsymbol{t}\, t} \models \boldsymbol{y} : \boldsymbol{U}$. Therefore, $f\boldsymbol{t}^* t\boldsymbol{y} \in S = [\![V]\!]^I_{\xi_{\boldsymbol{z}}^{\boldsymbol{S}}, \theta_{\boldsymbol{z} z}^{\boldsymbol{t}\, t}}$. Now, since S satisfies (R1), $f\boldsymbol{t}^* t\boldsymbol{y} \in \mathcal{SN}$ and $t \in \mathcal{SN}$.

(R2) Let $t \in R$ and $t' \in \to\!(t)$. We must prove that $t' \in R$, hence that $f\boldsymbol{t}^* t'\boldsymbol{y}\theta \in S = [\![V]\!]^I_{\xi_{\boldsymbol{z}}^{\boldsymbol{S}}, \theta_{\boldsymbol{z} z}^{\boldsymbol{t}\, t}}$. This follows from the fact that $f\boldsymbol{t}^* t\boldsymbol{y}\theta \in S$ (since $t \in R$) and S satisfies (R2).

(R3) Let t be a neutral term such that $\to\!(t) \subseteq R$. We must prove that $t \in R$, hence that $u = f\boldsymbol{t}^* t\boldsymbol{y}\theta \in S = [\![V]\!]^I_{\xi_{\boldsymbol{z}}^{\boldsymbol{S}}, \theta_{\boldsymbol{z} z}^{\boldsymbol{t}\, t}}$. Since u is neutral and S satisfies (R3), it suffices to prove that $\to\!(u) \subseteq S$. Since $\boldsymbol{y}\theta \in \mathcal{SN}$ by (R1), we proceed by induction on $\boldsymbol{y}\theta$ with $\to$ as well-founded ordering. The only difficult case could be when u is head-reducible, but this is not possible since t is neutral, hence not constructor-headed. □

5 Admissible Recursors

Since we changed the interpretation of constant predicate symbols, we must check several things in order to preserve the strong normalization result of [5].

- We must make sure that the interpretation of primitive types is still $\mathcal{SN}$ since this is used for proving the computability of first-order symbols and the interpretation of some defined predicate symbols (see Lemma 5).

[5] In Section 8, we give weaker conditions for dealing with inductive-recursive types.

- We must also prove that every symbol is computable.
  - For extended recursors, this follows from the definition of the interpretation for constant predicate symbols, and thus, does not require safety.
  - For first-order symbols, nothing is changed.
  - For higher-order symbols distinct from recursors, we must make sure that the accessible arguments of a computable constructor-headed term are computable.
  - For constructors, this does not follow from the interpretation for constant predicate symbols anymore. We therefore have to prove it.

We now define general conditions for these requirements to be satisfied.

Definition 4 (Admissible recursors). *Assume that every constructor is equipped with a set* $\mathrm{Acc}(c) \subseteq \{1, \dots, \alpha_c\}$ *of* accessible arguments. *Let* $C : (\boldsymbol{z} : \boldsymbol{V})\star$ *be a constant predicate symbol.* $\mathcal{R}ec(C)$ *is* complete w.r.t. accessibility *if, for all* $c : (\boldsymbol{x} : \boldsymbol{T})C\boldsymbol{v}$, $j \in \mathrm{Acc}(c)$, $\boldsymbol{x}\eta$ *and* $\boldsymbol{x}\sigma$, *if* $\eta \models \Gamma_c$, $\boldsymbol{v}\sigma \in \mathcal{SN}$ *and* $c\boldsymbol{x}\sigma \in [\![C\boldsymbol{v}]\!]_{\eta,\sigma}$ *then* $x_j\sigma \in [\![T_j]\!]_{\eta,\sigma}$.

A recursor $f : (\boldsymbol{z} : \boldsymbol{V})(z : C\boldsymbol{z})(\boldsymbol{y} : \boldsymbol{U})V$ *is* head-computable w.r.t *a constructor* $c : (\boldsymbol{x} : \boldsymbol{T})C\boldsymbol{v}$ *if, for all* $\boldsymbol{x}\eta$, $\boldsymbol{x}\sigma$, $\boldsymbol{y}\xi$, $\boldsymbol{y}\theta$, $\boldsymbol{S} = [\![\boldsymbol{v}]\!]_{\eta,\sigma}$, *if* $\eta, \sigma \models \Gamma_c$ *and* $\xi^{\boldsymbol{S}}_{\boldsymbol{z}}, \theta^{\boldsymbol{v}\sigma}_{\boldsymbol{z}}{}^{c\boldsymbol{x}\sigma}_{z} \models \boldsymbol{y} : \boldsymbol{U}$, *then every head-reduct of* $f\boldsymbol{v}\sigma(c\boldsymbol{x}\sigma)\boldsymbol{y}\theta$ *belongs to* $[\![V]\!]_{\xi^{\boldsymbol{S}}_{\boldsymbol{z}}, \theta^{\boldsymbol{v}\sigma}_{\boldsymbol{z}}{}^{c\boldsymbol{x}\sigma}_{z}}$. *A recursor is* head-computable *if it is head-computable w.r.t. every constructor.* $\mathcal{R}ec(C)$ *is* head-computable *if all its recursors are head-computable.*

$\mathcal{R}ec(C)$ *is* admissible *if it is head-computable and complete w.r.t. accessibility.*

We first prove that the interpretation of primitive types is $\mathcal{SN}$.

Lemma 5 (Primitive types). *Types equivalent to* C *are* primitive *if, for all* $D \simeq C$, $D : \star$ *and, for all* $d : (\boldsymbol{x} : \boldsymbol{T})D$, $\mathrm{Acc}(d) = \{1, \dots, \alpha_d\}$ *and every* T_j *is a primitive type* $E \leq C$. *Let* $C : \star$ *be a primitive symbol. If recursors are head-computable then* $I_C = \mathcal{SN}$.

Proof. By definition, $I_C \subseteq \mathcal{SN}$. We prove that, if $t \in \mathcal{SN}$ then $t \in I_C$, by induction on t with $\rightarrow \cup \rhd$ as well-founded ordering. Let $f : (z : C)(\boldsymbol{y} : \boldsymbol{U})V$ be a recursor, $\boldsymbol{y}\xi$ and $\boldsymbol{y}\theta$ such that $\xi, \theta^t_z \models \boldsymbol{y} : \boldsymbol{U}$. We must prove that $v = ft\boldsymbol{y}\theta \in S = [\![V]\!]_{\xi,\theta^t_z}$. Since v is neutral, it suffices to prove that $\rightarrow(v) \subseteq S$. We proceed by induction on $t\boldsymbol{y}\theta$ with $\rightarrow$ as well-founded ordering ($\boldsymbol{y}\theta \in \mathcal{SN}$ by R1). If the reduction takes place in $t\boldsymbol{y}\theta$, we can conclude by induction hypothesis. Assume now that v' is a head-reduct of v. By assumption on recursors (Definition 1), t is of the form $c\boldsymbol{u}$ with $c : (\boldsymbol{x} : \boldsymbol{T})C$. Since C is primitive, every u_j is accessible and every T_j is a primitive type $D \leq C$. By induction hypothesis, $u_j \in I_D$. Therefore, $\emptyset, \{\boldsymbol{x} \mapsto \boldsymbol{u}\} \models \Gamma_c$ and, since $\xi, \theta^t_z \models \boldsymbol{y} : \boldsymbol{U}$ and recursors are head-computable, $v' \in S$. $\square$

Theorem 6 (Strong normalization). *Assume that every constant predicate symbol* C *is equipped with an admissible set* $\mathcal{R}ec(C)$ *of extended recursors distinct from constructors. If* $\rightarrow$ *is confluent and strong recursors and symbols that are not recursors satisfy the conditions given in [5] then* $\beta\cup\mathcal{R}$ *is strongly normalizing.*

Proof. Let $\vdash_f$ (resp. $\vdash_f^<$) be the typing relation of the CAC whose symbols are (resp. strictly) smaller than f. By induction on f, we prove that, if $\Gamma \vdash_f t : T$ and $\xi, \theta \models \Gamma$ then $t\theta \in [\![T]\!]_{\xi,\theta}$. By (symb), if $g \le f$ and $\vdash_f g : \tau_g$ then $\vdash_f^< \tau_g : s_g$. Therefore, the induction hypothesis can be applied to the subterms of τ_g.

We first prove that recursors are computable. Let $f : (\boldsymbol{z} : \boldsymbol{V})(z : C\boldsymbol{z})(\boldsymbol{y} : \boldsymbol{U})V$ be a recursor and assume that $\xi, \theta \models \Gamma_f$. We must prove that $v = f\boldsymbol{z}\theta z\theta\boldsymbol{y}\theta \in S = [\![V]\!]_{\xi,\theta}$. Since v is neutral, it suffices to prove that $\rightarrow(v) \subseteq S$. We proceed by induction on $\boldsymbol{z}\theta z\theta\boldsymbol{y}\theta$ with $\rightarrow$ as well-founded ordering ($\boldsymbol{z}\theta z\theta\boldsymbol{y}\theta \in \mathcal{SN}$ by R1). If the reduction takes place in $\boldsymbol{z}\theta z\theta\boldsymbol{y}\theta$, we conclude by induction hypothesis. Assume now that we have a head-reduct v'. By assumption on recursors (Definition 1), $z\theta$ is of the form $c\boldsymbol{u}$ with $c : (\boldsymbol{x} : \boldsymbol{T})C\boldsymbol{v}$, and v' is a head-reduct of $v_0 = f\boldsymbol{z}\theta^* z\theta\boldsymbol{y}\theta$ where $\boldsymbol{z}\theta^*$ are the normal forms of $\boldsymbol{z}\theta$. Since $\xi, \theta \models \Gamma_f$, we have $z\theta = c\boldsymbol{u} \in [\![C\boldsymbol{z}]\!]_{\xi,\theta} = I_C(\boldsymbol{z}\theta, \boldsymbol{z}\xi)$. Therefore, $v_0 \in S$ and, by (R2), $v' \in S$.

We now prove that constructors are computable. Let $c : (\boldsymbol{x} : \boldsymbol{T})C\boldsymbol{v}$ be a constructor of $C : (\boldsymbol{z} : \boldsymbol{V})\star$, $\boldsymbol{x}\eta$ and $\boldsymbol{x}\sigma$ such that $\eta, \sigma \models \Gamma_c$. We must prove that $c\boldsymbol{x}\sigma \in [\![C\boldsymbol{v}]\!]_{\eta,\sigma} = I_C(\boldsymbol{v}\sigma, \boldsymbol{S})$ where $\boldsymbol{S} = [\![\boldsymbol{v}]\!]_{\eta,\sigma}$. By induction hypothesis, we have $\boldsymbol{v}\sigma \in \mathcal{SN}$. So, let $f : (\boldsymbol{z} : \boldsymbol{V})(z : C\boldsymbol{z})(\boldsymbol{y} : \boldsymbol{U})V$ be a recursor of C, $\boldsymbol{y}\xi$ and $\boldsymbol{y}\theta$ such that $\xi_{\boldsymbol{z}}^{\boldsymbol{S}}, \theta_{\boldsymbol{z}\ z}^{\boldsymbol{v}\sigma\, c\boldsymbol{x}\sigma} \models \boldsymbol{y} : \boldsymbol{U}$. We must prove that $v = f\boldsymbol{v}\sigma^*(c\boldsymbol{x}\sigma)\boldsymbol{y}\theta \in S = [\![V]\!]_{\xi_{\boldsymbol{z}}^{\boldsymbol{S}}, \theta_{\boldsymbol{z}\ z}^{\boldsymbol{v}\sigma\, c\boldsymbol{x}\sigma}}$. Since v is neutral, it suffices to prove that $\rightarrow(v) \subseteq S$. Since $\boldsymbol{y}\theta \in \mathcal{SN}$, we can proceed by induction on $\boldsymbol{y}\theta$ with $\rightarrow$ as well-founded ordering.

In the case of a reduction in $\boldsymbol{y}\theta$, we conclude by induction hypothesis. In the case of a head-reduction, we conclude by head-computability of f. And, in the case of a reduction in $c\boldsymbol{x}\sigma$, we conclude by the computability lemmas for function symbols in [5]: if the strong normalization conditions are satisfied and accessibility is correct w.r.t. computability, then every reduct of $c\boldsymbol{x}\sigma$ belongs to $[\![C\boldsymbol{v}]\!]_{\eta,\sigma}$. The fact that accessibility is correct w.r.t. computability follows from the completeness of the set of recursors w.r.t. accessibility. □

6 The Calculus of Inductive Constructions

As an example, we prove the admissibility of a large class of weak recursors for strictly positive types, from which Coq's recursors [22] can be easily derived. This can be extended to strong recursors and to some non-strictly positive types (see Section 7).

Definition 7. *Let $C : (\boldsymbol{z} : \boldsymbol{V})\star$ and $\boldsymbol{c}$ be* strictly positive *constructors of C, that is, if c_i is of type $(\boldsymbol{x} : \boldsymbol{T})C\boldsymbol{v}$ then either no type equivalent to C occurs in T_j or T_j is of the form $(\boldsymbol{\alpha} : \boldsymbol{W})C\boldsymbol{w}$ with no type equivalent to C occurring in $\boldsymbol{W}$. The* parameters *of C is the biggest sequence $\boldsymbol{q}$ such that $C : (\boldsymbol{q} : \boldsymbol{Q})(\boldsymbol{z} : \boldsymbol{V})\star$ and each c_i is of type $(\boldsymbol{q} : \boldsymbol{Q})(\boldsymbol{x} : \boldsymbol{T})C\boldsymbol{q}\boldsymbol{v}$ with $T_j = (\boldsymbol{\alpha} : \boldsymbol{W})C\boldsymbol{q}\boldsymbol{w}$ if C occurs in T_j.*

The canonical *weak recursor*[6] *of C w.r.t $\boldsymbol{c}$ is $rec_{\boldsymbol{c}}^* : (\boldsymbol{q} : \boldsymbol{Q})(\boldsymbol{z} : \boldsymbol{V})(z : C\boldsymbol{q}\boldsymbol{z})(P : (\boldsymbol{z} : \boldsymbol{V})C\boldsymbol{q}\boldsymbol{z} \Rightarrow \star)(\boldsymbol{y} : \boldsymbol{U})P\boldsymbol{z}z$ with $U_i = (\boldsymbol{x} : \boldsymbol{T})(\boldsymbol{x}' : \boldsymbol{T}')P\boldsymbol{v}(c_i\boldsymbol{q}\boldsymbol{x})$, $T_j' = (\boldsymbol{\alpha} : \boldsymbol{W})P\boldsymbol{w}(x_j\boldsymbol{\alpha})$ if $T_j = (\boldsymbol{\alpha} : \boldsymbol{W})C\boldsymbol{q}\boldsymbol{w}$, and $T_j' = T_j$ otherwise, defined*

[6] Strong recursors cannot be defined by taking $P : (\boldsymbol{z} : \boldsymbol{V})C\boldsymbol{q}\boldsymbol{z} \Rightarrow \Box$ instead since $(\boldsymbol{z} : \boldsymbol{V})C\boldsymbol{q}\boldsymbol{z} \Rightarrow \Box$ is not typable in CC. They must be defined for each P.

by the rules $rec^*_{\boldsymbol{c}}\boldsymbol{qz}(c_i\boldsymbol{q}'\boldsymbol{x})P\boldsymbol{y} \to y_i\boldsymbol{xt}'$ *where* $t'_j = [\boldsymbol{\alpha} : \boldsymbol{W}](rec^*_{\boldsymbol{c}}\boldsymbol{qw}(x_j\boldsymbol{\alpha})P\boldsymbol{y})$ *if* $T_j = (\boldsymbol{\alpha} : \boldsymbol{W})C\boldsymbol{qw}$*, and* $t'_j = x_j$ *otherwise.*[7]

Lemma 8. *The set of canonical recursors is complete w.r.t. accessibility.*[8]

Proof. Let $c = c_i : (\boldsymbol{q} : \boldsymbol{Q})(\boldsymbol{x} : \boldsymbol{T})C\boldsymbol{qv}$ be a constructor of $C : (\boldsymbol{q} : \boldsymbol{Q})(\boldsymbol{z} : \boldsymbol{V})\star$, $\boldsymbol{q}\eta$, $\boldsymbol{x}\eta$, $\boldsymbol{q}\sigma$ and $\boldsymbol{x}\sigma$ such that $\boldsymbol{q}\sigma\boldsymbol{v}\sigma \in \mathcal{SN}$ and $c\boldsymbol{q}\sigma\boldsymbol{x}\sigma \in [\![C\boldsymbol{qv}]\!]_{\eta,\sigma} = I_C(\boldsymbol{q}\sigma\boldsymbol{v}\sigma, \boldsymbol{q}\xi[\![\boldsymbol{v}]\!]_{\eta,\sigma})$. Let $\boldsymbol{a} = \boldsymbol{qx}$ and $\boldsymbol{A} = \boldsymbol{QT}$. We must prove that, for all j, $a_j\sigma \in [\![A_j]\!]_{\eta,\sigma}$. For the sake of simplicity, we assume that weak and strong recursors have the same syntax. Since $\boldsymbol{q}\sigma\boldsymbol{v}\sigma$ have normal forms, it suffices to find u_j such that $rec_{\boldsymbol{c}}\boldsymbol{qv}(c\boldsymbol{qx})P_ju_j \to u_j\boldsymbol{xt}' \to^*_\beta a_j$. Take $u_j = [\boldsymbol{x} : \boldsymbol{T}][\boldsymbol{x}' : \boldsymbol{T}']a_j$. □

Lemma 9. *Canonical recursors are head-computable.*

Proof. Let $f : (\boldsymbol{q} : \boldsymbol{Q})(\boldsymbol{z} : \boldsymbol{V})(z : C\boldsymbol{qz})(P : (\boldsymbol{z} : \boldsymbol{V})C\boldsymbol{qz} \Rightarrow \star)(\boldsymbol{y} : \boldsymbol{U})P\boldsymbol{z}z$ be the canonical weak recursor w.r.t. $\boldsymbol{c}$, $T = (\boldsymbol{z} : \boldsymbol{V})C\boldsymbol{qz} \Rightarrow \star$, $c = c_i : (\boldsymbol{q} : \boldsymbol{Q})(\boldsymbol{x} : \boldsymbol{T})C\boldsymbol{qv}$, $\boldsymbol{q}\eta$, $\boldsymbol{q}\sigma$, $\boldsymbol{x}\eta$, $\boldsymbol{x}\sigma$, $P\xi$, $P\theta$, $\boldsymbol{y}\xi$, $\boldsymbol{y}\theta$, $\boldsymbol{R} = [\![\boldsymbol{v}]\!]_{\eta,\sigma}$, $\xi' = \xi^{\boldsymbol{R}}_{\boldsymbol{z}}$ and $\theta' = \theta^{\boldsymbol{v}\sigma}_{\boldsymbol{z}}{}^{c\boldsymbol{x}\sigma}_{z}$, and assume that $\eta, \sigma \models \Gamma_c$ and $\eta\xi', \sigma\theta' \models P : T, \boldsymbol{y} : \boldsymbol{U}$. We must prove that $y_i\theta\boldsymbol{x}\sigma\boldsymbol{t}'\sigma\theta \in [\![P\boldsymbol{z}z]\!]_{\xi',\theta'}$.

We have $y_i\theta \in [\![U_i]\!]_{\xi',\theta'}$, $U_i = (\boldsymbol{x} : \boldsymbol{T})(\boldsymbol{x}' : \boldsymbol{T}')P\boldsymbol{v}(c\boldsymbol{qx})$ and $x_j\sigma \in [\![T_j]\!]_{\eta,\sigma} = [\![T_j]\!]_{\eta\xi',\sigma\theta'}$. We prove that $t'_j\sigma\theta \in [\![T'_j]\!]_{\eta\xi',\sigma\theta'}$. If $T'_j = T_j$ then $t'_j\sigma\theta = x_j\sigma$ and we are done. Otherwise, $T_j = (\boldsymbol{\alpha} : \boldsymbol{W})C\boldsymbol{qw}$, $T'_j = (\boldsymbol{\alpha} : \boldsymbol{W})P\boldsymbol{w}(x_j\boldsymbol{\alpha})$ and $t'_j = [\boldsymbol{\alpha} : \boldsymbol{W}]f\boldsymbol{qw}(x_j\boldsymbol{\alpha})P\boldsymbol{y}$. Let $\boldsymbol{\alpha}\zeta$ and $\boldsymbol{\alpha}\gamma$ such that $\eta\xi'\zeta, \sigma\theta'\gamma \models \boldsymbol{\alpha} : \boldsymbol{W}$. Let $t = x_j\sigma\boldsymbol{\alpha}\gamma$. We must prove that $v = f\boldsymbol{q}\sigma\boldsymbol{w}\sigma\gamma tP\theta\boldsymbol{y}\theta \in S = [\![P\boldsymbol{w}(x_j\boldsymbol{\alpha})]\!]_{\eta\xi'\zeta,\sigma\theta'\gamma}$. Since v is neutral, it suffices to prove that $\to(v) \subseteq S$.

We proceed by induction on $\boldsymbol{q}\sigma\boldsymbol{w}\sigma\gamma tP\theta\boldsymbol{y}\theta \in \mathcal{SN}$ with $\to$ as well-founded ordering (we can assume that $\boldsymbol{w}\sigma\gamma \in \mathcal{SN}$ since $\vdash^<_f \tau_f : s_f$). In the case of a reduction in $\boldsymbol{q}\sigma\boldsymbol{w}\sigma\gamma tP\theta\boldsymbol{y}\theta$, we conclude by induction hypothesis. Assume now that we have a head-reduct v'. By definition of recursors, v' is also a head-reduct of $v_0 = f\boldsymbol{q}\sigma^*\boldsymbol{w}\sigma\gamma^*tP\theta\boldsymbol{y}\theta$ where $\boldsymbol{q}\sigma^*\boldsymbol{w}\sigma\gamma^*$ are the normal forms of $\boldsymbol{q}\sigma\boldsymbol{w}\sigma\gamma$. If $v_0 \in S$ then, by (R2), $v' \in S$. So, let us prove that $v_0 \in S$.

By candidate substitution, $S = [\![P\boldsymbol{z}z]\!]_{\xi^{\boldsymbol{S}}_{\boldsymbol{z}},\theta^{\boldsymbol{w}\sigma\gamma}_{\boldsymbol{z}}{}^{t}_{z}}$ with $\boldsymbol{S} = [\![\boldsymbol{w}]\!]_{\eta\xi'\zeta,\sigma\theta'\gamma} = [\![\boldsymbol{w}]\!]_{\eta\xi\zeta,\sigma\theta\gamma}$ for $\mathrm{FV}(\boldsymbol{w}) \subseteq \{\boldsymbol{q}, P, \boldsymbol{x}, \boldsymbol{\alpha}\}$. Since $x_j\sigma \in [\![T_j]\!]_{\eta\xi',\sigma\theta'}$ and $\eta\xi'\zeta, \sigma\theta'\gamma \models \boldsymbol{\alpha} : \boldsymbol{W}$, $t \in [\![C\boldsymbol{qw}]\!]_{\eta\xi'\zeta,\sigma\theta'\gamma} = I_C(\boldsymbol{q}\sigma\boldsymbol{w}\sigma\gamma, \boldsymbol{q}\xi\boldsymbol{S})$. Since $\eta\xi', \sigma\theta' \models P : T, \boldsymbol{y} : \boldsymbol{U}$ and $\mathrm{FV}(T\boldsymbol{U}) \subseteq \{\boldsymbol{q}, P\}$, we have $\eta\xi, \sigma\theta \models P : T, \boldsymbol{y} : \boldsymbol{U}$ and $\eta\xi^{\boldsymbol{S}}_{\boldsymbol{z}}, \sigma\theta^{\boldsymbol{w}\sigma\gamma}_{\boldsymbol{z}}{}^{t}_{z} \models P : T, \boldsymbol{y} : \boldsymbol{U}$. Therefore, $v_0 \in S$. □

It follows that CAC essentially subsumes CIC as defined in [23]. Theorem 6 cannot be applied to CIC directly since CIC and CAC do not have the same syntax and the same typing rules. So, in [5], we defined a sub-system of CIC, called CIC$^-$, whose terms can be translated into a CAC. Without requiring inductive types to be *safe* and to satisfy (I6), we think that CIC$^-$ is essentially as powerful as CIC.

Theorem 10. *The system CIC*$^-$ *defined in [5] (Chapter 7) is strongly normalizing even though inductive types are unsafe and do not satisfy (I6).*

---

[7] We could erase the useless arguments $t^{\square}_j = x_j$ when $T^{\square}_j = T_j$.

[8] In [23] (Lemma 4.35), Werner proves a similar result.

7 Non-strictly Positive Types

We are going to see that the use of elimination-based interpretations allows us to have functions defined by recursion on non-strictly positive types too, while CIC has always been restricted to strictly positive types. An interesting example is given by Abel's formalization of first-order terms with continuations as an inductive type $trm : \star$ with the constructors [1]:

$$\begin{aligned} var &: nat \Rightarrow trm \\ fun &: nat \Rightarrow (list\ trm) \Rightarrow trm \\ mu &: \neg\neg trm \Rightarrow trm \end{aligned}$$

where $list : \star \Rightarrow \star$ is the type of polymorphic lists, $\neg X$ is an abbreviation for $X \Rightarrow \bot$ (in the next section, we prove that $\neg$ can be defined as a function), and $\bot : \star$ is the empty type. Its recursor $rec : (A : \star)(y_1 : nat \Rightarrow A)\ (y_2 : nat \Rightarrow list\ trm \Rightarrow listA \Rightarrow A)(y_3 : \neg\neg trm \Rightarrow \neg\neg A \Rightarrow A)(z : trm)A$ can be defined by:

$$\begin{aligned} rec\ A\ y_1\ y_2\ y_3\ (var\ n) &\rightarrow y_1\ n \\ rec\ A\ y_1\ y_2\ y_3\ (fun\ n\ l) &\rightarrow y_2\ n\ l\ (map\ trm\ A\ (rec\ A\ y_1\ y_2\ y_3)\ l) \\ rec\ A\ y_1\ y_2\ y_3\ (mu\ f) &\rightarrow y_3\ f\ [x : \neg A](f\ [y : trm](x\ (rec\ A\ y_1\ y_2\ y_3\ y))) \end{aligned}$$

where $map : (A : \star)(B : \star)(A \Rightarrow B) \Rightarrow list\ A \Rightarrow list\ B$ is defined by:

$$\begin{aligned} map\ A\ B\ f\ (nil\ A') &\rightarrow (nil\ B) \\ map\ A\ B\ f\ (cons\ A'\ x\ l) &\rightarrow cons\ B\ (f\ x)\ (map\ A\ B\ f\ l) \\ map\ A\ B\ f\ (app\ A'\ l\ l') &\rightarrow app\ B\ (map\ A\ B\ f\ l)\ (map\ A\ B\ f\ l') \end{aligned}$$

We now check that rec is an admissible recursor. Completeness w.r.t. accessibility is easy. For the head-computability, we only detail the case of mu. Let $f\sigma$, $t = mu\ f\sigma$, $A\xi$, $A\theta$ and $\boldsymbol{y}\theta$ such that $\emptyset, \sigma \models \Gamma_{mu}$ and $\xi, \sigma\theta_z^t \models \Gamma = A : \star$, $\boldsymbol{y} : \boldsymbol{U}$ where U_i is the type of y_i. Let $b = recA\theta\boldsymbol{y}\theta$, $c = [y : trm](x(by))$ and $a = [x : \neg A\theta](f\sigma c)$. We must prove that $y_3\theta f\sigma a \in [\![A]\!]_{\xi,\sigma\theta_z^t} = A\xi$.

Since $\xi, \sigma\theta_z^t \models \Gamma$, $y_3\theta \in [\![\neg\neg trm \Rightarrow \neg\neg A \Rightarrow A]\!]_{\xi,\theta}$. Since $\emptyset, \sigma \models \Gamma_{mu}$, $f\sigma \in [\![\neg\neg trm]\!]$. Thus, we are left to prove that $a \in [\![\neg\neg A]\!]_{\xi,\theta}$, that is, $f\sigma c\gamma \in I_\bot$ for all $x\gamma \in [\![\neg A]\!]_{\xi,\theta}$. Since $f\sigma \in [\![\neg\neg trm]\!]$, it suffices to prove that $c\gamma \in [\![\neg trm]\!]$, that is, $x\gamma(by\gamma) \in I_\bot$ for all $y\gamma \in I_{trm}$. This follows from the facts that $x\gamma \in [\![\neg A]\!]_{\xi,\theta}$ and $by\gamma \in A\xi$ since $y\gamma \in I_{trm}$.

8 Inductive-Recursive Types

In this section, we define new positivity conditions for dealing with *inductive-recursive type definitions* [12]. An inductive-recursive type C has constructors whose arguments have a type Ft with F defined by recursion on $t : C$, that is, a predicate F and its domain C are defined at the same time.

A simple example is the type $dlist : (A : \star)(\# : A \Rightarrow A \Rightarrow \star)\star$ of lists made of distinct elements thanks to the predicate $fresh : (A : \star)(\# : A \Rightarrow A \Rightarrow \star)A \Rightarrow (dlist\ A\,\#) \Rightarrow \star$ parametrized by a function $\#$ to test whether two elements are distinct. The constructors of $dlist$ are:

$$nil : (A:\star)(\#:A\Rightarrow A\Rightarrow\star)(dlist\,A\,\#)$$
$$cons : (A:\star)(\#:A\Rightarrow A\Rightarrow\star)(x:A)(l:dlist\,A\,\#)(fresh\ A\ \#\ x\ l)\Rightarrow(dlist\,A\,\#)$$

and the rules defining $fresh$ are:

$$\begin{array}{rcl} fresh\ A\ \#\ x\ (nil\ A') & \rightarrow & \top \\ fresh\ A\ \#\ x\ (cons\ A'\ y\ l\ h) & \rightarrow & x\#y \wedge fresh\ A\ \#\ x\ l \end{array}$$

where $\top$ is the proposition always true and $\wedge$ the connector "and". Other examples are given by Martin-Löf's definition of the first universe *à la* Tarski [12] or by Pollack's formalization of record types with manifest fields [21].

Definition 11 (Positive/negative positions). *Assume that every predicate symbol* $f : (\boldsymbol{x} : \boldsymbol{t})U$ *is equipped with a set* $\mathrm{Mon}^+(f) \subseteq \{i \le \alpha_f \mid x_i \in \mathcal{X}^\Box\}$ *of* monotone arguments *and a set* $\mathrm{Mon}^-(f) \subseteq \{i \le \alpha_f \mid x_i \in \mathcal{X}^\Box\}$ *of* anti-monotone arguments. *The sets of* positive positions $\mathrm{Pos}^+(t)$ *and* negative positions $\mathrm{Pos}^-(t)$ *in a term* t *are inductively defined as follows:*

- $\mathrm{Pos}^\delta(s) = \mathrm{Pos}^\delta(x) = \{\varepsilon \mid \delta = +\}$,
- $\mathrm{Pos}^\delta((x:U)V) = 1.\mathrm{Pos}^{-\delta}(U) \cup 2.\mathrm{Pos}^\delta(V)$,
- $\mathrm{Pos}^\delta([x:U]v) = 2.\mathrm{Pos}^\delta(v)$,
- $\mathrm{Pos}^\delta(tu) = 1.\mathrm{Pos}^\delta(t)$ *if* $t \ne f\boldsymbol{t}$,
- $\mathrm{Pos}^\delta(f\boldsymbol{t}) = \{1^{|\boldsymbol{t}|} \mid \delta = +\} \cup \bigcup\{1^{|\boldsymbol{t}|-i}2.\mathrm{Pos}^{\epsilon\delta}(t_i) \mid \epsilon \in \{-,+\},\ i \in \mathrm{Mon}^\epsilon(f)\}$,

where $\delta \in \{-,+\}$, $-+ = -$ *and* $-- = +$ *(usual rule of signs).*

Theorem 12 (Strong normalization). *Definition 1 is modified as follows. A pre-recursor* $f : (\boldsymbol{z} : \boldsymbol{V})(z : C\boldsymbol{z})W$ *is a* recursor *if:*

- *no* $F > C$ *occurs in* W,
- *every* $F \simeq C$ *occurs only positively in* W,
- *if* $i \in \mathrm{Mon}^\delta(C)$ *then* $\mathrm{Pos}(z_i, W) \subseteq \mathrm{Pos}^\delta(W)$.

Assume furthermore that, for every rule $F\boldsymbol{l} \rightarrow r$*:*

- *no* $G > F$ *occurs in* r,
- *for all* $i \in \mathrm{Mon}^\epsilon(F)$, $l_i \in \mathcal{X}^\Box$ *and* $\mathrm{Pos}(l_i, r) \subseteq \mathrm{Pos}^\epsilon(r)$.

Then, Theorem 6 is still valid.

Proof. For Theorem 6 to be still valid, we must make sure that φ (see Definition 2) is still monotone, hence has a least fixpoint. To this end, we need to prove that $[\![t]\!]^I_{\xi,\theta}$ is monotone (resp. anti-monotone) w.r.t. $x\xi$ if x occurs only positively (resp. negatively) in t, and that $[\![t]\!]^I_{\xi,\theta}$ is monotone (resp. anti-monotone) w.r.t. I_C if C occurs only positively (resp. negatively) in t. These results are easily extended to the new positivity conditions by reasoning by induction on the well-founded ordering used for defining the defined predicate symbols.

Let us see what happens in the case where $t = F\boldsymbol{t}$ with F a defined predicate symbol. Let $\le^+ = \le$ and $\le^- = \ge$. We want to prove that, if $\xi_1 \le_x \xi_2$ (*i.e.* $x\xi_1 \le x\xi_2$ and, for all $y \ne x$, $y\xi_1 = y\xi_2$) and $\mathrm{Pos}(x,t) \subseteq \mathrm{Pos}^\delta(t)$, then $[\![t]\!]^I_{\xi_1,\theta} \le^\delta [\![t]\!]^I_{\xi_2,\theta}$.

By definition of I_F, if the normal forms of $\boldsymbol{t}\theta$ matches the left hand-side of $F\boldsymbol{l} \to r$, then $[\![F\boldsymbol{t}]\!]^I_{\xi_i,\theta} = [\![r]\!]^I_{\xi'_i,\sigma}$ where σ is the matching substitution and, for all $y \in \mathrm{FV}(r)$, $y\xi'_i = [\![t_{\kappa_y}]\!]^I_{\xi_i,\theta}$ where κ_y is such that $l_{\kappa_y} = y$ (see [5] for details). Now, since $\mathrm{Pos}(x, F\boldsymbol{t}) \subseteq \mathrm{Pos}^\delta(F\boldsymbol{t})$, $\mathrm{Pos}(x, t_{\kappa_y}) \subseteq \mathrm{Pos}^{\epsilon\delta}(t_{\kappa_y})$ for some ϵ. Hence, by induction hypothesis, $\xi'_1 \leq^{\epsilon\delta}_y \xi'_2$. Now, since $\mathrm{Pos}(y, r) \subseteq \mathrm{Pos}^\epsilon(r)$, by induction hypothesis again, $[\![r]\!]_{\xi'_1,\sigma} \leq^{\epsilon^2\delta}=\leq^\delta [\![r]\!]_{\xi'_2,\sigma}$. □

For instance, in the positive type trm of Section 7, instead of considering $\neg\neg A$ as an abbreviation, one can consider $\neg$ as a predicate symbol defined by the rule $\neg A \to A \Rightarrow \bot$ with $\mathrm{Mon}^-(\neg) = \{1\}$. Then, one easily checks that A occurs negatively in $A \Rightarrow \bot$, and hence that trm occurs positively in $\neg\neg trm$ since $\mathrm{Pos}^+(\neg\neg trm) = \{1\} \cup 2.\mathrm{Pos}^-(\neg trm) = \{1\} \cup 2.2.\mathrm{Pos}^+(trm) = \{1, 2.2\}$.

9 Conclusion

By using an elimination-based interpretation for inductive types, we proved that the Calculus of Algebraic Constructions completely subsumes the Calculus of Inductive Constructions. We define general conditions on extended recursors for preserving strong normalization and show that these conditions are satisfied by a large class of recursors for strictly positive types and by non-strictly positive types too. Finally, we give general positivity conditions for dealing with inductive-recursive types.

Acknowledgments. I would like to thank C. Paulin, R. Matthes, J.-P. Jouannaud, D. Walukiewicz, G. Dowek and the anonymous referees for their useful comments on previous versions of this paper. Part of this work was performed during my stay at Cambridge (UK) thanks to a grant from the INRIA.

References

1. A. Abel. Termination checking with types. Technical Report 0201, Ludwig Maximilians Universität, München, Germany, 2002.
2. F. Barbanera, M. Fernández, and H. Geuvers. Modularity of strong normalization and confluence in the algebraic-λ-cube. In *Proceedings of the 9th IEEE Symposium on Logic in Computer Science*, 1994.
3. H. Barendregt. Lambda calculi with types. In S. Abramski, D. Gabbay, and T. Maibaum, editors, *Handbook of logic in computer science*, volume 2. Oxford University Press, 1992.
4. F. Blanqui. Definitions by rewriting in the Calculus of Constructions (extended abstract). In *Proceedings of the 16th IEEE Symposium on Logic in Computer Science*, 2001.
5. F. Blanqui. *Théorie des Types et Récriture.* PhD thesis, Université Paris XI, Orsay, France, 2001. Available in english as "Type Theory and Rewriting".
6. F. Blanqui. Definitions by rewriting in the Calculus of Constructions, 2003. Journal submission, 68 pages.

7. F. Blanqui. A short and flexible strong normalization proof for the Calculus of Algebraic Constructions with curried rewriting, 2003. Draft.
8. T. Coquand. Pattern matching with dependent types. In *Proceedings of the International Workshop on Types for Proofs and Programs, 1992*. `http://www.lfcs.informatics.ed.ac.uk/research/types-bra/proc/`.
9. T. Coquand and G. Huet. The Calculus of Constructions. *Information and Computation*, 76(2–3):95–120, 1988.
10. T. Coquand and C. Paulin-Mohring. Inductively defined types. In *Proceedings of the International Conference on Computer Logic*, Lecture Notes in Computer Science 417, 1988.
11. N. Dershowitz and J.-P. Jouannaud. Rewrite systems. In J. van Leeuwen, editor, *Handbook of Theoretical Computer Science*, volume B, chapter 6. North-Holland, 1990.
12. P. Dybjer. A general formulation of simultaneous inductive-recursive definitions in type theory. *Journal of Symbolic Logic*, 65(2):525–549, 2000.
13. J.-Y. Girard, Y. Lafont, and P. Taylor. *Proofs and Types*. Cambridge University Press, 1988.
14. R. Harper and J. Mitchell. Parametricity and variants of Girard's J operator. *Information Processing Letters*, 70:1–5, 1999.
15. J.-P. Jouannaud and M. Okada. Executable higher-order algebraic specification languages. In *Proceedings of the 6th IEEE Symposium on Logic in Computer Science*, 1991.
16. J. W. Klop, V. van Oostrom, and F. van Raamsdonk. Combinatory reduction systems: introduction and survey. *Theoretical Computer Science*, 121:279–308, 1993.
17. R. Matthes. *Extensions of System F by Iteration and Primitive Recursion on Monotone Inductive Types*. PhD thesis, Ludwig Maximilians Universität, München, Germany, 1998.
18. C. McBride. *Dependently typed functional programs and their proofs*. PhD thesis, University of Edinburgh, United Kingdom, 1999.
19. N. P. Mendler. *Inductive Definition in Type Theory*. PhD thesis, Cornell University, United States, 1987.
20. C. Paulin-Mohring. Personal communication, 2001.
21. R. Pollack. Dependently typed records in type theory. *Formal Aspects of Computing*, 13(3–5):341–363, 2002.
22. Coq Development Team. *The Coq Proof Assistant Reference Manual – Version 7.3*. INRIA Rocquencourt, France, 2002. `http://coq.inria.fr/`.
23. B. Werner. *Une Théorie des Constructions Inductives*. PhD thesis, Université Paris VII, France, 1994.

On Strong Normalization in the Intersection Type Discipline
(Extended Abstract)

Gérard Boudol

INRIA Sophia Antipolis
BP 93 – 06902, Sophia Antipolis Cedex, France

Abstract. We give a proof for the strong normalization result in the intersection type discipline, which we obtain by putting together some well-known results and proof techniques. Our proof uses a variant of Klop's extended λ-calculus, for which it is shown that strong normalization is equivalent to weak normalization. This is proved here by means of a finiteness of developments theorem, obtained following de Vrijer's combinatory technique. Then we use the standard argument, formalized by Lévy as "the creation of redexes is decreasing" and implemented in proofs of weak normalization by Turing, and Coppo and Dezani for the intersection type discipline, to show that a typable expression of the extended calculus is normalizing.

1 Introduction

In this paper we propose, by putting together some more or less well-known results and proof techniques, a proof of the classical result that λ-terms typable in the intersection type discipline are strongly normalizable. This typing discipline was introduced by Coppo and Dezani [5], and in later developments several variants were considered (see [1,4,6,12,19,23]). Since we are interested here into strong normalization, we do not consider the systems that include the type constant ω introduced by Sallé [25] to type any term. In their original paper [5], Coppo and Dezani established a weak normalization result, using a technique close to the one of Turing for the simply typed λ-calculus (see [7,9,10]). Regarding strong normalization, there are, as far as I know, only two kinds of proofs: one by Pottinger [23], and another one by Kfoury and Wells [14]. The technique used by Pottinger can be called, following Tait [28], the "realizability" technique, since it relies on an interpretation of types closely related to Kleene's original recursive realizability interpretation [17], the main ingredient of which is: f realizes $(\sigma \to \tau)$ if and only if for all x, if x realizes σ, then (fx) realizes τ. Other well-known names for similar concepts are "convertibility" (or "computability") [27], "candidats de réductibilité" [8,9], or "logical relations" [22].

The proof of strong normalization using the realizability technique is very elegant and "amazingly slick", as Klop says [18]. This technique also has the advantage of being very general, and applicable to many type systems (see [3,19]).

M. Hofmann (Ed.): TLCA 2003, LNCS 2701, pp. 60–74, 2003.

However, having seen such a proof, one may keep asking: "what is decreasing?" By contrast, a "syntactic", or combinatory proof of normalization, like the one of Kfoury and Wells [14], although perhaps less concise than a proof using the realizability technique, has the advantage of explaining "why" a typed expression normalizes, by exhibiting a measure that decreases with the reduction, or more accurately with a well-chosen reduction of the subject. This is the way we shall follow here.

To introduce our work, we first recall the observation on which relies Turing's proof for the simply typed λ-calculus, which is that *a created redex has a strictly smaller degree than its creator* (see [10] for instance). This was first stated formally by Lévy, who showed (see [20], Lemma 3 p. 31, and [2], Exercise 14.5.3) that there are only three ways to create a redex in the λ-calculus, namely – indicating also the (simple) type of the corresponding abstractions:

$$
\begin{array}{rcll}
(\lambda x\lambda y.M)NP & \underset{\beta}{\rightarrow} & (\lambda y\{x\mapsto N\}M)P & \qquad (\theta\rightarrow(\tau\rightarrow\sigma)) \text{ vs } (\tau\rightarrow\sigma)\\
(\lambda x\, x)(\lambda y M)N & \underset{\beta}{\rightarrow} & (\lambda y M)N & \qquad ((\tau\rightarrow\sigma)\rightarrow(\tau\rightarrow\sigma)) \text{ vs } (\tau\rightarrow\sigma)\\
(\lambda x.C[(xN)])\lambda y M & \underset{\beta}{\rightarrow} & C^{\ulcorner}[(\lambda y M)N^{\urcorner}] & \qquad ((\tau\rightarrow\sigma)\rightarrow\theta) \text{ vs } (\tau\rightarrow\sigma)
\end{array}
$$

We see that the degree of the redex, that is the size of the type of its abstraction part, strictly decreases. Then, if we define the measure of a typing proof as the pair (m, r) where m is the maximal degree of a redex, and r is the number of redexes of degree m, we can prove that this measure strictly decreases, with respect to the lexicographic ordering, by reducing an innermost redex of maximal degree in the subject of the typing. This is the way followed by Turing, and by Coppo and Dezani [5] (see also [9,10]). But obviously, this only proves weak normalization, and does not guarantee strong normalization, because, as it is well-known, the weak normalizability of each sub-expression of a term does not imply the strong normalizability of the whole expression, as shown by $(\lambda x.\mathsf{F}(x\varDelta))\varDelta$ for instance, where $\mathsf{F} = \lambda x\lambda y\, y$ and $\varDelta = \lambda z(zz)$. This is due to an anomaly of the λ-calculus, which allows to forget about some reductions. Clearly, if we are interested in strong normalization, we should better not forget anything to compute.

Then we slightly modify our aim: we shall show in fact the (strong) normalization result for an *extended* λ-calculus, where no redex may disappear. Specifically, we use a variant of Klop's calculus [18]. The idea of Klop's calculus is very simple and elegant: it is to introduce a new construction building a pair $[M, N]$ of expressions, where M is the "main" part, from the λ-calculus point of view, and N is an expression that would have been discarded by ordinary β-reduction. Typically, if $x \notin \mathsf{fv}(M)$, one reduces the β_K-redex $(\lambda x M N)$ into $[M, N]$, rather than into M. As a consequence, one only performs "non-erasing reductions" in Klop's calculus. There are many ways of formalizing "non-erasing reductions" in the λ-calculus – see the surveys in [13,26] –, which are related with each other. For instance, de Groote [10] introduces a new notion of reduction β_S, which, if we read $[M, N]$ as an irreducible β_K-redex $(\lambda x M N)$, where $x \notin \mathsf{fv}(M)$, is precisely a rule of Klop's calculus, namely $[M, N]P \rightarrow [MP, N]$. Here we shall treat this as a "structural equivalence" $[M, N]P \equiv [MP, N]$, in order to keep the usual notion of β-redex[1]. This is the main variation we introduce on Klop's calculus. As pointed out to me by Joe Wells, the calculus we introduce is in fact a notational variant (and a restriction) of a calculus by Kamareddine [13], which provides a very general framework for studying weak and strong normalization. Following the terminology of [13], our calculus is, if we identify $[M, N]$ with a β_K-redex $(\lambda x M N)$, the λ-calculus equipped with *generalized* β_I*-reduction* (looking through β_K-redexes only).

The proof technique used here is close to the one of Kfoury and Wells, since both use the "strong normalization from weak normalization" technique (see [26]). A difference between Kfoury and Wells' proof and the one presented in this paper is that we use a calculus enjoying a "uniformity" property – that is, weak and strong normalization coincide, see [16] –, while they introduce a new notion of reduction, called γ-reduction, and rely on a specific strategy for normalizing an expression. Another difference is that we then show in the standard way that the measure of typing is decreasing with a well-chosen reduction of the subject, where they use a rather elaborate notion of reduction for a Church style presentation of the intersection type discipline in an extended λ-calculus.

Our paper is organized as follows: in the first technical section, we introduce a variant of Klop's calculus, which is a restriction of Kamareddine's calculus. Next, we show the main property of this calculus, namely that strong normalizability is equivalent to weak normalizability. Our proof differs from Klop's and Kamareddine's ones, which both use a technique due to Nederpelt. Here we prove this by means of a "conservation theorem" (see [2,10]), which in turn relies on the "finiteness of developments". The latter is proved by adapting the combinatory method of de Vrijer [29]. Then we introduce the intersection type discipline, extended to Klop's calculus, establishing a connection between the "system $\mathcal{D}$" of Pottinger [19,23] and the original, syntax directed system of Coppo and Dezani [5]. Finally we prove the weak normalization property for the latter, adapting the proof in [5], that is using Lévy's observation that "created redexes are smaller than their creator," from which follows the strong normalization result. We conclude by some remarks on our proof, observing in particular that the strong normalizability result for the intersection type discipline is equivalent to the classical finiteness of developments property.

Note. There has been a lot of work in the past about weak and strong normalization, conservation, developments, etc., regarding the λ-calculus, and its generalizations (higher-order rewrite systems). I will try to cite some of the relevant references here, but there will certainly be missing items in the bibliography. I apologize to the missing authors, like Church for instance, for the oversight.

For lack of space, most of the proofs have been omitted from this extended abstract. They can be found in the full version of the paper, available from the author's web page.

---

[1] As noted by Kfoury and Wells [15], the degree of a created β_S-redex is not necessarily smaller than the one of its creator. This difficulty does not arise with Klop's calculus. Moreover, in our variant we only have to deal with "standard" β-redexes.

2 The Extended λ-Calculus

Our extended λ-calculus is basically Klop's one [18] (see also [26]), but differs from it on some points: as we said, we do not include the reduction $[M, N]P \to [MP, N]$, but we rather treat this as an equivalence. More precisely, we allow a λ-abstraction to be applied to an argument across discarded expressions. Moreover, we do not consider the reduction $[M, N] \to M$; instead we regard this as a projection onto the λ-calculus. Finally, instead of introducing an embedding ι of the λ-calculus determined by $\iota(\lambda x M) = \lambda x[\iota(M), x]$, we split the β-reduction in two parts, keeping the ordinary notion for β_I-redexes, and reducing β_K-redexes as explained in the Introduction. The syntax is as follows:

$$S,\, T \ldots \;::=\; x \mid \lambda x S \mid (ST) \mid [S, T] \qquad (\Lambda^*)$$

As usual, $M, N \ldots$ denote terms of the λ-calculus, belonging to Λ. The operation of (capture avoiding) substitution in Λ^*, denoted $\{x \mapsto T\}S$, is defined as usual. For the usual notational conventions regarding the λ-calculus, we refer to [2]. For any notion of reduction x, we denote by $\underset{x}{\to}$ the reduction relation determined by x, that is the least relation containing x and compatible with the constructions of the calculus, and we let $\mathcal{N}_x$, $\mathcal{WN}_x$ and $\mathcal{SN}_x$ be respectively the set of x-normal forms, weakly x-normalizing and strongly x-normalizing expressions. (We recall that an expression is weakly x-normalizing if it has a finite sequence of x-reductions to an x-normal form, while it its strongly x-normalizing if it has no infinite sequence of x-reductions.)

To define our *notion of reduction* κ on Λ^*, we allow n to be 0 in an expression $[\cdots[S, U_1]\cdots, U_n]$, in which case this denotes S. We abbreviate this as $[S, U_1, \ldots, U_n]$, which we could also denote, following Klop [18], by $S_{\boldsymbol{U}}$. The reduction $\underset{\kappa}{\to}$ is given by the two following axiom schemas:

$$\frac{x \in \mathsf{fv}(S)}{([\lambda x S, U_1, \ldots, U_n]T) \underset{\kappa}{\to} [\{x \mapsto T\}S, U_1, \ldots, U_n]}$$

$$\frac{x \notin \mathsf{fv}(S)}{([\lambda x S, U_1, \ldots, U_n]T) \underset{\kappa}{\to} [S, U_1, \ldots, U_n, T]}$$

One can see that an ordinary β-redex is also a κ-redex. Let us see an example – this is the Example 9.1.7 in [2] of an expression such that any sub-expression of any reduct of it is weakly β-normalizing, but which is not strongly β-normalizing. With $\mathsf{F} = \lambda x\,\mathsf{I}$ where $\mathsf{I} = \lambda y\, y$ we have $(\mathsf{F}X) \underset{\kappa}{\to} [\mathsf{I}, X]$. Then, for $M = \lambda z.\mathsf{F}(zz)$, we have $M \underset{\kappa}{\to} \lambda z[\mathsf{I}, (zz)] = N$, and therefore

$$(MM) \underset{\kappa}{\to} (MN) \underset{\kappa}{\to} (NN) \underset{\kappa}{\to} [\mathsf{I}, (NN)] \underset{\kappa}{\to} [\mathsf{I}, [\mathsf{I}, (NN)]] \underset{\kappa}{\to} \cdots$$

As remarked by Joe Wells, our calculus is just a notational variant of a calculus by Kamareddine [13] introducing *generalized β-reduction.* The calculus of Kamareddine is obtained by interpreting, as we suggested in the Introduction, the construct $[M, N]$ as a β_K-redex $(\lambda x M N)$, where $x \notin \mathsf{fv}(M)$. Then our notion of κ-reduction is a particular case of Kamareddine's generalized β-reduction, where we only use β_I-reduction. In the following technical developments, we could certainly use Kamareddine's notation – choosing one or the other is a matter of taste.

We define a projection from Λ^* onto Λ, denoted S^{hd}, which takes, recursively, the first component of the pairs $[S, T]$, which is the "main" component – the "proper part," following Klop's terminology:

$$\begin{aligned} x^{\mathsf{hd}} &= x \\ (\lambda x S)^{\mathsf{hd}} &= \lambda x S^{\mathsf{hd}} \\ (ST)^{\mathsf{hd}} &= (S^{\mathsf{hd}} T^{\mathsf{hd}}) \\ [S, T]^{\mathsf{hd}} &= S^{\mathsf{hd}} \end{aligned}$$

Clearly, $M^{\mathsf{hd}} = M$ for $M \in \Lambda$. Now we show that β-reduction is lifted to Λ^* by the mapping S^{hd}. We need the following

Lemma 2.1. $(\{x \mapsto T\} S)^{\mathsf{hd}} = \{x \mapsto T^{\mathsf{hd}} S^{\mathsf{hd}}\}$

Proof. by induction on S, trivial. □

Lemma 2.2. $S^{\mathsf{hd}} \underset{\beta}{\rightarrow} N \;\Rightarrow\; \exists T.\ S \underset{\kappa}{\rightarrow} T \;\&\; N = T^{\mathsf{hd}}$

Proof. by induction on S, and then by induction on the definition of $\underset{\beta}{\rightarrow}$. □

An immediate consequence is that if M has an infinite sequence of β-reductions, then it also has an infinite sequence of κ-reductions. By contraposition, a strongly κ-normalizing λ-expression is strongly β-normalizing, that is:

Corollary 2.3. $\Lambda \cap \mathcal{SN}_\kappa \subseteq \mathcal{SN}_\beta$

In the next section we show that, in fact, any *weakly* κ-normalizing λ-expression is *strongly* β-normalizing. To this end, we show a "conservation theorem" (*cf.* [2,10]).

3 The Conservation Theorem

Klop has shown that weak and strong normalizability are equivalent in his calculus. This result has been extended, by Khasidashvili & al. [16] for higher-order, non-orthogonal rewrite systems, and, as regards the λ-calculus, by Kamareddine [13], generalizing a proof by de Groote [10] that uses, as the one of Klop, a proof technique due to Nederpelt. Then our result below is a consequence of a more general conservation result of [13]. Nevertheless, we give a detailed proof of this result here, because the technique we use is different and, we hope, simpler. We derive here the property that $\mathcal{SN}_\kappa = \mathcal{WN}_\kappa$ from a standard conservation result, which in turn is a consequence of the Finiteness of Development Theorem. A nice proof of the latter can be found in [24], following some of the steps of the realizability technique. However, since our aim here is to provide a "syntactic" proof of strong normalization, we give a purely combinatory proof of this theorem, by adapting a technique of de Vrijer [29]. To define conveniently the notion of development, we follow [3], introducing a labelled version of the calculus. The labels are actually very simple here: they are only used to mark the initial redexes in an expression, and their residuals. A marked initial redex is thus $([\underline{\lambda} xS, S_1, \ldots, S_n]T)$, while an abstraction λxM which is not applied to an argument, but may appear in a created redex after some reductions, is never marked. The syntax is

$$\begin{array}{rcl} S,\, T \ldots & ::= & x \mid \lambda xS \mid (ST) \mid [S,T] \mid (FT) \qquad (\underline{\Lambda}^*) \\ F & ::= & \underline{\lambda} xS \mid [F,S] \end{array}$$

The relation $\underset{\underline{\kappa}}{\rightarrow}$ is defined as the κ-reduction, except that only marked redexes are allowed to be reduced. Then $\underline{\kappa}$-reduction is the notion of *development*, where we only reduce (residuals of) initial redexes. Our aim now is to show the Finiteness of Developments Theorem, that is, any term is strongly $\underline{\kappa}$-normalizing. We shall show this by adapting de Vrijer's proof, that is by exhibiting a measure on terms that strictly decreases along $\underline{\kappa}$-reduction. The idea is the following: since $\underline{\kappa}$-redexes cannot be created (nor discarded) by $\underline{\kappa}$-reduction, one can estimate the *multiplicity*, denoted $\mu(x, S)$, a variable x can have in a term S, which is a bound for the number of occurrences of this variable in any $\underline{\kappa}$-reduct S' of S. This is defined inductively as follows:

$$\mu(x,y) = \begin{cases} 1 \text{ if } x = y \\ 0 \text{ otherwise} \end{cases}$$

$$\mu(x,\lambda yS) = \begin{cases} \mu(x,S) \text{ if } x \neq y \\ 0 \qquad\quad \text{otherwise} \end{cases}$$

$$\mu(x,(ST)) = \mu(x,S) + \mu(x,T) \;=\; \mu(x,[S,T])$$

$$\mu(x,([\underline{\lambda} yS_0, S_1, \ldots, S_n]T)) = \begin{cases} \sum\limits_{0<i\leqslant n} \mu(x,S_i) + \mu(x,T)\cdot \mathrm{max}(1,\mu(y,S_0)) \text{ if } x = y \\ \sum\limits_{0\leqslant i\leqslant n} \mu(x,S_i) + \mu(x,T)\cdot \mathrm{max}(1,\mu(y,S_0)) \text{ otherwise} \end{cases}$$

The following should be obvious:

Remark 3.1. $\mu(x,S) = 0 \;\Leftrightarrow\; x \notin \mathsf{fv}(S)$

This implies that the definition of $\mu(x,S)$ is not affected if we assume that the bound variables of S are renamed so that x is not bound in S. It should also be clear that $\mu(x,S)$ is greater than the number of occurrences of x in S, and is exactly this number if $S \in \mathcal{N}_{\underline{\kappa}}$. As a preliminary result, we show:

Lemma 3.2. $x \neq y \;\Rightarrow\; \mu(y, \{x \mapsto T\}S) = \mu(y, S) + \mu(y, T) \cdot \mu(x, S)$

Proof. by induction on S. □

As we announced, we can prove that $\mu(x, S)$ is a bound for the number of occurrences of x in $\underline{\kappa}$-reducts of S:

Corollary 3.3. $S \underset{\underline{\kappa}}{\rightarrow} S' \;\Rightarrow\; \mu(x, S') = \mu(x, S)$

Now we define the *$\underline{\kappa}$-norm* of an expression S, which is intended to be a bound for the length of $\underline{\kappa}$-reduction sequences starting from S. The idea here is to exploit the multiplicity of the abstracted variable in a redex, say $\mu(x, S)$ if we are to reduce $([\underline{\lambda} xS, \ldots]T)$, to anticipate the duplication of reductions (from the argument T) that may occur when reducing this redex. Then the $\underline{\kappa}$-norm of S is the integer $|S|_{\underline{\kappa}}$ inductively defined as follows:

$$\begin{aligned} |x|_{\underline{\kappa}} &= 0 \\ |\lambda xS|_{\underline{\kappa}} &= |S|_{\underline{\kappa}} \\ |(ST)|_{\underline{\kappa}} &= |S|_{\underline{\kappa}} + |T|_{\underline{\kappa}} \;=\; |[S,T]|_{\underline{\kappa}} \\ |([\underline{\lambda} xS_0, S_1, \ldots, S_n]T)|_{\underline{\kappa}} &= 1 + \sum_i |S_i|_{\underline{\kappa}} + |T|_{\underline{\kappa}} \cdot \max(1, \mu(x, S_0)) \end{aligned}$$

Below we show that $|S|_{\underline{\kappa}}$ is indeed a bound for the length of $\underset{\underline{\kappa}}{\rightarrow}$-reduction sequences starting from S. In order to prove this, we need:

Lemma 3.4. $|\{x \mapsto T\}S|_{\underline{\kappa}} = |S|_{\underline{\kappa}} + |T|_{\underline{\kappa}} \cdot \mu(x, S)$

Proof. by induction on S. □

Lemma 3.5. $S \underset{\underline{\kappa}}{\rightarrow} S' \;\Rightarrow\; |S|_{\underline{\kappa}} > |S'|_{\underline{\kappa}}$

Proof. by induction on the definition of $S \underset{\underline{\kappa}}{\rightarrow} S'$. □

An obvious consequence of this is the "finiteness of developments" property:

Theorem 3.6. *(Finiteness of Developments)* $\mathcal{SN}_{\underline{\kappa}} = \underline{\Lambda}^*$

Now we prove the conservation theorem. To this end, for any expression S we define $S^\diamond$, which is the result of contracting, in an inside-out way, all (and only) the marked initial redexes in S:

(i) $x^\diamond = x$

(ii) $(\lambda xS)^\diamond = \lambda xS^\diamond$

(iii) $(ST)^\diamond = (S^\diamond T^\diamond)$

(iv) $[S,T]^\diamond = [S^\diamond, T^\diamond]$

(v) $$([\underline{\lambda} xS, S_1, \ldots, S_n]T)^\diamond = \begin{cases} [\{x \mapsto T^\diamond\}S^\diamond, S_1^\diamond, \ldots, S_n^\diamond] & \text{if } x \in \mathrm{fv}(S) \\ [S^\diamond, S_1^\diamond, \ldots, S_n^\diamond, T^\diamond] & \text{otherwise} \end{cases}$$

(This is denoted $\varphi(S)$ in [3,24].) The following is analogous to the Lemma 2.3.14 in [3] – notice, however, a difference in the last point:

Lemma 3.7. (i) $(\{x \mapsto T\}S)^\diamond = \{x \mapsto T^\diamond\}S^\diamond$
(ii) $S \underset{\underline{\kappa}}{\rightarrow} T \;\Rightarrow\; T^\diamond = S^\diamond$
(iii) $S \underset{\kappa}{\rightarrow} T \;\Rightarrow\; S^\diamond \overset{+}{\underset{\kappa}{\rightarrow}} T^\diamond$

The last property is specific of the non-erasing reductions: in the ordinary λ-calculus, we would have $S^\diamond \overset{*}{\underset{\beta}{\rightarrow}} T^\diamond$. In other words, in $\underline{\Lambda}^*$ any redex has a residual by a reduction – except obviously if it is the reduced redex. This is clearly the key property towards the conservation theorem. The proof of the latter will use the following consequence of the finiteness of developments – which does *not* hold for $\beta, \underline{\beta}$:

Lemma 3.8. $S^\diamond \in \mathcal{SN}_\kappa \;\Rightarrow\; S \in \mathcal{SN}_{\kappa,\underline{\kappa}}$

Proof. we show that if S has an infinite sequence of κ and $\underline{\kappa}$-reductions, then $S^\diamond$ has an infinite sequence of κ-reductions. Assume that

$$S = S_0 \underset{\kappa,\underline{\kappa}}{\longrightarrow} S_1 \;\cdots\; S_n \underset{\kappa,\underline{\kappa}}{\longrightarrow} S_{n+1} \;\cdots$$

Then by the Lemma 3.7 we have

$$S^\diamond = S_0^\diamond \overset{*}{\underset{\kappa}{\rightarrow}} S_1^\diamond \;\cdots\; S_n^\diamond \overset{*}{\underset{\kappa}{\rightarrow}} S_{n+1}^\diamond \;\cdots$$

where $S_{i+1}^\diamond = S_i^\diamond$ if $S_i \underset{\underline{\kappa}}{\rightarrow} S_{i+1}$, and $S_i^\diamond \overset{+}{\underset{\kappa}{\rightarrow}} S_{i+1}^\diamond$ if $S_i \underset{\kappa}{\rightarrow} S_{i+1}$. By the Theorem 3.6 for all i there exists $j > i$ such that $S_i \underset{\kappa}{\rightarrow} S_j$, and therefore the κ-reduction sequence above from $S^\diamond$ is indeed infinite. □

Now let us denote by $\overline{S}$ the mapping from $\underline{\Lambda}^*$ to Λ^* which consists in removing the underlining, transforming $\underline{\lambda}$ into λ. (This is denoted $|S|$ in [3].)

Lemma 3.9. $\overline{S} \underset{\kappa}{\rightarrow} T \;\Rightarrow\; \exists U.\, \overline{U} = T \;\&\; S \underset{\kappa,\underline{\kappa}}{\longrightarrow} U$ *(This is analogous to the Lemma 2.3.13 in [3]. The proof, by induction on S, is immediate.)*

Corollary 3.10. $S \in \mathcal{SN}_{\kappa,\underline{\kappa}} \;\Rightarrow\; \overline{S} \in \mathcal{SN}_\kappa$

Theorem 3.11. *(The Conservation Theorem)* $\forall S \in \Lambda^*.\; S \underset{\kappa}{\rightarrow} T \;\&\; T \in \mathcal{SN}_\kappa \;\Rightarrow\; S \in \mathcal{SN}_\kappa$

Proof. let $([\lambda x S_0, S_1, \ldots, S_n]U)$ be the redex, at the occurrence C, which is contracted in the reduction $S \underset{\kappa}{\rightarrow} T$, so that $S = C[([\lambda x S_0, S_1, \ldots, S_n]U)]$, and let $W = C[([\underline{\lambda} x S_0, S_1, \ldots, S_n]U)]$. We obviously have $\overline{W} = S$, and it is easy to see that $W^\diamond = T$. Then $W \in \mathcal{SN}_{\kappa,\underline{\kappa}}$ by the Lemma 3.8, and therefore $S \in \mathcal{SN}_\kappa$ by the Corollary 3.10. □

A consequence of this result is that any reduction strategy (choosing a redex when there is one) only normalizes the strongly normalizing expressions, in other words $\mathcal{SN}_\kappa = \mathcal{WN}_\kappa$. Then, by the Corollary 2.3, we have:

Corollary 3.12. $\Lambda \cap \mathcal{WN}_\kappa \subseteq \mathcal{SN}_\beta$

To conclude this section we notice that the Lemmas 3.7 and 3.9 can be used to show the Church-Rosser property, exactly as in [3]:

Theorem 3.13. *(Church-Rosser Property) If* $S \overset{*}{\underset{\kappa}{\rightarrow}} T$ *and* $S \overset{*}{\underset{\kappa}{\rightarrow}} U$ *then there exists* S' *such that* $T \overset{*}{\underset{\kappa}{\rightarrow}} S'$ *and* $U \overset{*}{\underset{\kappa}{\rightarrow}} S'$*.*

Notice that, by contrast with the technique of Klop [18] and Kamareddine [13] (and of de Groote [10]), we do not need the Church-Rosser property to establish the "uniformity" result, that is $\mathcal{SN}_\kappa = \mathcal{WN}_\kappa$. Klop, de Groote and Kamareddine all use an idea of Nederpelt, that strong and weak termination of a transition relation coincide if this relation has the Church-Rosser property and is "increasing" (that is, there exists a measure that strictly increases along the transitions).

4 Intersection Types

Regarding the intersection type discipline, strong normalization was first proved by Pottinger [23] for a system later called "system $\mathcal{D}$" by Krivine [19]. The types of this system are:

$$\tau,\ \sigma\ldots \ ::=\ t \mid (\tau \to \sigma) \mid (\tau \wedge \sigma)$$

where t is any type variable. We define the size $|\tau|$ of a type τ as follows:

$$\begin{aligned} |t| &= 0 \\ |(\tau \to \sigma)| &= 1 + \max(|\tau|, |\sigma|) \\ |(\tau \wedge \sigma)| &= \max(|\tau|, |\sigma|) \end{aligned}$$

In order to relate system $\mathcal{D}$ to the original system of Coppo and Dezani [5], we consider in fact a system which is slightly more generous than system $\mathcal{D}$. To define the typing rules of this system, we introduce the following axioms regarding the conjunction:

$$((\tau_0 \wedge \tau_1) \wedge \tau_2) = (\tau_0 \wedge (\tau_1 \wedge \tau_2)) \quad \text{(A)}$$

$$(\tau_0 \wedge \tau_1) = (\tau_1 \wedge \tau_0) \quad \text{(C)}$$

$$(\tau \wedge \tau) = \tau \quad \text{(I)}$$

$$(\tau \to (\sigma_0 \wedge \sigma_1)) = ((\tau \to \sigma_0) \wedge (\tau \to \sigma_1)) \quad \text{(D)}$$

We denote by $\approx$ the congruence on types generated by these axioms. As usual, typing contexts Γ, $\Delta\ldots$ are mappings from a finite set of variables to types. The typing of the construct $[S,T]$ should be obvious: since S is the "main expression" in $[S,T]$, the type of $[S,T]$ is the type of S. Then the typing rules are:

$$\frac{\Gamma(x) \approx \tau}{\Gamma \vdash x : \tau} \qquad \frac{\Gamma \vdash S : \sigma \quad \Gamma(x) = \tau}{\Gamma \backslash x \vdash \lambda x S : (\tau \to \sigma)} \qquad \frac{\Gamma \vdash S : (\tau \to \sigma) \quad \Gamma \vdash T : \tau}{\Gamma \vdash (ST) : \sigma}$$

$$\frac{\Gamma \vdash S : \sigma \quad \Gamma \vdash T : \tau}{\Gamma \vdash [S,T] : \sigma}$$

$$\frac{\Gamma \vdash S : \tau \quad \Gamma \vdash S : \sigma}{\Gamma \vdash S : (\tau \wedge \sigma)} \qquad \frac{\Gamma \vdash S : (\tau \wedge \sigma)}{\Gamma \vdash S : \tau} \qquad \frac{\Gamma \vdash S : (\tau \wedge \sigma)}{\Gamma \vdash S : \sigma}$$

(plus an implicit rule for α-conversion). We call this type system for (our variant of) Klop's extended λ-calculus the "system $\mathcal{D}^*$" – although the axiom is slightly more generous than the usual one. For instance, the following typing holds in system $\mathcal{D}^*$, but not in system $\mathcal{D}$:

$$\vdash \mathsf{I} : ((\tau \to \sigma_0) \wedge (\tau \to \sigma_1)) \to (\tau \to (\sigma_0 \wedge \sigma_1))$$

This system is a restriction of the system of Barendregt & al. [4] (see also [12]), which uses a subtyping relation containing $\approx$. Since we are interested in strong normalization, our system does not include the constant ω that is a type for any expression. To prove the strong normalization result however, it is more convenient to use a syntax directed, equivalent type system, namely the original system of Coppo and Dezani [5]. Before introducing this system, let us see a preliminary property of system $\mathcal{D}^*$. Let us denote by $\Gamma \vdash_{\mathcal{D}^*} S : \sigma$ the fact that the sequent $\Gamma \vdash S : \sigma$ is provable in system $\mathcal{D}^*$. Next, we extend the congruence $\approx$ to typing contexts in the obvious way: $\Gamma \approx \Delta$ if and only if $\Gamma(x) \approx \Delta(x)$ for all x. Then we have:

Lemma 4.1. $\Gamma \vdash_{\mathcal{D}^*} S : \sigma \;\&\; \Gamma \approx \Delta \;\Rightarrow\; \Delta \vdash_{\mathcal{D}^*} S : \sigma$.

(The proof, by induction on the inference of $\Gamma \vdash_{\mathcal{D}^*} S : \sigma$, is trivial.) The type system of Coppo and Dezani [5], that we extend to Klop's calculus and then call "system $\mathcal{E}^*$", uses a restricted form of types. Let us say that the type ϕ is *prime* if it is built according to the following syntax:

$$\begin{aligned} \phi, \psi \ldots \; &::= \; t \mid (\xi \to \phi) \\ \xi, \zeta \ldots \; &::= \; \phi \mid (\xi \wedge \zeta) \end{aligned}$$

We shall consider these types up to the congruence generated by the associativity law (A) (notice that this congruence preserves the size). Then we write in particular $(\phi_1 \wedge \cdots \wedge \phi_n \to \psi)$ since the exact bracketing of $\phi_1 \wedge \cdots \wedge \phi_n$ is immaterial. We shall in fact denote such a type $(\phi_1, \ldots, \phi_n \to \psi)$. With this notation, the syntax of prime types is simply

$$\phi, \psi \ldots \; ::= \; t \mid (\phi_1, \ldots, \phi_n \to \psi)$$

though one should not forget that these are types in the sense of system $\mathcal{D}^*$. Now for each type τ we define the set $\tau^\star$ of its *prime factors*, by induction on τ, as follows:

$$\begin{aligned} t^\star &= \{t\} \\ (\tau \wedge \sigma)^\star &= \tau^\star \cup \sigma^\star \\ (\tau \to \sigma)^\star &= \{\, (\phi_1, \ldots, \phi_n \to \psi) \mid \tau^\star = \{\phi_1, \ldots, \phi_n\} \ \&\ \psi \in \sigma^\star \,\} \end{aligned}$$

Notice that we could have assumed a canonical way to enumerate types, in order to build smaller sets $(\tau \to \sigma)^\star$.

Lemma 4.2. *If* $\tau^\star = \{\phi_1, \ldots, \phi_n\}$ *then* $\tau \approx \phi_1 \wedge \cdots \wedge \phi_n$.

Proof. by induction on the structure of types. For the case of $(\tau \to \sigma)$ we use the distributivity law (D). □

The rules of system $\mathcal{E}^*$ are as follows:

$$\frac{\phi \in \Gamma(x)^\star}{\Gamma \vdash x : \phi} \qquad \frac{\Gamma \vdash S : \psi \quad \Gamma(x)^\star = \{\phi_1, \ldots, \phi_n\}}{\Gamma \backslash x \vdash \lambda x S : (\phi_1, \ldots, \phi_n \to \psi)}$$

$$\frac{\Gamma \vdash S : (\phi_1, \ldots, \phi_n \to \psi) \quad \Gamma \vdash T : \phi_1 \ \cdots \ \Gamma \vdash T : \phi_n}{\Gamma \vdash (ST) : \psi} \qquad \frac{\Gamma \vdash S : \phi \quad \Gamma \vdash T : \psi}{\Gamma \vdash [S, T] : \phi}$$

Proposition 4.3. (i) $\Gamma \vdash_{\mathcal{E}^*} S : \psi \ \Rightarrow\ \Gamma \vdash_{\mathcal{D}^*} S : \psi$
(ii) $\Gamma \vdash_{\mathcal{D}^*} S : \sigma \ \&\ \phi \in \sigma^\star \ \Rightarrow\ \Gamma \vdash_{\mathcal{E}^*} S : \phi$

Proof. (i) by induction on the inference of $\Gamma \vdash_{\mathcal{E}^*} S : \psi$, using the Lemmas 4.2 and 4.1.

(ii) we proceed by induction on the inference of $\Gamma \vdash_{\mathcal{D}^*} S : \sigma$, and by cases on the last rule used in this inference. □

A similar, but more elaborate result is shown by Hindley [11] for the system of Barendregt, Coppo & Dezani [4] (see also [1]). An immediate consequence of this property is

Corollary 4.4. *A λ-expression is typable in system $\mathcal{D}^*$ if and only if it is typable in system $\mathcal{E}^*$.*

Obviously, the previous proposition provides a much more precise and effective statement. This result shows that, to prove that typable expressions are strongly normalizing, we may assume that "typable" means "typable in system $\mathcal{E}^*$".

5 Strong Normalization

In this section we prove that typable λ-expressions are strongly β-normalizing. The key property for this is a particular subject reduction property, the proof of which is most conveniently established using the presentation of the type system as a *natural deduction* system (see [6,9]). Expressed in the form of natural deduction, system $\mathcal{E}^*$ consists in the following rules:

$$\dfrac{\begin{array}{c}\cdots\ [x:\phi_1]\ \cdots\ [x:\phi_n]\ \cdots\\ \vdots\\ S:\psi\end{array}}{\lambda xS:(\phi_1,\ldots,\phi_n\to\psi)}\,(\lambda)\qquad \dfrac{\begin{array}{c}\vdots\\ S:(\phi_1,\ldots,\phi_n\to\psi)\end{array}\quad \begin{array}{c}\vdots\\ T:\phi_1\end{array}\ \cdots\ \begin{array}{c}\vdots\\ T:\phi_n\end{array}}{(ST):\psi}\,(@)$$

$$\dfrac{\begin{array}{c}\vdots\\ S:\phi\end{array}\quad \begin{array}{c}\vdots\\ T:\psi\end{array}}{[S,T]:\phi}\,[,]$$

where, in the (λ)-rule, $[\cdots]$ means as usual that all the assumptions regarding x made in proving $S:\psi$, with possible additions, are discharged. This involves implicit contraction (the same ϕ_i may appear several times as an assumption for x) and weakening, or "vacuous cancellations", see for instance [12]. A *proof* of typing is defined, rather informally, as a tree Π build according to these rules. Now we define the size of a proof. First, let us call *degree* of an occurrence of a redex[2] the size of the type of its function part (which is an abstraction). Then the *size* $|\Pi|$ of the proof Π is the pair (m,r) where the integer m is the maximum degree in Π, and r is the number of redexes of degree m. More formally, we define inductively the size of a proof as follows:

(i) if Π is the tree consisting in the leaf $x:\phi$ then $|\Pi|=(0,0)$.

(ii) if Π is the proof

$$\dfrac{\Phi}{\lambda xS:(\phi_1,\ldots,\phi_n\to\psi)}$$

where Φ is a proof of $S:\psi$, involving the assumptions $\phi_1,\ldots,\phi_n$ about x, then $|\Pi|=|\Phi|$.

(iii) if Π is the proof

$$\dfrac{\Phi\quad \Psi_1\ \cdots\ \Psi_n}{(ST):\psi}$$

[2] since an argument in an application may have to be typed with several types, a given redex of a term may have sereral occurrences in a proof of typing.

where Φ is a proof of $S:(\phi_1,\ldots,\phi_n \to \psi)$, and the Ψ_i's are proofs of the $T:\phi_i$'s, with $|\Phi| = (m_0, r_0)$ and $|\Psi_i| = (m_i, r_i)$ then $|\Pi| = (m, r)$ where, if we let $\sigma = (\phi_1,\ldots,\phi_n \to \psi)$:

$$m = \begin{cases} \max\{|\sigma|, m_0, m_1, \ldots, m_n\} & \text{if } S = [\lambda x U, U_1, \ldots, U_s] \\ \max\{m_0, m_1, \ldots, m_n\} & \text{otherwise} \end{cases}$$

$$r = \begin{cases} 1 + \sum\{\, r_i \mid 0 \leqslant i \leqslant n \;\&\; m_i = m \,\} & \text{if } S = [\lambda x U, U_1, \ldots, U_s] \;\&\; m = |\sigma| \\ \sum\{\, r_i \mid 0 \leqslant i \leqslant n \;\&\; m_i = m \,\} & \text{otherwise} \end{cases}$$

(iv) if Π is the proof

$$\frac{\Phi \quad \Psi}{[S,T]:\phi}$$

where Φ and Ψ are respectively proofs of $S:\phi$ and $T:\psi$, with $|\Phi| = (k,p)$ and $|\Psi| = (h,q)$, then $|\Pi| = (m,r)$ where $m = \max(k,h)$ and $r = p$ if $k = m > h$, else $r = q$ if $h = m > k$, or else $r = p + q$.

We observe that, by definition, $|\Pi| = (0,0)$ if and only if Π is a proof of typing of a κ-normal form. We denote by $\prec$ the strict lexicographic ordering on pairs of integers. Now we show a refined subject reduction property:

Proposition 5.1. *Let Π be a proof of $S:\psi$. If $S \underset{\kappa}{\to} T$ by contracting an innermost redex of maximal m degree in Π, then there exists a proof Φ of $T:\psi$ such that Φ contains strictly less redexes of degree m than Π, and therefore $|\Phi| \prec |\Pi|$.*

The proof is an adaptation of the one of Coppo and Dezani [5] for the similar property for the λ-calculus. The pairing construct of Klop's calculus does not introduce any particular difficulty. An immediate consequence is the classical result that λ-expressions typable in the intersection type discipline are strongly β-normalizing:

Theorem 5.2. *(Strong Normalization) Any λ-expression typable in system $\mathcal{D}$ is strongly β-normalizing.*

Proof. if M is typable in $\mathcal{D}$, it is also typable in system $\mathcal{E}^*$, by the Corollary 4.4. Then $M \in \mathcal{WN}_\kappa$ by the previous proposition, and therefore $M \in \mathcal{SN}_\beta$ by the Corollary 3.12. □

As one can see, the last part of the proof is just an easy adaptation of the standard proof of weak normalization for typable expressions. Then, regarding strong normalization, the real work is done in proving that in the extended λ-calculus weak and strong normalizability are equivalent. This in turn relies on the classical "finiteness of developments" property, which, as a matter of fact, is the only strong normalization result proper that we proved in this paper. Then we can conclude that "finiteness of developments" is the termination result that implies strong normalizability in the intersection type discipline. We notice that the two results are actually equivalent, since Parigot [21] has shown that the finiteness of developments can be derived from the strong normalization theorem for system $\mathcal{D}$ (this is also reported in [19], and the proof easily extends to the calculus we used here).

6 Conclusion

We have proved the strong normalization result for the intersection type discipline in a way which, we think, is quite natural and intuitive. First, the choice of Klop's calculus for dealing with strong normalization is a natural one, since it adds to the λ-calculus just what is needed to deal with β_K-redexes in an appropriate way, and this is the difficult part in "syntactic" proofs of termination. As we said above, we could as well use Kamareddine's calculus of generalized β-reduction [13], of which our variant of Klop's calculus is a restriction, where we only "look through" β_K-redexes. Second, we think that Lévy's argument, showing what is decreasing in the typing when the subject is reduced, is also quite natural and simple. The technique presented here, collecting some more or less well-known facts about reduction in the (typed) λ-calculus, is perhaps not as general as the realizability technique, however. For instance, we cannot immediately apply it to system $\mathcal{F}$ [8,9], since the property that created redexes are smaller than their creator does not hold in this system, where one can instantiate arbitrarily a quantified type $\forall t.\tau$, thus increasing the size of the type.

References

[1] S. van Bakel, *Intersection type assignment systems,* TCS Vol. 151 No. 2 (1995) 348–435.

[2] H. Barendregt, *The Lambda Calculus,* Studies in Logic 103, North-Holland, Revised Edition (1984).

[3] H. Barendregt, *Lambda Calculi with Types,* in Handbook of Logic in Computer Science, Vol. 2 (S. Abramsky, Dov M. Gabbay & T.S.E. Maibaum, Eds.), Oxford University Press (1992) 117–309.

[4] H. Barendregt, M. Coppo, M. Dezani-Ciancaglini, *A filter lambda model and the completeness of type assignment,* J. of Symbolic Logic 48 (1983) 931–940.

[5] M. Coppo, M. Dezani-Ciancaglini, *An extension of the basic functionality theory for the λ-calculus,* Notre Dame J. of Formal Logic 21 (1980) 685–693.

[6] M. Coppo, M. Dezani-Ciancaglini, B. Venneri, *Functional characters of solvable terms,* Zeit. Math. Logik Grund. 27 (1981) 45–58.

[7] R. Gandy, *An early proof of normalization by A.M. Turing,* In To H.B. Curry: Essays on Combinatory Logic, Lambda Calculus and Formalism (J.R. Hindley and J.P. Seldin, Eds.), Academic Press (1980) 453–455.

[8] J.-Y. Girard, *Une extension de l'interprétation de Gödel à l'analyse, et son application à l'élimination des coupures dans l'analyse et la théorie des types,* Second Scandinavian Logic Symposium (Ed. J.E. Fenstad), North-Holland (1971) 63–92.

[9] J.-Y. Girard, Y. Lafont, P. Taylor, *Proofs and Types*, Cambridge Tracts in Theoretical Computer Science 7, Cambridge University Press (1989).

[10] Ph. de Groote, *The conservation theorem revisited*, Typed Lambda Calculi and Applications, Lecture Notes in Comput. Sci. 664 (1993) 163–178.

[11] R. Hindley, *The simple semantics for Coppo-Dezani-Sallé types*, Intern. Symp. on Programming, Lecture Notes in Comput. Sci. 137 (1982) 212–226.

[12] R. Hindley, *Types with intersection: an introduction*, Formal Aspects of Computing Vol. 4 (1992) 470–486.

[13] F. Kamareddine, *Postponement, conservation and preservation of strong normalization for generalized reduction*, J. of Logic and Computation Vol. 10 No. 5 (2000) 721–738.

[14] A. Kfoury, J. Wells, *New notions of reduction and non-semantic proofs of strong β-normalization in typed λ-calculi*, LICS (1995) 311–321.

[15] A. Kfoury, J. Wells, *Addendum to "New notions of reduction and non-semantic proofs of strong β-normalization in typed λ-calculi"*, Tech. Rep. 95–007, Comput. Sci. Dept., Boston University (1995).

[16] Z. Khasidashvili, M. Ogawa, V. van Oostrom, *Uniform normalisation beyond orthogonality*, RTA, Lecture Notes in Comput. Sci. 2051 (2001) 122–136.

[17] S.C. Kleene, *On the interpretation of intuitionistic number theory*, J. of Symbolic Logic, Vol. 10 (1945) 109–124.

[18] J.W. Klop, *Combinatory Reduction Systems*, PhD Thesis, Utrecht University. Mathematical Centre Tracts Vol. 127, Mathematisch Centrum, Amsterdam (1980).

[19] J.-L. Krivine, *Lambda-Calcul: Types et Modèles*, Masson, Paris (1990). English translation "Lambda-Calculus, Types and Models", Ellis Horwood (1993).

[20] J.-J. Lévy, *Réductions correctes et optimales dans le lambda-calcul*, Thèse, Université Paris 7 (1978).

[21] M. Parigot, *Internal labellings in lambda-calculus*, MFCS, Lecture Notes in Comput. Sci. 452 (1990) 439–445.

[22] G. Plotkin, *Lambda-definability and logical relations*, Memo SAI-RM-4, University of Edinburgh (1973).

[23] G. Pottinger, *A type assignment for the strongly normalizable λ-terms*, In To H.B. Curry: Essays on Combinatory Logic, Lambda Calculus and Formalism (J.R. Hindley and J.P. Seldin, Eds.), Academic Press (1980) 561–577.

[24] F. van Raamsdonk, P. Severi, M.H. Sørensen, H. Xi, *Perpetual reductions in λ-calculus*, Information and Computation Vol. 149 No. 2 (1999) 173–225.

[25] P. Sallé, *Une extension de la théorie des types en λ-calcul*, ICALP, Lecture Notes in Comput. Sci. 62 (1978) 398–410.

[26] M.H. Sørensen, *Strong normalization from weak normalization in typed λ-calculi*, Information and Computation Vol. 133 No. 1 (1997) 35–71.

[27] W. Tait, *Intensional interpretations of functionals of finite type I*, J. of Symbolic Logic 32 (1967) 198–212.

[28] W. Tait, *A realizability interpretation of the theory of species*, Logic Colloquium, Lecture Notes in Mathematics 453 (1975) 240–251.

[29] R. de Vrijer, *A direct proof of the finite developments theorem*, J. of Symbolic Logic, Vol. 50 (1985) 339–343.

Relative Definability and Models of Unary PCF

Antonio Bucciarelli, Benjamin Leperchey, and Vincent Padovani

PPS, Université Denis Diderot
175 Rue du Chevaleret, 75013 Paris

Abstract. We show that the poset of degrees of relative definability in the Scott model of Unary PCF is non trivial, and that, nevertheless, the hierarchy of order extensional models of the language is reduced to a bottom element (the fully abstract model) and a top one (the Scott model itself).

1 Introduction

Finitary versions of PCF and related languages have been studied in the last decade, in order to settle the well-known "full abstraction problem" for the full language. In 1993, A. Jung and A. Stoughton [4] proposed the following crash test for a solution to that problem to be a "good" one: it should provide an algorithm for deciding observational equivalences for the finitary fragment of the language (*i.e.* the language whose unique ground type is `bool`, FPCF).

Shortly later, R. Loader proved that the observational equivalence of FPCF is undecidable [9]. An immediate corollary of this breakthrough result is that the problem of FPCF-definability in, say, the Scott model of FPCF, is undecidable. Hence, relative FPCF-definability in that model is also undecidable.

The poset of degrees of relative definability is somehow related to the existence of hierarchies of (order extensional) models. The idea is the following: given a big model of FPCF, w.r.t. a natural notion of embedding, defined via a logical relation (see Sect. 2.4) an undefinable element x of the model, and a logical relation R_x which proves that x is undefinable, one can:

- remove all elements of the model which are not invariant w.r.t. R_x,
- perform an extensional collapse on the remaining elements,

hence obtaining a new, smaller, model.

Let us consider, as an example, the following hierarchy of monotone functions $g_n : bool^n \rightarrow bool$ (generalizing a well-known example of first-order, stable and non sequential function, due to G. Berry)

$$g_n(x_1, ..., x_n) = \mathtt{tt} \Leftrightarrow \exists \boldsymbol{y} \in G_n \; \boldsymbol{x} \geq \boldsymbol{y}$$

where G_n is the set of circular permutations of $(\bot, \underbrace{\mathtt{tt}, ..., \mathtt{tt}}_{n-2}, \mathtt{ff})$

For all $n \geq 3$, g_n is FPCF-undefinable; moreover g_{n+1} is FPCF-definable relatively to g_n, and the converse does not hold.

M. Hofmann (Ed.): TLCA 2003, LNCS 2701, pp. 75–89, 2003.

Now, for all n, we define a logical relation R_n such that g_n (and hence all the g_{n-i}) is not invariant w.r.t R_n, and g_{n+1} (and hence all the g_{n+i}) is invariant w.r.t. R_n. This relation R_n is an instance of "sequentiality relation" (see definition 2), and in Sieber's terminology $R_n = S^{n+1}_{\{1,..,n\}\{1,..,n+1\}}$, *i.e.*, R_n contains, at ground type, the set of $(n+1)$-tuples which are either constant or does contain an occurrence of $\perp$ in one of the first n components.

Performing the two operations described above w.r.t. R_n, yields a model which does not contain g_n and contains g_{n+1}.

Hence, concerning FPCF, we have that:

- The observational equivalence is undecidable.
- The definability problem in the Scott model is undecidable.
- The relative definability problem (in the Scott model) is undecidable, and the poset of degrees of relative definability is infinite.
- There exist infinite hierarchies of standard, order extensional models.

Several authors have investigated restrictions on the syntax of FPCF which make observational equivalence decidable: V. Padovani has shown that this can be achieved by eliminating all non ground constants (the `if-then-else` in FPCF) [12]. Schmidt-Schauss [15] and independently, R. Loader [8], have proved that observational equivalence is decidable also for the "unary" version of PCF (a single ground type o with two constants $\perp$ and $\top$ and a "sequential convergence test" $\wedge : o \to o \to o$).

In a recent paper [5], J. Laird shows that Berry's model of bidomains is universal for UPCF (using the listing algorithm devised by Schmidt-Shauss).

In this paper, we address the following questions:

- Is the poset of degree of relative definability in the Scott model of UPCF trivial? (*i.e.* does it contain just the degree of definable functions, and the one of functions equivalent to the "parallel convergence test" $\vee$?)
- Is the hierarchy of (standard and order extensional) model of UPCF trivial? (*i.e.* does it contain just the Scott and the fully abstract model?)

Our first remark was that a positive answer to the first question implies a positive answer to the second one. Surprisingly enough, it turns out that the poset of degree is non trivial, and the one of models is trivial[1].

The point is that, when applying the "collapsing" technique described above to the Scott model of UPCF in order to eliminate, say, the degree of $\vee$, by picking up an appropriate logical relation (typically, Sieber's $S^3_{\{1,2\},\{1,2,3\}}$) then all the other degrees collapse too, either in the first phase (elimination of non-invariant elements) or in the second one (extensional collapse of the invariant elements).

[1] The fact that the poset of models is trivial has an alternative proof, simpler then ours, due to J. Laird [6], but less general. In fact we are able to apply our result also in order to reason about the hierarchy of models of FPCF (Sect. 4.4).

2 Preliminaries

We introduce the notion of relative definability, the language UPCF and its Scott model, and logical relations, that we use both to compare degrees of definability and models.

2.1 Degrees of Definability

Given an applied calculus L, a model $\mathcal{M}$ of L, and two elements $f \in \mathcal{M}^\tau$ and $g \in \mathcal{M}^\sigma$, we say that *f is smaller than g in the L-definability preorder of $\mathcal{M}$*, $f \preceq_L^{\mathcal{M}} g$, if there exists an L-term $M : \sigma \to \tau$ such that $[\![M]\!]^{\mathcal{M}} g = f$.

A *degree of L-definability in $\mathcal{M}$* is an equivalence class of the equivalence relation associated with the preorder above, and degrees are partially ordered by $\preceq_L^{\mathcal{M}}$. The poset of degrees always has a smallest element, namely the degree of definable elements.

Degrees of PCF-definability in the Scott-continuous model, often called *degrees of parallelism* have been studied for instance in [1], while degrees of PCF-definability in the model of strongly stable function (which could be called *degrees of intensionality*) have been investigated in [2,10,11].

Of course, if $\mathcal{M}$ has the definability property w.r.t. L (i.e. if any element of $\mathcal{M}$ is the denotation of some L-term), then the poset of degrees of L-definability in $\mathcal{M}$ is reduced to a singleton.

When the language and the model we refer to are clear from the context, we will omit "L" and "$\mathcal{M}$", in the definition and notations above, and we will speak of "degrees of definability", or even, when the model is order extensional, of "degrees of parallelism". Moreover, we use the same symbols for the constants of L and their denotations in $\mathcal{M}$.

In the rest of this section, we focus on Unary PCF. Nevertheless, all the definitions and results apply to FPCF too, changing appropriately the ground type and its standard interpretation.

2.2 Unary PCF

Unary PCF is an example of applied λ-calculus: its ground constants are $\bot, \top : o$, and the only first order constant is $\wedge : o^2 \to o$,

Let $\mathcal{M}$ be a standard model of UPCF, that is:

- $[\![o]\!] = \{\bot, \top\}$, $[\![\top]\!] = \top$ and $[\![\bot]\!] = \bot$,
- $[\![\sigma \to \tau]\!]$ is a subset of $[\![\tau]\!]^{[\![\sigma]\!]}$.

We can define an extensional order $\leq$ on $\mathcal{M}$:

- At type o, $x \leq_o y$ iff $x = \bot$ or $x = y$,
- At type $\sigma \to \tau$, $f \leq_{\sigma\to\tau} g$ iff $\forall x \in [\![\sigma]\!], fx \leq_\tau gx$.

Definition 1. *$\mathcal{M}$ is a standard order-extensional model if:*

- *$\mathcal{M}$ is a standard model,*
- *All functions are monotonic for the order $\leq$.*

The Scott model $\mathcal{E}$ of UPCF is the standard, order extensional model where $[\![\sigma \to \tau]\!] = \{f : [\![\sigma]\!] \to [\![\tau]\!] \mid f \text{ is } \leq\text{-monotonic}\}$, each $[\![\sigma]\!]$ being partially ordered by $\leq$.

2.3 Sequentiality Relations

This definition and the proposition are taken from [14].

Definition 2. *If $A \subseteq B \subseteq \{1, ..., n\}$, $S^n_{A,B}$ is the set of tuples $(x_1, ..., x_n)$ such that either $\exists i \in A, x_i = \bot$ or $\forall i, j \in B, x_i = x_j$.*

A Sieber relation is a logical relation which is an intersection of a number of $S^n_{A,B}$ at base type.

Proposition 3. *The Sieber relations are exactly the relations which contain the constants of UPCF.*

The fundamental lemma of logical relations ensures that the denotation of any UPCF term, in any model, is invariant w.r.t. all Sieber relations.

2.4 Hierarchies of Models

Definition 4. *Given two standard models $\mathcal{M}$ and $\mathcal{N}$, the relation $R_{\mathcal{M},\mathcal{N}}$ is the only logical relation which is the identity at the base type.*

Definition 5. *A logical relation R is functional if*

$$\forall \sigma, \forall f \in [\![\sigma]\!]^{\mathcal{M}}, \forall g, g' \in [\![\sigma]\!]^{\mathcal{N}}, fRg \wedge fRg' \Longrightarrow g = g'$$

If R is functional and $x \in [\![\sigma]\!]^{\mathcal{M}}$, we write $R(x)$ for the only $y \in [\![\sigma]\!]^{\mathcal{N}}$ such that xRy.

A logical relation R is onto if

$$\forall \sigma, \forall g \in [\![\sigma]\!]^{\mathcal{N}}, \exists f \in [\![\sigma]\!]^{\mathcal{M}}, fRg$$

Definition 6. *A model $\mathcal{M}$ is smaller than a model $\mathcal{N}$ if $R_{\mathcal{N},\mathcal{M}}$ is functional and onto. $\mathcal{M}$ and $\mathcal{N}$ are isomorphic if $\mathcal{M}$ is smaller than $\mathcal{N}$ and $\mathcal{N}$ is smaller than $\mathcal{M}$.*

Lemma 7. *Assume R^σ is functional and onto. Let $x, x' \in [\![\sigma]\!]^{\mathcal{E}}$, $y \in [\![\sigma]\!]^{\mathcal{M}}$. If $y = R(x)$ and $y = R(x')$ then $y = R(x \wedge x')$. If $y \leq R(x)$ and $y \leq R(x')$ then $y \leq R(x \wedge x')$.*

Proof. If $\sigma = o$, R is the identity. Assume σ is a functional type, and let a be an argument of x and x'. $R(xa) = yR(a) = R(x'a)$. By hypothesis, $R(xa \wedge x'a) = yR(a)$ so $R((x \wedge x')a) = yR(a)$. Finally, we get $R(x \wedge x') = y$. The proof is the same for $\leq$. □

Lemma 8. *If R is functional and onto, there exists a total monotonic map R^{-1} such that $\forall x \in [\![\alpha]\!]^{\mathcal{M}}, x = R(R^{-1}(x))$.*

Proof. Lemma 7 allows us to define $R^{-1}(y)$ as the smallest x such that $R(x) = y$. R^{-1} is monotonic: if $x \leq y$, $x \leq R(R^{-1}(x))$ and $x \leq R(R^{-1}(y))$. With lemma 7, we get $x \leq R(R^{-1}(x) \wedge R^{-1}(y))$, and the monotonicity of R gives $R(R^{-1}(x)) \geq R(R^{-1}(x) \wedge R^{-1}(y))$. This yields $x = R(R^{-1}(x) \wedge R^{-1}(y))$, so $R^{-1}(x) \leq R^{-1}(x) \wedge R^{-1}(y)$, and finally $R^{-1}(x) \leq R^{-1}(y)$. □

Proposition 9. *If $\mathcal{M}$ and $\mathcal{N}$ are standard order-extensional models of UPCF and $\mathcal{N}$ is fully abstract, $\mathcal{N}$ is smaller than $\mathcal{M}$*

Proof. We write $R = R_{\mathcal{M},\mathcal{N}}$. At ground type, R is a bijection. Assume that R is functional and onto at types σ and τ.

Assume $fR^{\sigma\to\tau}g$ and $fR^{\sigma\to\tau}g'$. Let $x \in [\![\sigma]\!]^{\mathcal{N}}$. Since R^{σ} is onto, there exists $y \in [\![\sigma]\!]^{\mathcal{M}}$ such that $R(y) = x$. We get $gx = g(R(y)) = R(fy) = g'(R(y)) = g'x$, which entails $g = g'$: $R^{\sigma\to\tau}$ is functional.

Let $g \in [\![\sigma \to \tau]\!]^{\mathcal{N}}$. Since N is fully abstract, there exists a closed term G such that $[\![G]\!]^{\mathcal{N}} = g$. By the fundamental lemma of logical relations

$$[\![G]\!]^{\mathcal{M}} R^{\sigma\to\tau} [\![G]\!]^{\mathcal{N}}$$

This yields $[\![G]\!]^{\mathcal{M}} R^{\sigma\to\tau} g$: $R^{\sigma\to\tau}$ is onto. □

3 Some Degrees in the Big Model of UPCF

We know that the poset of degrees of UPCF-definability in the Scott model has a smallest element $\bot_{deg}$ (the degree of definable elements). A biggest degree $\top_{deg}$ (the degree of the "parallel convergence test" $\underline{\lambda} xy.\ x \vee y$), also exists:

Lemma 10. *All elements of the Scott model of UPCF are definable relatively to $\underline{\lambda} xy.\ x \vee y$.*

Proof. We make an induction on the types.

Let $f : \sigma_1 \to ... \to \sigma_n \to o$. Let $(x_{i1}, ..., x_{in})_{i=1...m}$ be the trace of f (the smallest tuples such that f yields $\top$).

If $x : \sigma_i$, the function $\leq_x$ mapping y to $\top$ if $x \leq y$ and to $\bot$ otherwise is definable: if $(a_{i1}, ..., a_{il})_{i=1...k}$ is the trace of x, and $a_{ij} = [\![A_{ij}]\!]$ (by hypothesis), then $\lambda y.(yA_{11}...A_{1l}) \wedge ... \wedge (yA_{k1}...A_{kl})$ defines $\leq_x$.

One easily checks that f is defined by

$$\lambda y_1...y_n.((\leq_{x_{11}} y_1) \wedge ... \wedge (\leq_{x_{1n}} y_n)) \vee ... \vee ((\leq_{x_{m1}} y_1) \wedge ... \wedge (\leq_{x_{mn}} y_n))$$

□

In the Scott model of Unary PCF, one can easily show that any first order function belongs either to the smallest degree (*i.e.* it is definable), or to the biggest one (*i.e.* it allows to ($UPCF$-)define any other element of the model).

Lemma 11. *If $\psi : o^n \to o$ is undefinable, then there exists $u_1, \ldots, u_n$ where $u_i \in \{x, y, \top, \bot\}$, such that $[\![\lambda fxy.(fu_1 \ldots u_n)]\!]\,\psi = \vee$.*

Proof. By hypothesis, ψ is not a constant function. Suppose there is a unique minimal sequence $\boldsymbol{c} \in \{\top, \bot\}^n$ such that $\psi(\boldsymbol{c}) = \top$. Take $u_i = \top$ if $c_i = \bot$, x_i otherwise. Then $[\![\lambda x_1 \ldots x_n.u_1 \wedge \ldots \wedge u_n]\!] = \psi$, a contradiction.

Let $\boldsymbol{c}^1, \boldsymbol{c}^2 \in \{\top, \bot\}^n$ be minimal distinct sequences such that $\psi(\boldsymbol{c}^1) = \psi(\boldsymbol{c}^2) = \top$. Take $u_i = x$ if $(c_i^1, c_i^2) = (\top, \bot)$, y if $(c_i^1, c_i^2) = (\bot, \top)$, d if $c_i^1 = c_i^2 = [\![d]\!]$. Check that $[\![\lambda fxy.(fu_1 \ldots u_n)]\!]\,\psi = \vee$. □

In other words, all first order types possess the 2-DEG property. We show that 2-DEG is not preserved at higher types, by constructing two intermediate degrees.

Let $\phi \in \mathcal{E}^{(o^3 \to o) \to (o^3 \to o)}$ be the function defined by

$$\phi(f) = \begin{cases} \underline{\lambda} xyz.\ x \vee y \vee z \text{ if } f = \underline{\lambda} xyz.\ x \vee y \\ f \qquad\qquad\qquad \text{otherwise} \end{cases}$$

Proposition 12. $\bot_{deg} \prec \phi \prec \top_{deg}$.

Proof. First of all, ϕ is monotone since there is no element of $\mathcal{E}^{(o^3 \to o)}$ strictly in between $\underline{\lambda} xyz.\ x \vee y$ and $\underline{\lambda} xyz.\ x \vee y \vee z$.

Concerning $\bot_{deg} \prec \phi$, we are going to show that ϕ is non-invariant w.r.t. a particular $UPCF$-relation: $S^3_{\{1,2\}\{1,2,3\}}$. Let $f = \underline{\lambda} xyz.\ x \wedge y$, $g = \underline{\lambda} xyz.\ x \vee y$, and let us use S as a shorthand for $S^3_{\{1,2\}\{1,2,3\}}$.

First of all, it is easy to see that $(f, g, g) \in S$, since whenever (x_1, x_2, x_3), (y_1, y_2, y_3), $(z_1, z_2, z_3) \in S$ are such that $f\ x_1\ y_1\ z_1 = \top$ and $g\ x_2\ y_2\ z_2 = \top$, then either $x_1 = x_2 = \top$ or $y_1 = y_2 = \top$. In the former case we conclude that $x_3 = \top$, in the latter that $y_3 = \top$; in both cases, $g\ x_3\ y_3\ z_3 = \top$, and hence $(f, g, g) \in S$.

Next we prove that the ϕ-image of (f, g, g) is not in S, and hence that ϕ is not definable.

Let $h = \underline{\lambda} xyz.\ x \vee y \vee z$. The ϕ-image of (f, g, g) is (f, h, h); the following diagram shows that $(f, h, h) \notin S$:

$$\begin{array}{l} f\ h\ h \\ \top \bot \bot \in S \\ \top \bot \bot \in S \\ \underline{\bot \top \bot \in S} \\ \top \top \bot \notin S \end{array}$$

In order to show that $\phi \prec \top_{deg}$, we show that $\top_{deg} \not\preceq \phi$. We prove by induction on (n, m) that there is no normal η-long term M with n free occurrences of f, m occurrences of $\wedge$, such that $[\![\lambda fxy.M]\!]\,\phi = \vee$:

- $[\![\lambda fxy.x]\!]\,\phi$, $[\![\lambda fxy.y]\!]\,\phi \neq \vee$,
- if $M = M_1 \wedge M_2$, then:
  - either $[\![\lambda fxy.M_1]\!]\,\phi\ \bot\bot = \top$ and $[\![\lambda fxy.M_1]\!]\,\phi = \vee$,

- or $[\![\lambda fxy.M_2]\!]\,\phi\ \bot\bot = \top$ and $[\![\lambda fxy.M_2]\!]\,\phi = \vee$.

– if $M = (f(\lambda z_1 z_2 z_3.w))\, v_1\, v_2\, v_3$ where a $[\![\lambda fxy.v_i]\!]\,\phi$ is undefinable, then the latter equals the only undefinable of type $o \to (o \to o)$, that is, $\vee$.

– if $M = (f(\lambda z_1 z_2 z_3.w))\, N_1\, N_2\, N_3$ where no N_i contains an occurrence of f and $[\![(\lambda fxyz_1z_2z_3.w)]\!]\,\phi$ is undefinable, then by lemma 11 there exists $(t_1, t_2, u_1, u_2, u_3) \in \{x, y, \top, \bot\}^5$ such that:

$$[\![(\lambda fxy.w[t_1/x, t_2/y, u_1/z_1, u_2/z_2, u_3/z_3])]\!]\,\phi = \vee$$

Since all substituted terms are of ground type, the term in the left-hand side is in normal, η-long form, and contains $n-1$ occurrences of f.

– if $M = (f\, P\, N_1\, N_2\, N_3)$ where P, N_1, N_2, N_3 contain no occurrences of f, then for all $(t, u) \in \{\top, \bot\}^2$ and for $P' = P[t/x, u/y]$, $N_i' = N[t/x, u/y]$, we have:

$$\begin{aligned}
[\![\lambda fxy.M]\!]\,\phi\,[\![t]\!]\,[\![u]\!] &= \phi\,[\![P']\!]\,[\![N_1']\!]\,[\![N_2']\!]\,[\![N_3']\!] \\
&= [\![P']\!]\,[\![N_1']\!]\,[\![N_2']\!]\,[\![N_3']\!] \\
&= [\![(P'\, N_1'\, N_2'\, N_3')]\!] \\
&= [\![\lambda xy.(P\, N_1\, N_2\, N_3)]\!]\,[\![t]\!]\,[\![u]\!]
\end{aligned}$$

In other words $[\![\lambda fxy.M]\!]\,\phi = [\![\lambda fxy.(P\, N_1\, N_2\, N_3)]\!]\,\phi = \vee$ where the normal form of $(P\, N_1\, N_2\, N_3)$ contains no occurrence of f.

□

By using ϕ, we are now able to define a new degree, represented by a function Φ which moreover is S-invariant: let $\Phi \in \mathcal{E}^{((o^3 \to o)\to(o^3\to o))\to(o^2\to o)}$ be the function defined by

$$\Phi(\psi) = \begin{cases} \underline{\lambda} xy.\top & \text{if } \psi > \phi \\ \underline{\lambda} xy.x \vee y & \text{if } \psi = \phi \\ \underline{\lambda} xy.\bot & \text{if } \psi \not\geq \phi \end{cases}$$

Proposition 13. *The degrees of ϕ and Φ are incomparable.*

Proof. First of all, it is easy to see that Φ is actually an element of $\mathcal{E}$, i.e. a monotone function.

Concerning $\Phi \not\preceq \phi$, it is enough to remark that, if $\Phi \preceq \phi$, then $\underline{\lambda} xy.\, x \vee y \preceq \phi$.

In order to show that $\phi \not\preceq \Phi$, we prove that Φ is invariant w.r.t. $S^3_{\{1,2\}\{1,2,3\}} = S$ (and we conclude using the fundamental lemma of logical relations, since ϕ is not S-invariant).

This amounts to showing that whenever

$$\begin{array}{l}
\Phi\ \ \Phi\ \ \Phi \\
\psi_1\ \psi_2\ \psi_3 \\
x_1\ x_2\ x_3 \in S \\
\underline{y_1\ y_2\ y_3 \in S} \\
\top\ \ \top\ \ \bot
\end{array}$$

one has $(\psi_1, \psi_2, \psi_3) \notin S$, *i.e.* that $\psi_1, \psi_2 \geq \phi$ and $\psi_3 \not\geq \phi$ entail $(\psi_1, \psi_2, \psi_3) \notin S$.

Now, we decompose $\psi_3 \not\geq \phi$ in two (not mutually exclusive) cases, and prove $(\psi_1, \psi_2, \psi_3) \notin S$ for both of them:

- case 1: $\psi_3(\underline{\lambda} xyz.\ x \vee y) \leq \underline{\lambda} xyz.\ x \vee y \vee z$.
  let $f = \underline{\lambda} xyz.\ x \wedge y$, $g = \underline{\lambda} xyz.\ x \vee y$, as before.

$$\begin{array}{llll}
\psi_1 & \psi_2 & \psi_3 & \\
f & g & g & \in S \\
\top & \bot & \bot & \in S \\
\top & \bot & \bot & \in S \\
\bot & \top & \bot & \in S \\
\hline
\top & \top & \bot & \notin S
\end{array}$$

- case 2: there exist $f_0 \in \mathcal{E}^{o\to o\to o\to o}, x_0, y_0, z_0 \in \mathcal{E}^o$ such that $\psi_3\ f_0\ x_0\ y_0\ z_0 = \bot$ and $f_0\ x_0\ y_0\ z_0 = \top$ (remark that if both (case 1) and (case 2) do not hold, then $\psi_3 > \phi$). Let $f' \in \mathcal{E}^{o\to o\to o\to o}$ be the function defined by

$$f'\ x\ y\ z = \begin{cases} \top \text{ if } x \geq x_0,\ y \geq y_0 \text{ and } z \geq z_0 \\ \bot \text{ otherwise} \end{cases}$$

  Remark that f' is definable and $f' \leq f_0$. Showing that $(f', f', f_0) \in S$ is trivial. We can now conclude:

$$\begin{array}{llll}
\psi_1 & \psi_2 & \psi_3 & \\
f' & f' & f_0 & \in S \\
x_0 & x_0 & x_0 & \in S \\
y_0 & y_0 & y_0 & \in S \\
z_0 & z_0 & z_0 & \in S \\
\hline
\top & \top & \bot & \notin S
\end{array}$$

□

We can summarize the results of this section by the following diagram, showing a fragment of the poset of degrees:

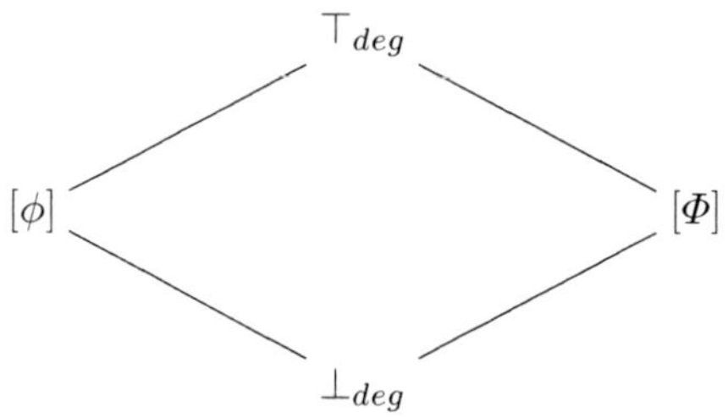

4 Standard Order-Extensional Models

In this section, we state a theorem about a fragment of the hierarchy of standard order-extensional models: all the models strictly greater than the bidomains model contain a weak version of the parallel or. We apply this result to UPCF, and give some clues about FPCF.

4.1 Bidomains

Definition 14. *We write $x \uparrow y$ if x and y are bounded. A dI-domain $(D, \sqsubseteq, \bot)$ is a Scott domain such that:*

- *Each compact element has finitely many lower bounds,*
- $\forall x, y, z \in D, y \uparrow z \Longrightarrow x \sqcap (y \sqcup z) = (x \sqcap y) \sqcup (x \sqcap z)$.

A stable function between two dI-domains X and Y is a Scott-continuous function f such that

$$\forall x, y \in X, x \uparrow y \Longrightarrow f(x \sqcap y) = f(x) \sqcap f(y)$$

Definition 15. *$(D, \leq, \sqsubseteq, \bot)$ is a bidomain if:*

- *$(D, \leq, \bot)$ is a Scott domain,*
- *$(D, \sqsubseteq, \bot)$ is a dI-domain,*
- *the identity between $(D, \leq, \bot)$ and $(D, \sqsubseteq, \bot)$ is continuous,*
- *if x and y are bounded in $(D, \sqsubseteq, \bot)$ then $x \wedge y = x \sqcap y$.*

Proposition 16. *The category of bidomains is cartesian closed.*

Proof. We define $D \Rightarrow D'$:

- $f \in D \Rightarrow D'$ if f is stable for $\sqsubseteq$ and continuous for $\leq$ and $\sqsubseteq$
- $f \leq g$ iff $\forall x \in D, fx \leq gx$
- $f \sqsubseteq g$ iff $\forall x, y \in D, x \sqsubseteq y \Longrightarrow fx = fy \sqcap gx$

□

Definition 17. *Let $t \in [\![\alpha_1 \to \alpha_2 \to \ldots \to o]\!]^{\mathcal{M}}$. $Tr(t)$ is the set of $((x_1, ..., x_n), y) \in ([\![\alpha_1]\!]^{\mathcal{M}} \times \ldots \times [\![\alpha_n]\!]^{\mathcal{M}}) \times [\![o]\!]^{\mathcal{M}}$ such that:*

- $y \neq \bot$
- $tx_1...x_n = y$,
- *For any $x'_i \in [\![\alpha_i]\!]^{\mathcal{M}}$, if $\forall i, x'_i \sqsubseteq x_i$ and $tx'_1...x'_n = y$, then $\forall i, x'_i = x_i$.*

Note that $Tr(t) \subseteq Tr(u)$ entails $t \sqsubseteq u$.

4.2 Bidomains and $\vee^-$

$\vee^-$ is a parallel function smaller than the usual parallel or:

Definition 18. *Let $\top \neq \bot$ be an element of the base domain. $\vee^-_\top$ is the function defined by:*

$$\vee^-_\top xy = \begin{cases} \top \text{ if } x = \top \text{ or } y = \top \\ \bot \text{ otherwise} \end{cases}$$

Note that if a model of FPCF contains $\vee^-_{\mathtt{t}}$ or $\vee^-_{\mathtt{f}}$, it contains them both. Thus we speak of $\vee^-$ meaning "any $\vee^-_{\mathtt{x}}$". For UPCF, $\vee^- = \underline{\lambda} xy.x \vee y$.

First, we state a useful lemma:

Lemma 19. *If domains of $\mathcal{M}$ and $\mathcal{B}$ are isomorphic for types smaller or equal to $\alpha \to \beta$, and $f : [\![\alpha]\!]^{\mathcal{M}} \to [\![\beta]\!]^{\mathcal{M}}$, is $\leq$-monotone and $\sqsubseteq$-stable, then $f \in [\![\alpha \to \beta]\!]^{\mathcal{M}}$.*

Proof. We write $R = R_{\mathcal{M},\mathcal{B}}$. If $a, b \in [\![\alpha]\!]^{\mathcal{M}}$, we define $a \sqsubseteq b \iff R(a) \sqsubseteq R(b)$

Let us now assume that f is $\leq$-continuous and $\sqsubseteq$-stable. Since the domains of $\mathcal{B}$ are isomorphic to the domains of $\mathcal{M}$, we can define a function $\hat{f}$ between domains of $\mathcal{B}$. $\hat{f}$ is also $\leq$-continuous and $\sqsubseteq$-stable: $\hat{f}$ is an element of $[\![\alpha \to \beta]\!]^{\mathcal{B}}$.

As $R^{\alpha\to\beta}$ is a bijection, we can use R^{-1}. By definition of $R^{\alpha\to\beta}$,

$$\forall x \in [\![\alpha]\!], (R^{-1}(\hat{f})x, \hat{f}R(x)) \in R^{\beta}$$

By definition of $\hat{f}$,

$$\forall x \in [\![\alpha]\!], (fx, \hat{f}R(x)) \in R^{\beta}$$

Since R^{β} is a bijection, $\forall x \in [\![\alpha]\!], fx = R^{-1}(f)x$. We get $f = R^{-1}(\hat{f})$, which means that $f \in [\![\alpha \to \beta]\!]^{\mathcal{M}}$. □

We can now proceed with the theorem:

Theorem 20. *If a standard order-extensional model is strictly greater than the bidomains model $\mathcal{B}$, then it contains $\vee^-$.*

Proof. Let $\mathcal{M}$ be a standard order-extensional model such that $\mathcal{B}$ is strictly smaller than $\mathcal{M}$. Let ω be a smallest type (for the type depth) such that $R_{\mathcal{M},\mathcal{B}}$ does not define an isomorphism between $[\![\omega]\!]^{\mathcal{M}}$ and $[\![\omega]\!]^{\mathcal{B}}$.

Since $\mathcal{B}$ is smaller than $\mathcal{M}$ and $[\![\omega]\!]^{\mathcal{M}} \neq [\![\omega]\!]^{\mathcal{B}}$, there is a $\phi \in [\![\omega]\!]^{\mathcal{M}}$ such that $\nexists \psi \in [\![\omega]\!]^{\mathcal{B}}, \phi R_{\mathcal{M},\mathcal{B}} \psi$. As the argument and result domains are isomorphic, we can write $\phi \notin [\![\sigma \to \tau]\!]^{\mathcal{B}}$. Of course, $\omega \neq o$. Let us write $\omega = \sigma \to \tau$. Since ϕ is continuous for $\leq$, we have that ϕ is not stable for $\sqsubseteq$. Either ϕ is not monotone for $\sqsubseteq$, or there exist f and g bounded in $([\![\sigma]\!], \sqsubseteq)$ such that $\phi(f \wedge g) \neq \phi f \wedge \phi g$. We prove that $\vee^-$ is definable in each case.

Let us assume that ϕ is not $\sqsubseteq$-monotone. We choose f and g such that $f \sqsubseteq g$ and $\phi f \not\sqsubseteq \phi g$. If $\tau = o$, we have $\phi f = \top$ and $\phi g = \bot$. But, since $f \sqsubseteq g$ entails $f \leq g$, ϕ is not $\leq$-monotone. Thus, τ is functional: let us write $\tau = \tau_1 \to .. \to \tau_n \to o$.

As $\phi f \not\sqsubseteq \phi g$, $Tr(\phi f) \not\subseteq Tr(\phi g)$. We choose $((y_1 ... y_n), \top) \in Tr(\phi f) \backslash Tr(\phi g)$ (where $\top$ can be any element in $[\![o]\!]$ except $\bot$) Since ϕ is $\leq$-monotone, $f \leq g$, and $\phi f y_1 ... y_n = \top$, $\phi g y_1 ... y_n = \top$. Therefore, there is a $((x_1 ... x_n), \top) \in Tr(\phi g)$ such that $\forall i, x_i \sqsubseteq y_i$. Since $((y_1, ..., y_n), \top) \notin Tr(\phi g)$, there is an i such that $x_i \sqsubset y_i$, which entails $\phi f x_1 ... x_n = \bot$. We have chosen f, g, x_i and y_i such that:

- $\phi f x_1 ... x_n = \bot$,
- $\phi g x_1 ... x_n = \top$,
- $\phi f y_1 ... y_n = \top$.

As $f \sqsubseteq g$, and by lemma 19, the function $m : o \to \sigma$ defined by $m\bot = f$ and $m\top = g$ is in $\mathcal{M}$. As $x_i \sqsubseteq y_i$, and by lemma 19, the function $\chi_i : o \to \sigma$ defined by $\chi_i\bot = x_i$ and $\chi_i\top = y_i$ is in $\mathcal{M}$. One easily checks that

$$\underline{\lambda} ab.\phi(\psi a)(\chi_1 b)...(\chi_n b) = \vee_{\top}^{-}$$

Let us now assume that ϕ is $\sqsubseteq$-monotone, and that there exist f, g, h such that $f \sqsubseteq h$, $g \sqsubseteq h$, and $\phi(f \sqcap g) \neq (\phi f) \sqcap (\phi g)$. Since ϕ is $\sqsubseteq$-monotone, $\phi(f \sqcap g) \sqsubseteq (\phi f) \sqcap (\phi g)$, and we get $\phi(f \sqcap g) \sqsubset (\phi f) \sqcap (\phi g)$. This yields:

$$\phi(f \sqcap g) < (\phi f) \sqcap (\phi g)$$

We can choose $x_1...x_n$ (where n can be 0) and $\top \in [\![o]\!]$ ($\top \neq \bot$) such that $\phi(f \sqcap g)x_1...x_n = \bot$ and $((\phi f) \sqcap (\phi g))x_1...x_n = \top$. We can define a stable function ψ by:

- $\psi\bot\bot = f \sqcap g$,
- $\psi\top\bot = f$,
- $\psi\bot\top = g$,
- $\psi\top\top = h$.

Being $\leq$-continuous and $\sqsubseteq$-stable, ψ is in $\mathcal{M}$. One easily checks that

$$\underline{\lambda} ab.\phi(\psi ab)x_1...x_n = \vee_{\top}^{-}$$

□

4.3 Unary PCF

We apply the theorem to the case of UPCF.

Proposition 21. *Any standard order-extensional model of UPCF is smaller than the Scott model $\mathcal{E}$.*

Proof. Let $\mathcal{M}$ be a standard order-extensional model. We write R for $R_{\mathcal{E},\mathcal{M}}$.

At type o, R is a bijection. Assume that R^σ and R^τ are functional and onto, and let us prove that $R^{\sigma\to\tau}$ is also functional and onto. Seeing R as a partial function, we write $R(x)$ for the only y such that xRy. R^{-1} is the function defined in lemma 8 at types σ and τ.

Let $f \in [\![\sigma \to \tau]\!]^{\mathcal{E}}$, $g, g' \in [\![\sigma \to \tau]\!]^{\mathcal{M}}$ such that $fR^{\sigma\to\tau}g$ and $fR^{\sigma\to\tau}g'$. Let $x \in [\![\sigma]\!]^{\mathcal{M}}$. By definition of $R^{\sigma\to\tau}$,

$$gx = g(R(R^{-1}(x))) = R(f(R^{-1}(x)) = g'(R(R^{-1}(x))) = g'x$$

Thus, $R^{\sigma\to\tau}$ is functional.

Let $g \in [\![\sigma \to \tau]\!]^{\mathcal{M}}$. If x has an image by R, we define $\hat{f}(x) = R^{-1}(g(R(x)))$. Since the domains of $\mathcal{E}$ are lattices[2], we can define

$$f(x) = \bigvee_{R^{-1}(y) \leq x} \hat{f}(R^{-1}(y))$$

The monotonicity of R and R^{-1} yields the monotonicity of f. If $x \in [\![\sigma]\!]^{\mathcal{E}}$, $R(fx) = g(R(x))$. As $R^\sigma = \{(x, R(x))\}$, $f R^{\sigma\to\tau} g$: $R^{\sigma\to\tau}$ is onto. □

Theorem 22. *The model $\mathcal{B}$ of bidomains of UPCF is fully abstract.*

Proof. See[5] □

The following is an easy consequence of lemma 10:

Lemma 23. *If $\vee^-$ is in $\mathcal{M}$, then $\mathcal{M} \approx \mathcal{E}$.*

Corollary 24. *There are only two standard order-extensional models of UPCF: $\mathcal{B}$ and $\mathcal{E}$.*

Proof. Combine these results with theorem 20. □

4.4 Finitary PCF

We show that there are infinitely many degrees above $[\vee^-]$ and infinitely many models smaller than $\mathcal{E}$ that contain $\vee^-$.

Let us call $\vee^n$ the function of type $o^n \to o$ defined by:

$$\vee^n x_1 \ldots x_n = \begin{cases} t & \text{if } n \text{ of } n-1 \text{ of the } x_i \text{ are equal to } t \\ f & \text{if } x_1 = \ldots = x_n = f \\ \bot & \text{otherwise} \end{cases}$$

We write $x \wedge y = if\ x\ then\ (if\ y\ then\ t\ else\ \bot)\ else\ (if\ y\ then\ \bot\ else\ f)$. One easily checks that $\vee^n = \lambda x_1 ... x_n.\ \vee^{n-1} (x_1 \wedge x_2) x_3 ... x_n$, which entails $\vee^n \preceq \vee^{n-1}$. Let us prove that $\vee^n \prec \vee^{n-1}$, with the n-ary relation S^n defined by

$$S^n = S^n_{\{1\ldots n\},\{1\ldots n\}} \cap \left(\bigcap_{A \subseteq \{1\ldots n-1\}} S^n_{A,A} \right)$$

As one can check, the tuples $(x_1, ..., x_n)$ in S^n are exactly such that:

- there is no $i, j < n$ such that $x_i = t$ and $x_j = f$
- the tuple is not $tt...tf$ nor $ff...ft$.

First, $\vee^n$ is not invariant by S^{n+1} as shown by:

[2] Apart from this, the proof would be valid for FPCF

$$\begin{array}{cccccc}
\vee^n & \vee^n & \vee^n & \dots & \vee^n & \vee^n \\
\bot & t & t & \dots & t & f \\
t & \bot & t & & t & f \\
t & t & \bot & & t & f \\
\vdots & & & \ddots & \vdots & \vdots \\
t & t & t & \dots & \bot & f \\
\hline
t & t & t & \dots & t & f
\end{array}$$

Let us now prove that $\vee^n$ is invariant by S^n. Then we conclude with the fundamental lemma of logical relations. Let $x_{i,j}$ and y_i such that:

$$\begin{array}{ccc}
\vee^n & \dots & \vee^n \\
x_{11} & \dots & x_{n1} \\
\vdots & & \vdots \\
x_{1n} & \dots & x_{nn} \\
\hline
y_1 & \dots & y_n
\end{array}$$

and $(y_1, ..., y_n) \notin S^n$.

- If there exist $i, i' < n$ such that $y_i = t$ and $y_{i'} = f$ then for all j but one, $x_{ij} = t$ and $x_{i'j} = f$, which entails $(x_{1j} ... x_{nj}) \notin S^n$.
- If $\forall i < n, y_i = t$ and $y_n = f$, one can check that there exists a j_0 such that $\forall i < n, x_{ij_0} = t$. Since for all j, $x_{nj} = f$, $(x_{1j_0}, ..., x_{nj_0}) \notin S^n$.
- If $\forall i < n, y_i = f$ and $y_n = t$, for all j but one that $\forall i < n, x_{ij} = f$ and $x_{nj} = t$, which entails $(x_{1j}, ..., x_{nj}) \notin S^n$.

Note that for $n \geq 3$, $\vee^- = \underline{\lambda} xy. \vee^n tt \dots txy$, hence we have defined a sequence of undefinable elements of $\mathcal{E}$:

$$\vee^- \prec ... \prec \vee^4 \prec \vee^3 \prec \vee^2 = \vee$$

These elements being first-order functions, they cannot vanish with an extensional collapse. Thus, we have defined an infinite hierarchy of standard order extensional models of FPCF:

$$\dots \subset \mathcal{E}^4 \subset \mathcal{E}^3 \subset \mathcal{E}$$

These models are not greater than $\mathcal{B}$, but there might be corresponding models above $\mathcal{B}$.

5 Conclusion

We have shown that the poset of degrees of parallelism in the Scott model of UPCF is non-trivial, and that the poset of extensional models of the language is reduced to the fully abstract and the Scott ones.

Some open questions arise naturally:

- Decidability of the definability (and relative definability) problem in the Scott model of UPCF.
- Existence of infinitely many degrees of relative definability in the Scott models of UPCF.

A broader framework for this work is the study of the three related issues below[3] for a given extensional, finitary applied λ-calculus L, in order to explore the boundary decidable/undecidable with respect to the set of constants of L.

(a) Decidability of the definability (and relative definability) problem in the Scott model.
(b) Existence of a (finitary and "non-syntactic") fully abstract model.
(c) Decidability of the observational equivalence.

For FPCF, we know that (c), and hence (a) and (b), are false [9].

For UPCF, (b) is true [5] and (a) open.

For the simply typed λ-calculus without constants (replacing "Scott model" with "full set-theoretic model") (c) is true [12], (a) false [7,3], (b) open.

For finitary, parallel PCF, (a) is true [13].

References

1. A. Bucciarelli. *Degrees of Parallelism in the Continuous Type Hierarchy.* Theoretical Computer Science 177:59–71, 1997.
2. T. Ehrhard. *A Relative Definability Result for Strongly Stable Functions and Some Corollaries.* Information and Computation, Academic Press, 152:111–137, 1999.
3. T. Joly *Encoding of the halting problem into the Monster type and applications.* This volume.
4. A. Jung, A. Stoughton. *Studying the Fully Abstract Model of PCF Within its Continuous Function Model.* in Typed Lambda-Calculi and Applications, Eds. M. Bezem and J.F. Groote, LNCS 664, Springer, 1993.
5. J. Laird. *A Fully Abstract Bidomain Model of Unary FPC.* Unpublished notes, available at www.cogs.susx.ac.uk/users/jiml. 2002.
6. J. Laird. Personal communication. 2002.
7. R. Loader. *The undecidability of lambda-definability.* in Logic, Meaning and Computation, Essays in Memory of Alonzo Church, KluwerAcademic Publishers, 2001.
8. R. Loader. *Unary PCF is decidable.* Theoretical Computer Science 206, 1998.
9. R. Loader. *Finitary PCF is not decidable.* Theoretical Computer Science 266, 2001.
10. J. Longley. *The sequentially realizable functionals.* To appear in Annals of Pure and Applied Logic.
11. J. Longley. *When is a functional program not a functional program?* Proceedings of Fourth ACM SIGPLAN International Conference on Functional Programming, Paris, September 1999. ACM Press.
12. V. Padovani. *Decidability of all minimal models.* in M. Coppo and S. Berardi eds, Types for Proofs and Programs, LNCS 1158, Springer 1996.
13. G. Plotkin. *LCF considerer as a programming language.* Theoretical Computer Science 5, 1977.

[3] The second one is stated informally; (c) is weaker than (b) and (b) weaker than (a).

14. K. Sieber. *Reasoning about Sequential Functions via Logical Relations.* Proc. LMS Symposium on Applications of Categories in Computer Science, M. Fourman, P. Johnstone, A. Pitts eds, LMS Lecture Note Series 177, Cambridge University Press, 1992.
15. M. Schmidt-Schauss. *Decidability of Behavioral Equivalence in Unary PCF.* Theoretical Computer Science 216:363–373, 1999.

Principal Typing in Elementary Affine Logic*

Paolo Coppola[1] and Simona Ronchi della Rocca[2]

[1] Dipartimento di Matematica e Informatica – Università di Udine
via delle Scienze 206, I-33100 Udine, Italy
coppola@dimi.uniud.it
[2] Dipartimento di Informatica – Università di Torino
Corso Svizzera 185, I-10149 Torino, Italy
ronchi@di.unito.it

Abstract. Elementary Affine Logic (EAL) is a variant of the Linear Logic characterizing the computational power of the elementary bounded Turing machines. The EAL Type Inference problem is the problem of automatically assigning to terms of λ-calculus EAL formulas as types. This problem is proved to be decidable, and an algorithm is showed, building, for every λ-term, either a negative answer or a finite set of *type schemata*, from which all and only its typings can be derived, through suitable operations.

Introduction

Starting from a seminal idea of Girard, who designed the first variant of Linear Logic normalizable in polynomial time, the Light Linear Logic [10], some logical systems have been proposed in the literature, characterizing different complexity classes. In this paper we deal with Elementary Affine Logic (EAL), which characterizes the computational power of the elementary bounded Turing machines: this result has been proved in [9] for its second order version and moreover it has been extended to the propositional fragment in [1]. In [4] a language has been designed, Λ^{EA}, starting from the implicational fragment of Elementary Affine Logic. The interest of such a language deals not only in its computational bound, but in the fact that EAL-typeable terms can be reduced with the *abstract* subset of Lamping's optimal reduction algorithm [11] obtaining excellent performances.

Λ^{EA} is untyped and it can be assigned types which are formulas of EAL through a type assignment system in natural deduction style (NEAL), proving statements (typings) of the shape $\Gamma \vdash_{\text{NEAL}} M : A$, where the context Γ assigns types to variables. However the language, as almost all designed in this way, is quite complex, both as syntax and as reduction rules. Indeed its syntax reflects the inference rules, and its reduction rules reflect all the normalization steps, which are quite involved. So it cannot be proposed as paradigm for a real programming language. Then it would be interesting to have a decidable procedure

* Paper partially supported by MIUR-cofin-2000 "Linear Logic and Beyond",EC-TMR project "Linear Logic",IST-2001-33477 DART Project,MIUR-cofin-2001 COMETA

M. Hofmann (Ed.): TLCA 2003, LNCS 2701, pp. 90–104, 2003.

for assigning EAL formulas as types to terms of the standard λ-calculus: so the user can write a program in the usual style, and the compiler automatically will check its typability, assuring, in case of positive result, its complexity bound and enabling it to be reduced with the abstract Lamping's algorithm. In this paper we define such a procedure.

The problem is not easy, since there is not a type inference system for the λ- calculus. But it is easy to define a correspondence between Λ^{EA} and the λ-calculus, which preserves the operational behavior of terms, just by performing the substitutions that are implicit in terms of Λ^{EA}. Then the typability of λ-terms can be defined in the following way: a term $M \in \Lambda$ can have a typing $\Gamma \vdash_{\mathrm{NEAL}} M : A$ if and only if there is a term M' of Λ^{EA} such that it corresponds to M, and $\Gamma \vdash_{\mathrm{NEAL}} M' : A$. But in order to reach our aim, which is to check typability in order to reduce λ-terms with the Lamping's abstract algorithm, we need to restrict the previous definition, by taking into account only *simple* terms of Λ^{EA}, i.e., terms whose proofnets correspond to those obtained as initial translation of λ-terms. But it is important to point out that the result of this paper would remain true even without this restriction. The number of terms of Λ^{EA} corresponding to a λ-term is in general infinite (thanks to the possibility of adding boxes just replacing variables by variables), and so at a first look the problem of typability for the λ-calculus seems at most semi-decidable.

In [4] a procedure has been designed for assigning to a λ-term, that can be typed in the Curry's type assignment system [5,6,7], a type of EAL: the procedure takes as input a pair of a term and a Curry's type that can be assigned to it, and produces as output either a negative result or a typing, assigning to the term itself an EAL type related to the Curry type in input, in the sense that it expresses the same functionality behavior.

In this paper we give a stronger solution of the problem, by designing a modular type inference procedure for λ-calculus. The procedure is given in three steps. First we introduce the language of abstract EA-terms, terms in a meta-language, to which EAL types can be assigned through a suitable type assignment system. Each abstract EA-term represents an infinite set of EA-terms, in such a way that it can be assigned all and only the typings of the EA-terms in the set. We prove that all typing information of the infinite set of EA-term corresponding to a given λ-term can be codified in a finite set of *canonical form*, i.e., abstract EA-terms in normal form with respect to a suitable reduction relation preserving typings. *Canonical forms* give information on the box structure of the typing. Such structure was only implicit in [4]. Moreover we prove that typing for abstract EA-terms is decidable, in the sense that every abstract EA-term has a principal *typing scheme*, from which all and only the typings for it can be derived, through suitable operations. So we can deduce that a λ-term has all and only the typings of the canonical forms corresponding to it, and, being this set finite, it has a finite number of principal typings. All the proofs are constructive, in the sense that we show two algorithms, the first one building, for every λ-term, all and only the canonical forms corresponding to it, and the second one giving, for every abstract EA-term, either a negative answer or its principal typing. The notion of principal typing we found is not standard. Indeed, the operations that allow to obtain types from schemata are substitutions under some *constraints*, represented by linear equations.

Another example of type derivation for the λ-calculus, in a bounded complexity logic, is that one of Baillot [2] and [3], assigning types in Light Affine Logic to λ-terms. Here the main problem is to deal with two different modal operators. Related works are in [12,8], in the field of linear decorations of intuitionistic derivations.

The paper is organized as follows. In Section 1 the λ^{EA}-calculus and the type assignment system $\vdash_{\text{NEAL}}$ are recalled. Section 2 presents the abstract EA-terms, their canonical forms, and their relation with Λ. In Section 3 the principal typing property for abstract EA-terms is proved. In Section 4 the principal typing property for the λ-calculus is stated. The appendix contains, in its section A.1, the algorithm building the set of canonical forms corresponding to a given λ-term, and in the section A.2 the principal typing algorithm.

1 The λ^{EA}-Calculus

In this section we will briefly recall the λ^{EA}-calculus, as defined in [4], its typing rules, its relation with the λ-calculus, and the problem of typability of λ-calculus.

Definition 1. *i) The set of terms of the λ^{EA}-calculus (Λ^{EA}) is generated by the grammar: $M ::= x \mid \lambda x.M \mid (M\ M) \mid !(M)\left[{}^{M}/x, \dots, {}^{M}/x\right] \mid [M]_{M=x,y}$ where all variables occur at most once. Λ^{EA} is ranged over by M, N, P, Q.*

ii) EA-types *are formulas of Elementary Affine Logic, and they are generated by the grammar $A ::= a \mid A \multimap A \mid !A$ where a belongs to a countable set of basic type constants. EA-types will be ranged over by A, B, C.*

iii) The type assignment system $\vdash_{\text{NEAL}}$ assigns EA-types to EA-terms, starting from a context, assigning EA-types to variables. The rules of the system are given in Table 1.

Table 1. Type assignment system for EA-terms.

$$\frac{}{\Gamma, x : A \vdash_{\text{NEAL}} x : A}\ ax \qquad \frac{\Gamma \vdash_{\text{NEAL}} M : !A \quad \Delta, x : !A, y : !A \vdash_{\text{NEAL}} N : B}{\Gamma, \Delta \vdash_{\text{NEAL}} [N]_{M=x,y} : B}\ contr$$

$$\frac{\Gamma, x : A \vdash_{\text{NEAL}} M : B}{\Gamma \vdash_{\text{NEAL}} \lambda x.M : A \multimap B}\ (\multimap I) \qquad \frac{\Gamma \vdash_{\text{NEAL}} M : A \multimap B \quad \Delta \vdash_{\text{NEAL}} N : A}{\Gamma, \Delta \vdash_{\text{NEAL}} (M\ N) : B}\ (\multimap E)$$

$$\frac{\Delta_1 \vdash_{\text{NEAL}} M_1 : !A_1 \quad \cdots \quad \Delta_n \vdash_{\text{NEAL}} M_n : !A_n \quad x_1 : A_1, \dots, x_n : A_n \vdash_{\text{NEAL}} N : B}{\Gamma, \Delta_1, \dots, \Delta_n \vdash_{\text{NEAL}} !(N)\left[{}^{M_1}/x_1, \dots, {}^{M_n}/x_n\right] : !B}\ !$$

The following definition relates terms of Λ^{EA} with the terms of the classical λ-calculus.

Definition 2. *i) The set of terms of the λ-calculus (Λ) are defined by the grammar $M ::= x \mid MM \mid \lambda x.M$.*
By abuse of notation, Λ will be ranged over by M, N, P, Q, as Λ^{EA}, being the different meaning clear from the context.
ii) The translation function $()^* : \Lambda^{EA} \to \Lambda$ is defined by induction on the structure of the EA-term as follows:

$$(x)^* = x \qquad (\lambda x.M)^* = \lambda x.M^* \qquad ((M_1\ M_2))^* = (M_1^*\ M_2^*)$$
$$(!(M)[^{M_1}/x_1, \cdots, ^{M_n}/x_n])^* = M^*\{M_1^*/x_1, \ldots, M_n^*/x_n\}$$
$$([M]_{N=x_1,x_2})^* = M^*\{N^*/x_1, N^*/x_2\}$$

In order to extend to Λ the notion of type assignment with EA-types, let us introduce a particular class of EA-terms, the simple ones, which correspond, if translated into proofnets, to contracting at most variables. We discard EA-terms contracting non trivial subterms otherwise we loose the possibility of reducing lambda terms with the abstract Lamping's algorithm.

Definition 3. *i) The* length *$L(M)$ of an EA-term M is defined inductively as follows:*
$L(x) = 0;\quad L(\lambda x.M) = 1 + L(M);\quad L((M\ N)) = 1 + L(M) + L(N);$
$L(!\,(M)\,[^{M_1}/x_1, \ldots, ^{M_n}/x_n]) = L(M) + \sum_{i=1}^{n} L(M_i);$
$L([M]_{N=x,y}) = L(M) + L(N).$
ii) An EA-term M is simple *if and only if*
1. M has no subterm of the form $[M_1]_{M_2=x,y}$ where $(M_2)^*$ is not a variable,
2. $L(M) = L((M)^*)$

Condition ii)2. is not redundant. Indeed $M \equiv !([(x_1(x_2y))]_{x=x_1,x_2})[^{(z\ w)}/x, ^k/y]$ does not violate condition ii)1. but $L(M) = L((M)^*) - 1$.

Definition 4. *Let $M \in \Lambda$ and let $EA_M = \{N \mid N \in \Lambda^{EA}$, N is simple and $(N)^* = M\}$. Then $\Gamma \vdash_{\mathrm{NEAL}} M : A$ if and only if $\Gamma \vdash_{\mathrm{NEAL}} P : A$, for some $P \in EA_M$. EA_M is the set of EA-terms* corresponding *to M.*

Remark 1. It can be easily checked that the condition to be simple is not preserved under reduction. Consequently, the notion of typability we give for the λ-calculus does not enjoy the subject reduction property. But this lack does not invalidate our result. As a matter of fact, we want to use the notion of typability just as an initial check: in fact, if a term can be typed, then it can be evaluated with the abstract Lamping's algorithm. Since the algorithm reduces untyped terms, subject reduction is not needed. Finally notice that if we relax the constraint about simple terms it is no more assured that the term can be reduced with the abstract Lamping's algorithm.

The cardinality of EA_M is always infinite, for all $M \in \Lambda$, thanks to the possibility of replacing variables by variables in (!) subterms. In the following example, an infinite set of terms of Λ^{EA}, all corresponding to the λ-term $\lambda xy.(x(xy))$, is showed. Note that the replacement of variables by variables in (!) subterms cannot be avoided, since it is necessary for writing EA-terms corresponding to non linear λ-terms.

Example 1. For every $n \geq 0$, the EA-term 2_n, defined as:

$$\lambda x.[\lambda y.\underbrace{!(!(...!}_{n+1}(x_3(x_4 y_1))[^{x_5}/x_3, {}^{x_6}/x_4, {}^{y_2}/y_1]...)[^{x_1}/x_{2n-1}, {}^{x_2}/x_{2n}, {}^{y}/y_n]]_{x=x_1,x_2}$$

belongs to $EA_{\lambda xy.(x(xy))}$.

So, at a first look, the problem of EA-typability for the λ-calculus seems to be semi-decidable. But we will see in the following Section that for every M, EA_M has a finite representation.

2 Abstract EA-Terms and Canonical Forms

In this section the notion of abstract EA-terms, and a type assignment for assigning EA-types to them are introduced. The notion of abstract EA-term allows for a finite representation of the set EA_M, for every $M \in \Lambda$. In fact, an abstract EA-term M describes an infinite set of EA-terms, which can differ from each other in the number of nested occurrences of both the ! constructor and the contraction constructor. Moreover a reduction relation on abstract EA-terms is defined, which allows to choose a representative of each equivalence class of abstract EA-terms, in such a way that the typings of the terms belonging to the set EA_M, for every M, are all and only the typings of a finite set of abstract EA-terms in normal forms (called canonical forms).

Definition 5. *The set* Abs^{EA} *of* abstract EA-terms *is generated by the following grammar:*

$$M ::= x \mid \lambda x.M \mid (M\ M) \mid\ [M]_{N_1 \to (\widehat{x_1}),\dots,N_k \to (\widehat{x_k})} \mid\ \nabla(M)[^{N_1}/x_1, \cdots, {}^{N_k}/x_n]$$

with the condition that every variable occurs at most once in a term.

The type assignment system assigning formulas of EAL to terms of Abs^{EA} is shown in Table 2. Note that the main differences between this system and that one for Λ^{EA} are that the rule $(!^m)$ is parametric, in the sense that it can assign a type having any number $m > 0$ of modalities, and the rule (contr) allows for contracting simultaneously any set of different variables.

The relation between abstract EA-terms and EA-terms is expressed by the two functions $emb : \Lambda^{EA} \to Abs^{EA}$ and $bme : Abs^{EA} \to \mathrm{P}(\Lambda^{EA})$ defined in the following.

$$\begin{aligned}
emb(x) &= x \\
emb(\lambda x.M) &= \lambda x.emb(M) \\
emb((M_1\ M_2)) &= (emb(M_1)\ emb(M_2)) \\
emb([M]_{N=x,y}) &= [emb(M)]_{emb(N) \to (x,y)} \\
emb(!(M)[^{N_1}/x_1, \cdots, {}^{N_k}/x_k]) &= \nabla(emb(M))[^{emb(N_1)}/x_1, \cdots, {}^{emb(N_k)}/x_k]
\end{aligned}$$

Table 2. Type assignment system for Abs^{EA}-terms.

$$\frac{}{\Gamma, x : A \vdash_{abs} x : A}\ ax$$

$$\frac{\Gamma, x : A \vdash_{abs} M : B}{\Gamma \vdash_{abs} \lambda x.M : A \multimap B}\ (\multimap I) \qquad \frac{\Gamma \vdash_{abs} M : A \multimap B \quad \Delta \vdash_{abs} N : A}{\Gamma, \Delta \vdash_{abs} M\ N : B}\ (\multimap E)$$

$$\frac{\begin{array}{ll} \Gamma_1 \vdash_{abs} N_1 :!A_1 & x_1^1 :!A_1, \ldots, x_{n_1}^1 :!A_1, \\ \vdots & \vdots \\ \Gamma_k \vdash_{abs} N_k :!A_k & \Delta, x_1^k :!A_k, \ldots, x_{n_k}^k :!A_k \vdash_{abs} M : B \end{array}}{\Gamma_1, \ldots, \Gamma_k, \Delta \vdash_{abs} [M]_{N_1 \blacksquare\ (x_1^1, \cdots, x_{n_1}^1), \ldots, N_k \blacksquare\ (x_1^k, \cdots, x_{n_k}^k)} : B}\ contr$$

$$\frac{\begin{array}{l} \Delta_1 \vdash_{abs} N_1 : \overbrace{!...!}^{m} A_1 \\ \vdots \\ \Delta_k \vdash_{abs} N_k : \overbrace{!...!}^{m} A_k \qquad x_1 : A_1, \ldots, x_k : A_k \vdash_{abs} M : B \quad m > 0 \end{array}}{\Gamma, \Delta_1, \ldots, \Delta_k \vdash_{abs} \nabla(M)[^{N_1}/x_1, \cdots, {}^{N_k}/x_n] : \underbrace{!...!}_{m} B}\ !^m$$

$$\begin{aligned} bme(x) &= \{x\} \\ bme(\lambda x.M) &= \{\lambda x.M' | M' \in bme(M)\} \\ bme((M_1\ M_2) &= \{(M_1'\ M_2') | M_1' \in bme(M_1) \wedge M_2' \in bme(M_2)\} \end{aligned}$$

$$bme\big(\nabla(M)[^{N_1}/x_1, \cdots {}^{N_k}/x_k]\big) = \{\overbrace{!(\cdots !(}^{n} M')[z_1^1/x_1, \ldots, z_k^1/x_k]\overbrace{)\cdots)}^{n-1}[^{N_1'}/z_1^n, \ldots$$
$$\ldots, {}^{N_k'}/z_k^n]\ |M' \in bme(M) \wedge N_i' \in bme(N_i) \wedge n > 0\}$$

$$bme([M]_{N \to (x_1^1, x_1^2, \ldots, x_1^{n_1}), \ldots}) = bme\left(\left[[M]_{y \to (x_1^1, x_1^2)}\right]_{N \to (y, x_1^3, \ldots, x_1^{n_1}), \cdots}\right)$$

y fresh variable

$$bme([M]_{N_1 \to (x_1^1, x_1^2), N_2 \to (\widehat{x_2}), \ldots}) = bme\left(\left[[M]_{N_1 \to (x_1^1, x_1^2)}\right]_{N_2 \to (\widehat{x_2}), \cdots}\right)$$

$$\begin{aligned} bme([M]_{N \to (x_1, x_2)}) &= \{[M']_{N' = x_1, x_2}\ |M' \in bme(M) \wedge N' \in bme(N)\} \\ bme([M]_{N \to (x)}) &= bme(M\{^N/x\}). \end{aligned}$$

The following lemma holds:

Lemma 1. *i)* $\forall M \in \Lambda^{EA} \quad \Gamma \vdash_{\mathrm{NEAL}} M : A \Rightarrow \Gamma \vdash_{abs} emb(M) : A$.
ii) $\forall M \in Abs^{EA}$

$$\Gamma \vdash_{abs} M : A \Rightarrow \exists M' \in bme(M)\ \Gamma \vdash_{\mathrm{NEAL}} M' : A.$$

Example 2. $bme(\lambda x.[\lambda y.(\nabla(x_3(x_4y_1)))[x_1/x_3, x_2/x_4, y/y_1]]_{x\to x_1,x_2}) = \{2_n \mid n \geq 1\}$, where 2_n is defined in Example 1.

The reduction relation $\to_{\mathsf{Can}}$, defined in Figure 1, is the operational counterpart of some commuting conversions on proofs in NEAL. In fact, let us consider the two abstract EA-terms:

$$X \equiv \lambda xy.[[(yx_1x_2x_3)]_{z\to(x_1,x_2)}]_{x\to(z,x_3)} \text{ and } Y \equiv \lambda xy.[(yx_1x_2x_3)]_{x\to(x_1,x_2,x_3)}.$$

X and Y represent the same set of EA-terms, and they have the same types. Indeed, type derivations for X and Y differ just in the fact that the contraction of x_1, x_2, x_3 into x is made in two steps for X and in one step for Y. So we choose Y as representative of both terms, and we put $X \to_{\mathsf{Can}} Y$. Moreover, let consider the two terms:

$$X^{\square} \equiv \nabla(\nabla(M)[^{y_1}/x_1, ..., ^{y_k}/x_k])[^{N_1}/y_1, ..., ^{N_k}/y_k] \text{ and } Y^{\square} \equiv \nabla(M)[^{N_1}/x_1, ..., ^{N_k}/x_k].$$

$bme(X') \subseteq bme(Y')$, and for every type derivation for X' there is a type derivation for Y', both deriving a type of the shape $\underbrace{!...!}_{m} A$, for some A. The two derivations differ just in the fact that, while the m modalities are introduced for Y in one application of the rule $!^m$, the derivation for X uses two applications of the same rule. So we put $X' \to_{\mathsf{Can}} Y'$. The other rules composing $\to_{\mathsf{Can}}$ are motivated by similar reasonings.

It is easy to prove that $\to_{\mathsf{Can}}$ is confluent.

Lemma 2 (Subject reduction). *Let $M \in Abs^{EA}$ and $M \to^*_{\mathsf{Can}} N$, then*

$$\Gamma \vdash_{abs} M : A \Rightarrow \Gamma \vdash_{abs} N : A.$$

Proof. By induction on the length of the reduction.

Let us call *canonical forms* the EA-terms in normal forms with respect to the reduction $\to_{\mathsf{Can}}$. Notice that the normal form of a simple EA-term is simple too.

Lemma 3. *For every $M \in \Lambda$, let $C(M) = \{N \mid N$ is canonic and simple and $(N)^* = M\}$. Then $B \vdash_{\mathrm{NEAL}} M : \sigma$ if and only if $B \vdash_{abs} P : \sigma$, for some $P \in C(M)$.*

Proof. (only if) $B \vdash_{\mathrm{NEAL}} M : \sigma$ implies, by definition, $B \vdash_{\mathrm{NEAL}} N : \sigma$, for some $N \in EA_M$, which in its turn implies, by Lemma 1.i), $B \vdash_{abs} emb(N) : \sigma$. So, if P is the canonical form of $emb(N)$, $B \vdash_{abs} P : \sigma$, by Lemma 2. By definition, $P \in C(M)$.

(if) $B \vdash_{abs} P : \sigma$, for some $P \in C(M)$, implies, by Lemma 1.ii), there is $N \in bme(P)$ such that $B \vdash_{\mathrm{NEAL}} N : \sigma$. Note that N is simple by definition of $C(M)$. So $B \vdash_{\mathrm{NEAL}} (N)^* : \sigma$, and, since $M = (N)^*$, the proof is given.

The following lemma proves the key property for the decidability of EA-typing for λ-terms.

Lemma 4. *For every $M \in \Lambda$, $C(M)$ is finite.*

The proof is constructive, since we will show an algorithm $\mathcal{C}$ such that, for every $M \in \Lambda$, $\mathcal{C}(M)$ build $C(M)$, and we prove that $\mathcal{C}$ is correct and complete. All this stuff is showed in the Appendix A.1.

3 Principal Typing for $\vdash_{abs}$

In this section we will prove that $\vdash_{abs}$ enjoys the principal typing property, i.e., every typing can be derived from a particular typing schemata by means of suitable operations. First of all, let us introduce the notion of *type scheme.*

Definition 6. *i)* Type schemata *are defined by the following grammar:*

$$\sigma ::= \alpha \mid \sigma \multimap \sigma \mid !^{p}(\sigma)$$

where the exponential p *is defined by the following grammar:*

$$p ::= n \mid p + p$$

where α belongs to a countable set of scheme variables, ranged over by α, β, γ, and n belongs to a countable set of literals; type schemata are ranged over by σ, τ, ρ and exponentials are ranged over by p, q, r. Let T denote the set of type schemata.

ii) *A* scheme substitution *is a function from type schemata to types, denoted by a pair of substitutions $< S, X >$, where S replaces scheme variables by types and X replaces literals by natural numbers. The application of $< S, X >$ to a type scheme is defined inductively as follows:*

$$< S, X > (\alpha) = S(\alpha);$$

$$< S, X > (\sigma \multimap \tau) =< S, X > (\sigma) \multimap < S, X > (\tau);$$

$$< S, X > (!^{n_1 + \ldots + n_i}\sigma) = \underbrace{!\ldots!}_{q} < S, X > (\sigma),$$

where $q = X(n_1) + \ldots + X(n_i)$.

$\equiv$ denotes the syntactical identity between both types and type schemes.

In order to define the principal typing, we need a unification algorithm for type schemes. But first some technical definitions are necessary.

First we introduce a function F collapsing all consecutive $!^{p}$ in a type scheme:

$$F(\alpha) = \alpha;\ F(\sigma \multimap \tau) = F(\sigma) \multimap F(\tau);$$

$$F(!^{p}(\sigma)) =!^{p}F(\sigma) \text{ if } F(\sigma) \text{ is not of the form } !^{q}\tau.$$

$$F(!^{p}(\sigma)) =!^{p+q}(\tau) \text{ if } F(\sigma) =!^{q}\tau.$$

In the following we assume that type schemes are already transformed by F, i.e. it is not the case that a subterm $!^p(!^q(\sigma))$ occurs in a type scheme.

Moreover, let $=_e$ be the relation between type schemes defined as follows:

$\alpha =_e \alpha$; $\sigma =_e \mu$ and $\tau =_e \nu$ imply $\sigma \multimap \tau =_e \mu \multimap \nu$; $\sigma =_e \tau$ implies $!^p\sigma =_e !^q\tau$. Roughly speaking, two type schemes are $=_e$ if and only if they are identical modulo the exponentials.

The unification algorithm, which we will present in SOS style, is a function U from $\mathrm{T} \times \mathrm{T}$ to pairs of the shape $< C, s >$, where C (the *modality* set) is a set of natural linear constraints, in the form $p = q$, where p and q are exponentials, and s is a substitution, replacing scheme variables by type schemes. A set C of linear constraints is *solvable* if there is a substitution X from literals to natural numbers such that, for every constraint $n_1 + ... + n_i = m_1 + ... + m_j$ in C, $X(n_1) + ... + X(n_i) = X(m_1) + ... + X(m_j)$. Clearly the solvability of a set of linear constraints is a decidable problem.

$$\frac{}{U(\alpha, \alpha) =< \emptyset, [] >}$$

$$\frac{\alpha \text{ is a scheme variable not occurring in } \tau}{U(\alpha, \tau) =< \emptyset, [\alpha \mapsto \tau] >}$$

$$\frac{\alpha \text{ is a scheme variable not occurring in } \sigma}{U(\sigma, \alpha) =< \emptyset, [\alpha \mapsto \sigma] >}$$

$$\frac{U(\rho, \mu) =< C, s >}{U(!^p\rho, !^q\mu) =< C \cup \{p = q\}, s >}$$

$$\frac{U(\sigma_1, \tau_1) =< C_1, s_1 > U(s_1(\sigma_2), s_1(\tau_2)) =< C_2, s_2 >}{U(\sigma_1 \multimap \sigma_2, \tau_1 \multimap \tau_2) =< C_1 \cup C_2, s_1 \circ s_2 >}$$

In all other cases, U is undefined: for example both $U(\alpha, \alpha \multimap \beta)$ and $U(!^p\alpha, \sigma \multimap \tau)$ are undefined.

The following lemma proves that U is the most general unifier for type schemes, with respect to $=_e$.

Lemma 5. *i) (correctness)* $U(\sigma, \tau) =< C, s >$ *implies* $s(\sigma) =_e s(\tau)$.
ii) (completeness) $s(\sigma) =_e s(\tau)$ *implies* $U(\sigma, \tau) =< C, s' >$ *and* $s = s' \circ s''$, *for some* s''.

Proof. (correctness) By induction on the rules deriving $U(\sigma, \tau) =< c, s >$.

(completeness) By induction on the pair (number of symbols of σ, number of scheme variables occurring in σ).

The following lemma extends the behavior of U to scheme substitutions: it is the key lemma for proving the correctness of our principal typing result.

Lemma 6. *Let* $< S, X >$ *be a scheme substitution such that* $< S, X > (\sigma) \equiv < S, X > (\tau)$. *Then* $U(\sigma, \tau) =< C, s >$, *and there is* $< S', X' >$ *such that* X' *is a solution of* C *and* $< S', X' > (s(\sigma)) \equiv < S, X > (\sigma)$.

Proof. By induction on the pair (number of symbols of σ, number of scheme variables occurring in σ).

Using U, we can now design an algorithm, PT, that, for every abstract EA-term M, either gives a negative answer or build a triple $< \Theta, \sigma, C >$, where Θ is a scheme context, i.e. a finite set of scheme assignments, σ is a type scheme and C is a set of constraints. The algorithm PT is showed in the Appendix A.2. For every term $M \in Abs^{EA}$, $PT(M)$, provided it exists, is the principal typing of M, in the sense that all and only the typings for M can be derived from it through scheme substitutions, as stated in the next theorem. In the statement of the theorem, we will denote by $< S, X > (\Theta)$, where Θ is a scheme context, the scheme context $\{x :< S, X > (\sigma) \mid x : \sigma \in \Theta\}$.

Theorem 1 (Principal typing for Abs^{EA}).
(correctness) $PT(M) =< \Theta, \sigma, C >$ *implies, for all scheme substitution* $< S, X >$ *such that* X *satisfies* C, $< S, X > (\Theta) \vdash_{abs} M :< S, X > (\sigma)$.
(completeness) $\Gamma \vdash_{abs} M : A$ *implies* $PT(M) =< \Theta, \sigma, C >$ *and there is a scheme substitution* $< S, X >$ *such that* X *satisfies* C, $\Gamma \supseteq< S, X > (\Theta)$ *and* $A \equiv< S, X > (\sigma)$.

Proof. (correctness) By induction on M, using the fact that the derivations are syntax directed.
(completeness) By induction on the derivation of $\Gamma \vdash_{abs} M : A$, using Lemma 6.

Example 3. Let M be the abstract term $\lambda x.[\lambda y.\nabla(x_1(x_2 t))\ [z_1/x_1, z_2/x_2, y/t]_{x \to (z_1,z_2)}$. $PT(M) =< \emptyset,\ !^n(\alpha \multimap \alpha) \multimap\ !^n\alpha \multimap !^n\ \alpha, \emptyset >$.

The principal typing property can be extended in an obvious way to all the terms of Λ^{EA}, in the sense that an EA-term M has the principal typing of the abstract term $emb(M)$ with the additional constraints $n = 1$ for any exponential n.

4 Type Inference for λ-Calculus

In this section we prove the main result of this paper, namely that a term of the λ-calculus is typeable in the type assignment system $\vdash_{\mathrm{NEAL}}$, if and only if it has a finite number of principal typing schemes. This result is an easy consequence of the two ones we proved in the previous sections: the principal typing property for the abstract EA-terms, and the fact the number of canonical forms associated to every λ-term is finite. So we can prove that a λ-term M has all and only the principal typings of the canonical forms belonging to the set $C(M)$.

Theorem 2 (Principal typing for Λ in EAL). $\forall M \in \Lambda$, $\Gamma \vdash_{\mathrm{NEAL}} M : A$ *if and only if* $PT(N) =< \Theta, \sigma, C >$ *and* $\Gamma \supseteq< S, X > (\Theta)$, $A =< S, X > (\sigma)$, *for some scheme substitution* $< S, X >$ *such that* X *satisfies* C *and for some* $N \in C(M)$.

Proof. The proof is an easy consequence of Lemma 3 and of Theorem 1.

Let us define the following erasure function E transforming EA-types (normalized by the function F defined in the previous section) into Curry's types:

$$E(\alpha) = \alpha;\ E(!^p\sigma) = E(\sigma);\ E(\sigma \multimap \tau) = E(\sigma) \to E(\tau).$$

Assume that a term $M \in \Lambda$ be typeable in the type assignment system $\vdash_{\mathrm{NEAL}}$, and consequently in the Curry's type assignment system $\vdash_C$. Since $\vdash_C$ has the principal typing property, let $< \Gamma, \sigma >$ be its principal typing in this system, where Γ is a context and σ a Curry's type. Then for each $< \Theta, \tau, C > \in PT(M)$, $E(\Theta) = \Gamma$ and $E(\tau) = \sigma$ (where $E(\Theta)$ is the distribution of E on all type schemes occurring in Θ).

Example 4. Let M be the λ-term $\lambda x.\lambda y.(x(x\,y))$. The set $\mathcal{C}(M)$ contains 24 elements and only six of them are typeable in EAL. They are:

- $\lambda x.[\lambda y.\nabla((x_3\,(x_4\,y_1)))[^{x_1}/_{x_3}, {}^{x_2}/_{x_4}, {}^{y}/_{y_1}]]_{x\to(x_1,x_2)}$ with
  $$PT = \langle \emptyset, !^n(\alpha \multimap \alpha) \multimap !^n\alpha \multimap !^n\alpha, \emptyset\rangle;$$
- $\lambda x.[\nabla(\lambda y.(x_3\,(x_4\,y)))[^{x_1}/_{x_3}, {}^{x_2}/_{x_4}]]_{x\to(x_1,x_2)}$ with
  $$PT = \langle \emptyset, !^n(\alpha \multimap \alpha) \multimap !^n(\alpha \multimap \alpha), \emptyset\}\rangle;$$
- $\lambda x.[\nabla(\lambda y.\nabla((x_5\,(x_6\,y_1)))[^{x_3}/_{x_5}, {}^{x_4}/_{x_6}, {}^{y}/_{y_1}])[^{x_1}/_{x_3}, {}^{x_2}/_{x_4}]]_{x\to(x_1,x_2)}$ with
  $$PT = \langle \emptyset, !^{n_1}(\alpha \multimap \alpha) \multimap !^{n_2}(!^{n_3}\alpha \multimap !^{n_3}\alpha), \{n_1 = n_2 + n_3\}\rangle;$$
- $\nabla(\lambda x.[\lambda y.\nabla((x_3\,(x_4\,y_1)))[^{x_1}/_{x_3}, {}^{x_2}/_{x_4}, {}^{y}/_{y_1}]]_{x\to(x_1,x_2)})[\,]$ with
  $$PT = \langle \emptyset, !^m(!^n(\alpha \multimap \alpha) \multimap !^n\alpha \multimap !^n\alpha), \emptyset\rangle;$$
- $\nabla(\lambda x.[\nabla(\lambda y.(x_3\,(x_4\,y)))[^{x_1}/_{x_3}, {}^{x_2}/_{x_4}]]_{x\to(x_1,x_2)})[\,]$ with
  $$PT = \langle \emptyset, !^m(!^n(\alpha \multimap \alpha) \multimap !^n(\alpha \multimap \alpha)), \emptyset\rangle;$$
- $\nabla(\lambda x.[\nabla(\lambda y.\nabla((x_5\,(x_6\,y_1)))[^{x_3}/_{x_5}, {}^{x_4}/_{x_6}, {}^{y}/_{y_1}])[^{x_1}/_{x_3}, {}^{,x_2}/_{x_4}]]_{x\to(x_1,x_2)})[\,]$ with
  $$PT = \langle \emptyset, !^m(!^{n_1}(\alpha \multimap \alpha) \multimap !^{n_2}(!^{n_3}\alpha \multimap !^{n_3}\alpha)), \{n_1 = n_2 + n_3\}\rangle;$$

References

1. Asperti, A., Coppola, P., Martini, S.: (Optimal) duplication is not elementary recursive. In: Proceedings of the 27th ACM SIGPLAN-SIGACT Symposium on Principles of Programming Languages (POLP-00), N.Y., ACM Press (2000) 96–107
2. Baillot, P.: Checking Polynomial Time Complexity with Types. In: Proc. of 2nd IFIP International Conference on Theoretical Computer Science (TCS 2002). Volume 223 of IFIP Conference Proceedings., Montréal, Québec, Canada (2002) 370–382
3. Baillot, P.: Type inference for polynomial time complexity via constraints on words. Prepublication LIPN Universite Paris Nord (2003)
4. Coppola, P., Martini, S.: Typing Lambda Terms in Elementary Logic with Linear Constraints. In Abramsky, S., ed.: Proc. of Typed Lambda Calculi and Applications, 5th International Conference, TLCA 2001. Volume 2044 of Lecture Notes in Computer Science., Springer (2001) 76–90
5. Curry, H.B.: Functionality in combinatory logic. In: Proc. Nat. Acad. Science USA. Volume 20. (1934) 584–590

6. Curry, H.B., Feys, R.: Combinatory Logic, Volume I. Studies in Logic and the Foundations of Mathematics. North-Holland, Amsterdam (1958)
7. Curry, H.B., Hindley, J.R., Seldin, J.P.: Combinatory Logic, Volume II. Studies in Logic and the Foundations of Mathematics. North-Holland, Amsterdam (1972)
8. Danos, V., Joinet, J.B., Schellinx, H.: On the linear decoration of intuitionistic derivations. In: Archive for Mathematical Logic. Volume 33. (1995) 387–412
9. Danos, V., Joinet, J.B.: Linear Logic and Elementary Time. ICC'99, to appear in Information & Computation (1999)
10. Girard, J.Y.: Light linear logic. Information and Computation **204** (1998) 143–175
11. Lamping, J.: An algorithm for optimal lambda calculus reduction. In ACM, ed.: POPL '90. Proceedings of the seventeenth annual ACM symposium on Principles of programming languages, January 17–19, 1990, San Francisco, CA, New York, NY, USA, ACM Press (1990) 16–30
12. Schellinx, H.: The Noble Art of Linear Decorating. PhD thesis, Institute for Logic, Language and Computation, University of Amsterdam (1994)

A Appendix

A.1 The Algorithm $\mathcal{C}$

Notation. Let $\mathbb{L}(M)$ be the *linearization* of M with respect to all its free variables and let $\mathbb{L}_x(M)$ be the set of fresh variables generated by $\mathbb{L}$ during the linearization of x in M. I.e. if $M = (x(xyy))$ then $\mathbb{L}(M) = (x_1(x_2y_1y_2))$ and $\mathbb{L}_x(M) = \{x_1, x_2\}$, $\mathbb{L}_y(M) = \{y_1, y_2\}$ and $\mathbb{L}_z(M) = \emptyset$ for any other variable z. Moreover, if T_1 and T_2 are two sets of terms, then $T_1 @ T_2$ is the set of all terms obtained by applying a term of T_1 to a term of T_2, and, if R is any term constructor, RT_1 is the set of terms obtained by applying R to every term in T_1.

The algorithm uses three auxiliary functions $\mathbb{T}, \mathbb{F}, \mathbb{F}' : \Lambda \to \mathrm{P}(Abs^{EA})$ and is defined by the following equations:

$$\begin{aligned}
\mathcal{C}(M) &= \texttt{if } \exists x_1, \ldots, x_k \in_{i>1} \texttt{FV}(M) \\
&\quad \texttt{then } \big[\mathbb{T}(\mathbb{L}(M)) \cup \mathbb{F}(\mathbb{L}(M))\big]_{x_1 \to (\mathbb{L}_{x_1}(M)), \ldots, x_k \to (\mathbb{L}_{x_k}(M))} \\
&\quad \texttt{else } \mathbb{T}(M) \cup \mathbb{F}(M) \\
\mathbb{T}(x) &= \{x\} \\
\mathbb{T}(\lambda x.M) &= \texttt{if } x \in_{i>1} \texttt{FV}(M) \texttt{ then } \lambda x.\big[\mathbb{T}(\mathbb{L}(M)) \cup \mathbb{F}(\mathbb{L}(M))\big]_{x \to (\mathbb{L}_x(M))} \\
&\quad \texttt{else } \lambda x.(\mathbb{T}(M) \cup \mathbb{F}(M)) \\
\mathbb{T}((M\ N)) &= \mathbb{T}(M) @ \Big(\mathbb{T}(N) \cup \mathbb{F}(N)\Big) \\
\mathbb{F}(x) &= \mathbb{F}'(x) = \emptyset \\
\mathbb{F}(M \neq x) &= \mathbb{F}'(M) \ \cup\ \nabla\big(\mathbb{F}'(M\{\widehat{z}/\texttt{FV}(M)\}) \cup \mathbb{T}(M\{\widehat{z}/\texttt{FV}(M)\})\big)[\texttt{FV}(M)/\widehat{z}] \\
\mathbb{F}'(M \neq x) &= \forall A_1, \ldots, A_n, P, n > 0 \text{ s.t. } M =_\alpha P\{A_1/y_1, \cdots, A_n/y_n\} \\
&\qquad \{y_1, \ldots, y_{n+k}\} = \texttt{FV}(P)\ P \neq x\ \forall 1 \leq i \leq n\ A_i = (A_{i_1}\ A_{i_2}) \\
&\quad \nabla\big(\mathbb{T}(P\{\widehat{z_1^k}/\widehat{y_{n+1}^{n+k}}\}) \cup (\mathbb{F}'(P\{\widehat{z_1^k}/\widehat{y_{n+1}^{n+k}}\})\big)\big[\mathbb{T}(A_1)/y_1, \cdots, \mathbb{T}(A_n)/y_n, \widehat{y_{n+1}^{n+k}}/\widehat{z_1^k}\big]
\end{aligned}$$

$$\square\ \overset{\square}{\square}(M)[y_1/x_1,\cdots,y_n/x_n]\ \overset{\square}{[}{}^{M_1}/y_1,\cdots,{}^{M_n}/y_n]\square_{\mathsf{Can}}\ \square(M)\left[{}^{M_1}/x_1,\ldots,{}^{M_n}/x_n\right]$$

$$\overset{\square}{[M]}_{\ldots,x_i\to(\widehat{y^{n_i}}),\ldots\ \overset{\square}{\ldots},N_j\to(z^j_1,\ldots,z^j_{k-1},x_i,z^j_{k+1},\ldots,z^j_{m_j}),\ldots}\ \square_{\mathsf{Can}}\ \overset{\square}{[M]}_{\ldots\ \overset{\square}{\ldots},N_j\to(z^j_1,\ldots,z^j_{k-1},\widehat{y^{n_i}},z^j_{k+1},\ldots,z^j_{m_j}),\ldots}$$

$$\overset{\square}{[M]}_{N_1\to(\widehat{x^1}),\ldots,N_i\to(\widehat{x^{n_i}}),\ldots\ \overset{\square}{Slist}}\ \square_{\mathsf{Can}}\ \overset{\square}{[M]}_{N_1\to(\widehat{x^1}),\ldots,N_{i-1}\to(\widehat{x^{i-1}}),N_{i+1}\to(\widehat{x^{i+1}}),\ldots\ \overset{\square}{Slist},N_i\to(\widehat{x^{n_i}})}$$

if $\mathtt{FV}(N_i)\ \square\ \mathtt{CV}(Slist)=\square$.

$$\square(M)[{}^{M_1}/x_1,\cdots,{}^{\nabla(N)\left[{}^{P_1}/y_1,\ldots,{}^{P_m}/y_m\right]}/x_i,\cdots,{}^{M_n}/x_n]\square_{\mathsf{Can}}\ \square(M\{N/x_i\})[{}^{M_1}/x_1,\cdots,{}^{P_1}/y_1,\cdots,{}^{P_m}/y_m,\ldots{}^{M_n}/x_n]$$

$$\square(M)[\cdots,{}^{[M_i]_{N_1\to(\widehat{y^1}),\ldots,N_k\to(\widehat{y^k})}}/x_i,\cdots]\square_{\mathsf{Can}}\ \left[\square(M)[\cdots,{}^{M_i\{\widehat{z^1}/\widehat{y^1},\ldots,\widehat{z^k}/\widehat{y^k}\}}/x_i,\cdots]\right]_{N_1\to(\widehat{z^1}),\ldots,N_k\to(\widehat{z^k})}$$

$$\square([M]_{\ldots,x\to(\widehat{x}),\ldots})[\cdots,y/x,\cdots]\square_{\mathsf{Can}}\ \overset{\square}{\square}([M]_{\ldots})[\cdots,\widehat{y}/\widehat{x},\cdots]\overset{\square}{\ }_{y\to(\widehat{y})}$$

$$[M]_{\ldots,[N]_{P_1\to(\widehat{y^1}),\ldots,P_k\to(\widehat{y^k})}\to(\widehat{x^i}),\ldots}\ \square_{\mathsf{Can}}\ \left[[M]_{\ldots,N\{\widehat{z^1}/\widehat{y^1},\cdots,\widehat{z^k}/\widehat{y^k}\}\to(\widehat{x^i}),\ldots}\right]_{P_1\to(\widehat{z^1}),\ldots,P_k\to(\widehat{z^k})}$$

$$([M]_{M_1\to(\widehat{x^1}),\ldots,M_k\to(\widehat{x^k})}\ N)\square_{\mathsf{Can}}\left[(M\{\widehat{y^1}/\widehat{x^1},\ldots,\widehat{y^k}/\widehat{x^k}\}\ N)\right]_{M_1\to(\widehat{y^1}),\ldots,M_k\to(\widehat{y^k})}$$

$$(M\ [N]_{N_1\to(\widehat{x^1}),\ldots,N_k\to(\widehat{x^k})})\square_{\mathsf{Can}}\left[(M\ N\{\widehat{y^1}/\widehat{x^1},\ldots,\widehat{y^k}/\widehat{x^k}\})\right]_{N_1\to(\widehat{y^1}),\ldots,N_k\to(\widehat{y^k})}$$

$\lambda x.[M]_{Slist}\square_{\mathsf{Can}}[\lambda y.M']_{Slist'}$ where $x\not\square\ \mathtt{SFV}(Slist)\ \square\ \mathtt{SSV}(Slist)$ and $[M']_{Slist'}=[M]_{Slist}\{y/x\}$

$$\square(x)[{}^{M}/x]\square_{\mathsf{Can}}M$$

$\square(M)[\cdots,{}^{N}/x,\cdots]\square_{\mathsf{Can}}\square(M)[\cdots]$ where $x\not\square\mathtt{FV}(M)$

$$[x]_{\ldots,N\to(\ldots,x,\ldots),\ldots}\square_{\mathsf{Can}}N$$

$[M]_{\ldots,N\to(\ldots,x,\ldots),\ldots}\square_{\mathsf{Can}}[M]_{\ldots,N\to(\ldots),\ldots}$ where $x\not\square\mathtt{FV}(M)$

$$[M]_{\ldots,N\to(x),\ldots}\square_{\mathsf{Can}}[M\{N/x\}]_{\ldots}$$

Fig. 1. Reduction relation $\to_{\mathsf{Can}}$ for Abs^{EA} terms.

where $\widehat{z}$ and $\widehat{z_1^k}$ are fresh variables. $\widehat{z_1^k}/\widehat{y_{n+1}^{n+k}}$ stands for $z_1/y_{n+1},\cdots,z_k/y_{n+k}$, $\widehat{z}/\mathtt{FV}(M)$ stands for the complete renaming of free variables of M with fresh ones, and $\mathtt{FV}(M)/\widehat{z}$ stands for the inverse substitution.

Fact 1. $\forall\mathbb{C}\in\mathcal{C}(M)\qquad(\mathbb{C})^*=M$

The proof of soundness and completeness of the algorithm $\mathcal{C}$ is based on the fact that the simple canonical forms can be generated by the grammar given in the next definition. The sets of free variables FV, banged variables BV and contracted variables CV used by the grammar, are defined in Figure 2.

Definition 7 **(CC^{EA}).** *The set of* simple canonical EAL-terms CC^{EA} *is generated by the following grammar (CC is the starting symbol):*

$$
\begin{aligned}
CC &::= [K]_{Clist} \ \textit{where}\ \mathtt{CV}(Clist) \subseteq \mathtt{FV}(K) \mid K \\
Clist &::= y \to (\widehat{x}) \mid y \to (\widehat{x}), Clist \ \textit{where}\ |\widehat{x}| \geq 2 \\
K &::= \nabla(B)[\widehat{x}/\widehat{y}] \ \textit{where}\ \mathtt{BV}(\widehat{x}/\widehat{y}) = \mathtt{FV}(B) \mid B \mid x \\
B &::= \nabla\,(B)\,[L] \ \textit{where}\ \mathtt{BV}(L) = \mathtt{FV}(B) \mid R \\
L &::= A/x \mid y/x, L \mid A/x, L \\
R &::= \lambda x.[K]_{x \to (x_1,\ldots,x_n)} \ \textit{where}\ \{x_1, \ldots, x_n\} \subseteq \mathtt{FV}(K) \ \mid \lambda x.K \mid A \\
A &::= (R\ K) \mid (x\ K)
\end{aligned}
$$

where all variables are linear, $\widehat{x}$ *stands for* $x_1, \ldots, x_n$ *and* $n > 0$.

Notice that side condition $|\widehat{x}| \geq 2$ in the production of $Clist$ implies $[x]_{Clist}$ is not a possible term in CC^{EA} by side condition $\mathtt{CV}(Clist) \subseteq \mathtt{FV}(K)$ in production of CC and by $\{x_1, \ldots, x_n\} \subseteq \mathtt{FV}(K)$ in production of R.

$$
\begin{array}{ll}
\mathtt{FV}(x) = \{x\} & \mathtt{FV}(\square\ (M)[^{N_1}/x_1, \cdots, {}^{N_k}/x_k]) = \bigcup_{i=1}^{k} \mathtt{FV}(N_i) \\
\mathtt{FV}(\lambda x.M) = \mathtt{FV}(M) \setminus \{x\} & \mathtt{FV}(x \square\ (x_1, \ldots, x_n), Clist) = \{x\} \square\ \mathtt{FV}(Clist) \\
\mathtt{FV}((M_1\ M_2)) = \mathtt{FV}(M_1) \square\ \mathtt{FV}(M_2) & \mathtt{CV}(x \square\ (x_1, \ldots, x_n), Clist) = \{x_1, \ldots, x_n\} \square\ \mathtt{CV}(Clist) \\
\mathtt{FV}([M]_{Clist}) = (\mathtt{FV}(M) \setminus \mathtt{CV}(Clist)) \square\ \mathtt{FV}(Clist) & \mathtt{BV}(M/x, Blist) = \{x\} \square\ \mathtt{BV}(Blist)
\end{array}
$$

Fig. 2. Free, Contracted and Banged Variables for EA-terms.

It is easy, but boring, to check that the grammar of definition 7 is correct and complete, in the sense that it generates all and only the simple canonical forms.

Lemma 7. *i)* $\forall M \in \Lambda \quad \mathcal{C}(M) \subseteq CC^{EA}$.
ii) $N \in CC^{EA}$ *and* $(N)^* = M$ *implies* $N \in \mathcal{C}((M))$.

Theorem 3 (Soundness and Completeness of $\mathcal{C}$). $\forall M \in \Lambda$
1. $\mathcal{C}(M) \subseteq C(M)$;
2. $N \in C(M)$ *implies* $N \in \mathcal{C}(M)$.

Proof. 1. By Lemma 7.i) for any $\mathbb{C} \in \mathcal{C}(M)$ we have $\mathbb{C} \in CC^{EA}$. Moreover $(\mathbb{C})^* = M$ by Fact 1. Then $\mathbb{C}$ is in $C(M)$.
2. By Lemma 7.ii).

A.2 The Algorithm *PT*

Let Let

$$
\begin{aligned}
U(\sigma_1, \sigma_2, \ldots, \sigma_n) = \ & \mathtt{let}\ U(\sigma_1, \sigma_2) = < C_1, s_1 > \\
& \mathtt{and\ let}\ U(s_1 \circ \cdots \circ s_i(\sigma_{i+1}), s_1 \circ \cdots \circ s_i(\sigma_{i+2})) = < C_{i+1}, s_{i+1} > \\
& \qquad \mathtt{in}\ < C_1 \cup \cdots \cup C_n, s_1 \circ \cdots \circ s_n > .
\end{aligned}
$$

The function *PT* is defined as follows.

- $PT(x) =< \{x : \alpha\}, \alpha, \emptyset >$, where α is fresh;
- $PT(\lambda x.M) =$ `let` $PT(M) =< B, \sigma, C >$ `in`
  `if` $B = B' \cup \{x : \tau\}$ `then` $< B', \tau \multimap \sigma, C >$
  `else` $< B, \alpha \multimap \sigma, C >$ where α is fresh;
- $PT(M\ N) =$ `let` $PT(M) =< B_1, \sigma_1, C_1 >$ `and` $PT(N) =< B_2, \sigma_2, C_2 >$
  `and let` they be disjoint `in let` $U(\sigma_1, \sigma_2 \multimap \alpha) =< C, s >$ (α fresh)
  `in` $< s(B_1 \cup B_2), s(\alpha), C \cup C_1 \cup C_2 >$;
- $PT([M]_{N_1 \to (\widehat{x_1}), \ldots, N_k \to (\widehat{x_k})}) =$ `let` $PT(M) =< B, \sigma, C >$ `and`
  $PT(N_i) =< B_i, \sigma_i, C_i >$ (all disjoint and disjoint from $PT(M)$)
  `and let` $U(B(x^1_1), \ldots, B(x^1_{n_1}), \sigma_1, !^{m_1}\alpha_1) =< C'_1, s_1 >$
  `and` $U(s_1 \circ \cdots \circ s_i(B(x^{i+1}_1)), \ldots, s_1 \circ \cdots \circ s_i(B(x^{i+1}_{n_{i+1}})),$
  $s_1 \circ \cdots \circ s_i(\sigma_{i+1}), !^{m_{i+1}}\alpha_{i+1}) =< C'_{i+1}, s_{i+1} >$
  `and` $s = s_1 \circ \cdots \circ s_k$
  `in`$< s(B/\widehat{x_1}, ..., \widehat{x_k} \cup \bigcup_{1 \leq j \leq k} B_j), s(\sigma), C \cup \bigcup_{1 \leq j \leq k} C_j \cup \bigcup_{1 \leq j \leq k} C'_j >$.
  (where α_i and m_i are fresh and $B/\widehat{x_1}, ..., \widehat{x_k}$ denotes the context obtained from B by deleting the assignments to variables in $\widehat{x_i}$ $(1 \leq i \leq k)$)
- $PT(\nabla(M)[N_1/x_1, \ldots, N_m/x_m]) =$ `let` $PT(M) =< B, \sigma, C >$
  `and` $PT(N_i) =< B_i, \sigma_i, C_i >$ (all disjoint and disjoint from $PT(M)$)
  `and let` $U(!^n B(x_1), \sigma_1) =< C'_1, s_1 >$
  `and` $U(!^n(s_1 \circ \cdots \circ s_i(B(x_{i+1}))), s_1 \circ \cdots \circ s_i(\sigma_{i+1})) =< C'_{i+1}, s_{i+1} >$
  `and` $s = s_1 \circ \cdots \circ s_m$
  `in` $<s(\bigcup_{1 \leq j \leq m} B_j), !^n(s(\sigma)), C \cup \bigcup_{1 \leq j \leq m} C_j \cup \bigcup_{1 \leq j \leq m} C'_j >$ (n fresh).
  where x^i_j denotes the j-th component of $\widehat{x_i}$.

Note that PT is defined modulo names of type variables.

A Logical Framework with Dependently Typed Records

Thierry Coquand [1], Randy Pollack [2], and Makoto Takeyama [1]

[1] Chalmers Tekniska Högskola, Sweden, {coquand,makoto}@cs.chalmers.se
[2] Edinburgh University, U.K., rap@inf.ed.ac.uk

1 Introduction

Our long term goal is a system to formally represent complex structured mathematical objects, and proofs and computation on such objects; e.g. a foundational computer algebra system. Our approach is informed by the long development of module systems for functional programming based on dependent record types as signatures [20]. For our logical purposes, however, we want a dependently typed base language. In this paper we propose an extension of Martin-Löf's logical framework [23,19] with dependently typed records, and present the semantic foundation and the typechecking algorithm of our system. Some of the work is formally checked in Coq [7].[1] We have also implemented and experimented with several related systems. Our proposal combines a semantic foundation, provably sound typechecking, good expressiveness (e.g. subtyping, sharing) and first-class higher-order modules.

The development of functional programming modules has addressed many aspects of the problem, such as use of *manifest* or *transparent* fields to control the information available in a signature, signature *strengthening*, type abstraction, sharing and subtyping [17,12,18]. The problem of modularity is not, however, closed, with much current research into first-class higher-order modules, recursive modules and mixins.

There has also been work on dependently typed records over dependent base languages. A first practical implementation is described in [3], however without semantic foundation. An original extension of Martin-Löf's logical framework is given in [6], however it lacks manifest fields to express sharing, and lacks metamathematical analysis. A general approach to adding modules on top of a Pure Type System is described in [9], but these modules are not first-class. [24] reviews much of this work, and gives an interpretation of dependently typed records in a type theory with inductive-recursive definition; however this approach does not address subtyping, and is infeasible in practice. An approach in the system Nuprl is given in [15]. Related work on equality in logical frameworks is found in [13].

For many examples of the kinds of modular constructions we want our system to support we point the reader to [6,5,3,18,12,9,24].

[1] The formal development covers sections 2-4, is slightly different than that in this paper, and uses several unproved assumptions. It is available from http://homepages.inf.ed.ac.uk/rap/export/TLCA03.v.

M. Hofmann (Ed.): TLCA 2003, LNCS 2701, pp. 105–119, 2003.

General approach. We begin with a model; then a variety of concrete expression languages may be interpreted in that model. This simplifies understanding of, and proofs about the expression languages and their typechecking. Although the expression language may have many typechecking rules, they can be checked individually for soundness w.r.t. the model.

Our starting model is quite close to a fragment of the PER model of NuPrl [1]. However, unlike the closed terms of [1] we use open terms, for decidability of (definitional) equality and of type-checking. A further idea is needed, suggested by the work of Miquel [22] on semantics of subtyping: this is a general form of η-expansions (also see [11]). The main result says this η-expansion interacts well with the PER semantics: two terms are related by the PER interpreting a type iff their η-expanded forms are β-convertible.

Outline of the paper. We describe stepwise our PER model, first for a core logical framework, then with singleton types and then with record types. In each case, we present a possible syntax for expressions and judgements, for which our model provides a realisability semantics, and for which we can prove termination of type-checking. We end by suggesting possible extensions and further work.

2 Model for a Logical Framework

We present a model for a logical framework in which, up to "η-expansion", well typed objects are normalising and have a decidable notion of equality. This model will be used in later sections to interpret more concrete frameworks.

2.1 Values

Objects and types. Let x, y, range over an infinite decidable set of *identifiers*, $\mathcal{I}$. We will be informal about variables and binding. The *objects* are usual *untyped* λ-terms.

$$M, N \ ::= \ x \mid M\ M \mid \lambda x.M$$

Let $\mathcal{O}$ be the set of objects. The equality on objects is β*-conversion*, which we write as $\simeq$. We write $M[N/x]$ for capture avoiding substitution of N for x in M.

We define (weakly) *normalisable* as usual, and say that a term is *neutral* iff it is normalisable and of the form

$$\nu \ ::= \ x \mid \nu\ M.$$

Every normal object application is neutral.

We consider also the category of syntactic *types*

$$A, B \ ::= \ El\, M \mid \mathsf{fun}\, A\, x.B \mid \star$$

Capture avoiding substitution, $A[N/x]$ is defined on types. When context makes clear what variable is being replaced, we may write $A[N]$ for $A[N/x]$. We write $A_1 \rightarrow A_2$ for $\mathsf{fun}\, A_1\, x.A_2$ when x is not free in A_2.

2.2 Categorical Judgement: Type

We will interpret types as partial equivalence relations (per) over objects. A *per* on set D is a binary relation $\mathcal{A}$ on D which is symmetric and transitive. We may write $u_1 = u_2 : \mathcal{A}$ for $\mathcal{A}(u_1, u_2)$ and $u : \mathcal{A}$ for $u = u : \mathcal{A}$. Write $[\mathcal{A}]$ for the set of all $u \in D$ such that $u : \mathcal{A}$, and $\mathsf{per}(D)$ for the set of all pers on D. If $\mathcal{A} \in \mathsf{per}(D)$ then $Fam(\mathcal{A})$ is the set of all functions $\mathcal{F} : [\mathcal{A}] \to \mathsf{per}(D)$ such that $\mathcal{F}(u_1) = \mathcal{F}(u_2)$ (extensionally equal) whenever $u_1 = u_2 : \mathcal{A}$.

If $\mathcal{A} \in \mathsf{per}(\mathcal{O})$ and $\mathcal{F} \in Fam(\mathcal{A})$ we can form $\Pi(\mathcal{A}, \mathcal{F}) \in \mathsf{per}(\mathcal{O})$, defined by $w_1 = w_2 : \Pi(\mathcal{A}, \mathcal{F})$ iff $u_1 = u_2 : \mathcal{A}$ implies $w_1\ u_1 = w_2\ u_2 : \mathcal{F}(u_1)$. Notice that we could have written $w_1\ u_1 = w_2\ u_2 : \mathcal{F}(u_2)$ as well, since $\mathcal{F}(u_1) = \mathcal{F}(u_2)$ in this case.

Being a type. We define inductively a relation "$=$" of *intensional equality* on the set of types. (It will turn out to be a per.) Simultaneously for each A and B such that $A = B$ we define pers on objects, written $\overline{A}$ and $\overline{B}$. For generality this definition is parameterised by the interpretations of $\star$ and El.

We say that $\mathcal{A} \in \mathsf{per}(\mathcal{O})$ is *saturated* iff

- $\nu : \mathcal{A}$ for every neutral ν,
- if $u : \mathcal{A}$ then u is normalisable, and
- if $u_1 = u_2 : \mathcal{A}$ then $u_1 \simeq u_2$.

E.g. the relation "*M_1 and M_2 are neutral and equal*" is saturated.

Definition 1. Let $\overline{\star} \in \mathsf{per}(\mathcal{O})$ be saturated, and $\mathcal{E} \in Fam(\overline{\star})$ s.t. $\mathcal{E}(M)$ is saturated for $M : \overline{\star}$. The clauses defining intensional type equality are as follows.

- $\star = \star$.
- $El\, M = El\, N$ whenever $M = N : \overline{\star}$. $\overline{El\, M}$ is $\mathcal{E}(M)$.
- $\mathsf{fun}\, A_1\, x_1.B_1 = \mathsf{fun}\, A_2\, x_2.B_2$ whenever
  - $A_1 = A_2$,
  - $M \mapsto \overline{B_i[M]} \in Fam(\overline{A_i})$ (i=1,2), and
  - $M_1 = M_2 : \overline{A_1} \implies B_1[M_1] = B_2[M_2]$.

  $\overline{\mathsf{fun}\, A\, x.B}$ is $\Pi(\overline{A}, M \mapsto \overline{B[M]})$.

This definition respects α, β equality on syntactic types. It is *intensional* in the sense that $\overline{A} = \overline{B}$ (extensionally) does not imply $A = B$.

We sometimes write $A \in \mathbf{Type}$ for $A = A$, and $M : \overline{A}$ for $M = M : \overline{A}$. Writing $\overline{A}$ always implies $A \in \mathbf{Type}$.

Lemma 2. – *If $A = B$ then $\overline{A}$ and $\overline{B}$ are extensionally equal.*
– *The relation $A = B$ is a per on types.*

Having proved lemma 2, we can now give a simpler equivalent formulation of type equality.

Lemma 3. $\mathsf{fun}\, A_1\, x_1.B_1 = \mathsf{fun}\, A_2\, x_2.B_2$ *iff*
$A_1 = A_2$ *and* $M_1 = M_2 : \overline{A_1} \implies B_1[M_1] = B_2[M_2]$.

2.3 Normalisation and Decidability

We define an operation of η-expansion at type A (written $\eta\{A\}$) such that $M : \overline{A}$ implies $\eta\{A\}M$ is normalising.

Definition 4. $\eta\{\_\}$ is defined by recursion on the syntactic class of types.[2]

$$\begin{aligned}\eta\{\star\} &= \lambda x.x \\ \eta\{El\, M\} &= \lambda x.x \\ \eta\{\mathsf{fun}\, A\, x.B\} &= \lambda u.\lambda z.\eta\{B[\widehat{z}]\}\ (u\ \widehat{z}) \quad \text{where } \widehat{z} = \eta\{A\}z\end{aligned}$$

where u and z are distinct and not free in $\eta\{A\}$ or B. □

The *base types* are those of the form $\star$ or $El\, M$. If, for instance, A is a base type, we have

$$\eta\{(A \to A) \to A\} = \lambda u.\lambda z.u\ (\lambda x.z\ x).$$

We emphasise that we do not have η-conversion at the level of objects, but in the type system to be presented later there is no observable difference between η-convertible expressions.

Theorem 5. *Let* $A \in \mathbf{Type}$.

1. $\eta\{A\}\nu : \overline{A}$, *where* ν *is neutral.*
2. *If* $M : \overline{A}$ *then* $\eta\{A\}M$ *is normalisable.*
3. *If* $M : \overline{A}$ *then* $M = \eta\{A\}M : \overline{A}$.
4. *If* $M_1 = M_2 : \overline{A}$ *then* $\eta\{A\}M_1 \simeq \eta\{A\}M_2$.

Proof. This proof is checked in Coq [7]. The four parts are proved simultaneously by induction on the proof that $A \in \mathbf{Type}$. It is direct if A is a base type.

Consider now a $\mathbf{Type}$ of the form $\mathsf{fun}\, A\, y.B$. We have

$$\eta\{\mathsf{fun}\, A\, y.B\}\ t \simeq \lambda x.\eta\{B[\eta\{A\}x]\}\ (t\ (\eta\{A\}x)) \tag{1}$$

We do part 1. as an example. Assuming that t is neutral and that $M_1 = M_2 : \overline{A}$, we must show

$$\eta\{\mathsf{fun}\, A\, y.B\}\ t\ M_1 = \eta\{\mathsf{fun}\, A\, y.B\}\ t\ M_2 : \overline{B[M_1]}.$$

Writing $\widehat{M_i}$ for $\eta\{A\}M_i$ $(i = 1, 2)$, and using (1), we need

$$\eta\{B[\widehat{M_1}]\}\ (t\ \widehat{M_1}) = \eta\{B[\widehat{M_2}]\}\ (t\ \widehat{M_2}) : \overline{B[M_1]}.$$

By induction hypothesis, $\widehat{M_1} \simeq \widehat{M_2}$, so it suffices to show

$$\eta\{B[\widehat{M_1}]\}\ (t\ \widehat{M_1}) : \overline{B[M_1]}.$$

By IH, $M_1 = \widehat{M_1} : \overline{A}$ (so $\overline{B[\widehat{M_1}]} = \overline{B[M_1]}$) and $\widehat{M_1}$ is normalisable. Thus $t\,\widehat{M_1}$ is neutral, and by IH $\eta\{B[\widehat{M_1}]\}\ (t\ \widehat{M_1}) : \overline{B[\widehat{M_1}]}$. □

Corollary 6. *Let* $A \in \mathbf{Type}$, $M_1 : \overline{A}$ *and* $M_2 : \overline{A}$.

1. *If* $\eta\{A\}M_1 \simeq \eta\{A\}M_2$ *then* $M_1 = M_2 : \overline{A}$.
2. *The relation* $M_1 = M_2 : \overline{A}$ *is decidable.*

[2] The number of funs is reduced on each recursive call.

Remark on eta. Using $\beta\eta$-conversion for equality on objects (instead of β-conversion) would simplify the statements of theorem 5 and corollary 6, since our operator $\eta\{-\}$ is no longer needed. This simplification explains why [8] can model a type theory with one universe and η-conversion. Such a simplification however does not seem possible in presence of singleton types, or even unit types, a difficulty noticed in [13].

2.4 Hypothetical Judgements

Contexts. We consider contexts

$$C \ ::= \ \triangledown \mid C, x{:}A \qquad (\triangledown \text{ is the empty context.})$$

We say $x{:}A$ *in* C iff $C = C', x'{:}A'$ and $x{:}A = x'{:}A'$ or $x{:}A$ *in* C'. We also say $x \in C$ if $x{:}A$ *in* C for some A. In writing $C, x{:}A$ we assume $x \notin C$.

Environments. An environment, ρ, is a function of type $\mathcal{I} \to \mathcal{O}$. ρ_0 is the identity environment, $\rho_0(x) = x$. $M\rho$ (resp. $A\rho$) is the simultaneous substitution of $\rho(x)$ for all free occurrences of x in M (resp. A). The composition of environments is defined by $(\rho_1 \circ \rho_2)x = (\rho_1\, x)\rho_2$. Notice $M(\rho_1 \circ \rho_2) \simeq (M\rho_1)\rho_2$. We write $(\rho, x{=}M)$ for the *update* of ρ, defined by

$$(\rho, x{=}M)(x) = M, \qquad (\rho, x{=}M)(y) = \rho(y) \text{ if } y \neq x.$$

Lemma 7. *For A a type and ρ an environment, $(\eta\{A\})\rho \simeq \eta\{A\rho\}$. Hence for any object M, $(\eta\{A\}\, M)\rho \simeq \eta\{A\rho\}\,(M\rho)$.*

Definition 8. We inductively define a judgement of form $\rho_1 = \rho_2 : C$ by the rules:

$$\frac{}{\rho_1 = \rho_2 : \triangledown} \qquad \frac{\rho_1 = \rho_2 : C \qquad A\rho_1 \in \textbf{Type} \qquad \rho_1 x = \rho_2 x : \overline{A\rho_1}}{\rho_1 = \rho_2 : C, x{:}A}$$

We may write $\rho : C$ for $\rho = \rho : C$.

Lemma 9. *If $\rho_1 = \rho_2 : C$ and $y \notin C$ then for any M, $\rho_1 = (\rho_2, y{=}M) : C$.*

Hypothetical judgements. We define simultaneously three judgement forms, C valid, $A_1 = A_2\, [C]$ and $M_1 = M_2 : A\, [C]$. (We may write A type $[C]$ for $A = A\, [C]$, and $M : A\, [C]$ for $M = M : A\, [C]$.)

$$\frac{}{\triangledown \ \mathsf{valid}} \qquad \frac{x \notin C \qquad A \ \mathsf{type}\ [C]}{C, x{:}A \ \mathsf{valid}} \tag{2}$$

$$\frac{C \ \mathsf{valid} \qquad \forall \rho_1, \rho_2 \,.\, \rho_1 = \rho_2 : C \implies A_1\rho_1 = A_2\rho_2}{A_1 = A_2\, [C]}$$

$$\frac{A \ \mathsf{type}\ [C] \qquad \forall \rho_1, \rho_2 \,.\, \rho_1 = \rho_2 : C \implies M_1\rho_1 = M_2\rho_2 : \overline{A\rho_1}}{M_1 = M_2 : A\, [C]} \tag{3}$$

For the well-formedness of rule (3), note that the first premise guarantees that $\overline{A\rho_1}$ is defined.

Lemma 10. *Let* C *be* valid.

- *The following relations are pers*

$$\begin{array}{rcl} \rho_1, \rho_2 & \mapsto & \rho_1 = \rho_2 : C, \\ A_1, A_2 & \mapsto & A_1 = A_2\,[C], \\ M_1, M_2 & \mapsto & M_1 = M_2 : A\,[C] \end{array}$$

- *If* $\rho_1 = \rho_2 : C$ *and* $y \notin C$ *then for any* M, $(\rho_1, y{=}M) = (\rho_2, y{=}M) : C$.

Theorem 11. *The following implications hold. We write them as rules (consider also rules (2)) to invite comparison with the rules of Martin–Löf's logical framework.*

type formation and type equality

$$\frac{C \text{ valid}}{\star \text{ type } [C]} \qquad \frac{M = N : \star\,[C]}{El\,M = El\,N\,[C]} \qquad \frac{A_1 = A_2\,[C] \qquad B_1 = B_2\,[C, x{:}A_1]}{\mathsf{fun}\,A_1\,x.B_1 = \mathsf{fun}\,A_2\,x.B_2\,[C]} \tag{4}$$

objects

$$\frac{C, x{:}A \text{ valid}}{x : A\,[C, x{:}A]} \qquad \frac{M : B\,[C, x{:}A]}{\lambda x.M : \mathsf{fun}\,A\,x.B\,[C]} \qquad \frac{M : \mathsf{fun}\,A\,x.B\,[C] \qquad N : A\,[C]}{M\,N : B[N]\,[C]}$$

type conversion

$$\frac{M = N : A\,[C] \qquad A = B\,[C]}{M = N : B\,[C]}$$

weakening

$$\frac{B_1 = B_2\,[C] \qquad C, x{:}A \text{ valid}}{B_1 = B_2\,[C, x{:}A]} \qquad \frac{M = N : B\,[C] \qquad C, x{:}A \text{ valid}}{M = N : B\,[C, x{:}A]}$$

Proof. All parts are proved directly from the meanings of the definitions using the lemmas 2, 9 and 10. However there are a lot of details to be handled. We have checked this theorem in Coq [7]. □

Decidability. We give conditions for $M_1 = M_2 : A\,[C]$ to be decidable.

Definition 12. Eta-expansion of environment ρ at context C is defined by recursion on the structure of C.

$$\begin{array}{rcl} \eta\{\nabla\}\rho & = & \rho \\ \eta\{C, x{:}A\}\rho & = & (\rho',\, x{=}\eta\{A\rho'\}(\rho x)) \quad \text{where } \rho' = \eta\{C\}\rho \end{array}$$

Write ρ_C for $\eta\{C\}\rho_0$, where ρ_0 is the identity environment.

Lemma 13. *1.* $\rho_C \circ \rho = \eta\{C\}\rho$.
2. If C valid *then* $\rho_C : C$.
3. If C valid *and* $\rho : C$ *then* $\rho = \eta\{C\}\rho : C$.
4. If $M_1 : A\,[C]$ *and* $M_2 : A\,[C]$ *then* $M_1 = M_2 : A\,[C]$ *is equivalent to* $M_1\rho_C = M_2\rho_C : \overline{A\rho_C}$.

Corollary 14. *If* $M_1 : A\,[C]$ *and* $M_2 : A\,[C]$ *then* $M_1 = M_2 : A\,[C]$ *is equivalent to* $(\eta\{A\}M_1)\rho_C \simeq (\eta\{A\}M_2)\rho_C$, *which is decidable.*

2.5 "Syntactic" Type Equality

We now define a syntactic relation of shape $C \vdash A_1 = A_2$ which is decidable and sound for the semantic relation.

Definition 15. The rules of syntactic type equality are as follows.

$$\frac{}{C \vdash \star = \star} \qquad \frac{M_1 = M_2 : \star\,[C]}{C \vdash \mathit{El}\, M_1 = \mathit{El}\, M_2} \qquad \frac{C \vdash A_1 = A_2 \qquad C, x{:}A_1 \vdash B_1 = B_2}{C \vdash \mathsf{fun}\, A_1\, x.B_1 = \mathsf{fun}\, A_2\, x.B_2}$$

Lemma 16. *1. If* $A_1\,\mathsf{type}\,[C]$ *and* $A_2\,\mathsf{type}\,[C]$ *then* $C \vdash A_1 = A_2$ *is decidable.*
2. If C valid *and* $C \vdash A_1 = A_2$ *then* $A_1 = A_2\,[C]$.

3 A Logical Framework in Syntax

We give concrete syntax for a core LF, including rules for typechecking that interpret expressions in the model. This interpretation is semantically sound.

3.1 Expressions and Judgements

The syntax of expressions and expression contexts is defined by

$$\begin{array}{rcl} e & ::= & z \mid e\,e \mid \texttt{[}z\texttt{:}e\texttt{]}e \mid \texttt{*} \mid \texttt{El}\,e \mid \{z\texttt{:}e\}e \mid e \texttt{->} e \\ \Gamma & ::= & \blacktriangledown \mid \Gamma, x\texttt{:}e \end{array}$$

(As usual, we omit $\blacktriangledown$, the notation for the empty context, in concrete syntax.)

For simplicity and definiteness in this paper, expressions and contexts are taken concretely. Not even α-renaming is assumed. There is no substitution defined on expressions; that is delegated to the interpretation. Thus, given an implementation of pure lambda terms, our typechecking rules are directly implementable as presented below. However, more conventional languages can also be interpreted by our model.

Two judgement forms are defined simultaneously,

- $C \vdash e \Rightarrow A$, meaning that expression e is interpreted in C as type A.
- $C \vdash e \Rightarrow M : A$, meaning that expression e is interpreted in C as object M which has type A.

while a third can be defined afterwards.

- $\Gamma \Rightarrow C$, meaning that Γ is interpreted as the valid context C.

This organization is typical for implementations of type theory such as Lego [16] and Coq [7] that maintain a "current checked environment" representing the mathematics developed so far, since it avoids repeatedly rechecking that current context [21].

3.2 Typechecking

The rules are: type formation

$$\frac{}{C \vdash \texttt{*} \Rightarrow \star} \qquad \frac{C \vdash e \Rightarrow M : \star}{C \vdash \texttt{El}\, e \Rightarrow El\, M} \qquad \frac{C \vdash e_1 \Rightarrow A \qquad C \vdash e_2 \Rightarrow B}{C \vdash e_1 \texttt{->} e_2 \Rightarrow A \to B}$$

$$\frac{C \vdash e_1 \Rightarrow A \qquad x \notin C \qquad C, x{:}A \vdash e_2 \Rightarrow B}{C \vdash \{x{:}e_1\}e_2 \Rightarrow \mathsf{fun}\, A\, x.B}$$

objects

$$\frac{x{:}A \ in\ C}{C \vdash x \Rightarrow x : A} \qquad \frac{C \vdash e_1 \Rightarrow A \qquad x \notin C \qquad C, x{:}A \vdash e_2 \Rightarrow M : B}{C \vdash [x{:}e_1]e_2 \Rightarrow \lambda x.M : \mathsf{fun}\, A\, x.B} \qquad (5)$$

$$\frac{C \vdash e_1 \Rightarrow M_1 : \mathsf{fun}\, A_1\, x.B \qquad C \vdash e_2 \Rightarrow M_2 : A_2 \qquad C \vdash A_1 = A_2}{C \vdash e_1\, e_2 \Rightarrow M_1\, M_2 : B[M_2]} \qquad (6)$$

validity

$$\frac{}{\blacktriangledown \Rightarrow \triangledown} \qquad \frac{\Gamma \Rightarrow C \qquad C \vdash e \Rightarrow A \qquad x \notin C}{\Gamma, x{:}e \Rightarrow C, x{:}A}$$

In these rules, as in the rest of the paper, semantic values are taken up to α, β-equality as usual. However, as mentioned above, expressions are taken concretely. Thus some "good" expressions do not typecheck for reasons of variable clash. For example `[x:*][x:El x]x` is rejected, while `[y:*][y:El x]y` is accepted. The reason we include e_1 `->` e_2 as an expression, rather than as an abbreviation, is so that nested arrows do not fail due to variable clash.

η-conversion in the expression language. η-convertible expressions are indistinguishable by typechecking. For example, consider `x:{a:*}El a` and its η-expansion, `[a:*](x a)`. The only way these expressions might be compared during the typechecking of an expression is by the third premise of the rule for applications, (6), which calls definition 15. Since `x` and `[a:*](x a)` are objects, they can only occur in types inside the `El` constructor, hence will both be η-expanded by the premise of the second rule of definition 15 (via corollary 14). These η-expansions are β-convertible, so considered equal.

3.3 Correctness and Termination of Typechecking

Theorem 17. – *If* $C \vdash a \Rightarrow A$ *and* C valid *then* A type $[C]$.
– *If* $C \vdash e \Rightarrow M : A$ *and* C valid *then* $M : A\,[C]$.

Proof. By simultaneous induction on derivations. Straightforward using the definition of valid (rule (2)), theorem 11 and lemma 16. We do two cases.

Pi type The derivation ends with

$$\frac{C \vdash a \Rightarrow A \qquad x \notin C \qquad C, x{:}A \vdash b \Rightarrow B}{C \vdash \{x{:}a\}b \Rightarrow \mathsf{fun}\, A\, x.B}$$

Assume C valid. By first IH, A type $[C]$, so $C, x{:}A$ valid. Hence by second IH, B type $[C, x{:}A]$. By theorem 11, fun $A\ x.B$ type $[C]$, as required.

Application The derivation ends with

$$\frac{C \vdash e_1 \Rightarrow M_1 : \mathsf{fun}\, A_1\ x.B \qquad C \vdash e_2 \Rightarrow M_2 : A_2 \qquad C \vdash A_1 = A_2}{C \vdash e_1\, e_2 \Rightarrow M_1\, M_2 : B[M_2]}$$

Assume C valid. By IH, $M_1 : \mathsf{fun}\, A_1\ x.B\ [C]$ and $M_2 : A_2\ [C]$. By lemma 16, $A_1 = A_2\ [C]$. $M_1\, M_2 : B[M_2]\ [C]$ follows from theorem 11. □

Corollary 18. *If* $\Gamma \Rightarrow C$ *then* C valid.

Corollary 19. *1. If* C valid *then* $\exists A\ .\ C \vdash e \Rightarrow A$ *and* $\exists M, A\ .\ C \vdash e \Rightarrow M : A$ *are decidable.*
2. $\exists C\ .\ \Gamma \Rightarrow C$ *is decidable.*
3. $\exists C, A\ .\ \Gamma \Rightarrow C\ \wedge\ C \vdash e \Rightarrow A$ *and* $\exists C, M, A\ .\ \Gamma \Rightarrow C\ \wedge\ C \vdash e \Rightarrow M : A$ *are decidable.*

4 Singleton Types and Definitions

We extend the semantics with singleton types in order to interpret local and global definitions (this section) and *manifest fields* in signatures (section 5). In this section we do not need subtyping for singletons. In this paper we do not consider singletons in the expression language itself. For other work on singletons see [2,10,25,14].

4.1 Singleton Types in the Model

The category of syntactic types is extended with a notation $A/_M$ for the singleton of object M at type A.

$$A, B\ ::=\ El\, M \mid \mathsf{fun}\, A\ x.B \mid \star \mid A/_M.$$

If $\mathcal{A} \in \mathsf{per}(D)$ and $u : \mathcal{A}$ we form the singleton $\mathcal{A}/u \in \mathsf{per}(D)$ which is defined by $u_1 = u_2 : \mathcal{A}/u$ iff $u_1 = u : \mathcal{A}$ and $u_2 = u : \mathcal{A}$. Definition 1 is extended with a new clause

- $A_1/_{M_1} = A_2/_{M_2}$ iff $A_1 = A_2$ and $M_1 = M_2 : \overline{A_1}$.
  $M_1 = M_2 : \overline{A/_N}$ iff $M_1 = M_2 = N : \overline{A}$.

Lemma 2 holds of this extended definition.

Definition 4 is extended with the clause

$$\eta\{A/_M\}\ =\ \lambda u.\eta\{A\}M$$

where u is not free in $\eta\{A\}M$. Theorem 5 holds for $A/_M \in$ **Type**.

The following implications hold, extending theorem 11.

$$\frac{M : A\ [C]}{M : A/_M\ [C]} \qquad \frac{M : A/_N\ [C]}{M = N : A\ [C]} \qquad \frac{M : B\ [C, x{:}A/_N]}{M[N] : B[N]\ [C]} \tag{7}$$

Lemma 16 holds in the extended system.

4.2 Typechecking Definitions

Extend the syntax of expressions and expression contexts with local and global definitions respectively.

$$\begin{array}{lcl} e & ::= & z \mid e\,e \mid \texttt{[}z\texttt{:}e\texttt{]}e \mid \texttt{*} \mid \texttt{El}\,e \mid \{z\texttt{:}e\}e \mid e \texttt{->} e \mid \texttt{[}z\texttt{=}e\texttt{]}e \\ \Gamma & ::= & \blacktriangledown \mid \Gamma\texttt{,}x\texttt{:}e \mid \Gamma\texttt{,}x\texttt{=}e \end{array}$$

The typechecking rule for variables (first rule (5)) works unchanged for global definitions because variable x of type $A/_N$ η-expands to $\eta\{A\}N$. Thus global definitions are expanded only for equality testing, and heuristics to avoid some definition expansion are possible in pragmatic implementatons.

We have new typechecking rules for local definitions in types and in objects.

$$\frac{C \vdash e_1 \Rightarrow N : A \qquad x \notin C \qquad C, x{:}A/_N \vdash e_2 \Rightarrow B}{C \vdash \texttt{[}x\texttt{=}e_1\texttt{]}e_2 \Rightarrow B[N]}$$

$$\frac{C \vdash e_1 \Rightarrow N : A \qquad x \notin C \qquad C, x{:}A/_N \vdash e_2 \Rightarrow M : B}{C \vdash \texttt{[}x\texttt{=}e_1\texttt{]}e_2 \Rightarrow M[N] : B[N]}$$

and for validity with global definitions.

$$\frac{\Gamma \Rightarrow C \qquad C \vdash e \Rightarrow N : A \qquad x \notin C}{\Gamma\texttt{,}x\texttt{=}e \Rightarrow C, x{:}A/_N}$$

Semantical soundness and termination of typechecking follow as expected, using rules (7). Notice that no singleton ever occurs on the RHS of judgement forms $C \vdash e \Rightarrow M : A$ or $C \vdash e \Rightarrow A$, so no new syntactic type equality rules (definition 15) are needed.

5 Records, Signatures, and Subtyping

The model extends to both left associating and right associating signatures [24]. After some wrangling between the authors, we decided to present the more standard right associating signatures. Convenient use of signatures requires subtyping of some kind. Here we describe *structural subtyping* for signatures. There are many interesting subtyping rules for singletons [2], but we include only what is needed for manifest signatures.

5.1 Signatures and Records in the Model

We use identifiers as field labels, but often write l, k, instead of x for labels to make an informal distinction. The category of syntactic types is extended with a unit type, $\mathbf{1}$ (which will serve as a top type, and as the empty signature), and with a signature extension constructor.

$$A, B, S \;::=\; \mathit{El}\,M \mid \mathsf{fun}\,A\,x.B \mid \star \mid A/_M \mid \mathbf{1} \mid \langle l{:}A,\, x.S\rangle.$$ [3]

[3] The notation $\langle l{:}A,\, x.S\rangle$ is analogous to $\{l \triangleright x{:}A, S\}$ originally proposed in [12].

We say $l \in A$ iff $A = <k{:}A',\ x.B'>$ and $l = k$ or $l \in B'$. In writing $<l{:}A,\ x.B>$ we assume $l \notin B$.

To the category of objects we introduce $()$ (which serves as the empty record and the canonical element of $\mathbf{1}$) and objects for record extension and field projection.

$$M, N \ ::= \ x \mid M\ M \mid \lambda x.M \mid () \mid (l{=}M,\ M) \mid M.l$$

There are new reduction rules for *dot reduction* (record elimination)

$$(l{=}M_1,\ M_2).l \succ M_1, \qquad\qquad (k{=}M_1,\ M_2).l \succ M_2.l \quad (l \neq k).$$

Clearly, every object has a unique dot-normal form, so $\succ$ is Church-Rosser. Also $\succ$ commutes with β, so by the Hindley-Rosen lemma [4], $\succ\cup\beta$ is Church-Rosser. Equality on objects is defined as the closure of $\succ \cup \beta$. This equality does not include surjective pairing or record field permutation, but because of generalised η-expansion, $\eta\{A\}$, the interpreted expression language does.

Definition 1 is extended with clauses for $\mathbf{1}$ and $<l{:}A,\ x.B>$.

- $\mathbf{1} = \mathbf{1}$. $\overline{\mathbf{1}}$ is $\mathcal{O} \times \mathcal{O}$.
- $<l_1{:}A_1,\ x_1.B_1> = <l_2{:}A_2,\ x_2.B_2>$ iff
  - $l_1 = l_2$
  - $A_1 = A_2$,
  - $M \mapsto \overline{B_i[M]} \in Fam(\overline{A_i})$ (i=1,2), and
  - $M_1 = M_2 : \overline{A_1} \implies B_1[M_1] = B_2[M_2]$.

  $M_1 = M_2 : \overline{<l{:}A,\ x.B>}$ iff $M_1.l = M_2.l : \overline{A} \ \wedge\ M_1 = M_2 : \overline{B[M_1.l]}$.

Lemma 2 holds of this extended definition, and an analogue of lemma 3 shows the definition can be simplified.

Definition 4 is extended with new clauses.

$$\begin{aligned} \eta\{\mathbf{1}\} &= \lambda u.() \\ \eta\{<l{:}A,\ x.B>\} &= \lambda z.(l{=}\hat{z},\ \eta\{B[\hat{z}]\}z) \quad \text{where } \hat{z} = \eta\{A\}(z.l) \end{aligned}$$

where z is not free in $\eta\{A\}$ or B. Theorem 5 holds for the extended notion of **Type**.

The following implications are derivable, extending theorem 11.

$$\frac{C \ \mathsf{valid}}{\mathbf{1}\ \mathsf{type}\ [C]} \qquad \frac{l_1 = l_2 \qquad A_1 = A_2\ [C] \qquad B_1 = B_2\ [C, x{:}A_1]}{<l_1{:}A_1,\ x.B_1> = <l_2{:}A_2,\ x.B_2>\ [C]}$$

$$\frac{C \ \mathsf{valid}}{M = N : \mathbf{1}\ [C]} \qquad \frac{M_A : A\ [C] \qquad M_B : B[M_A]\ [C] \qquad B\ \mathsf{type}\ [C, x{:}A]}{(l{=}M_A,\ M_B) : <l{:}A,\ x.B>\ [C]}$$

$$\frac{M : <l{:}A,\ x.B>\ [C]}{M.l : A\ [C]} \qquad \frac{M : <l{:}A,\ x.B>\ [C] \qquad l \neq k \quad k \in B}{M.k : B[M.l]\ [C]}$$

5.2 Subtyping

For $A, B \in \mathbf{Type}$ we define the semantic notion *categorical subtype*, written $A \sqsubseteq B$, to be $\overline{A} \subseteq \overline{B}$ (extensionally subrelation); this is adequate for the claims in this paper.

The notion of *hypothetical subtype* is defined by

$$A \sqsubseteq B\,[C] \quad = \quad A\,\mathsf{type}\,[C] \;\wedge\; B\,\mathsf{type}\,[C] \;\wedge\; \forall \rho\,.\,\rho : C \implies A\rho \sqsubseteq B\rho$$

Lemma 20. *Analogous to the type conversion rule (theorem 11) we have.*

$$\frac{M = N : A\,[C] \qquad A \sqsubseteq B\,[C]}{M = N : B\,[C]}$$

Syntactic subtyping. We define a relation of *syntactic subtyping*, which replaces syntactic type equality, definition 15.

Definition 21. The rules of syntactic subtyping are as follows.

$$\frac{}{C \vdash \star/_M \sqsubseteq \star} \qquad \frac{M_1 = M_2 : \star\,[C]}{C \vdash (El\,M_1)/_M \sqsubseteq El\,M_2} \qquad \frac{}{C \vdash A/_M \sqsubseteq \mathbf{1}}$$

$$\frac{C, x{:}A_2 \vdash A_2/_x \sqsubseteq A_1 \qquad C, x{:}A_2 \vdash B_1/_{(M\,x)} \sqsubseteq B_2}{C \vdash (\mathsf{fun}\,A_1\,x.B_1)/_M \sqsubseteq \mathsf{fun}\,A_2\,x.B_2}$$

$$\frac{C \mid\; A/_M \sqsubseteq B \qquad M = N : B\,[C]}{C \vdash A/_M \sqsubseteq B/_N} \tag{8}$$

$$\frac{l{:}A'/_N \;in\; S/_M \qquad C \vdash A'/_N \sqsubseteq A \qquad C \vdash S/_M \sqsubseteq B[N]}{C \vdash S/_M \sqsubseteq <l{:}A,\, x.B>} \tag{9}$$

where the relation $l{:}A \;in\; S$ is defined by the following rules.

$$\frac{}{l{:}A/_N \;in\; <l{:}A/_N,\, x.B>/_M} \qquad \frac{A \text{ not singleton}}{l{:}A/_{M.l} \;in\; <l{:}A,\, x.B>/_M} \tag{10}$$

$$\frac{l{:}A/_N \;in\; B[M.l']/_M \qquad l \neq l'}{l{:}A/_N \;in\; <l'{:}A',\, x.B>/_M}$$

This formulation computes the principal type of a record directly, rather than using an auxiliary operation of *signature strengthening*, as in [18,9]. This is most clearly seen in the second rule (10).

Syntactic subtyping is decidable and sound.

Lemma 22. *If* $A_1\,\mathsf{type}\,[C]$ *and* $A_2\,\mathsf{type}\,[C]$ *then* $C \vdash A_1 \sqsubseteq A_2$ *is decidable and implies* $A_1 \sqsubseteq A_2\,[C]$.

5.3 Typechecking with Signatures, Records, and Subtyping

Extend the syntax of expressions.

$$\begin{array}{rl} e ::= & z \mid e\,e \mid \texttt{[}z\texttt{:}e\texttt{]}e \mid \texttt{*} \mid \texttt{El}\,e \mid \{z\texttt{:}e\}e \mid e \texttt{->} e \mid \texttt{[}z\texttt{=}e\texttt{]}e \mid \\ & \texttt{<>} \mid \texttt{()} \mid \texttt{<}l\texttt{:}e\texttt{,}\ e\texttt{>} \mid \texttt{<}l\texttt{=}e\texttt{,}\ e\texttt{>} \mid \texttt{(}l\texttt{=}e\texttt{,}\ e\texttt{)}. \end{array}$$

$\texttt{<>}$ is the empty signature; $\texttt{()}$ the empty record. $\texttt{<}l\texttt{:}e\texttt{,}\ e\texttt{>}$ (resp. $\texttt{<}l\texttt{=}e\texttt{,}\ e\texttt{>}$) is an *opaque* (resp. *manifest*) signature extension. $\texttt{(}l\texttt{=}e\texttt{,}\ e\texttt{)}$ is record extension.

In order to use the syntactic subtyping relation in typechecking, we replace the typechecking rule for applications, equation (6), by

$$\frac{C \vdash e_1 \Rightarrow M_1 : \mathsf{fun}\, A_1\, x.B \qquad C \vdash e_2 \Rightarrow M_2 : A_2 \qquad C \vdash A_2/_{M_2} \sqsubseteq A_1}{C \vdash e_1\, e_2 \Rightarrow M_1\, M_2 : B[M_2]} \tag{11}$$

Note how the third premise of this rule calls syntactic subtyping with a type-correct singleton LHS. This goes some way towards clarifying the rules of definition 21.

We have new typechecking rules for signature formation

$$\frac{}{C \vdash \texttt{<>} \Rightarrow \mathbf{1}} \qquad \frac{C \vdash e_1 \Rightarrow A \qquad l \notin C \qquad C, l{:}A \vdash e_2 \Rightarrow B}{C \vdash \texttt{<}l\texttt{:}e_1\texttt{,}\ e_2\texttt{>} \Rightarrow \texttt{<}l{:}A,\ l.B\texttt{>}}$$

$$\frac{C \vdash e_1 \Rightarrow M : A \qquad l \notin C \qquad C, l{:}A/_M \vdash e_2 \Rightarrow B}{C \vdash \texttt{<}l\texttt{=}e_1\texttt{,}\ e_2\texttt{>} \Rightarrow \texttt{<}l{:}A/_M,\ l.B\texttt{>}}$$

introduction

$$\frac{}{C \vdash \texttt{()} \Rightarrow () : \mathbf{1}} \qquad \frac{C \vdash e_1 \Rightarrow M_1 : A \qquad C \vdash e_2 \Rightarrow M_2 : B \qquad x \notin C}{C \vdash \texttt{(}l\texttt{=}e_1\texttt{,}\ e_2\texttt{)} \Rightarrow (l{=}M_1,\ M_2) : \texttt{<}l{:}A/_{M_1},\ x.B\texttt{>}}$$

and elimination.

$$\frac{C \vdash e \Rightarrow M : S \qquad l{:}B/_N\ \mathit{in}\ S/_M}{C \vdash e.l \Rightarrow N : B}$$

Semantical soundness and termination of typechecking follow as expected.

Signature strengthening. Suppose $\texttt{x:<a:*>}$ is declared. Can $\texttt{x}$ be used where an expression of type $\texttt{<a=x.a>}$ is expected? We have $C \vdash \texttt{<a:*>} \Rightarrow \texttt{<}a{:}\star\texttt{>}$ and $C \vdash \texttt{<a=x.a>} \Rightarrow \texttt{<}a{:}\star/_{x.a}\texttt{>}$ By rule (11), the question reduces to deciding $C \vdash \texttt{<}a{:}\star\texttt{>}/_x \sqsubseteq \texttt{<}a{:}\star/_{x.a}\texttt{>}$. This is derivable by rules (9) and (10).

6 Conclusion and Further Work

There is some way to go before our system is a usable modular logic. What we can say now is that the proposal in this paper is metamathematically well founded and directly implementable. It has most of the features of functional language module systems. Here are some of the issues we intend to work on.

Type Families and Kinds. The present system does not accept the expression `[a:*]El a -> *` (for the set of unary predicates over arbitrary type `a`) although for any particular `a:*`, `El a -> *` is well typed. In order to allow such expressions to be typed, we introduce new syntactic classes of *type families* and *kinds.*

$$\begin{array}{rcl} F & ::= & A \mid \lambda x.F \mid F\,M \\ K & ::= & \square \mid \mathsf{fun}\,A\,x.K \end{array}$$

Analogous to definition 1, we say what it means to be a categorical kind. Then the development proceeds for type families classified by kinds, just as it did for objects classified by types. Type family application is clearly normalising, and no new problems arise.

The `with` *notation.* A convenient feature of many functional programming module systems is the ability to add new manifest constraints to existing signatures. For example if `grpSig` and `monSig` are the signatures of groups and monoids respectively, perhaps from an existing library, one wants to define

```
ringSig = <G:grpSig, M:monSig with crr=G.crr, ...>
```

where the `with` clause states that the group and monoid share the same carrier. We have formalised and implemented this in earlier versions of our system. The model we have given can already interpret this notation.

Inductive sets. For practical applcation we must be able to extend the framework with new inductively defined sets: booleans, natural numbers, lists, ... It is clear that definitions such as the naturals can be added to our model. Let N, 0, S and rec be new objects. Add the expected computation rules for these new objects. Now choose $\overline{\star}$ such that $N = N : \overline{\star}$ (definition 1 is parameterised by saturated $\overline{\star}$) and define $El\,N$ appropriately.

Proof irrelevance. Adding a type Prop (analogous to $\star$) with a type family $Proof\ M$ (analogous to $El\,M$), where $\eta\{Proof\ M\} = \lambda u.()$, we get a model for a type theory with proof irrelevance and decidability of equality.

What type theory is it? Our typechecking rules have an unusual form. Can our model interpret the official rules of Martin-Löf's framework [23], or Luo's variation [19]?

Information hiding. One important aspect we have not yet addressed is information hiding, so that a module can be replaced by any other having the same signature.

References

1. Stuart Allen. *A Non-Type-Theoretic Semantics For Type-Theoretic Language.* PhD thesis, Cornell Univ., 1987. Report TR87-866.

2. David Aspinall. Subtyping with singleton types. In *Proc. Computer Science Logic, CSL'94*, volume 933 of *LNCS*, 1995.
3. Lennart Augustsson. Cayenne – a language with dependent types. In *ICFP '98*, volume 34(1) of *SIGPLAN Notices*, pages 239–250. ACM, June 1999.
4. H. P. Barendregt. *The Lambda Calculus: Its Syntax and Semantics*, volume 103 of *Studies in Logic and the Foundations of Mathematics*. North-Holland, revised edition, 1984.
5. G. Betarte. *Dependent Record Types and Formal Abstract Reasoning*. PhD thesis, Chalmers Univ. of Technology, 1998.
6. G. Betarte and A. Tasistro. Extension of Martin-Löf's type theory with record types and subtyping. In G. Sambin and J. Smith, editors, *Twenty Five Years of Constructive Type Theory*. Oxford Univ. Press, 1998.
7. Coq. The Coq Project, 2002. http://coq.inria.fr/.
8. Thierry Coquand. An algorithm for testing conversion in type theory. In G. Huet and G.D. Plotkin, editors, *Logical Frameworks*. Cambridge Univ. Press, 1991.
9. J. Courant. MC: A module calculus for Pure Type Systems. Technical Report 1217, CNRS Université Paris Sud 8623: LRI, June 1999.
10. Judicaël Courant. Strong normalization with singleton types. In Stefan Van Bakel, editor, *Second Workshop on Intersection Types and Related Systems*, volume 70 of *Electronic Notes in Computer Science*. Elsevier, 2002.
11. Jean-Yves Girard. Locus solum: From the rules of logic to the logic of rules. *Mathematical Structures in Computer Science*, 11(3):301–506, 2001.
12. R. Harper and M. Lillibridge. A type-theoretic approach to higher-order modules with sharing. In *POPL'94*. ACM Press, 1994.
13. Robert Harper and Frank Pfenning. On equivalence and canonical forms in the LF type theory. (Submitted for publication.), October 2001.
14. S. Hayashi. Singleton, union and intersection types for program extraction. *Information and Computation*, 109:174–210, 1994.
15. Alexei Kopylov. Dependent intersection: A new way of defining records in type theory. Technical Report TR2000-1809, Cornell University, 2000.
16. LEGO. The LEGO Proof Assistant, 2001. `www.dcs.ed.ac.uk/home/lego/`.
17. X. Leroy. Manifest types, modules, and separate compilation. In *POPL'94*, New York, NY, USA, 1994. ACM Press.
18. X. Leroy. A syntactic theory of type generativity and sharing. *Journal of Functional Programming*, 6(5):667–698, September 1996.
19. Z. Luo. *Computation and Reasoning: A Type Theory for Computer Science*. International Series of Monographs on Computer Science. Oxford Univ. Press, 1994.
20. D. MacQueen. Using dependent types to express modular structure. In *POPL'86*, 1986.
21. J. McKinna and R. Pollack. Some lambda calculus and type theory formalized. *Journal of Automated Reasoning*, 23(3–4), November 1999.
22. Alexandre Miquel. The implicit calculus of constructions. In S. Abramsky, editor, *5th Conf. on Typed Lambda Calculi and Applications, TLCA'01*, volume 2044 of *LNCS*. Springer-Verlag, 2001.
23. B. Nordström, K. Petersson, and J. Smith. Martin-löf's type theory. In Abramsky, Gabbai, and Maibaum, editors, *Handbook of Logic in Computer Science*, volume 5. Oxford Univ. Press, 2001.
24. Robert Pollack. Dependently typed records in type theory. *Formal Aspects of Computing*, 13:386–402, 2002.
25. Christopher Stone. *Singleton Kinds and Singleton Types*. PhD thesis, Carnegie Mellon University, 2000. Report CMU-CS-00-153.

A Sound and Complete CPS-Translation for $\lambda\mu$-Calculus

Ken-etsu Fujita

Department of Mathematics and Computer Science, Shimane University,
Matsue 690-8504, Japan
fujiken@cis.shimane-u.ac.jp
http://www.cis.shimane-u.ac.jp/~fujiken

Abstract. We investigate injectivity of the novel CPS-translation with surjective pairing which is originally introduced by Hofmann-Streicher. It is syntactically proved that the CPS-translation is sound and complete not only for the λ-calculus but also for the extensional $\lambda\mu$-calculus. The injective CPS-translation reveals a Church-Rosser fragment of the λ-calculus with surjective pairing and a neat connection to C-monoids.

1 Introduction

Parigot [13,14] introduced the $\lambda\mu$-calculus from the viewpoint of classical logic, and established an extension of the Curry-Howard isomorphism [10,7,12]. From the motivation of a universally computational point of view, we investigate type free $\lambda\mu$-calculus with extensionality [17,2,5].

In terms of a category of continuations, it is proved that for any $\lambda\mu$-theory a continuation semantics of typed $\lambda\mu$-calculus is sound and complete by Hofmann and Streicher [9]. Selinger [16] proposed the control category to establish an isomorphism between call-by-name and call-by-value $\lambda\mu$-calculi with conjunction and disjunction types. In Streicher and Reus [17], the category of negated domains is applied for a model of type free $\lambda\mu$-calculus. They remarked that the traditional CPS-translation naïvely based on Plotkin [15] cannot validate (η)-rule. All of the work [9,16,17] introduced a novel CPS-translation which requires, at least, products as a primitive notion, so that (η)-rule can be validated by the surjective pairing, as observed in [4].

In this paper, we show that the novel CPS-translation with surjective pairing is injective. It is syntactically proved that the CPS-translation is sound and complete not only for the extensional λ-calculus but also for the extensional $\lambda\mu$-calculus. As a corollary the injective CPS-translation reveals a Church-Rosser fragment of the λ-calculus with surjective pairing, which is not Church-Rosser as proved by Klop [1]. Moreover there exists an injection from $\lambda\mu$-calculus into C-monoids, since one has a one-to-one correspondence between the λ-calculus with surjective pairing and C-monoids as proved by Curien [3] and by Lambek and Scott [11]. Along the line of Plotkin [15], this work can also be regarded as a call-by-value simulation of call-by-name $\lambda\mu$-calculus with (η)-rule. It is remarked that the completeness in [15] has been proved by the essential use of

M. Hofmann (Ed.): TLCA 2003, LNCS 2701, pp. 120–134, 2003.

the Church-Rosser property of the target calculus. However, our target calculus is not Church-Rosser as stated above. In order to define an inverse translation, we introduce a context-free grammar which describes the image of the CPS-translation. Although this paper handles only type free $\lambda\mu$-calculus, our main theorems are still valid under typed $\lambda\mu$-calculus.

2 CPS-Translation of λ-Calculus into λ-Calculus with Surjective Pairing

2.1 Extensional λ-Calculus and CPS-Translation

First we show a preliminary result that the novel CPS-translation is sound and complete for type free λ-calculus with (η)-rule.

Definition 1 (λ-Calculus Λ). $\Lambda \ni M ::= x \mid \lambda x.M \mid MM$

(β) $(\lambda x.M_1)M_2 \to M_1[x := M_2]$
(η) $\lambda x.Mx \to M$ *if* $x \notin FV(M)$

Definition 2 (λ-Calculus with surjective pairing $\Lambda^{\langle\rangle}$).
$\Lambda^{\langle\rangle} \ni M ::= x \mid \lambda x.M \mid MM \mid \langle M, M\rangle \mid \pi_1(M) \mid \pi_2(M)$

(β) $(\lambda x.M_1)M_2 \to M_1[x := M_2]$
(η) $\lambda x.Mx \to M$ *if* $x \notin FV(M)$
(π) $\pi_i\langle M_1, M_2\rangle \to M_i \quad (i = 1, 2)$
(sp) $\langle \pi_1(M), \pi_2(M)\rangle \to M$

$FV(M)$ stands for the set of free variables in M. The one step reduction relation is denoted by $\to_R$ where R is (β), (η), $(\beta) + (\eta)$, $\lambda^{\langle\rangle}(= (\beta) + (\eta) + (\pi) + (\text{sp}))$, etc. We write $\to_R^+$ and $\to_R^*$ to denote the transitive closure and the reflexive and transitive closure of $\to_R$, respectively. We employ the notation $=_R$ to indicate the symmetric, reflexive and transitive closure of $\to_R$. The binary relation $\equiv$ denotes the syntactic identity.

Definition 3 (CPS-Translation : $\Lambda \to \Lambda^{\langle\rangle}$).

(i) $[\![x]\!] = x$
(ii) $[\![\lambda x.M]\!] = \lambda a.(\lambda x.[\![M]\!])(\pi_1 a)(\pi_2 a)$
(iii) $[\![M_1 M_2]\!] = \lambda a.[\![M_1]\!]\langle [\![M_2]\!], a\rangle$

Example 1. It is instructive to calculate the following where $m, n \geq 0$:

$$[\![\lambda x_1 \dots x_m.xM_1 \cdots M_n]\!] \to_\beta^+ \lambda a.x\langle [\![M_1]\!], \dots \langle [\![M_n]\!], \pi_2^m a\rangle \dots\rangle$$
$$[x_1 := \pi_1 a, x_2 := \pi_1(\pi_2 a), \dots, x_m := \pi_1(\pi_2^{m-1} a)]$$

Proposition 1 (Soundness). *Let* $M_1, M_2 \in \Lambda$.
If we have $M_1 \to_{\beta\eta} M_2$ *then* $[\![M_1]\!] \to_{\lambda^{\langle\rangle}}^+ [\![M_2]\!]$.

Proof. By induction on the derivation of $M_1 \to_{\beta\eta} M_2$. □

It is remarked that Proposition 1 holds true even under the call-by-value computation as follows:

$$V ::= x \mid \lambda x.M \mid \langle V, V\rangle \mid \pi_1(V) \mid \pi_2(V)$$

(β_v) $(\lambda x.M)V \to M[x := V]$
(η_v) $\lambda x.Vx \to V$
(π_v) $\pi_i\langle V_1, V_2\rangle \to V_i \quad (i = 1, 2)$
(sp_v) $\langle \pi_1 V, \pi_2 V\rangle \to V$

Hence this work can be regarded as a call-by-value simulation of call-by-name $\lambda\mu$-calculus with (η)-rule.

2.2 Universe of the CPS-Translation

We will give a definition of the inverse translation to each element of the universe of the CPS-translation:

$$Univ_\lambda \stackrel{\text{def}}{=} \{P \in \Lambda^{\langle\rangle} \mid [\![M]\!] \to^*_{\lambda\langle\rangle} P \text{ for some } M \in \Lambda\}$$

Every element in the universe will be generated by the following context-free grammar which is essentially applied in this work:

$$\mathcal{R} ::= x \mid \pi_1\mathcal{K} \mid (\lambda x.\mathcal{R})\mathcal{R} \mid \lambda a.\mathcal{R}\mathcal{K}$$
$$\mathcal{K} ::= a \mid \pi_2\mathcal{K} \mid \langle \mathcal{R}, \mathcal{K}\rangle$$

Lemma 1 (Subject reduction property).
The categories $\mathcal{R}$ and $\mathcal{K}$ are closed under the following reductions:
(β_x) $(\lambda x.R_1)R_2 \to R_1[x := R_2]$
(β_a) $(\lambda a.RK_1)K_2 \to RK_1[a := K_2]$
(η_a) $\lambda a.Ra \to R$ *if* $a \notin FV(R)$
$(\pi_{R,K})$ $\pi_1\langle R, K\rangle \to R$ *and* $\pi_2\langle R, K\rangle \to K$
(sp_K) $\langle \pi_1(K), \pi_2(K)\rangle \to K$

Proof. Because we have that $R_1[x := R_2] \in \mathcal{R}, K[x := R] \in \mathcal{K}$; and that $R[a := K] \in \mathcal{R}, K_1[a := K_2] \in \mathcal{K}$. □

Proposition 2. *$Univ_\lambda \subseteq \mathcal{R}$, i.e., $Univ_\lambda$ is generated by $\mathcal{R}$.*

Proof. From definition 3, $[\![M]\!] \in \mathcal{R}$ for any $M \in \Lambda$. Moreover, from Lemma 1, $\mathcal{R}$ and $\mathcal{K}$ are closed under the reductions, and hence $Univ_\lambda \subseteq \mathcal{R}$. □

There uniquely exists a projection normal form by the sole use of $(\pi_{R,K})$, and the projection normal form of K is in the following form:

$$K_{nf} ::= \pi_2^n a \mid \langle R_{nf}, K_{nf}\rangle$$

where $n \geq 0$. For a technical reason, an occurrence of a single variable $a \in \mathcal{K}$, i.e., $\pi_2^n a$ where $n = 0$ is handled as an (sp_K)-expansion form; $\langle \pi_1 a, \pi_2 a\rangle$. Under this consideration, K can be supposed to be in the form of $\langle R_1, \ldots, \langle R_m, \pi_2^n a\rangle \ldots\rangle$ with $m \geq 0, n \geq 1$.

Lemma 2 (π-normal form). *Let $m \geq 0$ and $n \geq 1$.*
Then every element in the universe Univ_λ is one of the following forms up to $(\pi_{R,K})$-reductions and (sp_K)-expansions:

(1) x
(2) $\pi_1(\pi_2^i a)$ *for some* $i \geq 0$
(3) $(\lambda x.R)R_1$
(4) $\lambda a.R\langle R_1, \ldots, \langle R_m, \pi_2^n a\rangle \ldots\rangle$ *for some* $m \geq 0$ *and* $n \geq 1$

where R and R_i $(1 \leq i \leq m)$ are in the form of (1), (2), (3), or (4) above.
We call the occurrence of $(\pi_2^n a)$ in the case of (4) above a tail with the variable a.
Moreover, the following property is satisfied under renaming of bound variables:

(i) *For each λa, there exists a unique occurrence of a tail $\pi_2^n a$ for some $n \geq 1$;*
(ii) *For any $\pi_1(\pi_2^i a)$ as a proper subterm where $i \geq 0$, there exists the least $n \geq 1$ and a unique tail $\pi_2^n a$ with the variable a such that the condition $i + 1 \leq n$ holds true.*

Proof. First obtain a projection normal form only by the use of $(\pi_{R,K})$-reductions, and then check whether the form has the condition (ii). The application of (sp_K)-expansion guarantees that (ii) holds true. □

2.3 Inverse Translation $\natural$

π-normal forms above play a role of representatives of the image Univ_λ under the translation $[\![-]\!]$. We give the definition of the inverse translation $\natural$ to every element in

$$\pi\text{-nf}(\mathit{Univ}_\lambda) \stackrel{\mathrm{def}}{=} \{\pi\text{-normal}(P) \in \Lambda^{\langle\rangle} \mid [\![M]\!] \to^*_{\lambda\langle\rangle} P \text{ for some } M \in \Lambda\}.$$

That is, $P^\natural = (\pi\text{-normal}(P))^\natural$ for any $P \in \mathit{Univ}_\lambda$.

Definition 4 (Inverse Translation $\natural : \pi$-nf(Univ_λ) $\to \Lambda$).

(1) $x^\natural = x$
(2) $(\pi_1(\pi_2^i a))^\natural = a_{i+1} \quad (i \geq 0)$
(3) $((\lambda x.R)R_1)^\natural = (\lambda x.R^\natural)R_1^\natural$
(4) $(\lambda a.R\langle R_1, \ldots \langle R_m, \pi_2^n a\rangle \ldots\rangle)^\natural = \lambda a_1 \ldots a_n.R^\natural R_1^\natural \cdots R_m^\natural \ (m \geq 0, n \geq 1)$

Lemma 3. *For any $M \in \Lambda$, we have $[\![M]\!]^\natural \to^*_\eta M$.*

Proof. By induction on the structure of $M \in \Lambda$. □

Lemma 4. *Let $R, R_1, \ldots, R_n \in \mathcal{R}$.*

(1) $R^\natural[x := R_1^\natural] = (R[x := R_1])^\natural$
(2) $(R[b := \langle R_1, \ldots, \langle R_m, \pi_2^n a\rangle \ldots\rangle])^\natural$
$= R^\natural[b_1 := R_1^\natural, \ldots, b_m := R_m^\natural, b_{m+1} := a_{n+1}, b_{m+2} := a_{n+2}, \cdots]$
under the simultaneous substitution where $m \geq 0$ and $n \geq 1$.

Proof. By straightforward induction on the structure of R. We show the base case for (2).

Case of $i+1 \leq m$:
$(\pi_1(\pi_2^i b)[b := \langle R_1, \ldots, \langle R_m, \pi_2^n a\rangle \ldots\rangle])^\natural$
$= (\pi_1(\pi_2^i \langle R_1, \ldots, \langle R_m, \pi_2^n a\rangle \ldots\rangle))^\natural$
$= R_{i+1}^\natural = (\pi_1(\pi_2^i b))^\natural [b_{i+1} := R_{i+1}^\natural]$

Case of $i+1 > m$:
$(\pi_1(\pi_2^i b)[b := \langle R_1, \ldots, \langle R_m, \pi_2^n a\rangle \ldots\rangle])^\natural$
$= (\pi_1(\pi_2^i \langle R_1, \ldots, \langle R_m, \pi_2^n a\rangle \ldots\rangle))^\natural$
$= (\pi_1(\pi_2^{n+i-m} a))^\natural = a_{n+i-m+1}$
$= (\pi_1(\pi_2^i b))^\sharp [b_{m+1} := a_{n+1}, \ldots]$
$= b_{i+1}[b_{m+1} := a_{n+1}, \ldots] = a_{i+1-m+n}$ □

Proposition 3 (Completeness). *Let* $P, Q \in Univ_\lambda$.

(1) *If* $P \to_{\beta_x} Q$ *then* $P^\natural \to_\beta Q^\natural$.
(2) *If* $P \to_{\beta_a} Q$ *then* $P^\natural \to_\beta^+ Q^\natural$.
(3) *If* $P \to_{\eta_a} Q$ *then* $P^\natural \to_\eta Q^\natural$.
(4) *If* $P \to_{\pi_{R,K}} Q$ *then* $P^\natural \equiv Q^\natural$.
(5) *If* $P \to_{\mathrm{sp}_K} Q$ *then* $P^\natural \to_\eta^* Q^\natural$.

Proof. By induction on the derivations. We show one case for (2), and other cases are straightforward.

Let K be $\langle S_1, \ldots, \langle S_q, \pi_2^p b\rangle \ldots\rangle$ with $q \geq 0, p \geq 1$, and K' be $\langle R_1, \ldots, \langle R_m, \pi_2^n a\rangle \ldots\rangle$ with $m \geq 0, n \geq 1$. Let θ be $[b := K']$. Now we prove the case P of $\lambda a.(\lambda b.RK)K'$:

$$(\lambda a.(\lambda b.RK)K')^\natural = \lambda a_1 \ldots a_n.(\lambda b_1 \ldots b_p.R^\natural S_1^\natural \cdots S_q^\natural) R_1^\natural \cdots R_m^\natural$$

Case of $p+1 \leq m$:
$\lambda a.(\lambda b.RK)K')^\natural$
$\to_\beta^+ \lambda a_1 \ldots a_n.(R^\natural S_1^\natural \cdots S_q^\natural)[b_1 := R_1^\natural, \ldots, b_p := R_p^\natural] R_{p+1}^\natural \cdots R_m^\natural$
$= \lambda a_1 \ldots a_n.(R^\natural S_1^\natural \cdots S_q^\natural)$
$[b_1 := R_1^\natural, \ldots, b_m := R_m^\natural, b_{m+1} := a_{m+1}, b_{m+2} := a_{m+2}, \ldots] R_{p+1}^\natural \cdots R_m^\natural$
since none of $b_{p+1}, b_{p+2}, \ldots$ appears in $R^\natural, S_1^\natural, \ldots, S_q^\natural$
$= \lambda a_1 \ldots a_n.((R\theta)^\natural (S_1\theta)^\natural \cdots (S_q\theta)^\natural) R_{p+1}^\natural \cdots R_m^\natural$
by Lemma 4
$= (\lambda a.R\theta\langle S_1\theta, \ldots, \langle S_q\theta, \langle R_{p+1}, \ldots, \langle R_m, \pi_2^n a\rangle \ldots\rangle\rangle \ldots\rangle)^\natural$
$= (\lambda a.R\theta\langle S_1\theta, \ldots, \langle S_q\theta, \pi_2^p\langle R_1, \ldots, \langle R_m, \pi_2^n a\rangle \ldots\rangle\rangle \ldots\rangle)^\natural$
$= (\lambda a.RK[b := K'])^\natural$

Another case of $p+1 > m$ can be verified similarly. □

Theorem 1. *Let* $M_1, M_2 \in \Lambda$. $M_1 =_{\beta\eta} M_2$ *if and only if* $[\![M_1]\!] =_{\lambda^{\langle\rangle}} [\![M_2]\!]$.

Proof. From Propositions 1 and 3 and Lemma 3. □

Corollary 1. *Let* $[\![\Lambda]\!] \stackrel{\text{def}}{=} \{[\![M]\!] \in \Lambda^{\langle\rangle} \mid M \in \Lambda\}$. *Then* $[\![\Lambda]\!]$ *is a Church-Rosser subset, in the sense that if* $P \to_{\lambda^{\langle\rangle}}^* P_1$ *and* $P \to_{\lambda^{\langle\rangle}}^* P_2$ *where* $P, P_1, P_2 \in [\![\Lambda]\!]$ *then there exists some* $Q \in [\![\Lambda]\!]$ *such that* $P_1 \to_{\lambda^{\langle\rangle}}^* Q$ *and* $P_2 \to_{\lambda^{\langle\rangle}}^* Q$.

3 Type Free $\lambda\mu$-Calculus

3.1 Extensional $\lambda\mu$-Calculus and CPS-Translation

We give the definition of type free $\lambda\mu$-calculus with (η). The syntax of the $\lambda\mu$-terms is defined from variables, λ-abstraction, application, or μ-abstraction over names denoted by α, where a term in the form of $[\alpha]M$ is called a named term.

Definition 5 ($\lambda\mu$-calculus).

$$\Lambda\mu \ni M ::= x \mid \lambda x.M \mid MM \mid \mu\alpha.N \qquad\qquad N ::= [\alpha]M$$

(β) $(\lambda x.M_1)M_2 \to M_1[x := M_2]$
(η) $\lambda x.Mx \to M$ *if* $x \notin FV(M)$
(μ) $(\mu\alpha.N)M \to \mu\alpha.N[\alpha \Leftarrow M]$
(μ_β) $\mu\alpha.[\beta](\mu\gamma.N) \to \mu\alpha.N[\gamma := \beta]$
(μ_η) $\mu\alpha.[\alpha]M \to M$ *if* $\alpha \notin FN(M)$

$FN(M)$ stands for the set of free names in M. The $\lambda\mu$-term $N[\alpha \Leftarrow M]$ denotes a term obtained by replacing each subterm of the form $[\alpha]M'$ in N with $[\alpha](M'M)$. This operation is inductively defined as follows:

1. $x[\alpha \Leftarrow M] = x$
2. $(\lambda x.M_1)[\alpha \Leftarrow M] = \lambda x.M_1[\alpha \Leftarrow M]$
3. $(M_1M_2)[\alpha \Leftarrow M] = (M_1[\alpha \Leftarrow M])(M_2[\alpha \Leftarrow M])$
4. $(\mu\beta.N)[\alpha \Leftarrow M] = \mu\gamma.N[\beta := \gamma][\alpha \Leftarrow M]$ where γ is a fresh name
5. $([\beta]M_1)[\alpha \Leftarrow M] = \begin{cases} [\beta]((M_1[\alpha \Leftarrow M])M), & \text{for } \alpha \equiv \beta \\ [\beta](M_1[\alpha \Leftarrow M]), & \text{otherwise} \end{cases}$

The term $M[\alpha \Leftarrow M_1, \ldots, M_n]$ denotes $M[\alpha \Leftarrow M_1] \cdots [\alpha \Leftarrow M_n]$.

The binary relation $=_{\lambda\mu}$ over $\Lambda\mu$ denotes the symmetric, reflexive and transitive closure of the one step reduction relation, i.e., the equivalence relation induced from the reduction rules.

Fact 1 **(1)** *For* $\mu\alpha.N \in \Lambda\mu$, $\mu\alpha.N =_{\lambda\mu} \lambda x.\mu\alpha.N[\alpha \Leftarrow x]$ *where* x *is fresh.*
(2) *Let* $M_1, M_2 \in \Lambda\mu$, *and* $x_i \notin FV(M_1M_2)$ *for* $1 \le i \le n$.
If we have $M_1[\alpha \Leftarrow x_1, \ldots, x_n] =_{\lambda\mu} M_2[\alpha \Leftarrow x_1, \ldots, x_n]$ *then* $M_1 =_{\lambda\mu} M_2$.

Proof. Here we show a proof only for (2):

$$\begin{aligned}
& M_1[\alpha \Leftarrow x_1, \ldots, x_n] =_{\lambda\mu} M_2[\alpha \Leftarrow x_1, \ldots, x_n] \\
& \Longrightarrow \mu\alpha.[\beta](M_1[\alpha \Leftarrow x_1, \ldots, x_n]) =_{\lambda\mu} \mu\alpha.[\beta](M_2[\alpha \Leftarrow x_1, \ldots, x_n]) \\
& \qquad \text{where } \beta \text{ is a fresh name} \\
& \Longrightarrow \lambda x_1 \ldots x_n.\mu\alpha.[\beta](M_1[\alpha \Leftarrow x_1, \ldots, x_n]) \\
& \qquad\qquad =_{\lambda\mu} \lambda x_1 \ldots x_n.\mu\alpha.[\beta](M_2[\alpha \Leftarrow x_1, \ldots, x_n]) \\
& \Longleftrightarrow \mu\alpha.[\beta]M_1 =_{\lambda\mu} \mu\alpha.[\beta]M_2 \qquad \text{by (1) above} \\
& \Longrightarrow \mu\beta.[\alpha](\mu\alpha.[\beta]M_1) =_{\lambda\mu} \mu\beta.[\alpha](\mu\alpha.[\beta]M_2) \\
& \Longleftrightarrow M_1 =_{\lambda\mu} M_2
\end{aligned}$$

□

Definition 6 (CPS-Translation : $\Lambda\mu \to \Lambda^{\langle\rangle}$).

(i) $[\![x]\!] = x$
(ii) $[\![\lambda x.M]\!] = \lambda a.(\lambda x.[\![M]\!])(\pi_1 a)(\pi_2 a)$
(iii) $[\![M_1 M_2]\!] = \lambda a.[\![M_1]\!]\langle[\![M_2]\!], a\rangle$
(iv) $[\![\mu a.[b]M]\!] = \lambda a.[\![M]\!]b$

Proposition 4 (Soundness). *Let $M_1, M_2 \in \Lambda\mu$. If we have $M_1 =_{\lambda\mu} M_2$ then $[\![M_2]\!] =_{\lambda^{\langle\rangle}} [\![M_2]\!]$.*

Proof. By induction on the derivation of $M_1 =_{\lambda\mu} M_2$, together with the following facts:

$$[\![M_1]\!][x := [\![M_2]\!]] \;=\; [\![M_1[x := M_2]]\!]$$
$$[\![M_1[\alpha \Leftarrow M_2]]\!] \to^*_\beta [\![M_1]\!][\alpha := \langle[\![M_2]\!], \alpha\rangle]$$ □

3.2 Universe of the CPS-Translation and Inverse Translation ♯

Following the idea of π-normal forms, we consider terms in the universe:

$$Univ_\mu \stackrel{\text{def}}{=} \{P \in \Lambda^{\langle\rangle} \mid [\![M]\!] \to^*_{\lambda^{\langle\rangle}} P \text{ for some } M \in \Lambda\mu\}$$

up to projections and (sp_K)-expansions.

If $\pi_1(\pi_2^i a)$ with $i \geq 0$ appears in R or R_j $(1 \leq j \leq m)$ of $\lambda a.R\langle R_1, \ldots, \langle R_m, \pi_2^n b\rangle \ldots\rangle$, then we can assume that there always exists the corresponding tail $\pi_2^n a$ with $0 \leq i < n$. Otherwise, an application of (β_a)-expansion makes this property hold true. Suppose that $\pi_1(\pi_2^i a)$ with $i \geq 0$ appears in R or R_j $(1 \leq j \leq m)$ but there is no corresponding tail $(a \not\equiv b)$. Then the following (β_a)-expansion produces the corresponding tail, and moreover the successive applications of $(\mathrm{sp_K})$-expansions guarantee that the condition $i + 1 \leq p$ holds for the tail $\pi_2^p a$:

$$\lambda a.R\langle R_1, \ldots, \langle R_m, \pi_2^n b\rangle \ldots\rangle =_\beta \lambda a.(\lambda c.R\langle R_1, \ldots, \langle R_m, \pi_2^n b\rangle \ldots\rangle)(\pi_2 a) \in \mathcal{R}$$

where c is fresh and $a \not\equiv b$.
For instance, the term

$$[\![\lambda y.\mu\alpha.[\alpha](y(\lambda x.\mu\beta.[\alpha]x))]\!] =_\beta \lambda a.\pi_1 a\langle\lambda b.\underline{\pi_1 b}(\pi_2 a), \pi_2 a\rangle$$

contains a subterm in the form of $\underline{\pi_1 b}$ which has no corresponding tail. The term above is handled as the following form with a fresh variable c:

$$\lambda a.\pi_1 a\langle\lambda b.(\lambda c.\pi_1 b(\pi_2 a))(\pi_2 b), \pi_2 a\rangle$$

Under this consideration of π-normal forms and the tail-producing (β_a)-expansions, the inverse translation will be defined for the following form of elements in $Univ_\mu$, denoted by π-$\mathcal{NF}$ $(Univ_\mu)$.

Lemma 5 (π-normal form). *Let $m \geq 0$ and $n \geq 1$. Then every element in the universe $\{P \in \Lambda^{\langle\rangle} \mid [\![M]\!] \to^*_{\lambda^{\langle\rangle}} P$ for some $M \in \Lambda\mu\}$ is one of the following forms up to $(\pi_{R,K})$-reductions, (sp_K)-expansions, and (β_a)-expansions:*

(1) x
(2) $\pi_1(\pi_2^i a)$ *for some* $i \geq 0$
(3) $(\lambda x.R)R_1$
(4) $\lambda a.R\langle R_1, \ldots, \langle R_m, \pi_2^n b\rangle \ldots\rangle$ *for some* $m \geq 0$ *and* $n \geq 1$

where R *and* R_i $(1 \leq i \leq m)$ *are in the form of* (1), (2), (3), *or* (4) *above. Moreover, for any* $\pi_1(\pi_2^i a)$ *as a proper subterm where* $i \geq 0$, *we have the corresponding and a unique tail* $\pi_2^n a$ *with the condition* $i + 1 \leq n$.

The essential distinction between Lemmata 2 and 5 is that the variable b of a tail $\pi_2^n b$ may be different from that of the binder λa in the case of (4) above.

Proof. For any $M \in \Lambda\mu$, $[\![M]\!]$ is in the form of (1), (4), or (4) with (3) up to (sp_K)-expansions. The forms above are closed under the reductions from Lemma 1. The applications of (β_a)-expansion and (sp_K)-expansions guarantee that the condition on tails holds true. Moreover the application of (sp_K)-expansion gives a unique tail. Suppose that we have two tails $\pi_2^n a$ and $\pi_2^{n+k} a$ with $k > 0$. Then we can adopt the tail $\pi_2^{n+k} a$, and by the use of (sp_K)-expansion, $\pi_2^n a$ can be replaced with $\langle \pi_1(\pi_2^n a), \ldots, \langle \pi_1(\pi_2^{n+k-1} a), \pi_2^{n+k} a\rangle \ldots\rangle \in \mathcal{K}$. □

Definition 7 (Inverse Translation $\sharp : \pi\text{-}\mathcal{NF}(Univ_\mu) \to \Lambda\mu$).

(1) $x^\sharp = x$
(2) $(\pi_1(\pi_2^i a))^\sharp = a_{i+1} \quad (i \geq 0)$
(3) $((\lambda x.R)R_1)^\sharp = (\lambda x.R^\sharp)R_1^\sharp$
(4) $(\lambda a.R\langle R_1, \ldots \langle R_m, \pi_2^n b\rangle \ldots\rangle)^\sharp = \mu a.[b](\lambda b_1 \ldots b_n.R^\sharp R_1^\sharp \cdots R_m^\sharp)$
$(m \geq 0, n \geq 1)$

For instance, the previous example $[\![\lambda y.\mu\alpha.[\alpha](y(\lambda x.\mu\beta.[\alpha]x))]\!]$ is decompiled as follows, where $R \equiv \pi_1 a$ and $R_1 \equiv \lambda b.(\lambda c.\pi_1 b(\pi_2 a))(\pi_2 b)$ in the case of Lemma 5 (4):

$$
\begin{aligned}
&(\lambda a.\pi_1 a\langle \lambda b.\pi_1 b(\pi_2 a), \pi_2 a\rangle)^\sharp\\
&= \mu a.[a](\lambda a_1.a_1(\lambda b.\pi_1 b(\pi_2 a))^\sharp)\\
&= \mu a.[a](\lambda a_1.a_1(\lambda b.(\lambda c.\pi_1 b(\pi_2 a))(\pi_2 b))^\sharp)\\
&= \mu a.[a](\lambda a_1.a_1(\mu b.[b](\lambda b_1.\mu c.[a](\lambda a_1.b_1))))\\
&\to_{\mu_\eta} \mu a.[a](\lambda a_1.a_1(\lambda b_1.\mu c.[a](\lambda a_1.b_1)))\\
&=_{\lambda\mu} \lambda y.\mu a.([a](\lambda a_1.a_1(\lambda b_1.\mu c.[a](\lambda a_1.b_1))))[a \Leftarrow y]\\
&\to_\beta^+ \lambda y.\mu a.[a](y(\lambda b_1.\mu c.[a]b_1))
\end{aligned}
$$

Lemma 6. (1) *Let* $M \in \Lambda\mu$. $[\![M]\!]^\sharp \to_{\eta\mu_\eta}^* M$
(2) *Let* $R, R_1, \ldots, R_m \in \mathcal{R}$. *Let* $m \geq 0$ *and* $n \geq 1$.

$$
\begin{aligned}
&(R[a := \langle R_1, \ldots, \langle R_m, \pi_2^n a\rangle \ldots\rangle])^\sharp[a \Leftarrow a_1, \ldots, a_n]\\
&=_\beta R^\sharp[a \Leftarrow R_1^\sharp, \ldots, R_m^\sharp]\\
&\qquad [a_1 := R_1^\sharp, \ldots, a_m := R_m^\sharp, a_{m+1} := a_{n+1}, a_{m+2} := a_{n+2}, \ldots]
\end{aligned}
$$

where $R_1, \ldots, R_m$ *have no occurrences of* a.

Proof. **(1)** By induction on the structure of M.

(2) By induction on the structure of R.

The case R of $\pi_1(\pi_2^i b)$ is the same as the proof given in Lemma 4. We show one of the other cases:

$$(\lambda a.R\langle S_1,\ldots,\langle S_p,\pi_2^q c\rangle\ldots\rangle\theta)^\sharp\Theta' =_\beta (\lambda a.R\langle S_1,\ldots,\langle S_p,\pi_2^q c\rangle\ldots\rangle)^\sharp\Theta\theta'$$

where $\theta \equiv [b := \langle R_1,\ldots,\langle R_m,\pi_2^n b\rangle\ldots\rangle]$,

$\theta' \equiv [b_1 := R_1^\sharp,\ldots,b_m := R_m^\sharp, b_{m+1} := b_{n+1}, b_{m+2} := b_{n+2},\ldots]$,

$\Theta \equiv [b \Leftarrow R_1^\sharp,\ldots,R_m^\sharp]$, and

$\Theta' \equiv [b \Leftarrow b_1,\ldots,b_n]$.

Since the case of $c \not\equiv b$ is straightforward, we show the case of $c \equiv b$ where $a \not\equiv b$:

$(\lambda a.R\langle S_1,\ldots,\langle S_p,\pi_2^q b\rangle\ldots\rangle)^\sharp[b \Leftarrow R_1^\sharp,\cdots,R_m^\sharp]\theta'$
$= (\mu a.[b](\lambda b_1\ldots b_q.R^\sharp S_1^\sharp\cdots S_p^\sharp))\Theta\theta'$
$= \mu a.[b](\lambda b_1\ldots b_q.R^\sharp\Theta S_1^\sharp\Theta\cdots S_p^\sharp\Theta)R_1^\sharp\cdots R_m^\sharp\theta'$

Case of $q+1 \le m$:

$\mu a.[b](\lambda b_1\ldots b_q.(R^\sharp\Theta)(S_1^\sharp\Theta)\cdots(S_p^\sharp\Theta))R_1^\sharp\cdots R_m^\sharp\theta'$
$\to_\beta^+ \mu a.[b]((R^\sharp\Theta)(S_1^\sharp\Theta)\cdots(S_p^\sharp\Theta))[b_1 := R_1^\sharp,\ldots,b_q := R_q^\sharp]R_{q+1}^\sharp\cdots R_m^\sharp\theta'$
$= \mu a.[b](R^\sharp\Theta\theta')(S_1^\sharp\Theta\theta')\cdots(S_p^\sharp\Theta\theta')R_{q+1}^\sharp\cdots R_m^\sharp$
since R_i has no occurrences of b and no free variables b_j ($j > q$) appear
$=_\beta \mu a.[b]((R\theta)^\sharp\Theta')((S_1\theta)^\sharp\Theta')\cdots((S_p\theta)^\sharp\Theta')R_{q+1}^\sharp\cdots R_m^\sharp$
by the induction hypotheses
$=_\beta (\mu a.[b](\lambda b_1\ldots b_n.(R\theta)^\sharp(S_1\theta)^\sharp\cdots(S_p\theta)^\sharp R_{q+1}^\sharp\cdots R_m^\sharp))\Theta'$
$= (\lambda a.R\theta\langle S_1\theta,\cdots,\langle S_p\theta,\langle R_{q+1},\cdots,\langle R_m,\pi_2^n b\rangle\ldots\rangle\rangle\ldots\rangle)^\sharp\Theta'$
$= (\lambda a.R\theta\langle S_1\theta,\cdots,\langle S_p\theta,\pi_2^q\langle R_1,\cdots,\langle R_m,\pi_2^n b\rangle\ldots\rangle\rangle\ldots\rangle)^\sharp\Theta'$
$= (\lambda a.R\langle S_1,\cdots,\langle S_p,\pi_2^q b\rangle\ldots\rangle\theta)^\sharp\Theta'$

Case of $q+1 > m$:

$\mu a.[b](\lambda b_1\ldots b_q.(R^\sharp\Theta)(S_1^\sharp\Theta)\cdots(S_p^\sharp\Theta))R_1^\sharp\cdots R_m^\sharp\theta'$
$\to_\beta^+ \mu a.[b](\lambda b_{m+1}\ldots b_q.(R^\sharp\Theta)(S_1^\sharp\Theta)\cdots(S_p^\sharp\Theta)[b_1 := R_1^\sharp,\ldots,b_m := R_m^\sharp])\theta'$
$= \mu a.[b](\lambda b_{n+1}\ldots b_{q-m+n}.(R^\sharp\Theta\theta')(S_1^\sharp\Theta\theta')\cdots(S_p^\sharp\Theta\theta'))$
since R_i has no occurrences of b and no free variables b_j ($j \ge n+1$) appear
$=_\beta \mu a.[b](\lambda b_{n+1}\ldots b_{q-m+n}.((R\theta)^\sharp\Theta')((S_1\theta)^\sharp\Theta')\cdots((S_p\theta)^\sharp\Theta'))$
by the induction hypotheses
$=_\beta (\mu a.[b](\lambda b_1\ldots b_n b_{n+1}\ldots b_{q-m+n}.(R\theta)^\sharp(S_1\theta)^\sharp\cdots(S_p\theta)^\sharp))\Theta'$
$= (\lambda a.R\theta\langle S_1\theta,\ldots,\langle S_p\theta,\pi_2^{q-m+n}b\rangle\ldots\rangle)^\sharp\Theta'$
$= (\lambda a.R\theta\langle S_1\theta,\ldots,\langle S_p\theta,\pi_2^q\langle R_1,\ldots,\langle R_m,\pi_2^n b\rangle\ldots\rangle\rangle\ldots\rangle)^\sharp\Theta'$
$= (\lambda a.R\langle S_1,\ldots,\langle S_p,\pi_2^q b\rangle\ldots\rangle\theta)^\sharp\Theta'$ □

Proposition 5 (Completeness). *Let* $P, Q \in \mathit{Univ}_\mu$.

(1) *If* $P \to_{\beta_x} Q$ *then* $P^\sharp \to_\beta Q^\sharp$.
(2) *If* $P \to_{\beta_a} Q$ *then* $P^\sharp =_{\lambda\mu} Q^\sharp$.
(3) *If* $P \to_{\eta_a} Q$ *then* $P^\sharp \to^+_{\eta\mu_\eta} Q^\sharp$.
(4) *If* $P \to_{\pi_{R,K}} Q$ *then* $P^\sharp \equiv Q^\sharp$.
(5) *If* $P \to_{\mathrm{sp}_K} Q$ *then* $P^\sharp \to^*_\eta Q^\sharp$.

Proof. By induction on the derivations. We show one case for (2).

Let K be $\langle S_1, \ldots, \langle S_q, \pi_2^p c\rangle \ldots\rangle$ with $q \geq 0, p \geq 1$, and K' be $\langle R_1, \ldots, \langle R_m, \pi_2^n d\rangle \ldots\rangle$ with $m \geq 0, n \geq 1$. Let $\Theta, \Theta', \Theta'', \theta, \theta'$ be as follows:

$$\begin{aligned}
\Theta &\equiv [b \Leftarrow R_1^\sharp, \ldots, R_m^\sharp];\\
\Theta' &\equiv [b \Leftarrow b_1, \ldots, b_n];\\
\Theta'' &\equiv [d \Leftarrow d_1, \ldots, d_n];\\
\theta' &\equiv [b_1 := R_1^\sharp, \ldots, b_m := R_m^\sharp, b_{m+1} := b_{n+1}, b_{m+2} := b_{n+2}, \ldots], \text{ and}\\
\theta &\equiv [b := \langle R_1, \ldots, \langle R_m, \pi_2^n b\rangle \ldots\rangle].
\end{aligned}$$

Now we prove the case P of $\lambda a.(\lambda b.RK)K'$:

$$\begin{aligned}
&(\lambda a.(\lambda b.RK)K')^\sharp\\
&\qquad = \mu a.[d](\lambda d_1 \ldots d_n.(\mu b.[c](\lambda c_1 \ldots c_p.R^\sharp S_1^\sharp \cdots S_q^\sharp))R_1^\sharp \cdots R_m^\sharp)
\end{aligned}$$

(i) We show the case of $b \equiv c$:

$$\begin{aligned}
&(\lambda a.(\lambda b.RK)K')^\sharp\\
&\qquad = \mu a.[d](\lambda d_1 \ldots d_n.(\mu b.[b](\lambda b_1 \ldots b_p.R^\sharp S_1^\sharp \cdots S_q^\sharp))R_1^\sharp \cdots R_m^\sharp)
\end{aligned}$$

(i-1) Case of $p+1 \leq m$:

(i-1-1) Subcase of $a \equiv d$:

$$\begin{aligned}
&(\lambda a.(\lambda b.RK)K')^\sharp\\
&\to^+_\mu \mu a.[d](\lambda d_1 \ldots d_n.\mu b.[b]((\lambda b_1 \ldots b_p.(R^\sharp\Theta)(S_1^\sharp\Theta) \cdots (S_q^\sharp\Theta))\\
&\qquad\qquad R_1^\sharp \cdots R_m^\sharp))\\
&\to^+_\beta \mu a.[d](\lambda d_1 \ldots d_n.\mu b.[b]((R^\sharp\Theta)(S_1^\sharp\Theta) \cdots (S_q^\sharp\Theta))\\
&\qquad\qquad [b_1 := R_1^\sharp, \ldots, b_p := R_p^\sharp]R_{p+1}^\sharp \cdots R_m^\sharp)\\
&= \mu a.[d](\lambda d_1 \ldots d_n.\mu b.[b]((R^\sharp\Theta\theta')(S_1^\sharp\Theta\theta') \cdots (S_q^\sharp\Theta\theta'))R_{p+1}^\sharp \cdots R_m^\sharp)\\
&\qquad \text{since we have no free occurrences of } b_{p+1}, b_{p+2}, \ldots\\
&=_\beta \mu a.[d](\lambda d_1 \ldots d_n.\mu b.[b](((R\theta)^\sharp\Theta')((S_1\theta)^\sharp\Theta') \cdots ((S_q\theta)^\sharp\Theta'))\\
&\qquad\qquad R_{p+1}^\sharp \cdots R_m^\sharp)\\
&\qquad \text{by Lemma 6 (2)}\\
&=_{\lambda\mu} \lambda d_1 \ldots d_n.\mu a.([d](\lambda d_1 \ldots d_n.\mu b.[b](((R\theta)^\sharp\Theta')((S_1\theta)^\sharp\Theta') \cdots\\
&\qquad\qquad \cdots ((S_q\theta)^\sharp\Theta'))R_{p+1}^\sharp \cdots R_m^\sharp))\Theta''\\
&\qquad \text{by Fact 1 (1)}\\
&\to^+_\beta \lambda d_1 \ldots d_n.\mu a.[d]\mu b.[b]((((R\theta)^\sharp\Theta')((S_1\theta)^\sharp\Theta') \cdots ((S_q\theta)^\sharp\Theta')\\
&\qquad\qquad R_{p+1}^\sharp \cdots R_m^\sharp)\Theta'')\\
&\to_{\mu_\beta} \lambda d_1 \ldots d_n.\mu a.[d]((((R\theta)^\sharp\Theta')((S_1\theta)^\sharp\Theta') \cdots ((S_q\theta)^\sharp\Theta')\\
&\qquad\qquad R_{p+1}^\sharp \cdots R_m^\sharp)\Theta''[b := d])\\
&= \lambda d_1 \ldots d_n.\mu a.[d](((R\theta)^\sharp(S_1\theta)^\sharp \cdots (S_q\theta)^\sharp R_{p+1}^\sharp \cdots R_m^\sharp)\Theta'\Theta''[b := d])
\end{aligned}$$

$= \lambda d_1 \ldots d_n.\mu a.[d](((R\theta[b := d])^\sharp (S_1\theta[b := d])^\sharp \cdots (S_q\theta[b := d])^\sharp$
$R^\sharp_{p+1} \cdots R^\sharp_m)\Theta'')$
$=_{\lambda\mu} \mu a.[d](\lambda d_1 \ldots d_n.(R\theta[b := d])^\sharp (S_1\theta[b := d])^\sharp \cdots (S_q\theta[b := d])^\sharp$
$R^\sharp_{p+1} \cdots R^\sharp_m)$
$= (\lambda a.R\theta[b := d]\langle S_1\theta[b := d], \ldots \langle S_q\theta[b := d], \langle R_{p+1}, \ldots,$
$\langle R_m, \pi_2^n d\rangle \ldots\rangle\rangle \ldots\rangle)^\sharp$
$= (\lambda a.R\theta[b := d]\langle S_1\theta[b := d], \ldots \langle S_q\theta[b := d], \pi_2^p K'\rangle \ldots\rangle)^\sharp$
$= (\lambda a.(RK)\theta[b := d])^\sharp$
$= (\lambda a.(RK)[b := K'])^\sharp$

(i-1-2) Subcase of $a \not\equiv d$:
Following the similar pattern to the case (i-1-1);
$(\lambda a.(\lambda b.RK)K')^\sharp$
$=_{\lambda\mu} \mu a.[d](\lambda d_1 \ldots d_n.\mu b.[b](((R\theta)^\sharp\Theta')((S_1\theta)^\sharp\Theta') \cdots ((S_q\theta)^\sharp\Theta'))R^\sharp_{p+1}$
$\cdots R^\sharp_m)$
by Lemma 6 (2)
$\stackrel{\text{def}}{=} M_1$
On the other hand;
$(\lambda a.(RK)[b := K'])^\sharp$
$= \mu a.[d](\lambda d_1 \ldots d_n.(R\theta[b := d])^\sharp (S_1\theta[b := d])^\sharp \cdots (S_q\theta[b := d])^\sharp R^\sharp_{p+1}$
$\cdots R^\sharp_m)$
$\stackrel{\text{def}}{=} M_2$
Now we verify that $M_1[d \Leftarrow d_1, \ldots, d_n] =_{\lambda\mu} M_2[d \Leftarrow d_1, \ldots, d_n]$:
$M_1\Theta''$
$\to^+_\beta \mu a.[d]\mu b.[b]((((R\theta)^\sharp\Theta')((S_1\theta)^\sharp\Theta') \cdots ((S_q\theta)^\sharp\Theta'))R^\sharp_{p+1} \cdots R^\sharp_m\Theta'')$
$\to_{\mu_\beta} \mu a.[d]((((R\theta)^\sharp\Theta')((S_1\theta)^\sharp\Theta') \cdots ((S_q\theta)^\sharp\Theta'))R^\sharp_{p+1} \cdots R^\sharp_m\Theta''$
$[b := d])$
$= \mu a.[d]((R\theta)^\sharp(S_1\theta)^\sharp \cdots (S_q\theta)^\sharp R^\sharp_{p+1} \cdots R^\sharp_m\Theta'\Theta''[b := d])$
By contrast;
$M_2\Theta''$
$\to^+_\beta \mu a.[d]((R\theta[b := d])^\sharp(S_1\theta[b := d])^\sharp \cdots (S_q\theta[b := d])^\sharp R^\sharp_{p+1} \cdots$
$\cdots R^\sharp_m\Theta'')$
Then $M_1\Theta'' =_{\lambda\mu} M_2\Theta''$ can be confirmed, and hence from fact 1 (2) we obtain the property:
$(\lambda a.(\lambda b.RK)K')^\sharp =_{\lambda\mu} M_1 =_{\lambda\mu} M_2 =_{\lambda\mu} (\lambda a.(RK)[b := K'])^\sharp$

(i-2) Case of $p + 1 > m$:

(i-2-1) Subcase of $a \equiv d$:
$(\lambda a.(\lambda b.RK)K')^\sharp$
$\to^+_\mu \mu a.[d](\lambda d_1 \ldots d_n.\mu b.[b]((\lambda b_1 \ldots b_p.(R^\sharp\Theta)(S_1^\sharp\Theta) \cdots (S_q^\sharp\Theta))R_1^\sharp \cdots$
$\cdots R^\sharp_m))$
$\to^+_\beta \mu a.[d](\lambda d_1 \ldots d_n.\mu b.[b](\lambda b_{m+1} \ldots b_p.(R^\sharp\Theta)(S_1^\sharp\Theta) \cdots (S_q^\sharp\Theta))$
$[b_1 := R_1^\sharp, \ldots, b_m := R_m^\sharp])$
$= \mu a.[d](\lambda d_1 \ldots d_n.\mu b.[b](\lambda b_{n+1} \ldots b_{n+p-m}.(R^\sharp\Theta\theta')(S_1^\sharp\Theta\theta') \cdots$
$\cdots (S_q^\sharp\Theta\theta')))$

$$=_\beta \mu a.[d](\lambda d_1 \ldots d_n.\mu b.[b](\lambda b_{n+1} \ldots b_{n+p-m}.((R\theta)^\sharp\Theta')((S_1\theta)^\sharp\Theta') \cdots \cdots ((S_q\theta)^\sharp\Theta')))$$

by Lemma 6 (2)

$$=_{\lambda\mu} \lambda d_1 \ldots d_n.\mu a.([d](\lambda d_1 \ldots d_n.\mu b.[b](\lambda b_{n+1} \ldots b_{n+p-m}.((R\theta)^\sharp\Theta') ((S_1\theta)^\sharp\Theta') \cdots ((S_q\theta)^\sharp\Theta'))))\Theta''$$

by Fact 1 (1)

$$\to_\beta^+ \lambda d_1 \ldots d_n.\mu a.[d]\mu b.[b](\lambda b_{n+1} \ldots b_{n+p-m}.((R\theta)^\sharp\Theta')((S_1\theta)^\sharp\Theta') \cdots \cdots ((S_q\theta)^\sharp\Theta')\Theta'')$$

$$\to_{\mu_\beta} \lambda d_1 \ldots d_n.\mu a.[d](\lambda b_{n+1} \ldots b_{n+p-m}.((R\theta)^\sharp\Theta')((S_1\theta)^\sharp\Theta') \cdots \cdots ((S_q\theta)^\sharp\Theta')\Theta''[b := d])$$

$$= \lambda d_1 \ldots d_n.\mu a.[d](\lambda b_{n+1} \ldots b_{n+p-m}.(R\theta)^\sharp(S_1\theta)^\sharp \cdots \cdots (S_q\theta)^\sharp\Theta'\Theta''[b := d])$$

$$= \lambda d_1 \ldots d_n.\mu a.[d](\lambda d_{n+1} \ldots d_{n+p-m}.(R\theta[b := d])^\sharp(S_1\theta[b := d])^\sharp \cdots \cdots (S_q\theta[b := d])^\sharp\Theta'')$$

$$=_{\lambda\mu} \mu a.[d](\lambda d_1 \ldots d_{n+p-m}.(R\theta[b := d])^\sharp(S_1\theta[b := d])^\sharp \cdots \cdots (S_q\theta[b := d])^\sharp)$$

$$= (\lambda a.R\theta[b := d]\langle S_1\theta[b := d], \ldots \langle S_q\theta[b := d], \pi_2^{n+p-m} d\rangle \ldots\rangle)^\sharp$$

$$= (\lambda a.R\theta[b := d]\langle S_1\theta[b := d], \ldots \langle S_q\theta[b := d], \pi_2^{p} K'\rangle \ldots\rangle)^\sharp$$

$$= (\lambda a.(RK)\theta[b := d])^\sharp$$

$$= (\lambda a.(RK)[b := K'])^\sharp$$

(i-2-2) Subcase of $a \not\equiv d$**:**

Following the similar pattern to the case (i-1-2);

$(\lambda a.(\lambda b.RK)K')^\sharp$

$$=_{\lambda\mu} \mu a.[d](\lambda d_1 \ldots d_n.\mu b.[b](\lambda b_{n+1} \ldots b_{n+p-m}.((R\theta)^\sharp\Theta')((S_1\theta)^\sharp\Theta') \cdots \cdots ((S_q\theta)^\sharp\Theta')))$$

$\stackrel{\text{def}}{=} M_3$

$(\lambda a.(RK)\theta[b := d])^\sharp$

$$= \mu a.[d](\lambda d_1 \ldots d_{n+p-m}.(R\theta[b := d])^\sharp(S_1\theta[b := d])^\sharp \cdots \cdots (S_q\theta[b := d])^\sharp)$$

$\stackrel{\text{def}}{=} M_4$

Here we prove that $M_3[d \Leftarrow d_1, \ldots, d_n] =_{\lambda\mu} M_4[d \Leftarrow d_1, \ldots, d_n]$:

$M_3\Theta''$

$$\to_\beta^+ \mu a.[d]\mu b.[b](\lambda b_{n+1} \ldots b_{n+p-m}.((R\theta)^\sharp\Theta')((S_1\theta)^\sharp\Theta') \cdots \cdots ((S_q\theta)^\sharp\Theta')\Theta'')$$

$$\to_{\mu_\beta} \mu a.[d](\lambda b_{n+1} \ldots b_{n+p-m}.(R\theta)^\sharp(S_1\theta)^\sharp \cdots (S_q\theta)^\sharp\Theta'\Theta''[b := d])$$

On the other hand;

$M_4\Theta''$

$$\to_\beta^+ \mu a.[d](\lambda d_{n+1} \ldots d_{n+p-m}.(R\theta[b := d])^\sharp(S_1\theta[b := d])^\sharp \cdots \cdots (S_q\theta[b := d])^\sharp\Theta'')$$

From fact 1 (2), $M_3\Theta'' =_{\lambda\mu} M_4\Theta''$ implies $M_3 =_{\lambda\mu} M_4$, and hence we have

$(\lambda a.(\lambda b.RK)K')^\sharp =_{\lambda\mu} (\lambda a.(RK)\theta[b := d])^\sharp$.

(ii) We show another case of $b \not\equiv c$:

(ii-1) Case of $a \equiv d$:

$(\lambda a.(\lambda b.RK)K')^\sharp$
$= \mu a.[d](\lambda d_1 \dots d_n.(\mu b.[c](\lambda c_1 \dots c_p.R^\sharp S_1^\sharp \cdots S_q^\sharp))R_1^\sharp \cdots R_m^\sharp)$
$\rightarrow_\mu^+ \mu a.[d](\lambda d_1 \dots d_n.\mu b.[c](\lambda c_1 \dots c_p.(R^\sharp \Theta)(S_1^\sharp \Theta) \cdots (S_q^\sharp \Theta)))$
$\rightarrow_\mu^+ \mu a.[d](\lambda d_1 \dots d_n.\mu b.[c](\lambda c_1 \dots c_p.(R^\sharp \Theta\theta')(S_1^\sharp \Theta\theta') \cdots (S_q^\sharp \Theta\theta')))$
since no free variables b_i appear
$=_\beta \mu a.[d](\lambda d_1 \dots d_n.\mu b.[c](\lambda c_1 \dots c_p.(R\theta)^\sharp \Theta'((S_1\theta)^\sharp \Theta') \cdots (S_q\theta)^\sharp \Theta'))$
by Lemma 6
$=_{\lambda\mu} \lambda d_1 \dots d_n.\mu a.[d]\mu b.([c](\lambda c_1 \dots c_p.((R\theta)^\sharp \Theta')((S_1\theta)^\sharp \Theta') \cdots$
$\cdots ((S_q\theta)^\sharp \Theta')))\Theta''$
by Fact 1 (1)
$\rightarrow_{\mu_\beta} \lambda d_1 \dots d_n.\mu a.([c](\lambda c_1 \dots c_p.(R\theta)^\sharp (S_1\theta)^\sharp \cdots (S_q\theta)^\sharp))\Theta'\Theta''[b := d]$
since $c \not\equiv b$
$= \lambda d_1 \dots d_n.\mu a.([c](\lambda c_1 \dots c_p.(R\theta[b := d])^\sharp (S_1\theta[b := d])^\sharp \cdots$
$\cdots (S_q\theta[b := d])^\sharp))\Theta''$
$=_{\lambda\mu} \mu a.[c](\lambda c_1 \dots c_p.(R\theta[b := d])^\sharp (S_1\theta[b := d])^\sharp \cdots (S_q\theta[b := d])^\sharp)$
by Fact 1 (1)
$= (\lambda a.R\theta[b := d]\langle S_1\theta[b := d], \dots, \langle S_q\theta[b := d], \pi_2^p c\rangle \dots\rangle)^\sharp$
$= (\lambda a.RK[b := K'])^\sharp$

(ii-2) Case of $a \not\equiv d$:

Following the similar pattern to the case (ii-1);
$(\lambda a.(\lambda b.RK)K')^\sharp$
$=_{\lambda\mu} \mu a.[d](\lambda d_1 \dots d_n.\mu b.[c](\lambda c_1 \dots c_p.(R\theta)^\sharp \Theta'((S_1\theta)^\sharp \Theta') \cdots (S_q\theta)^\sharp \Theta'))$
$\stackrel{\text{def}}{=} M_5$
$(\lambda a.RK[b := K'])^\sharp$
$= \mu a.[c](\lambda c_1 \dots c_p.(R\theta[b := d])^\sharp (S_1\theta[b := d])^\sharp \cdots (S_q\theta[b := d])^\sharp)$
$\stackrel{\text{def}}{=} M_6$
Now we verify that $M_5[d \Leftarrow d_1, \dots, d_n] =_{\lambda\mu} M_6[d \Leftarrow d_1, \dots, d_n]$:
$M_5\Theta''$
$\rightarrow_\beta^+ \mu a.[d]\mu b.([c](\lambda c_1 \dots c_p.(R\theta)^\sharp (S_1\theta)^\sharp \cdots (S_q\theta)^\sharp))\Theta'\Theta''$
$\rightarrow_{\mu_\beta} \mu a.([c](\lambda c_1 \dots c_p.(R\theta)^\sharp (S_1\theta)^\sharp \cdots (S_q\theta)^\sharp))\Theta'\Theta''[b := d]$
By contrast;
$M_6\Theta''$
$= \mu a.([c](\lambda c_1 \dots c_p.(R\theta[b := d])^\sharp (S_1\theta[b := d])^\sharp \cdots (S_q\theta[b := d])^\sharp))\Theta''$
Then $M_5\Theta'' =_{\lambda\mu} M_6\Theta''$ implies $M_5 =_{\lambda\mu} M_6$ by Fact 1 (2), and hence we have
$(\lambda a.(\lambda b.RK)K')^\sharp =_{\lambda\mu} M_5 =_{\lambda\mu} M_6 =_{\lambda\mu} (\lambda a.RK[b := K'])^\sharp$. □

Theorem 2. *Let* $M_1, M_2 \in \Lambda\mu$. $M_1 =_{\lambda\mu} M_2$ *if and only if* $[\![M_1]\!] =_{\lambda\langle\rangle} [\![M_2]\!]$.

Proof. From Propositions 4 and 5 and Lemma 6 (1). □

4 Concluding Remarks and Further Work

(1) $\Lambda^{\langle\rangle}$ is not Church-Rosser by Klop [1]. As stated in Corollary 1, Theorem 1 reveals that $[\![\Lambda]\!] \stackrel{\text{def}}{=} \{[\![M]\!] \in \Lambda^{\langle\rangle} \mid M \in \Lambda\}$ is a confluent fragment, in the sense that if $P_1 =_{\lambda\langle\rangle} P_2$ for $P_1, P_2 \in [\![\Lambda]\!]$ then there exists some $P \in [\![\Lambda]\!]$ such that $P_1 \to^*_{\lambda\langle\rangle} P$ and $P_2 \to^*_{\lambda\langle\rangle} P$.

(2) There exists a one-to-one correspondence between C-monoids and $\Lambda^{\langle\rangle}$ by Lambek and Scott [11] and by Curien [3]:
($\mathrm{C}_{\mathrm{Ass}}$) $(x \circ y) \circ z = x \circ (y \circ z)$
($\mathrm{C}_{\mathrm{Idl}}$) $1 \circ x = x$
($\mathrm{C}_{\mathrm{Idr}}$) $x \circ 1 = x$
($\mathrm{C}_{\mathrm{Fst}}$) $\pi_1 \circ \langle x, y\rangle = x$
($\mathrm{C}_{\mathrm{Snd}}$) $\pi_2 \circ \langle x, y\rangle = y$
(C_{SP}) $\langle \pi_1 \circ x, \pi_2 \circ x\rangle = x$
($\mathrm{C}_{\mathrm{App}}$) $\mathtt{App} \circ \langle \mathtt{Cur}(x) \circ \pi_1, \pi_2\rangle = x$
($\mathrm{C}_{\mathrm{S}\Lambda}$) $\mathtt{Cur}(\mathtt{App} \circ \langle x \circ \pi_1, \pi_2\rangle) = x$
Theorem 2 means that there also exists an injection from $\Lambda\mu$ into C-monoids, and hence C-monoids become an extensional model of $\lambda\mu$-calculus with (η).

(3) The recursive domains

$$U \cong U \times U \cong [U \to U]$$

give a model of the λ-calculus with surjective pairing [8]. The domains $U \cong U \times U \cong [U \to U]$ can provide continuation denotational semantics of the extensional $\lambda\mu$-calculus as well. From Theorem 2 the completeness of the continuation denotational semantics of $\Lambda\mu$ depends on that of the direct denotational semantics of $\Lambda^{\langle\rangle}$. See also [6] for a simulation relation over cpo's between the continuation denotational semantics of $\Lambda\mu$ and the CPS-translation followed by the direct denotational semantics of $\Lambda^{\langle\rangle}$.

(4) Even in the case of typed $\lambda\mu$-calculus, the novel CPS-translation works well as a negative translation from classical logic into intuitionistic logic consisting of $\neg$ and $\wedge$:
If we have $\Gamma \vdash M : A; \Delta$ in typed $\lambda\mu$-calculus, then $\Gamma^k, \Delta^* \vdash [\![M]\!] : A^k$ in typed λ-calculus,
where

$$\begin{cases} (A_1 \Rightarrow A_2)^k = \neg(A_1^k \wedge A_2^*); \\ \qquad\quad A^k = \neg\neg A \quad \text{if } A \text{ is atomic; and} \\ \qquad\quad A^* = B \quad \text{where } \neg B \equiv A^k. \end{cases}$$

Hofmann and Streicher [9] proved that in terms of a category of continuations, a call-by-name continuation semantics of typed $\lambda\mu$-calculus is sound and complete for every $\lambda\mu$-theory. Roughly speaking, A^k corresponds to the type of denotations of type A, i.e., $R^{[\![A]\!]}$ in terminology of [9], and A^* to the space of continuations of type A together with $(A_1 \Rightarrow A_2)^* \Leftrightarrow (A_1^k \wedge A_2^*)$. Their result implies our Theorem 2 in the simply typed case. Our syntactic method will also be applied to second order $\lambda\mu$-calculus [13,14] in an extended version of this paper.

Acknowledgements. I am grateful to Horai-Takahashi Masako for helpful discussions. The author is also grateful to Martin Hofmann and anonymous referees for their constructive comments and suggestions. This research has been supported by Grants-in-Aid for Scientific Research (C)(2)14540119, Japan Society for the Promotion of Science and by the Kayamori Foundation of Informational Science Advancement.

References

1. H. P. Barendregt: *The Lambda Calculus, Its Syntax and Semantics* (revised edition), North-Holland, 1984.
2. K. Baba, S. Hirokawa, and K. Fujita: Parallel Reduction in Type-Free $\lambda\mu$-Calculus, *Electronic Notes in Theoretical Computer Science*, Vol. 42, pp. 52–66, 2001.
3. P. -L. Curien: *Categorical Combinators, Sequential Algorithms and Functional Programming*, Pitman, London/John Wiley&Sons, 1986.
4. K. Fujita: Simple Models of Type Free $\lambda\mu$-Calculus, *The 18th Conference Proceedings Japan Society for Software Science and Technology*, 2001 (*in Japanese*).
5. K. Fujita: An interpretation of $\lambda\mu$-calculus in λ-calculus, *Information Processing Letters*, Vol. 84, No. 5, pp. 261–264, 2002.
6. K. Fujita: Continuation Semantics and CPS-translation of $\lambda\mu$-Calculus, *Scientiae Mathematicae Japonicae*, Vol. 57, No.1, pp. 73–82, 2003.
7. T. G. Griffin: A Formulae-as-Types Notion of Control, *Proc. the 17th Annual ACM Symposium on Principles of Programming Languages*, pp. 47–58, 1990.
8. C. A. Gunter and D. S. Scott: Semantic Domains, in: *Handbook of Theoretical Computer Science Vol. B : Formal Models and Semantics*, Elsevier Science Publishers B. V., 1990.
9. M. Hofmann and T. Streicher: Continuation models are universal for $\lambda\mu$-calculus, *Proc. the 12th Annual IEEE Symposium on Logic in Computer Science*, pp. 387–395, 1997.
10. W. Howard: The Formulae-as-Types Notion of Constructions, in: *To H.B.Curry: Essays on combinatory logic, lambda-calculus, and formalism*, Academic Press, pp. 479–490, 1980.
11. J. Lambek and P. J. Scott: *Introduction to higher order categorical logic*, Cambridge University Press, 1986.
12. C. R. Murthy: An Evaluation Semantics for Classical Proofs, *Proc. the 6th Annual IEEE Symposium on Logic in Computer Science*, pp. 96–107, 1991.
13. M. Parigot: $\lambda\mu$-Calculus: An Algorithmic Interpretation of Classical Natural Deduction, Lecture Notes in Computer Science 624, pp. 190–201, 1992.
14. M. Parigot: Proofs of Strong Normalization for Second Order Classical Natural Deduction, *J. Symbolic Logic*, Vol. 62, No. 4, pp. 1461–1479, 1997.
15. G. Plotkin: Call-by-Name, Call-by-Value and the λ-Calculus, *Theoretical Computer Science*, Vol. 1, pp. 125–159, 1975.
16. P. Selinger: Control Categories and Duality: on the Categorical Semantics of the Lambda-Mu Calculus, *Math. Struct. in Compu. Science*, Vol. 11, pp. 207–260, 2001.
17. T. Streicher and B. Reus: Classical Logic, Continuation Semantics and Abstract Machines, *J. Functional Programming*, Vol. 8, No. 6, pp. 543–572, 1998.

Abstraction Barrier-Observing Relational Parametricity

Jo Hannay

Department of Software Engineering,
Simula Research Laboratory, Pb. 134, NO-1325 Lysaker, Norway
jo@simula.no

Abstract. A concept of relational parametricity is developed taking into account the encapsulation mechanism inherent in universal types. This is then applied to data types and refinement, naturally giving rise to a notion of simulation relations that compose for data types with higher-order operations, and whose existence coincides with observational equivalence. The ideas are developed syntactically in lambda calculus with a relational logic. The new notion of relational parametricity is asserted axiomatically, and a corresponding parametric *per*-semantics is devised.

1 Introduction

Information hiding is essential in software development. It allows on the one hand, clients of a module to disregard irrelevant detail of the module; this is the *abstraction* aspect of information hiding, and on the other hand it protects the module from unauthorised access to inner workings; this is the *encapsulation* aspect. Thus, information hiding is a prerequisite for the plugability and interchangeability of software components. Information hiding is achieved through interfaces by raising appropriate *abstraction barriers*, and the fundamental theoretical device for this is the concept of *data type*, consisting of an encapsulated data representation and operations on that data representation.

A measure for interchangeability is *observational equivalence*; two data types are interchangeable if they exhibit equivalent observable behaviours. Observational equivalence is generally hard to show, and a simpler strategy is to show the existence of a *simulation relation*, *i.e.*, a relation between the data representations of the data types, which the respective data-type operations preserve. The existence of a simulation relation must coincide with observational equivalence for this to be a valid method, and for the sake of constructive stepwise refinement, simulation relations should compose. For data types with first-order operations, this is the case, but when higher-order operations are involved, both these properties break down in general. Several remedies to this have been found; on the semantic level we have *pre-logical relations* [14,13], *lax logical relations* [26,16], and *L-relations* [15]. On the logical level we have *abstraction barrier-observing* simulation relations [9,10]. The latter, developed in the setting of System F and a relational logic [25] with the assertion of *relational parametricity* [28], are directly motivated by the information-hiding mechanism in data types, in that they

M. Hofmann (Ed.): TLCA 2003, LNCS 2701, pp. 135–152, 2003.

reflect the way data-type operations are used by clients. Ultimately, these techniques apply in stepwise refinement processes producing certified components, *e.g.*, [30,13,1,23,10,7,8].

Abstraction barrier-observing simulation relations were designed specifically to solve the above issues regarding observational equivalence, and to remedy the lack of composability. However, the main ingredient in this new notion of simulation relation, namely respecting an abstraction barrier; and in fact the abstraction barrier itself, both originate from considerations on universal types.

This paper therefore goes back to basic principles and develops abstraction barrier-observing relations fundamentally at universal types. The first step is to define the appropriate family of relations, mirroring how polymorphic functionals may be used according to the abstraction barrier inherent in universal types. We argue that this is the correct relational concept for universal types. Then we go on and use this concept in asserting *abstraction barrier-observing relational parametricity*. If we insist on using abstraction barrier-observing relations for universal types, we must also assert the corresponding notion of relational parametricity, if we are to keep the parametricity principal or have any hopes of retaining proof power. The third step is to give a model for the logic and the abstraction barrier-observing relational parametricity axiom schema.

With this in place, abstraction barrier-observing simulation relations are a direct application of the abstraction barrier-observing relational concept for universal types, mirroring data-type abstraction barriers.

Besides the aesthetical benefit gained from developing the new notion of simulation relations as a natural consequence of basic considerations, we get a substantial simplification of formalism compared to [9,10], where one only asserts specialised instances of abstraction barrier-observing relational parametricity. In [9,10], the logical language and the calculus is augmented in order to present a non-syntactic model for these instances. With abstraction barrier-observing relational parametricity this is not necessary.

2 Type Theory and Logic

We review relevant formal aspects. For full accounts, see [3,21,6,25,2].

The second-order lambda-calculus F_2, or System F, has abstract syntax

$$\begin{array}{ll}\text{(types)}\ T & ::= X \mid (T \to T) \mid (\forall X.T) \\ \text{(terms)}\ t & ::= x \mid (\lambda x{:}T.t) \mid (tt) \mid (\Lambda X.t) \mid (tT)\end{array}$$

where X and x range over type and term variables respectively. For simplicity, we obey *Barendregt's variable convention*; bound variables are chosen to differ in name from free variables in any type or term. The main syntactic feature of System F is polymorphic functionals. For example, a general function-composition operator is achieved by $comp \stackrel{def}{=} \Lambda X, Y, Z.\lambda f : X \to Y.\lambda g : Y \to Z.\lambda x : X.g(fx)$. One can also define inductive types [4], *e.g.*, $\mathsf{Nat} \stackrel{def}{=} \forall X.X \to (X \to X) \to X$. The inductive inhabitants are the closed terms. It is easy to program constructors, as well as destructors and conditionals, since iteration is built in. We shall use inductive types Nat, Bool, and List_U, and various self-explanatory terms such as 0, succ, true, false, cons, cond $:\forall Y.\mathsf{Bool}\to Y\to Y\to Y$, isnil, *etc.* Binary products are encoded in System F as inductive types by $U\times V \stackrel{def}{=} \forall X.((U\to V\to X)\to X)$, where U and V have no free X, with constructor $\mathsf{pair}_{U,V}:U\to V\to U\times V$, and also destructors $\mathsf{proj}_{1U,V}: U\times V\to U$ and $\mathsf{proj}_{2U,V}: U\times V\to V$, defined by $\mathsf{pair}_{U,V}uv \stackrel{def}{=} \Lambda X.\lambda f: U\to V\to X.fuv$, $\mathsf{proj}_{1U,V}(x) \stackrel{def}{=} xU(\lambda x: U.\lambda y: V.x)$, and $\mathsf{proj}_{2U,V}(x) \stackrel{def}{=} xV(\lambda x{:}U.\lambda y{:}V.y)$. This generalises to n-ary products.

Consider a computation $f = \Lambda X.\lambda x{:}U[X].t[X,x]$. We refer to X as a *virtual data representation*, x as a collection of *virtual operations*, and the whole computation as a *virtual computation.* Any instance $(f\,A\,a)$ of the computation with an *actual data representation* A, and *actual operations* a then gives an *actual computation.* A crucial observation which will determine future development, is now embodied in the following obvious statement.

***Abs-Bar1*:** A virtual client computation $\Lambda X.\lambda x: T[X].t[X,x]$ cannot have free variables of types involving the virtual data representation X.

For data types, encapsulation is provided in the style of [22] by the following encoding of existential (abstract) types and pack and unpack combinators.

$$\exists X.T[X] \stackrel{def}{=} \forall Y.(\forall X.(T[X]\to Y)\to Y), \qquad Y \text{ not free in } T$$

$$\mathsf{pack}_{T[X]}:\forall X.(T[X]\to\exists X.T[X])$$
$$\mathsf{pack}_{T[X]}(A)(opns) \stackrel{def}{=} \Lambda Y.\lambda f{:}\forall X.(T[X]\to Y).f(A)(opns)$$

$$\mathsf{unpack}_{T[X]}:(\exists X.T[X])\to\forall Y.(\forall X.(T[X]\to Y)\to Y)$$
$$\mathsf{unpack}_{T[X]}(package)(B)(client) \stackrel{def}{=} package(B)(client)$$

Operationally, pack packages a data representation and an implementation of operations on that data representation to give a data type of the existential type. The resulting package is a polymorphic functional that given a client computation and its result domain, instantiates the client with the particular elements of the package. The unpack combinator is merely the application operator for pack. An abstract type for stacks of natural numbers could be $\exists X.(X\times(\mathsf{Nat}\to X\to X)\times(X\to X)\times(X\to\mathsf{Nat}))$. A data type of this type is, *e.g.*, $(\mathsf{pack}\,\mathsf{List}_{\mathsf{Nat}}\,\mathfrak{l})$, where $(\mathsf{proj}_1\,\mathfrak{l}) = \mathsf{nil}$, $(\mathsf{proj}_2\,\mathfrak{l}) = \mathsf{cons}$, $(\mathsf{proj}_3\,\mathfrak{l}) = \lambda l : \mathsf{List}_{\mathsf{Nat}}.\ (\mathsf{cond}\,\mathsf{List}_{\mathsf{Nat}}\,(\mathsf{isnil}\,l)\,\mathsf{nil}\,(\mathsf{cdr}\,l))$, and $(\mathsf{proj}_4\,\mathfrak{l}) = \lambda l : \mathsf{List}_{\mathsf{Nat}}.(\mathsf{cond}\,\mathsf{Nat}\,(\mathsf{isnil}\,l)\,0(\mathsf{car}\,l))$. For convenience we use a labelled product notation; $\exists X.T_{\mathsf{STACK}_{\mathsf{Nat}}}[X]$, where $T_{\mathsf{STACK}_{\mathsf{Nat}}}[X] \stackrel{def}{=} (\mathsf{empty}: X, \mathsf{push}: \mathsf{Nat}\to X\to X, \mathsf{pop}{:}X\to X, \mathsf{top}{:}X\to\mathsf{Nat})$. We call each $f_i{:}T_i[X]$ a *profile* of the existential type.

Existential types together with the pack and unpack combinators embody an abstraction barrier composed of three components.

First of all, ***Abs-Bar1*** says that a client computation $\Lambda X.\lambda x: T[X].t[X,x]$ cannot have free variables of types involving the virtual data representation X, *e.g.*, by Barendregt's variable convention, $y{:}X \,\triangleright\, \Lambda X.\lambda\mathfrak{x}{:}T_{\mathsf{STACK}_{\mathsf{Nat}}}.(\mathfrak{x}.\mathsf{push}\,n\,y)$ is ill-formed. A client computation may compute over types containing the virtual data representation X only by accessing virtual operations in the supplied collection $\mathfrak{x}$, as in *e.g.*, $\Lambda X.\lambda\mathfrak{x}{:}T_{\mathsf{STACK}_{\mathsf{Nat}}}.\ \mathfrak{x}.\mathsf{top}(\mathfrak{x}.\mathsf{push}\,n\,\mathfrak{x}.\mathsf{empty})$.

In large, up to $\beta\eta$-equivalence, package operations may only be applied to arguments definable by package operations. This aspect is essential in restricting access to the underlying data representation, which may contain invalid values. Note that a supplied package operation might itself make calls involving arguments of types over the actual data representation, which are not expressed in terms of package operations. In this case, it is not true that package operations will only be applied to definable arguments, but crucially, this is at the discretion of the package implementor, and not due to direct user application.

There are two other aspects of the abstraction barrier inherent in the encoding of existential types. The next one, a prerequisite for plugability, is the already stated condition in the definition of the encoding of existential types.

***Abs-Bar2*:** The type Y does not occur in T in the encoding

$$\exists X.T[X] \stackrel{def}{=} \forall Y.(\forall X.(T[X] \to Y) \to Y).$$

This ensures that data types do not depend on their environment of usage, other than via explicitly given parameters if $\exists X.T[X]$ has free types.

The third aspect arises from the fact that when using a data type, *i.e.*, a package of existential type, the user must provide the result type of the client computation. This entails the following.

***Abs-Bar3*:** Client computations $f : \forall X.(T[X] \to Y)$ cannot have a result type containing the virtual data representation, *i.e.*, the bound type variable X.

For example, $f \stackrel{def}{=} \Lambda X.\lambda\mathfrak{x} : T_{\mathsf{STACK}_{\mathsf{Nat}}}.(\mathfrak{x}.\mathsf{push}\, n\, \mathfrak{x}.\mathsf{empty}) : \forall X.(T_{\mathsf{STACK}_{\mathsf{Nat}}} \to X)$ is not a possible client; in order to use a package $(\mathsf{pack} Aa)$ in f, we must do $(\mathsf{pack} Aa)(C)(f)$, where C is the result type of f, which in this case is the inaccessible virtual data representation X. Due to ***Abs-Bar3***, the only way a package can be used is via client computations adhering to the above. This then fixes how packages may be used in actual computations, since all actual computations are uniform in that they inherit the form of the virtual computation from which they stem. We will refer to ***Abs-Bar1***, ***Abs-Bar2***, and ***Abs-Bar3*** jointly as ***Abs-Bar***.

The logic we shall use as a starting point is the logic for parametric polymorphism due to [25]. This logic is essentially a second-order logic augmented with relation symbols and a syntax for relation definition, and the assertion of relational parametricity as an axiom schema. See also [18,33].

The logic in [25] has formulae built using the standard connectives, but now basic predicates are not only equations, but also relation membership statements,

$$\phi ::= (t =_A u) \mid t\ R\ u \mid \cdots \mid \forall R \subset A \times B.\phi \mid \exists R \subset A \times B.\phi$$

where R ranges over relation variables. We write $\alpha[\xi]$ to indicate possible occurrences of variable ξ in type, term or formula α, and write $\alpha[\beta]$ for the substitution $\alpha[\beta/\xi]$, following the appropriate rules regarding capture.

Judgements for formula formation involve relation symbols, so contexts are augmented with relation variables, depending on type context, and obeying standard conventions. Relation definition is accommodated by the following syntax,

$$\frac{\Gamma, x\!:\!A, y\!:\!B \rhd \phi}{\Gamma \rhd (x\!:\!A, y\!:\!B)\ .\ \phi \quad \subset A \times B}$$

where ϕ is a formula. For example $\mathsf{eq}_A \stackrel{def}{=} (x\!:\!A, y\!:\!A).(x =_A y)$.

We now build complex relations using type formers. We get the *arrow-type relation* $\rho \to \rho' \subset (A \to A') \times (B \to B')$ from $\rho \subset A \times B$ and $\rho' \subset A' \times B'$ by

$$(\rho \to \rho') \stackrel{def}{=} (f : A \to A', g : B \to B') \ . \ (\forall x : A . \forall y : B \ . \ (x \rho y \ \Rightarrow \ (fx)\rho'(gy)))$$

i.e., f and g are related if they map ρ-related arguments to ρ'-related values. The *universal-type relation* $\forall(Y, Z, R \subset Y \times Z)\rho[R] \ \subset \ (\forall Y.A[Y]) \times (\forall Z.B[Z])$ is defined from $\rho[R] \subset A[Y] \times B[Z]$, where Y, Z and $R \subset Y \times Z$ are free, by

$$\begin{aligned}&\forall(Y, Z, R \subset Y \times Z)\rho[R] \stackrel{def}{=} \\ &\qquad (y : \forall Y.A[Y], z : \forall Z.B[Z]) \ . \ (\forall Y . \forall Z . \forall R \subset Y \times Z \ . \ ((yY)\rho[R](zZ)))\end{aligned}$$

i.e., two y and z are related at universal type if all instances yY and zZ are related in $\rho[R]$, whenever R relates Y and Z.

One defines the *action* of types on relations, by substituting relations for type variables in types. For $\boldsymbol{X} = X_1, \ldots, X_n$, $\boldsymbol{A} = A_1, \ldots, A_n$, $\boldsymbol{B} = B_1, \ldots, B_n$ and $\boldsymbol{\rho} = \rho_1, \ldots, \rho_n$, where $\rho_i \subset A_i \times B_i$, we get $T[\boldsymbol{\rho}] \subset T[\boldsymbol{A}] \times T[\boldsymbol{B}]$, the action of $T[\boldsymbol{X}]$ on $\boldsymbol{\rho}$, defined by cases on $T[\boldsymbol{X}]$ as follows:

$$\begin{aligned}T[\boldsymbol{X}] &= X_i : & T[\boldsymbol{\rho}] &= \rho_i \\ T[\boldsymbol{X}] &= T'[\boldsymbol{X}] \to T''[\boldsymbol{X}] : & T[\boldsymbol{\rho}] &= T'[\boldsymbol{\rho}] \to T''[\boldsymbol{\rho}] \\ T[\boldsymbol{X}] &= \forall X'.T'[\boldsymbol{X}, X'] : & T[\boldsymbol{\rho}] &= \forall(Y, Z, R \subset Y \times Z)T'[\boldsymbol{\rho}, R]\end{aligned}$$

The proof system is natural deduction, intuitionistic style, over formulae now involving relation symbols, and is augmented with inference rules for relation symbols in the obvious way. There are standard axioms for equational reasoning and $\beta\eta$-equalities. These imply extensionality for arrow and universal types.

Parametric polymorphism promotes all instances of a polymorphic functional to exhibit a uniform behaviour. Of various notions [32,2]; we adopt *relational parametricity* [28,17], where uniformity is defined by saying that if a polymorphic functional is instantiated at two related domains, the resulting instances should be related as well. In [25] this is directly asserted by the axiom schema,

$$\text{PARAM}: \ \forall \boldsymbol{Z} . \forall u : (\forall X . U[X, \boldsymbol{Z}]) \ . \ u \ (\forall X . U[X, \mathsf{eq}_{\boldsymbol{Z}}]) \ u$$

The logic with PARAM is sound w.r.t. the parametric *per*-model of [2] and also w.r.t. the syntactic parametric models of [11].

The assumption of PARAM yields interesting results. The following *Identity Extension Lemma* is fundamental and follows from PARAM and extensionality.

$$\forall \boldsymbol{Z} . \forall u, v : U[\boldsymbol{Z}] \ . \ (u \ U[\mathsf{eq}_{\boldsymbol{Z}}] \ v \ \Leftrightarrow \ (u =_{U[\boldsymbol{Z}]} v))$$

Weak versions of constructs such as products, sums, initial and final (co-)algebras are encodable in System F [4]. With PARAM, these constructs become provably universal constructions. For example, we can derive with PARAM,

$$\forall U, V . \forall z : U \times V \ . \ \mathsf{pair}(\mathsf{proj}_1 z)(\mathsf{proj}_2 z) = z$$

$$\begin{aligned}&\forall u : A \times A', v : B \times B', \rho \subset A \times B, \rho' \subset A' \times B' \ . \\ &\qquad u(\rho \times \rho')v \ \Leftrightarrow \ (\mathsf{fst}(u) \ \rho \ \mathsf{fst}(v) \wedge \mathsf{snd}(u) \ \rho' \ \mathsf{snd}(v))\end{aligned}$$

where $\rho \times \rho'$ is obtained by the action $(X \times X')[\rho, \rho']$.

3 Abstraction Barrier-Observing Relations

Abs-Bar says that function arguments in computations are bounded by formal parameters, *e.g.*, in the closed computation $f \stackrel{def}{=} \Lambda X.\lambda x\!:\!X\lambda s\!:\!X \to X.t[x,s]$, s will only be applied to arguments built from formal parameters x and s. This transfers to instances fA and fB. Relationally for $R \subset A \times B$, we would like *e.g.*, $s_A\ (R \to R)\ s_B$ to reflect this, in that only those x, y are considered for the antecedent $x\ R\ y$, that are admissible in the computations; we want something like $\forall x\!:\!A, y\!:\!B\ .\ x\ R\ y \wedge \mathsf{Dfnbl}(x,y)\ \Rightarrow\ s_Ax\ R\ s_By$. Although the idea is based on observing closed terms, we neither can, nor wish to be that specific formalistically. Thus if A, B, a, b, s_A, s_B are actual parameters to f, we set $\mathsf{Dfnbl}(x,y)$ to stand for $\exists f_X\!:\!\forall X.X \to (X \to X) \to X\ .\ f_X A a s_A = x \wedge f_X B b s_B = y$. For closed computations $\Lambda X.t$ in particular, this mirrors precisely application in t.

The *abo*-relation we now define formalises this idea. The full definition is complicated by universal-type nesting, and by the need to keep track of which actual parameters have (not) been received. Above, a, b, s_A, s_B are all received at the stage when $\mathsf{Dfnbl}(x,y)$ is defined, but consider what needs to be done for $\Lambda X.\lambda s\!:\!(X \to X).\lambda x\!:\!X.t[s,x]$. The various rôles universal types may assume complicate matters further. The instrumental aspect is however simple argument definability. Example 1 below illustrates *abo*-relations. In Sect. 4 we use *abo*-relations to devise abstraction barrier-observing simulation relations.

Definition 1 (*abo*-Relation). *For any type* $U[\boldsymbol{Z}]$*, we define the* abstraction barrier-observing relation $U[\mathbf{eq}_{\boldsymbol{Z}}]^{\mathsf{abo}}$ *as follows.*

$$U[\mathbf{eq}_{\boldsymbol{Z}}]^{\mathsf{abo}} \stackrel{def}{=} U[\mathbf{eq}_{\boldsymbol{Z}}] \qquad \text{if } U \text{ is not a universal type.}$$

$$(\forall X.U[X,\mathbf{eq}_{\boldsymbol{Z}}])^{\mathsf{abo}} \stackrel{def}{=} (\forall X.U[X,\mathbf{eq}_{\boldsymbol{Z}}])^{\sigma_0 l_0 \mathsf{abo}}$$

where $(\forall X.U[X,\mathbf{eq}_{\boldsymbol{Z}}])^{\sigma_0 l_0 \mathsf{abo}}$ *is given by the following. Regard* $\forall X.U$ *according to outermost arrow structure and universal quantifier nesting. For the* m^{th} *nesting, write* $\forall X_m.U_m = \forall X_m.U_{m1} \to \cdots \to U_{mn_m} \to U_{mc}$*, where* U_{mc} *is not an arrow type, but possibly a universal type* $\forall X_{m+1}.U_{m+1}$*. Setting* $\forall X.U \stackrel{def}{=} \forall X_1.U_1$ *as the first nesting, let* N *be the number of nestings in* $\forall X.U$*. Define*

$$\begin{array}{l}(\forall X_m.U_m[\boldsymbol{R}, X_m, \mathbf{eq}_{\boldsymbol{Z}}])^{\sigma_{m-1} l_{m-1} \mathsf{abo}} \stackrel{def}{=} \\ \qquad (f\!:\!\forall X_m.U_m[\boldsymbol{A}, X_m],\ g\!:\!\forall X_m.U_m[\boldsymbol{B}, X_m])\ . \\ \quad (\forall A_m, B_m, R_m \subset A_m \times B_m\ .\ fA_m\ (U_m[\boldsymbol{R}, R_m, \mathbf{eq}_{\boldsymbol{Z}}]_{\top}^{\sigma_m l_m \langle A_m, B_m\rangle})\ gB_m)\end{array}$$

Define sequences σ_m *and* l_m *as follows. Set* $\sigma_0 = l_0 \stackrel{def}{=} \varepsilon$*. If* U_m *has no free variables from* σ_{m-1}*, then* $\sigma_m \stackrel{def}{=} \langle X_j, \langle U_{mi}\rangle_{i=1}^{n_m}\rangle_{j=m}^{q}$*,* $q \leq N$*, where* $\forall X_{q+1}.U_{q+1}$ *is the first nesting within the* m^{th} *with no free occurrences of any* X_j*,* $1 \leq j \leq q$*. Set* $l_m \stackrel{def}{=} \varepsilon$*. If on the other hand,* U_m *has free variables from* σ_{m-1} *then* $\sigma_m \stackrel{def}{=} \sigma_{m-1}$*, and* $l_m \stackrel{def}{=} l_{m-1}$*. The sequence* l_m *will now be the basis for sequences* l_{mk} *at each* U_{mk} *with further items. These will have the format* $\langle\langle A_j, B_j\rangle\langle a_{j_i}, b_{j_i}\rangle_{i=1}^{k_j}\rangle_{j=p}^{m}$*, where* p *is the index of the first item in* σ_m*, and* $1 \leq k_j \leq n_j$*. We write* $l\langle\alpha,\beta\rangle$ *to indicate the implicit insertion of* $\langle\alpha,\beta\rangle$ *format-correctly in* l*. In general, we expand* $U[\boldsymbol{R}, \mathbf{eq}_{\boldsymbol{Z}}]_{\top}^{\sigma l}$ *through the following definitions.*

$$(U'[\boldsymbol{R},\mathbf{eq}_{\boldsymbol{Z}}]\to U''[\boldsymbol{R},\mathbf{eq}_{\boldsymbol{Z}}])^{\sigma l}_{\top} \stackrel{def}{=}$$
$$(f{:}U'[\boldsymbol{A}]\to U''[\boldsymbol{A}],\ g{:}U'[\boldsymbol{B}]\to U''[\boldsymbol{B}])\ .$$
$$(\forall a{:}U'[\boldsymbol{A}],\ b{:}U'[\boldsymbol{B}]\ .\ a\ (U'[\boldsymbol{R},\mathbf{eq}_{\boldsymbol{Z}}])^{\sigma l\langle a,b\rangle}\ b$$
$$\Rightarrow\ (fa)\ (U''[\boldsymbol{R},\mathbf{eq}_{\boldsymbol{Z}}])^{\sigma l\langle a,b\rangle}_{\top}\ (gb))$$

$U[\boldsymbol{R},\mathbf{eq}_{\boldsymbol{Z}}]^{\sigma l}_{\top} \stackrel{def}{=} U[\boldsymbol{R},\mathbf{eq}_{\boldsymbol{Z}}]^{\sigma l\mathsf{abo}}$, *if U is a universal type*

$U[\boldsymbol{R},\mathbf{eq}_{\boldsymbol{Z}}]^{\sigma l}_{\top} \stackrel{def}{=} U[\boldsymbol{R},\mathbf{eq}_{\boldsymbol{Z}}]^{\sigma l}$, *otherwise*

Then, $U[\boldsymbol{R},\mathbf{eq}_{\boldsymbol{Z}}]^{\sigma l}$, is defined by

$$R_i{}^{\sigma l} \stackrel{def}{=} R_i$$

$$\mathsf{eq}_{Z_i}{}^{\sigma l} \stackrel{def}{=} \mathsf{eq}_{Z_i}$$

$$(U'[\boldsymbol{R},\mathbf{eq}_{\boldsymbol{Z}}]\to U''[\boldsymbol{R},\mathbf{eq}_{\boldsymbol{Z}}])^{\sigma l} \stackrel{def}{=} (f{:}U'[\boldsymbol{A}]\to U''[\boldsymbol{A}],\ g{:}U'[\boldsymbol{B}]\to U''[\boldsymbol{B}])\ .$$
$$(\forall_{\sigma-l}.\forall x{:}U'[\boldsymbol{A}],\ y{:}U'[\boldsymbol{B}]\ .$$
$$(x\ U'[\boldsymbol{R},\mathbf{eq}_{\boldsymbol{Z}}]^{\sigma l(\sigma-l)}\ y\ \wedge\ \mathsf{Dfnbl}^{\sigma l(\sigma-l)}_{U'}(x,y))$$
$$\Rightarrow\ (fx)\ U''[\boldsymbol{R},\mathbf{eq}_{\boldsymbol{Z}}]^{\sigma l(\sigma-l)}\ (gy))$$

where $\forall_{\sigma-l} \stackrel{def}{=} \forall\{A_j,B_j,a_{j_i}{:}U_{j_i}[\boldsymbol{A}],b_{j_i}{:}U_{j_i}[\boldsymbol{B}]\ \mid$
$1\le j\le N,\ 1\le i\le n_j,\ \ \langle A_j,B_j\rangle,a_{j_i},b_{j_i}\notin l\wedge X_j,U_{j_i}\in\sigma\}$

and $l(\sigma-l)$, is l appended format-correctly with the $\forall_{\sigma-l}$ quantified items, and

$$\mathsf{Dfnbl}^{\sigma l(\sigma-l)}_{U'}(x,y) \stackrel{def}{=}$$
$$\exists f_{U'}{:}\forall X_p.(U_{p1}\to\cdots\to U_{p_{n_p}}\to\cdots\to(\forall X_q.U_{q1}\to\cdots\to U_{q_{n_q}}\to U'))\ .$$
$$(f_{U'}A_pa_{p_1}\cdots a_{p_{n_p}}\cdots A_qa_{q_1}\cdots a_{q_{n_q}})=x\ \wedge$$
$$(f_{U'}B_pb_{p_1}\cdots b_{p_{n_p}}\cdots B_qb_{1_q}\cdots b_{1_{n_q}})=y$$

where p is the index of the first item in σ. The functional $f_{U'}$ receives as arguments in proper order, the types and terms in l, and then in place of missing actual arguments, the "hypothetical" arguments given by the quantification $\forall_{\sigma-l}$.

Finally, the relation $(\forall Y.U'[\boldsymbol{R},Y,\mathbf{eq}_{\boldsymbol{Z}}])^{\sigma l}$ at universal type within some U_{mk} reflects various rôles terms in this position may play. First we define, if $\forall Y.U'$ is internal, *i.e., is not equal to some $\forall X_r.U_r$, $m<r\le q$, nor closed, then*

$$(\forall Y.U'[\boldsymbol{R},Y,\mathbf{eq}_{\boldsymbol{Z}}])^{\sigma l} \stackrel{def}{=} (f'{:}\forall Y.U'[\boldsymbol{A}],g'{:}\forall Y.U'[\boldsymbol{B}])\ .$$
$$(\textstyle\bigwedge_{V[X]}\ \ f'V[\boldsymbol{A}]\ \ (U'[\boldsymbol{R},V[\boldsymbol{R},\mathbf{eq}_{\boldsymbol{Z}}],\mathbf{eq}_{\boldsymbol{Z}}])^{\sigma l}\ \ g'V[\boldsymbol{B}]$$

and if $\forall Y.U'$ is not internal, and occurs only negatively within some U_{mk},

$$(\forall Y.U'[\boldsymbol{R},Y,\mathbf{eq}_{\boldsymbol{Z}}])^{\sigma l} \stackrel{def}{=} (\forall Y.U'[\boldsymbol{R},Y,\mathbf{eq}_{\boldsymbol{Z}}]))^{\sigma l\mathsf{abo}}$$

and otherwise,

$$(\forall Y.U'[\boldsymbol{R},Y,\mathbf{eq}_{\boldsymbol{Z}}])^{\sigma l} \stackrel{def}{=} (f'{:}\forall Y.U'[\boldsymbol{A}],g'{:}\forall Y.U'[\boldsymbol{B}])\ .$$
$$(\textstyle\bigwedge_{V[X]}\ \ f'V[\boldsymbol{A}]\ \ (U'[\boldsymbol{R},V[\boldsymbol{R},\mathbf{eq}_{\boldsymbol{Z}}],\mathbf{eq}_{\boldsymbol{Z}}])^{\sigma l}\ \ g'V[\boldsymbol{B}]\ \wedge$$
$$f'\ (\forall Y.U'[\boldsymbol{R},Y,\mathbf{eq}_{\boldsymbol{Z}}]))^{\sigma l\mathsf{abo}}\ g')$$

This concludes the definition.

Example 1. Consider the type $\mathsf{Nat} \stackrel{\mathit{def}}{=} \forall X.X \to (X \to X) \to X$. Observing the abstraction barrier in this universal type, we have for $n\!:\!\mathsf{Nat}$,

$$n\ (\forall X.X \to (X \to X) \to X)^{\mathsf{abo}}\ n \\ \stackrel{\mathit{def}}{\Leftrightarrow}\ \forall A, B, R \subset A \times B\ .\ nA\ (R \to (R \to R) \to R)_{\top}^{\langle A,B\rangle}\ nB$$

omitting the superscript $\sigma = X, X, (X \to X)$. This expands to

$$\begin{array}{l} \forall A, B, R \subset A \times B\ . \\ \quad \forall a\!:\!A, b\!:\!B\ .\ a\ R^{\langle A,B\rangle\langle a,b\rangle}\ b\ \Rightarrow \\ \qquad \forall s\!:\!A \to A, s'\!:\!B \to B\ .\ s\ (R \to R)^{\langle A,B\rangle\langle a,b\rangle\langle s,s'\rangle}\ s'\ \Rightarrow\ nAas\ R\ nBbs' \end{array}$$

Here, $R^{\langle A,B\rangle\langle a,b\rangle}$ is simply R, and $s\ (R \to R)^{\langle A,B\rangle\langle a,b\rangle\langle s,s'\rangle}\ s'$ expands to

$$\begin{array}{l} \forall x\!:\!A, y\!:\!B\ .\ x\ R\ y\ \wedge \\ \qquad (\exists f_X\!:\!\forall X.X \to (X \to X) \to X\ .\ f_X Aas = x \wedge f_X Bbs' = y) \\ \hfill \Rightarrow\ sx\ R\ s'y \end{array}$$

The definability clause here reflects that s and s' can be applied only to x and y admissible in virtual computations. For instance, if n were a closed computation, this would mean x and y built from a, s, and b, s'. Moreover, x and y are uniform, since they arise from the same virtual computation n.

Now, consider the type $\mathsf{Nat}' \stackrel{\mathit{def}}{=} \forall X.(X \to X) \to X \to X$, where the successor argument is received prior to the zero argument. Now we have

$$n\ (\forall X.(X \to X) \to X \to X)^{\mathsf{abo}}\ n \\ \stackrel{\mathit{def}}{\Leftrightarrow}\ \forall A, B, R \subset A \times B\ .\ nA\ ((R \to R) \to R \to R)_{\top}^{\langle A,B\rangle}\ nB$$

omitting the superscript $\sigma = X, (X \to X), X$. This expands to

$$\begin{array}{l} \forall A, B, R \subset A \times B\ . \\ \quad \forall s\!:\!A \to A, s'\!:\!B \to B\ .\ s\ (R \to R)^{\langle A,B\rangle\langle s,s'\rangle}\ s'\ \Rightarrow \\ \qquad\qquad \forall a\!:\!A, b\!:\!B\ .\ a\ R^{\langle A,B\rangle\langle s,s'\rangle\langle a,b\rangle}\ b\ \Rightarrow\ nAsa\ R\ nBs'b \end{array}$$

Here, $R^{\langle A,B\rangle\langle s,s'\rangle\langle a,b\rangle}$ is simply R, but $s\ (R \to R)^{\langle A,B\rangle\langle s,s'\rangle}\ s'$ expands to

$$\begin{array}{l} \forall a\!:\!A, b\!:\!B\ .\ \forall x\!:\!A, y\!:\!B\ . \\ \quad x\ R\ y\ \wedge (\exists f_X\!:\!\forall X.(X \to X) \to X \to X\ .\ f_X Asa = x \wedge f_X Bs'b = y) \\ \hfill \Rightarrow\ sx\ R\ s'y \end{array}$$

At the point when the successor arguments s and s' are received by n there is no knowledge of what the zero arguments will be. Therefore, the definability clause for the successor arguments is prepared for arbitrary zero arguments a, b.

To illustrate nesting, consider

$$f\ (\forall X.X \to (X \to X) \to (\forall Y.(Y \to X) \to (X \to Y) \to X))^{\mathsf{abo}}\ f$$

Omitting $\sigma = X, X, (X \to X), Y, (Y \to X), (X \to Y)$, this expands to

$$\begin{array}{l}
\forall A, B, R \subset A \times B\ . \\
\quad \forall a{:}A, b{:}B\ .\ a\ R^{\langle A,B\rangle\langle a,b\rangle}\ b\ \Rightarrow \\
\quad \forall s{:}A \to A, s'{:}B \to B\ .\ s\ (R \to R)^{\langle A,B\rangle\langle a,b\rangle\langle s,s'\rangle}\ s'\ \Rightarrow \\
\ \ \forall A', B', R' \subset A' \times B'\ . \\
\qquad \forall r{:}A' \to A, r'{:}B' \to B\ .\ r\ (R' \to R)^{\langle A,B\rangle\langle a,b\rangle\langle s,s'\rangle\langle A',B'\rangle\langle r,r'\rangle}\ r'\ \Rightarrow \\
\qquad \forall q{:}A \to A', q'{:}B \to B'\ .\ q\ (R \to R')^{\langle A,B\rangle\langle a,b\rangle\langle s,s'\rangle\langle A',B'\rangle\langle r,r'\rangle\langle q,q'\rangle}\ q' \\
\qquad\qquad \Rightarrow\ fAasA'rq\ R^{\langle A,B\rangle\langle a,b\rangle\langle s,s'\rangle\langle A',B'\rangle\langle r,r'\rangle\langle q,q'\rangle}\ fBbs'B'r'q'
\end{array}$$

Here, $s\ (R \to R)^{\langle A,B\rangle\langle a,b\rangle\langle s,s'\rangle}\ s'$ expands to

$$\begin{array}{l}
\forall A', B'. \forall r{:}A' \to A, r'{:}B' \to B, q{:}A \to A', q'{:}B \to B'\ .\ \forall x{:}A, y{:}B\ . \\
\quad x\ R\ y\ \wedge (\exists f_X{:}\forall X.X \to (X \to X) \to (\forall Y.(Y \to X) \to (X \to Y) \to X)\ . \\
\qquad\qquad\qquad\qquad f_X AasA'rq = x \wedge f_X Bbs'B'r'q' = y) \\
\qquad\qquad\qquad \Rightarrow\ sx\ R\ s'y
\end{array}$$

This reflects that arguments to s and s' may be defined also in terms of future r, q, r', and q', besides a, b, s, s'. For example if f were closed, we could have $f = \Lambda X.\lambda x{:}X, s{:}X \to X.\Lambda Y.\lambda r{:}Y \to X, q{:}X \to Y\ .\ s(r(qx))$.

To illustrate the situation for universal types within some U_{mk}, consider

$$f\ (\forall X.X \to (\forall Y.Y \to X) \to (X \to \mathsf{Bool} \to \mathsf{Nat}) \to \mathsf{Nat})^{\mathsf{abo}}\ f$$

We get for $\sigma = X, X, (\forall Y.Y \to X), (X \to \mathsf{Bool} \to \mathsf{Nat})$,

$$\begin{array}{l}
\forall A, B, R \subset A \times B\ .\ \forall a, b\ .\ a\ R\ b\ \Rightarrow \\
\ \ \forall q, q'\ .\ q\ (\forall Y.Y \to R)^{\sigma\langle A,B\rangle\langle a,b\rangle\langle q,q'\rangle}\ q'\ \Rightarrow \\
\quad \forall r, r'\ .\ r\ (R \to \mathsf{Bool} \to \mathsf{Nat})^{\sigma\langle A,B\rangle\langle a,b\rangle\langle q,q'\rangle\langle r,r'\rangle}\ r' \\
\qquad\qquad\qquad \Rightarrow\ fAaqr\ \mathsf{Nat}^{\sigma\langle A,B\rangle\langle a,b\rangle\langle q,q'\rangle\langle r,r'\rangle\mathsf{abo}}\ fBbq'r'
\end{array}$$

Here, $\forall Y.Y \to X$ is inner, *i.e.*, does not occur as a nesting, and is not closed. So for every type V, $V \to X$ is just a local profile within the type of f. Thus, $q\ (\forall Y.Y \to R)^{\sigma\langle A,B\rangle\langle a,b\rangle\langle q,q'\rangle}\ q'$ says $\bigwedge_V qV[A]\ (V[R] \to R)^{\sigma\langle A,B\rangle\langle a,b\rangle\langle q,q'\rangle}\ q'V[B]$, reflecting that q, q' may only be used internally. In contrast, Nat is not inner, since it is indeed a nesting, and is also closed. In a computation, terms of this type may be used in two ways; internally, and also externally; the use of $fAaqr$ and $fBbq'r'$ of type Nat is no longer bound by arguments to f. Thus, we have

$$\begin{array}{l}
x\ \ \mathsf{Nat}^{\sigma\langle A,B\rangle\langle a,b\rangle\langle q,q'\rangle\langle r,r'\rangle}\ \ y \overset{def}{\Leftrightarrow} \\
\bigwedge_{V[X]}\ xV[A]\ \ ((V[R] \to (V[R] \to V[R]) \to V[R])^{\sigma\langle A,B\rangle\langle a,b\rangle\langle q,q'\rangle\langle r,r'\rangle}\ \ yV[B] \\
\qquad\qquad\qquad\qquad\qquad \wedge\ x\ \ \mathsf{Nat}^{\sigma\langle A,B\rangle\langle a,b\rangle\langle q,q'\rangle\langle r,r'\rangle\mathsf{abo}}\ \ y
\end{array}$$

Since Nat has no free occurrences of X, $n\ \mathsf{Nat}^{\sigma\langle A,B\rangle\langle a,b\rangle\langle q,q'\rangle\langle r,r'\rangle\mathsf{abo}}\ m$ expands to the same as $\mathsf{Nat}^{\mathsf{abo}}$. Finally, Bool occurs only negatively in $X \to \mathsf{Bool} \to \mathsf{Nat}$ and the relation is here only $\mathsf{Bool}^{\mathsf{abo}}$, since any term in this position will not be used the internal way. All this reflects ***Abs-Bar1***. Consider for example the closed computation $f = \Lambda X.\lambda x{:}X, q{:}\forall Y.Y \to X, r{:}X \to \mathsf{Bool} \to \mathsf{Nat}\ .\ r(qXx)\mathsf{true}$. ○

Example 1 illustrates *abo*-relations reflexively. If two computations, $n, m : \mathsf{Nat}$ for the example above, are involved, the uniformity aspect is not appropriate. The reason Def. 1 ignores this, is that it is geared toward *abo*-identity extension. Thus, if for example $n\ (\forall X.X \to (X \to X) \to X)^{\mathsf{abo}}\ m$, then we shall get $n = m$, so in this context, there is only one computation involved after-all.

The abstraction barrier-observing formulation of relational parametricity is now given by the following axiom schema.

Definition 2 (*abo*-Parametricity).

$$abo\text{-}\textsc{Param}:\ \forall \boldsymbol{Z}.\forall u{:}(\forall X.U[X,\boldsymbol{Z}])\ .\ u\ (\forall X.U[X,\mathbf{eq}_{\boldsymbol{Z}}])^{\mathsf{abo}}\ u$$

The *abo*-version of the identity extension lemma does not follow from *abo*-Param, because we can no longer use extensionality. Nevertheless, in the spirit of observing abstraction barriers, we argue that in virtual computations, it suffices to consider extensionality only w.r.t. function arguments that will actually occur. The simplest way to capture this is in fact by asserting identity extension.

Definition 3 (*abo*-Identity Extension for Universal Types).

$$abo\text{-}\textsc{Iel}:\ \forall \boldsymbol{Z}.\forall u{:}(\forall X.U[X,\boldsymbol{Z}])\ .\ u\ (\forall X.U[X,\mathbf{eq}_{\boldsymbol{Z}}])^{\mathsf{abo}}\ v\ \Leftrightarrow\ u = v$$

Both *abo*-Param and *abo*-Iel will be shown to hold in the *abo*-parametric *per*-model. Regular parametricity, Param, will not hold in this model; in fact any logic containing both of Param and *abo*-Param is inconsistent. Note that *abo*-Iel implies *abo*-Param. Nevertheless, we choose to display both. We get,

Theorem 1 (*abo*-Identity Extension). *With abo-*Iel*, we have*

$$\forall \boldsymbol{Z}.\forall u, v{:}U[\boldsymbol{Z}]\ .\ u\ U[\mathbf{eq}_{\boldsymbol{Z}}]^{\mathsf{abo}}\ v\ \Leftrightarrow\ (u =_{U[\boldsymbol{Z}]} v)$$

Proof: Easy induction. □

There is an inner aspect of *abo*-identity extension as well.

Theorem 2 (*abo*-Identity Extension (Inner Aspect)). *Let U be a type with no occurrences of X_j in σ. Then we derive with abo-*Iel,

$$\forall u, v{:}U[\boldsymbol{Z}]\ .\ u\ U[\mathbf{eq}_{\boldsymbol{Z}}]^{\sigma l}\ v\ \Leftrightarrow\ u =_{U[\boldsymbol{Z}]} v$$

Proof: Induction on the structure of U. □

We also get

Theorem 3. *With abo-*PARAM *and abo-*IEL*, we derive*

$$\forall U, V.\forall z{:}U \times V \;.\; \mathsf{pair}(\mathsf{proj}_1 z)(\mathsf{proj}_2 z) = z$$

Proof: First show $\forall U, V, z{:}U \times V \;.\; z = z(U \times V)\mathsf{pair}$. Use Theorem 2. □

Theorem 4. *With abo-*PARAM *and abo-*IEL*, we derive*

$$\begin{array}{l}\forall u, v{:}T_1 \times T_2 \;.\; u\ (T_1[\mathbf{eq}_Z] \times T_2[\mathbf{eq}_Z])^{\mathsf{abo}}\ v \\ \quad \Leftrightarrow\ (\mathsf{proj}_1 u)\ T_1[\mathbf{eq}_Z]^{\mathsf{abo}}\ (\mathsf{proj}_1 v)\ \wedge\ (\mathsf{proj}_2 u)\ T_2[\mathbf{eq}_Z]^{\mathsf{abo}}\ (\mathsf{proj}_2 v)\end{array}$$

$$\begin{array}{l}\forall A, B, R \subset A \times B \;.\; \forall u{:}T_1[A] \times T_2[A],\ v{:}T_1[B] \times T_2[B] \;. \\ \quad u\ (T_1[R, \mathbf{eq}_Z] \times T_2[R, \mathbf{eq}_Z])^{T[X]\langle A,B\rangle\langle u,v\rangle}\ v \\ \qquad \Leftrightarrow\ (\mathsf{proj}_1 u)\ T_1[R, \mathbf{eq}_Z]^{T[X]\langle A,B\rangle\langle u,v\rangle}\ (\mathsf{proj}_1 v)\ \wedge \\ \qquad\qquad (\mathsf{proj}_2 u)\ T_2[R, \mathbf{eq}_Z]^{T[X]\langle A,B\rangle\langle u,v\rangle}\ (\mathsf{proj}_2 v)\end{array}$$

Proof: Use Theorem 3. □

4 Refinement

Consider now the issue of when two packages are interchangeable in a program. We view observational equivalence as the conceptual description of interchangeability, and simulation relations as part of a method for showing observational equivalence. We want the two notions to be equivalent. For data types with first-order operations, this equivalence is a fact under relational parametricity. At higher-order this is not the case. Also, the composability of simulation relations fails at higher order, compromising the constructive composition of refinement steps. We now show that by using *abo*-relations and *abo*-parametricity, we get the equivalence as well as composability, at any order.

To each refinement stage, a set *Obs* of *observable types* is associated, assumed to contain closed inductive types, such as Bool or Nat, and also any parameters. Two data types are interchangeable if their observable properties are indistinguishable, *i.e.*, packages should be observationally equivalent if it makes no difference which one is used in computations with observable result types. Thus,

Definition 4 (Observational Equivalence). *For* A, B, $\mathfrak{a}{:}T[A], \mathfrak{b}{:}T[B]$, *Obs*,

$$\bigwedge_{D \in Obs} \forall f{:}\forall X.(T[X] \to D) \;.\; (f A\, \mathfrak{a}) = (f B\, \mathfrak{b})$$

For example, an observable computation on natural-number stacks could be $\Lambda X.\lambda \mathfrak{x}{:}T_{\mathsf{STACK}_{\mathsf{Nat}}}[X] \;.\; \mathfrak{x}.\mathsf{top}(\mathfrak{x}.\mathsf{push}\, n\ \mathfrak{x}.\mathsf{empty})$.

Simulation relations arise from the concept of *data refinement* [12,5] and the use of relations to show *representation independence* [19,31,28,27], leading to *logical relations* for lambda calculus [20,21,34,24]. In our logic one can use the action of types on relations to define a syntactic mirror of the above ideas. Two data types are related by a simulation relation if there exists a relation on their data representations that is preserved by their corresponding operations.

Definition 5 (Simulation Relation). *For* A, B, $\mathfrak{a}:T[A], \mathfrak{b}:T[B]$,

$$\exists R \subset A \times B \;.\; \mathfrak{a}(T[R, \mathbf{eq}_Z])\mathfrak{b}$$

For specification refinement one wants to establish observational equivalence using simulation relations. The problem at higher order is that there might not exist a simulation relation, even in the presence of observational equivalence.

Example 2. Consider $Sig_{\mathsf{SetCE}} \stackrel{def}{=} \exists X.T_{\mathsf{SetCE}}[X]$, where

$$T_{\mathsf{SetCE}}[X] \stackrel{def}{=} (\mathsf{empty}:X,\ \mathsf{add}:\mathsf{Nat}\to X\to X,\ \mathsf{remove}:\mathsf{Nat}\to X\to X,\\ \mathsf{in}:\mathsf{Nat}\to X\to \mathsf{Bool},\ \mathsf{crossover}:(\mathsf{Nat}\to X\to X)\to\mathsf{Nat}\to\mathsf{Bool})$$

and consider $(\mathsf{pack}\ \mathsf{List}_{\mathsf{Nat}}\ \mathfrak{a}):Sig_{\mathsf{SetCE}}$ and $(\mathsf{pack}\ \mathsf{List}_{\mathsf{Nat}}\ \mathfrak{b}):Sig_{\mathsf{SetCE}}$, where

$$\begin{aligned}
\mathfrak{a} \stackrel{def}{=} (&\mathsf{empty} = \mathsf{nil},\\
&\mathsf{add} = \textit{cons-uniquesorted} \stackrel{def}{=} \lambda x:\mathsf{Nat}.\lambda l:\mathsf{List}_{\mathsf{Nat}}\ .\\
&\qquad \textit{return } l' \textit{ that is } l \textit{ with } x \textit{ uniquely inserted before first } y > x,\\
&\mathsf{remove} = \textit{del-first} \stackrel{def}{=} \lambda x:\mathsf{Nat}.\lambda l:\mathsf{List}_{\mathsf{Nat}}\ .\\
&\qquad \textit{return } l' \textit{ that is } l \textit{ with first occurrence of } x \textit{ removed},\\
&\mathsf{in} = \textit{in} \stackrel{def}{=} \lambda x:\mathsf{Nat}.\lambda l:\mathsf{List}_{\mathsf{Nat}}\ .\ \textit{return } \mathsf{true} \textit{ if } x \textit{ occurs in } l,\ \mathsf{false} \textit{ otherwise},\\
&\mathsf{crossover} \stackrel{def}{=} \lambda f:(\mathsf{Nat}\to\mathsf{List}_{\mathsf{Nat}}\to\mathsf{List}_{\mathsf{Nat}}).\lambda n:\mathsf{Nat}\ .\\
&\qquad \mathsf{in}(n)(f(n)(1 :: 0 :: \mathsf{nil})))\ )
\end{aligned}$$

Here we use the infix symbol :: to denote cons. Furthermore,

$$\begin{aligned}
\mathfrak{b} \stackrel{def}{=} (&\mathsf{empty} = \mathsf{nil},\\
&\mathsf{add} = \textit{cons-uniquesorted},\\
&\mathsf{remove} = \textit{del-all} \stackrel{def}{=} \lambda x:\mathsf{Nat}.\lambda l:\mathsf{List}_{\mathsf{Nat}}\ .\\
&\qquad \textit{return } l' \textit{ that is } l \textit{ with all occurrences of } x \textit{ removed},\\
&\mathsf{in} = \textit{in},\\
&\mathsf{crossover} \stackrel{def}{=} \lambda f:(\mathsf{Nat}\to\mathsf{List}_{\mathsf{Nat}}\to\mathsf{List}_{\mathsf{Nat}}).\lambda n:\mathsf{Nat}\ .\\
&\qquad \mathsf{in}(n)(f(n)(1 :: 1 :: 0 :: \mathsf{nil})))\ )
\end{aligned}$$

In the parametric minimal model of [11], all elements of the interpretation of $\mathsf{List}_{\mathsf{Nat}}$ and Bool are in correspondence with closed normal forms, and the ω-rule holds. Then observational equivalence holds between $\mathfrak{a}$ and $\mathfrak{b}$, but the existence of a simulation relation does not, because any simulation relation R demands

$$\mathfrak{a}.\mathsf{crossover}\ \ ((\mathsf{eq}_{\mathsf{Nat}}\to R\to R)\to\mathsf{eq}_{\mathsf{Nat}}\to\mathsf{eq}_{\mathsf{Bool}})\ \ \mathfrak{b}.\mathsf{crossover}$$

meaning that $\mathfrak{a}.\mathsf{crossover}(\textit{del-all})(1)$ and $\mathfrak{b}.\mathsf{crossover}(\textit{del-first})(1)$ are to be considered. Note that *del-all* really belongs to $\mathfrak{b}$ and *del-first* belongs to $\mathfrak{a}$.

For failure of composability, in addition to $(\mathsf{pack}\ \mathsf{List}_{\mathsf{Nat}}\ \mathfrak{a})$ and $(\mathsf{pack}\ \mathsf{List}_{\mathsf{Nat}}\ \mathfrak{b})$ above, consider $(\mathsf{pack}\ \mathsf{List}_{\mathsf{Nat}}\ \mathfrak{d}):Sig_{\mathsf{SetCE}}$, where

$$\mathfrak{d} \stackrel{def}{=} (\mathsf{empty} = \mathsf{nil}, \quad \mathsf{add} = cons, \quad \mathsf{remove} = del\text{-}all \quad \mathsf{in} = in, \quad \mathsf{crossover} \stackrel{def}{=} \lambda f : (\mathsf{Nat} \to \mathsf{List}_{\mathsf{Nat}} \to \mathsf{List}_{\mathsf{Nat}}).\lambda n : \mathsf{Nat} \, . \, \mathsf{in}(n)(f(n)(1 :: 1 :: 0 :: \mathsf{nil}))) \,)$$

Then the existence of simulation relations holds between $\mathfrak{a}$ and $\mathfrak{d}$, and between $\mathfrak{d}$ and $\mathfrak{b}$, but as before, not between $\mathfrak{a}$ and $\mathfrak{b}$. ○

Thus the standard concept of simulation relation is not adequate. For us, the reason for this is rooted in the encapsulation issue for universal types. The interchangeability of data types in a program translates to interchangeability of packages in virtual computations. Hence, the relevant context is that of any given computation, and therefore simulation relations should account for the encapsulation inherent in such computations. We therefore define,

Definition 6 (*abo*-Simulation Relation). *For* A, B, $\mathfrak{a}{:}T[A], \mathfrak{b}{:}T[B]$,

$$\exists R \subset A \times B \, . \, \mathfrak{a} \, (T[R, \mathbf{eq}_{\boldsymbol{Z}}])^{T[X]\langle A,B\rangle\langle \mathfrak{a},\mathfrak{b}\rangle} \, \mathfrak{b}$$

Example 2. (continued) We now have

$$\mathfrak{a}.\mathsf{crossover} \; ((\mathsf{eq}_{\mathsf{Nat}} \to R \to R) \to \mathsf{eq}_{\mathsf{Nat}} \to \mathsf{eq}_{\mathsf{Bool}})^{T_{\mathsf{SetCE}}\langle \mathsf{List}_{\mathsf{Nat}}, \mathsf{List}_{\mathsf{Nat}}\rangle\langle \mathfrak{a},\mathfrak{b}\rangle} \; \mathfrak{b}.\mathsf{crossover}$$

Only $\gamma, \delta : \mathsf{Nat} \to \mathsf{List}_{\mathsf{Nat}} \to \mathsf{List}_{\mathsf{Nat}}$ such that $\mathsf{Dfnbl}^{T_{\mathsf{SetCE}}\langle \mathsf{List}_{\mathsf{Nat}}, \mathsf{List}_{\mathsf{Nat}}\rangle\langle \mathfrak{a},\mathfrak{b}\rangle}_{\mathsf{Nat} \to X \to X}(\gamma, \delta)$, *i.e.*,

$$\exists f : \forall X. T_{\mathsf{SetCE}}[X] \to (\mathsf{Nat} \to X \to X) \, . \, (f \, \mathsf{List}_{\mathsf{Nat}} \, \mathfrak{a}) = \gamma \wedge (f \, \mathsf{List}_{\mathsf{Nat}} \, \mathfrak{b}) = \delta$$

are considered. This excludes *e.g.*, the cross-over pair (*del-all*, *del-first*). ○

In the following, using *abo*-simulation relations we establish the essential properties that do not hold for standard simulation relations. In the context of data types, we adhere to the following reasonable assumption.

HADT$_{Obs}$**:** Every profile $T_i[X] = T_{i1}[X] \to \cdots \to T_{n_i}[X] \to T_{c_i}[X]$ of an abstract type $\exists X.T[X]$ is such that $T_{c_i}[X]$ is either X or some $D \in Obs$.

Theorem 5. *Assuming* ***HADT***$_{Obs}$ *for* $T[X]$, *we get with abo*-PARAM *and abo*-IEL,

$$\forall A, B. \forall \mathfrak{a}{:}T[A], \mathfrak{b}{:}T[B] \, . \, \exists R \subset A \times B \, . \, \mathfrak{a} \; T[R, \mathbf{eq}_{\boldsymbol{Z}}]^{T[X]\langle A,B\rangle\langle \mathfrak{a},\mathfrak{b}\rangle} \; \mathfrak{b} \;\Leftrightarrow\; \textstyle\bigwedge_{D \in Obs} \forall f : \forall X.(T[X] \to D) \, . \, (f A \, \mathfrak{a}) = (f B \, \mathfrak{b})$$

Proof: $\Rightarrow$: By *abo*-PARAM $f \; (\forall X.T[X, \mathbf{eq}_{\boldsymbol{Z}}] \to D)^{\mathsf{abo}} \; f$ and Theorem 2.

$\Leftarrow$: Let $s \stackrel{def}{=} T[X]\langle A, B\rangle\langle \mathfrak{a}, \mathfrak{b}\rangle$. We must derive $\exists R \subset A \times B \, . \, \mathfrak{a}(T[R, \mathbf{eq}_{\boldsymbol{Z}}]^s)\mathfrak{b}$. We exhibit $R \stackrel{def}{=} (a : A, b : B) \, . \, (\mathsf{Dfnbl}^s_X(a, b))$. By Theorem 4, we must for every component $g : U_1 \to \cdots \to U_l \to U_c$ in $T[X]$, show the derivability of

$$\forall x_1 : U_1[A], \ldots, x_l : U_l[A] \;.\; \forall y_1 : U_1[B], \ldots, y_l : U_l[B] \;.$$
$$\textstyle\bigwedge_{1 \le i \le l} (x_i \; U_i[R, \mathbf{eq_Z}]^s \; y_i \;\wedge\; \mathsf{Dfnbl}^s_{U_i}(x_i, y_i))$$
$$\Rightarrow \quad (\mathfrak{a}.g\, x_1 \cdots x_l) \; U_c[R, \mathbf{eq_Z}]^s \; (\mathfrak{b}.g\, y_1 \cdots y_l)$$

We get $\exists f_{U_i} : \forall X.(T[X] \to U_i[X]) \;.\; (f_{U_i} A\, \mathfrak{a}) = x_i \wedge (f_{U_i} B\, \mathfrak{b}) = y_i$ from the antecedent $\mathsf{Dfnbl}^s_{U_i}(x_i, y_i)$. Let $f \stackrel{def}{=} \Lambda X.\lambda \mathfrak{x} : T[X] \;.\; (\mathfrak{x}.g(f_{U_1} X \mathfrak{x}) \cdots (f_{U_l} X \mathfrak{x}))$.

$U_c = D \in Obs$: By Theorem 2 it suffices to derive $\mathfrak{a}.g\, x_1 \cdots x_l =_D \mathfrak{b}.g\, y_1 \cdots y_l$. The assumption gives $(f A\, \mathfrak{a}) =_D (f B\, \mathfrak{b})$, which gives the result.

Suppose $U_c = X$: We must then derive

$$\exists f : \forall X.(T[X] \to U_c[X]) \;.\; (f A\, \mathfrak{a}) = (\mathfrak{a}.g\, x_1 \cdots x_l) \wedge (f B\, \mathfrak{b}) = (\mathfrak{b}.g\, y_1 \cdots y_l)$$

For this we display f above. □

Theorem 6. *Assuming* HADT_{Obs} *for* $T[X]$*, we get with abo-*PARAM *and abo-*IEL*,*

$$\forall A, B, C,\; R \subset A \times B,\; S \subset B \times C,\; \mathfrak{a} : T[A], \mathfrak{b} : T[B], \mathfrak{c} : T[C].$$
$$\mathfrak{a}(T[R, \mathbf{eq_Z}]^{T[X]\langle A,B\rangle\langle \mathfrak{a},\mathfrak{b}\rangle})\mathfrak{b} \;\wedge\; \mathfrak{b}(T[S, \mathbf{eq_Z}]^{T[X]\langle B,C\rangle\langle \mathfrak{b},\mathfrak{c}\rangle})\mathfrak{c}$$
$$\Rightarrow \; \mathfrak{a}(T[S \circ R, \mathbf{eq_Z}]^{T[X]\langle A,C\rangle\langle \mathfrak{a},\mathfrak{c}\rangle})\mathfrak{c}$$

Proof: The goal is to derive for every component $g : U_1 \to \cdots \to U_l \to U_c$ in T,

$$\forall x_1 : U_1[A], \ldots, x_l : U_l[A] \;.\; \forall z_1 : U_1[C], \ldots, z_l : U_l[C] \;.$$
$$\textstyle\bigwedge_{1 \le i \le l} (x_i \; U_i[S \circ R, \mathbf{eq_Z}]^{T[X]\langle A,C\rangle\langle \mathfrak{a},\mathfrak{c}\rangle} \; z_i \;\wedge\; \mathsf{Dfnbl}^{T[X]\langle A,C\rangle\langle \mathfrak{a},\mathfrak{c}\rangle}_{U_i}(x_i, z_i))$$
$$\Rightarrow \quad (\mathfrak{a}.g\, x_1 \cdots x_l) \; U_c[S \circ R, \mathbf{eq_Z}]^{T[X]\langle A,C\rangle\langle \mathfrak{a},\mathfrak{c}\rangle} \; (\mathfrak{c}.g\, z_1 \cdots z_l)$$

$\mathsf{Dfnbl}^{T[X]\langle A,C\rangle\langle \mathfrak{a},\mathfrak{c}\rangle}_{U_i}(x_i, z_i)$ gives an $f \stackrel{def}{=} \Lambda X.\lambda \mathfrak{x} : T[X] \;.\; (\mathfrak{x}.g(f_{U_1} X \mathfrak{x}) \cdots (f_{U_l} X \mathfrak{x}))$.

$U_c = D \in Obs$: By assumption and Theorem 5, $(f A\, \mathfrak{a}) = (f B\, \mathfrak{b}) = (f C \mathfrak{c})$, and $\mathfrak{a}.g\, x_1 \cdots x_l = (f A\, \mathfrak{a})$ and $(f C \mathfrak{c}) = \mathfrak{c}.g\, z_1 \cdots z_l$.

$U_c = X$: We must show $\exists b : U_c[B, \boldsymbol{Z}] \;.\; (\mathfrak{a}.g\, x_1 \cdots x_l) \; R \; b \wedge b \; S \; (\mathfrak{c}.g\, z_1 \cdots z_l)$. Exhibit $f B\, \mathfrak{b} = (\mathfrak{b}.g(f_{U_1} B \mathfrak{b}) \cdots (f_{U_l} B \mathfrak{b}))$ for b. □

Conventional simulation relations and relational parametricity do not give corresponding results to Theorem 5; the coincidence of the existence of simulation relations with observational equivalence, and Theorem 6; the composability of simulation relations, except for data types with first-order operations.

5 *abo*-Semantics

We present a *per*-model for the relational logic with *abo*-PARAM and *abo*-IEL. Just as the parametric *per*-model [2] is defined directly according to PARAM, the *abo*-parametric *per*-model will be defined according to *abo*-PARAM.

We use an obvious shorthand notation and simply write things like $U[\mathcal{A}, \mathcal{Z}]$ instead of $[\![A, \boldsymbol{Z} \rhd U[A, \boldsymbol{Z}]]\!]_{[A \mapsto \mathcal{A}, \boldsymbol{Z} \mapsto \mathcal{Z}]}$, and also things like $U[\mathcal{R}, \mathcal{Z}]^{\sigma\langle \mathcal{A},\mathcal{B}\rangle}$ for $[\![A, B, R \subset A \times B, \boldsymbol{Z} \rhd U[R, \mathbf{eq_Z}]^{\sigma\langle A,B\rangle}]\!]_{[A \mapsto \mathcal{A}, B \mapsto \mathcal{B}, \boldsymbol{Z} \mapsto \mathcal{Z}, R \mapsto \mathcal{R}]}$.

First, the semantics of a universal type $\forall X.U$ in the pure parametric *per*-model, *e.g.*, [29], is $(\cap_{\mathcal{A}\in\text{PER}}[\![U[\mathcal{A}]]\!])$. This gives Strachey's parametricity. In the relational parametric *per*-model, the semantics $(\cap_{\mathcal{A}} U[\mathcal{A}, \mathcal{Z}])^\flat$ is obtained by including only those elements which satisfy relational parametricity, thus

$$\begin{aligned} n\ (\cap_{\mathcal{A}} U[\mathcal{A}, \mathcal{Z}])^\flat\ m \ \stackrel{def}{\Leftrightarrow}\ & \forall \mathcal{A}, \mathcal{B} \in \text{PER},\ saturated\ \mathcal{R} \subset Dom(\mathcal{A}) \times Dom(\mathcal{B})\ . \\ & n\ U[\mathcal{A}, \mathcal{Z}]\ m \wedge n\ U[\mathcal{B}, \mathcal{Z}]\ m \wedge \\ & n\ U[\mathcal{R}, \mathcal{Z}]\ n \wedge m\ U[\mathcal{R}, \mathcal{Z}]\ m \end{aligned}$$

Alternatively, we may give the following definition, since relational parametricity is equivalent to identity extension at universal type.

$$\begin{aligned} n\ (\cap_{\mathcal{A}} U[\mathcal{A}, \mathcal{Z}])^\flat\ m \ \stackrel{def}{\Leftrightarrow}\ & \forall \mathcal{A}, \mathcal{B} \in \text{PER},\ saturated\ \mathcal{R} \subset Dom(\mathcal{A}) \times Dom(\mathcal{B})\ . \\ & n\ U[\mathcal{R}, \mathcal{Z}]\ m \end{aligned}$$

A model satisfying *abo*-Param and *abo*-Iel is then obtained by defining

$$\begin{aligned} n\ (\cap_{\mathcal{A}} U[\mathcal{A}, \mathcal{Z}])^{\flat^{\mathsf{abo}}}\ m \ \stackrel{def}{\Leftrightarrow}\ & \forall \mathcal{A}, \mathcal{B} \in \text{PER},\ saturated\ \mathcal{R} \subset Dom(\mathcal{A}) \times Dom(\mathcal{B})\ . \\ & n\ U[\mathcal{R}, \mathcal{Z}]_{\top}^{\sigma\langle \mathcal{A}, \mathcal{B}\rangle}\ m \end{aligned}$$

where σ depends on U according to Def. 1. The rest of the semantics is standard *per*-semantics, see *e.g.*, [29,2]. We must now show that this *abo*-parametric structure is a model for System F; every closed term has an interpretation. Intuitively, this should be so, since *abo*-relations are defined according to ***Abs-Bar1***, which in particular captures what closed polymorphic functionals look like.

Theorem 7. *Every closed term of System F has an interpretation in the abo-parametric per-structure.*

Proof: The interesting part of this is the semantics for universal types. Thus, show for every closed $f : \forall X.U$, that $f\ (\forall X.U[X])^{\mathsf{abo}}\ f$ holds in the structure. □

6 Final Remarks

We have provided a notion of relational parametricity that takes into account the abstraction barrier-observing (*abo*) mechanism in universal types. We showed how this then gives the notion of *abo*-simulation relation from [9,10], which resolves serious well-known problems for simulation relations in the framework of stepwise refinement. Furthermore, we introduced, asserted, and validated *abo*-relational parametricity. This massively simplifies both our task as formalists, as well as decreases the amount of reasoning to which a developer needs to commit, when using *abo*-simulation relations for refinement.

There is a link under *abo*-relational parametricity between *abo*-simulation relations and equality at existential type. This link exists for standard simulation relations under standard relational parametricity [25], but in that case there is a slightly annoying circularity issue. This issue is not present in the *abo* case. This is a consequence of Theorem 5. Also, relational parametricity gives induction, *e.g.,* on Nat, and so does *abo*-relational parametricity.

There are lots of further questions, *e.g.,* does the *abo*-parametric *per*-model have polymorphic functionals that are not closed-term denotable, and what else can be shown using *abo*-parametricity that cannot be analogously shown using regular parametricity; and *vice versa,* in particular w.r.t. universal constructions. Present research includes a comparison of *abo*-simulation relations to the prelogical relations of [14].

Acknowledgements. Don Sannella, Martin Hofmann, Uday Reddy, and David Aspinall have given essential feed-back on early versions of ideas in this paper. The anonymous referees have given further detailed and valuable comments.

References

1. R.-J. Back and J. Wright. *Refinement Calculus, A Systematic Introduction.* Graduate Texts in Computer Science. Springer Verlag, 1998.
2. E.S. Bainbridge, P.J. Freyd, A. Scedrov, and P.J. Scott. Functorial polymorphism. *Theoretical Computer Science*, 70:35–64, 1990.
3. H.P. Barendregt. Lambda calculi with types. In S. Abramsky, D.M. Gabbay, and T.S.E. Maibaum, *eds., Handbook of Logic in Computer Science*, volume 2, pages 118–309. Oxford University Press, 1992.
4. C. Böhm and A. Berarducci. Automatic synthesis of typed λ-programs on term algebras. *Theoretical Computer Science*, 39:135–154, 1985.
5. O.-J. Dahl. *Verifiable Programming, Revised version 1993.* Prentice Hall Int. Series in Computer Science; C.A.R. Hoare, Series Editor. Prentice-Hall, UK, 1992.
6. J.-Y. Girard, P. Taylor, and Y. Lafont. *Proofs and Types.* Cambridge Tracts in Theoretical Computer Science. Cambridge University Press, 1990.
7. J. Hannay. Specification refinement with System F. In *Computer Science Logic. Proc. of CSL'99*, vol. 1683 of *Lecture Notes in Comp. Sci.*, pages 530–545. Springer Verlag, 1999.
8. J. Hannay. Specification refinement with System F, the higher-order case. In *Recent Trends in Algebraic Development Techniques. Selected Papers from WADT'99*, volume 1827 of *Lecture Notes in Comp. Sci.*, pages 162–181. Springer Verlag, 1999.
9. J. Hannay. A higher-order simulation relation for System F. In *Foundations of Software Science and Computation Structures. Proc. of FOSSACS 2000*, vol. 1784 of *Lecture Notes in Comp. Sci.*, pages 130–145. Springer Verlag, 2000.
10. J. Hannay. *Abstraction Barriers and Refinement in the Polymorphic Lambda Calculus.* PhD thesis, Laboratory for Foundations of Computer Science (LFCS), University of Edinburgh, 2001.
11. R. Hasegawa. Parametricity of extensionally collapsed term models of polymorphism and their categorical properties. In *Theoretical Aspects of Computer Software. Proc. of TACS'91*, vol. 526 of *Lecture Notes in Comp. Sci.*, pages 495–512. Springer Verlag, 1991.
12. C.A.R. Hoare. Proofs of correctness of data representations. *Acta Informatica*, 1:271–281, 1972.

13. F. Honsell, J. Longley, D. Sannella, and A. Tarlecki. Constructive data refinement in typed lambda calculus. In *Foundations of Software Science and Computation Structures. Proc. of FOSSACS 2000*, vol. 1784 of *Lecture Notes in Comp. Sci.*, pages 161–176. Springer Verlag, 2000.
14. F. Honsell and D. Sannella. Prelogical relations. *Information and Computation*, 178:23–43, 2002.
15. Y. Kinoshita, P.W. O'Hearn, J. Power, M. Takeyama, and R.D. Tennent. An axiomatic approach to binary logical relations with applications to data refinement. In *Theoretical Aspects of Computer Software. Proc. of TACS'97*, vol. 1281 of *Lecture Notes in Comp. Sci.*, pages 191–212. Springer Verlag, 1997.
16. Y. Kinoshita and J. Power. Data refinement for call-by-value programming languages. In *Computer Science Logic. Proc. of CSL'99*, vol. 1683 of *Lecture Notes in Comp. Sci.*, pages 562–576. Springer Verlag, 1999.
17. Q. Ma and J.C. Reynolds. Types, abstraction and parametric polymorphism, part 2. In *Mathematical Foundations of Programming Semantics, Proc. of MFPS*, vol. 598 of *Lecture Notes in Comp. Sci.*, pages 1–40. Springer Verlag, 1991.
18. H. Mairson. Outline of a proof theory of parametricity. In *Functional Programming and Computer Architecture. Proc. of the 5th acm Conf.*, vol. 523 of *Lecture Notes in Comp. Sci.*, pages 313–327. Springer Verlag, 1991.
19. R. Milner. An algebraic definition of simulation between programs. In *Joint Conferences on Artificial Intelligence, Proc. of JCAI*, pages 481–489. Morgan Kaufman Publishers, 1971.
20. J.C. Mitchell. On the equivalence of data representations. In V. Lifschitz, *ed.*, *Artificial Intelligence and Mathematical Theory of Computation: Papers in Honor of John McCarthy*, pages 305–330. Academic Press, 1991.
21. J.C. Mitchell. *Foundations for Programming Languages.* Foundations of Computing. MIT Press, 1996.
22. J.C. Mitchell and G.D. Plotkin. Abstract types have existential type. *ACM Transactions on Programming Languages and Systems*, 10(3):470–502, 1988.
23. C. Morgan. *Programming from Specifications, 2nd ed.* Prentice Hall International Series in Computer Science; C.A.R. Hoare, Series Editor. Prentice-Hall, UK, 1994.
24. P.W. O'Hearn and R.D. Tennent. Relational parametricity and local variables. In *20th SIGPLAN-SIGACT Symposium on Principles of Programming Languages, Proceedings*, pages 171–184. ACM Press, 1993.
25. G.D. Plotkin and M. Abadi. A logic for parametric polymorphism. In *Typed Lambda Calculi and Applications. Proc. of TLCA'93*, vol. 664 of *Lecture Notes in Comp. Sci.*, pages 361–375. Springer Verlag, 1993.
26. G.D. Plotkin, J. Power, D. Sannella, and R.D. Tennent. Lax logical relations. In *Automata, Languages and Programming. Proc. of ICALP 2000*, vol. 1853 of *Lecture Notes in Comp. Sci.*, pages 85–102. Springer Verlag, 2000.
27. J.C. Reynolds. *The Craft of Programming.* Prentice-Hall International, 1981.
28. J.C. Reynolds. Types, abstraction and parametric polymorphism. In *Information Processing 83, Proc. of the IFIP 9th World Computer Congress*, pages 513–523. Elsevier Science Publishers B.V. (North-Holland), 1983.
29. E. Robinson. Notes on the second-order lambda calculus. Report 4/92, University of Sussex, School of Cognitive and Computing Sciences, 1992.
30. D. Sannella and A. Tarlecki. Essential concepts of algebraic specification and program development. *Formal Aspects of Computing*, 9:229–269, 1997.
31. O. Schoett. Behavioural correctness of data representations. *Science of Computer Programming*, 14:43–57, 1990.

32. C. Strachey. Fundamental concepts in programming languages. Lecture notes from the Int. Summer School in Programming Languages, Copenhagen, 1967.
33. I. Takeuti. An axiomatic system of parametricity. *Fundamenta Informaticae*, 20:1–29, 1998.
34. R.D. Tennent. Correctness of data representations in Algol-like languages. In A.W. Roscoe, *ed.*, *A Classical Mind: Essays in Honour of C.A.R. Hoare*. Prentice Hall International, 1997.

Encoding of the Halting Problem into the Monster Type and Applications

Thierry Joly

Équipe Preuves, Programmes et Systèmes, Université Paris VII,
2, place Jussieu, 75251 Paris cedex 05, France**
joly@pps.jussieu.fr

Abstract. The question whether a given functional of a full type structure (FTS for short) is λ-definable by a closed λ-term, raised by G. Huet in [Hue75] and known as the Definability Problem, was proved to be undecidable by R. Loader in 1993. More precisely, R. Loader proved that the problem is undecidable for every FTS over at least 7 ground elements (cf [Loa01]).

We solve here the remaining non trivial cases and show that the problem is undecidable whenever there are more than one ground element. The proof is based on a direct encoding of the Halting Problem for register machines into the Definability Problem restricted to the functionals of the Monster type $\mathbb{M} = (((o \to o) \to o) \to o) \to (o \to o)$. It follows that this new restriction of the Definability Problem, which is orthogonal to the ones considered so far, is also undecidable. Another consequence of the latter fact, besides the result stated above, is the undecidability of the β-Pattern Matching Problem, recently established by R. Loader in [Loa03].

1 Introduction

Notations. We consider here the simply typed λ-calculus λ^τ *à la Church* over a single atomic type o. The λ-terms and the variables of any type A are respectively denoted by capital letters and small letters: M^A, x^A. Their type may be omitted if it is of no importance or clear from the context.
The λ-terms are considered up to the renaming of their bound variables, so that $=$ actually denotes α-equivalence. The other kinds of λ-terms equalities are always explicitly written: $=_\beta$, $=_{\beta\eta}$, ...
Λ (resp. $\Lambda^\varnothing$) is the set of the λ-terms (resp. of the closed λ-terms) of λ^τ.
$\mathbf{I}^{A\to A}$ denotes the combinator $\lambda x^A.x$ and $M \circ N$ the λ-term $\lambda x.M(Nx)$.
Types are implicitely parenthesized to the right, *i.e.* $A_1 \to \ldots \to A_n \to B$ stands for $A_1 \to (\ldots \to (A_n \to B)\ldots)$. We also use for concision the type notations: $1 = o \to o$, $2 = 1 \to o, \ldots, n{+}1 = n \to o, \ldots$ and $1_2 = o \to 1, \ldots, 1_{n+1} = o \to 1_n, \ldots$

Recall that a hereditarily finite full type structure of λ^τ (FTS for short) is a collection $\mathcal{F} = (\mathcal{F}^A)_{A \in \mathrm{Type}(\lambda^\tau)}$ such that $\mathcal{F}^o \neq \varnothing$ is finite and $\mathcal{F}^{A\to B} = (\mathcal{F}^B)^{\mathcal{F}^A}$

** The present work was done at the University of Nijmegen, The Netherlands.

M. Hofmann (Ed.): TLCA 2003, LNCS 2701, pp. 153–166, 2003.

for all A, B. Recall also that a functional $\varphi \in \mathcal{F}^A$ is said λ-*definable* whenever for some closed λ-term M^A we have $[\![M]\!] = \varphi$, where $[\![M]\!]$ denotes the interpretation of M in $\mathcal{F}$ defined in the usual way (see e.g. [Fri75]). The basic question:

DP: *"Given a functional φ of any FTS, is φ λ-definable?"*

is known as the Definability Problem, and the most natural restrictions of this very general problem are with no doubt:

$\mathrm{DP}_{\mathcal{F}_n}$: *"Given a functional φ of $\mathcal{F}_n$, is φ λ-definable?"*

where $\mathcal{F}_n$ ($n \geqslant 1$) denotes the unique FTS (up to isomorphism) such that $\mathcal{F}_n^o$ has n elements. We may also consider the following restrictions of DP, which are somehow orthogonal to the previous ones:

DP_A: *"Given a functional $\varphi \in \mathcal{F}_n^A$ of any FTS $\mathcal{F}_n$, is φ λ-definable?"*

The Statman conjecture was that DP is decidable (*cf* [Sta82]) and was refuted by R. Loader, who proved that $\mathrm{DP}_{\mathcal{F}_n}$ is undecidable for every $n \geqslant 7$ (*cf* [Loa01]). The Definability Problem is related to another famous one, the Higher Order Pattern Matching Problem, raised by Huet in [Hue75]:

MP: *"Given $P, Q \in \Lambda^{\varnothing}$, is there $X \in \Lambda^{\varnothing}$ such that $P\,X =_{\beta\eta} Q$?"*

It is well known that the decidability of MP, which is still an open question, would have been a consequence of the Statman conjecture (*cf* [Sta82]). DP is actually equivalent to the following generalization of MP:

MP^+: *"Given a type substitution $+ = [o{:=}1_n]$ and $P, Q \in \Lambda^{\varnothing}$, is there a closed λ-term X such that $P\,X^+ =_{\beta\eta} Q$?"*

More precisely, for every type A the problem DP_A proves to be equivalent to:

MP^+_A: *"Given $P^{A^+\to o^+}, Q^{o^+} \in \Lambda^{\varnothing}$ where $+$ is any substitution $[o{:=}1_n]$, is there a closed λ-term X^A such that $P\,X^+ =_{\beta\eta} Q$?"*

We encode below (Section 2) the Halting Problem of a simple kind of register machines into functionals of the Monster type $\mathbb{M} = 3 \to 1$. This yields the undecidability of new restricted cases of the problems DP, MP^+, namely $\mathrm{DP}_{\mathbb{M}}$ and $\mathrm{MP}^+_{\mathbb{M}}$, which prove to be fruitful. E.g. from the undecidability of $\mathrm{MP}^+_{\mathbb{M}}$, we derive at once the undecidability of the following other generalization[1] of MP:

β-MP: *"Given $P, Q \in \Lambda^{\varnothing}$, is there $X \in \Lambda^{\varnothing}$ such that $P\,X =_{\beta} Q$?"*

which has already been established by R. Loader in [Loa03]. This alternative argument (Section 3), where the restriction of the equality of the calculus to $=_\beta$

[1] MP is indeed the particular case of β-MP where the given λ-terms P, Q are in long normal form, since for such P, Q, the equations $PX =_\beta Q$ and $PX =_{\beta\eta} Q$ have the same solutions $X \in \Lambda^{\varnothing}$.

is introduced at the latest, enlights the crucial role played by this restriction and the fact that the question whether MP is decidable is still wide open.

We also derive from the undecidability of $\mathrm{MP}^+_{\mathbb{M}}$ the one of $\mathrm{DP}_{\mathcal{F}_n}$ for every $n \geqslant 2$ (Sections 4,5). This closes the question of the decidability of $\mathrm{DP}_{\mathcal{F}_n}$ for all n, since $\mathrm{DP}_{\mathcal{F}_1}$ is equivalent to the Type Inhabitation Problem and is therefore easily decided (indeed $\varphi \in \mathcal{F}_1^A$ is λ-definable iff there is a closed λ-term M^A).

More applications of the encoding of the Halting Problem below will be detailed in another paper, where we will complete the present study of DP and establish for the larger λ-calculus $\lambda^{\to}$ with infinitely many atomic types that the decidability of DP_A is equivalent to several other statements about A. One of these statements (*"all the closed normal forms of type A can be written from a finite supply of variables"*) yields a procedure deciding for any given A whether DP_A is decidable. Another one (*"all the closed λ-terms of type A can be generated from a finite set of combinators"*) extends to $\lambda^{\to}$ the solution for λ^{τ} (cf [Jol01]) of the Finite Generation Problem raised by R. Statman in [Sta85].

2 The Definability Problem at the Monster Type

Every register machine M, as considered here, is completely determined by the (finite) number $p \geqslant 2$ of its registers, a finite set of states Σ, the choice of two different elements s, h in Σ and for all $q \in \Sigma \setminus \{h\}$, an instruction Instr$(q)$. It is intended that the machine starts computations *with blank registers* from state s on, halts whenever it reaches state h and that Instr(q) is the next instruction to be executed when the machine is in state q. For all $q \in \Sigma \setminus \{h\}$, Instr$(q)$ may only have one of the two following forms:

- Inc(k, q): Increment register n° k and go to state q.
- Jzd(k, q_1, q_2) (Jump if Zero & Decrement): If register n° k has content 0, then go to state q_1 else decrement register n° k and go to state q_2.

The complete configuration of the machine is described at any time by a tuple $(q, n_1, \ldots, n_p)$, where q is the current state and n_i $(1 \leqslant i \leqslant p)$ the current content of register n° i. More formally:

Definition I. Let $\mathrm{M} = (p, \Sigma, s, h, \mathrm{Instr})$ be any register machine.

(i) The sequence of the configurations $\mathbf{c}_n \in \Sigma \times \mathbb{N}^p$ of M during its computation is defined by $\mathbf{c}_0 = (s, 0, \ldots, 0)$ and, in the case where say $\mathbf{c}_n = (q, n_1, \ldots, n_p)$:

- If Instr$(q) = $ Inc(k, q'), then $\mathbf{c}_{n+1} = (q', n_1, \ldots, n_{k-1}, n_k + 1, n_{k+1}, \ldots, n_p)$.
- If Instr$(q) = $ Jzd(k, q_1, q_2), then:
  - $\mathbf{c}_{n+1} = (q_1, n_1, \ldots, n_p)$ in case $n_k = 0$,
  - $\mathbf{c}_{n+1} = (q_2, n_1, \ldots, n_{k-1}, n_k - 1, n_{k+1}, \ldots, n_p)$ otherwise.
- If $q = h$, then the machine halts and $\mathbf{c}_{n+1}$ is consequently not defined.

(ii) Let P_n be the λ-terms of type o inductively defined by $P_0 = e^o$ and:

- If $\mathbf{c}_n = (q, \ldots)$ and Instr$(q) = $ Inc$(k, \ldots)$, then $P_{n+1} = x_k^1 P_n$.

– If $\mathbf{c}_n = (q, \ldots)$ and $\mathrm{Instr}(q) = \mathrm{Jzd}(k, \ldots)$, then $P_{n+1} = l^3(\lambda v^1.Q)$, where the λ-term Q is obtained from P_n by replacing its rightmost occurrence of x_k^1 with v if any, and $Q = P_n$ otherwise.
– If $\mathbf{c}_n = (h, \ldots)$, then the machine halts and P_{n+1} is not defined.

(iii) let $\mathcal{F}$ be the FTS of λ^τ over the ground domain:

$$\mathcal{F}^o = (\Sigma \cup \{\bot, \top\}) \times \{0,1\}^p.$$

The basic role played by the terms P_n in the sequel is to encode the sequence of the instructions performed by M. It will be the duty of some valuation ρ into $\mathcal{F}$ to check from the interpretation $[\![P]\!]_\rho$ of any λ-term P^o whether P is a correct encoding of the computations of M until some step n (*i.e.* $P = P_n$) and whether the halting state h is reached at this step. All this works so far in a similar way as in the former refutation of the Statman conjecture by R. Loader (*cf* [Loa01]), where the Word Problem is encoded, λ-terms are used to denote the sequences of rewritings and where a FTS checks whether a λ-term denotes a sequence of rewritings reaching a given word.

The λ-terms P_n have to play here another important role. Since the computations are encoded whithin the single type $\mathbb{M}$ (note that the terms P_n are just subterms of closed normal λ-terms $N^{\mathbb{M}}$), the FTS $\mathcal{F}$ may only keep track of a finitely bounded information at every step of the computation (this is not the case in R. Loader's encoding, which uses infinitely many types instead of $\mathbb{M}$). Therefore, the terms P_n have to memorize what the FTS $\mathcal{F}$ cannot, namely the content of the registers of M. It is so by virtue of the following.

Remark. For every n such that $\mathbf{c}_n = (q, n_1, \ldots, n_p)$ is defined, n_i is the number of occurrences of x_i in P_n. This can be checked by a straightforward induction on n.

But the main compensation for the restriction to the single type $\mathbb{M}$ is the essential fact that $\mathcal{F}$ depends on the register machine M, whereas the FTS used in R. Loader's encoding is quite independent of the rewriting system considered. The first component of an element of $\mathcal{F}^o$ recalls the current state, $\bot$ is for rejection (in case the term read is not one of the terms P_n) and the additional state $\top$ is used to test the occurrences of the variable v in a term Q such that $Q[v := x_k] = P_n$, in order to check whether $l^3(\lambda v^1.Q) = P_{n+1}$. The other components are flags recalling whether the p registers currently contain 0 or not.

Lemma 1. *There is a valuation ρ in $\mathcal{F}$ which is recursively defined from the register machine* M *and such that for every normal λ-term P^o with free variables among $l^3, e^o, x_1^1, \ldots, x_p^1$ and interpretation say $[\![P]\!]_\rho = (c, f_1, \ldots, f_p)$:*

$$c \in \Sigma \text{ iff for some } n,\ P = P_n,\ \mathbf{c}_n = (c, n_1, \ldots, n_p) \text{ and } \forall i\ f_i = 0 \Leftrightarrow n_i = 0.$$

Proof. Let INC_i and TEST_i $(1 \leqslant i \leqslant p)$ be the functionals of $\mathcal{F}^1$ such that:

$$\mathrm{INC}_i(a, f_1, \ldots, f_p) = (c, f_1, \ldots, f_{i-1}, 1, f_{i+1}, \ldots, f_p),$$
$$\text{where } c = \begin{bmatrix} q & \text{if } a \in \Sigma \setminus \{h\} \text{ and } \mathrm{Instr}(a) = \mathrm{Inc}(k, q), \\ \top & \text{if } a = \top, \\ \bot & \text{otherwise.} \end{bmatrix}$$
$$\mathrm{TEST}_i(a, f_1, \ldots, f_p) = (c, f_1, \ldots, f_p), \text{ where } c = \begin{bmatrix} \top & \text{if } a \notin \{\bot, \top\} \text{ and } f_i = 0, \\ \bot & \text{otherwise.} \end{bmatrix}$$

Let us define a functional JZD of $\mathcal{F}^3$ by the following. For any $\phi \in \mathcal{F}^2$, let $\phi \,\mathrm{Id}_{\mathcal{F}^o} = (a, f_1, \ldots, f_p)$ and for all i, let $\phi \,\mathrm{INC}_i = (a_i, \ldots)$, $\phi \,\mathrm{TEST}_i = (b_i, \ldots)$. Then $\mathrm{JZD}\, \phi = (c, f_1, \ldots, f_p)$ where:

$$c = \begin{bmatrix} q_1 & \text{if } a_1 = \ldots = a_p = q \in \Sigma \setminus \{h\},\ \mathrm{Instr}(q) = \mathrm{Jzd}(k, q_1, q_2) \text{ and } f_k = 0, \\ q_2 & \text{if for some } k,\ \mathrm{Instr}(a_k) = \mathrm{Jzd}(k, q_1, q_2),\ \forall i \neq k\ \ a_i = \bot \text{ and } b_k = \top, \\ \top & \text{if } a_k = \top \text{ for some } k, \\ \bot & \text{if none of the previous cases applies.} \end{bmatrix}$$

At last, let $\mathcal{S}$ be the set of the normal λ-terms of type o whose free variables are among $l^3, e^o, x_1^1, \ldots, x_p^1, z_1^1, \ldots, z_p^1$ and let ρ be a valuation such that $\rho(l^3) = \mathrm{JZD}$, $\rho(e^o) = (s, 0, \ldots, 0)$, $\rho(x_i^1) = \mathrm{INC}_i$ and $\rho(z_i^1) = \mathrm{TEST}_i$ $(1 \leqslant i \leqslant p)$.

Let us first prove for every $P \in \mathcal{S}$ with interpretation $[\![P]\!]_\rho = (c, f_1, \ldots, f_p)$:

$$f_i = 1 \;\Leftrightarrow\; x_i^1 \text{ is free in } P \tag{1}$$

by induction on P. Since it comes at once from the definitions of $\rho(e^o)$, INC_k, TEST_k if $P = e^o$, $P = x_k Q$ or $P = z_k Q$, we only consider here the remaining case: $P = l^3(\lambda v^1.Q^o)$. The λ-term $(\lambda v^1.Q)(\lambda y^o.y)$ reduces to a λ-term P_0 of $\mathcal{S}$ having exactly the same free variables x_i as P and by induction hypothesis, $[\![P_0]\!]_\rho$ has the form $(a, g_1, \ldots, g_p)$ where $g_i = 1$ iff x_i is free in P_0, so that we only have to check: $f_i = g_i$ for all i. But this follows at once from the definition of JZD, because $[\![P]\!]_\rho = \mathrm{JZD}\, [\![\lambda v^1.Q]\!]_\rho$ and $[\![\lambda v^1.Q]\!]_\rho \mathrm{Id}_{\mathcal{F}^o} = [\![(\lambda v^1.Q)(\lambda y^o.y)]\!]_\rho = [\![P_0]\!]_\rho$.

In case $P = P_n$ and $\mathbf{c}_n = (c, n_1, \ldots, n_p)$, we get from (1) and the remark above: $f_i = 0 \Leftrightarrow n_i = 0$ for all i.

Let us now prove that the first component of $[\![P]\!]_\rho$ is given by:

$$c = \begin{bmatrix} q \in \Sigma & \text{if for some } n,\ P = P_n \text{ and } \mathbf{c}_n = (q, \ldots), \\ \top & \text{if } P \in \mathcal{T}, \\ \bot & \text{in every other case.} \end{bmatrix} \tag{2}$$

where $\mathcal{T} \subset \mathcal{S}$ is the set of the terms M such that for some k, M has exactly one free occurrence of z_k, none of the other variables z_i and such that the largest subterm in which z_k is not free can be given the form P_n where $x_k \notin P_n$ by substituting variables $x_1, \ldots, x_p$ for its free variables that are bound in M.

Here again, the induction on $P \in \mathcal{S}$ is straightforward from the above definitions whenever $P = e^o$, $P = x_k Q$ or $P = z_k Q$, and we just detail the case where $P = l^3(\lambda v^1.Q^o)$. Let $\phi = [\![\lambda v^1.Q]\!]_\rho$ and let us put like in the definition of JZD: $\phi\,\mathrm{Id}_{\mathcal{F}^o} = [\![Q]\!]_{\rho[v\leftarrow \mathrm{Id}_{\mathcal{F}^o}]} = (a, f_1, \ldots, f_p)$, $\phi\,\mathrm{INC}_i = [\![Q[v{:=}x_i]]\!]_\rho = (a_i, \ldots)$, $\phi\,\mathrm{TEST}_i = [\![Q[v{:=}z_i]]\!]_\rho = (b_i, \ldots)$.

By the definition of $\mathcal{T}$, we have $P \in \mathcal{T}$ iff $Q[v := x_i] \in \mathcal{T}$ for some i, i.e. by induction hypothesis, iff $a_i = \top$ for some i; moreover, the latter is equivalent to $c = \top$. We may therefore assume $P \notin \mathcal{T}$, $Q[v{:=}x_i] \notin \mathcal{T}$ and $a_i \in \Sigma \cup \{\bot\}$ for all i. In the case where no λ-term $Q[v{:=}x_i]$ is of the form P_n, neither is P and $a_i = \bot$ for all i by induction hypothesis, so that $[\![P]\!]_\rho = \bot$ and we are done again. Let us then put $Q[v{:=}x_k] = P_n$; by induction hypothesis, it follows $\mathbf{c}_n = (a_k, \ldots)$. If $a_k = h$ or $\mathrm{Instr}(a_k)$ has not the form $\mathrm{Jzd}(k, q_1, q_2)$, then P is not one of the terms P_i and the definition of JZD yields $[\![P]\!]_\rho = \bot$, hence (2) holds. We may therefore assume $\mathrm{Instr}(a_k) = \mathrm{Jzd}(k, q_1, q_2)$.

- If v does not occur free in Q, then we obtain $a_1 = \ldots = a_p = [\![Q]\!]_\rho$. In the case where x_k is not free either in $Q = P_n$, we get $f_k = n_k = 0$ by (1), hence $[\![P]\!]_\rho = q_1$ and $P = P_{n+1}$. Otherwise, $f_k = 1$, $[\![P]\!]_\rho = \bot$ and P has not the form P_i. This yields (2) in both cases.
- If v occurs free in Q, then there is no λ-term $Q[v{:=}x_i]$ of the form P_m except $Q[v{:=}x_k] = P_n$; hence by induction hypothesis, we get $a_i = \bot$ for all $i \neq k$.
  - If Q is obtained from $Q[v{:=}x_k]$ by replacing its rightmost occurrence of x_k, then $Q[v{:=}z_k] \in \mathcal{T}$ by the definition $\mathcal{T}$ and by induction hypothesis, $b_k = \top$. Hence, $[\![P]\!]_\rho = q_2$ and (2) follows since in this case, $P = P_{n+1}$.
  - Otherwise, there are several occurrences of v in Q or x_k occurs free on the right of v, so that P has not the form P_i and $Q[v := z_k] \notin \mathcal{T}$. The induction hypothesis gives then $b_k \neq \top$, $[\![P]\!]_\rho = \bot$ and therefore (2).

□

Proposition 2. *For every register machine* M *with m states and p registers say, there are* JZD $\in \mathcal{F}_k^3$, S $\in \mathcal{F}_k^o$ *and* H $\in \mathcal{F}_k^o$ *where* $k = 2^p(m+2)$ *which are recursively defined from* M *and such that:*

$$\mathrm{M}\ \textit{halts with cleared registers} \quad \Leftrightarrow \quad \exists X^{\mathrm{M}} \in \Lambda^{\varnothing}\ \ [\![X]\!]\,\mathrm{JZD}\,\mathrm{S} = \mathrm{H}.$$

Proof. The FTS $\mathcal{F}_k$ where $k = 2^p(m+2)$ may be identified with the FTS $\mathcal{F}$ defined above and for any register machine M with m states and p registers, we may consider a valuation ρ in $\mathcal{F}_k$ as in the lemma. From the latter and the remark made above, it follows that M halts with cleared registers iff there is a normal λ-term P^o with free variables among l^3, e^o such that $[\![P]\!]_\rho = (h, 0, \ldots, 0)$. We may therefore take $\mathrm{JZD} = \rho(l^3)$, $\mathrm{S} = \rho(e^o)$ and $\mathrm{H} = (h, 0, \ldots, 0)$. □

The point of halting with cleared registers is not a serious modification of the Halting Problem. Indeed, we may transform uniformly every machine M into a machine M′ which halts with cleared registers iff M halts, *e.g.* by replacing the halting state of M with new states $q_1, \ldots, q_{p+1}$ such that $\mathrm{Instr}(q_i) = \mathrm{Jzd}(i, q_{i+1}, q_i)$ $(1 \leqslant i \leqslant p)$, where p is the number of registers of M and q_{p+1} the new halting state. In the same way, the fact that we consider only machines starting with clear registers is not a restriction, since such a single-computation machine can be obtained for every more general machine and every input data by inserting instructions Inc at the initial state of the latter machine. Therefore, we get from the previous proposition:

Proposition 3. *The following problem is undecidable.*

> *"Given elements $s, h \in \mathcal{F}_n^o$ and $\varphi \in \mathcal{F}_n^3$ of any full type structure $\mathcal{F}_n$, is there a closed λ-term $X^{\mathbb{M}}$ such that $[\![X]\!]\,\varphi\, s = h$?"*

Theorem 4. $\mathrm{DP}_{\mathbb{M}}$ *is undecidable.*

Proof. Indeed, otherwise we could check out for any $s, h \in \mathcal{F}_n^o$, $\varphi \in \mathcal{F}_n^3$ whether one of the finitely many $\psi \in \mathcal{F}_n^{\mathbb{M}}$ such that $\psi\varphi s = h$ is λ-definable, but this contradicts Proposition 3.[2] □

3 Around the Higher Order Pattern Matching Problem

Let us explain briefly the equivalence of MP_A^+ and DP_A stated in the introduction. Let $PX^+ =_{\beta\eta} Q$ $(+ = [o{:=}1_n])$ be any instance of MP_A^+. By virtue of a famous theorem by R. Statman (*cf* [Sta82]), we may compute k such that for all $M \in \Lambda^\varnothing$, $\mathcal{F}_k \models M = Q \;\Leftrightarrow\; M =_{\beta\eta} Q$, so that the question is whether some functional of $\mathcal{E} = \{\varphi \in \mathcal{F}_k^{A^+} ; [\![P]\!]^{\mathcal{F}_k}\varphi = [\![Q]\!]^{\mathcal{F}_k}\}$ has the form $[\![X^+]\!]^{\mathcal{F}_k}$ $(X \in \Lambda^\varnothing)$. Moreover, we may identify $\mathcal{F}_k^{1_n}$ with the ground domain $\mathcal{F}_{k^{k^n}}^o$ and check that we have $[\![X^+]\!]^{\mathcal{F}_k} = [\![X]\!]^{\mathcal{F}_{k^{k^n}}}$ for all $X \in \Lambda^\varnothing$, so that the question is now whether some $\varphi \in \mathcal{E} \subseteq \mathcal{F}_{k^{k^n}}^A$ is λ-definable. Therefore, if DP_A is decidable, then so is MP_A^+. The converse comes from a well known reconstruction of the FTS $\mathcal{F}_n$ as the type structure $\mathcal{S}/{\sim} = \bigcup_A(\mathcal{S}^A/{\sim})$ where $\mathcal{S}^A$ is the set of the closed λ-terms of type A^+ $(+ = [o{:=}1_n])$ and where $\sim$ is the observational equivalence defined by $M \sim N \Leftrightarrow \forall C^{o^+}[\;]\; C[M] =_{\beta\eta} C[N]$. Indeed, the usual argument by induction on the reduction length of $C[M]$ establishes $\forall N \in \mathcal{S}^A\; MN \sim M'N \;\Rightarrow\; M \sim M'$, and therefore the extensionality of the structure $\mathcal{S}/{\sim}$. Moreover, we may easily build by a single induction on the type A λ-terms $R_\varphi \in \mathcal{S}^A$, $P_\varphi \in \mathcal{S}^{A\to o}$ for all $\varphi \in \mathcal{F}_n^A$ such that $\{R_i\,;\, i \in \mathcal{F}_n^o\} = \mathcal{S}^o$, $R_\varphi R_\psi \sim R_{\varphi\psi}$ and $P_\varphi R_\psi \sim R_1 \;\Leftrightarrow\; R_\varphi \sim R_\psi$; it follows that we get an isomorphism of $\mathcal{F}_n$ onto $\mathcal{S}/{\sim}$ by mapping every φ to the observational class of R_φ. At last, we may check that the interpretation of any $X \in \Lambda^\varnothing$ in $\mathcal{S}/{\sim}$ is just the observational class of X^+. Hence for all $X^A \in \Lambda^\varnothing$, $[\![X]\!] = \varphi \;\Leftrightarrow\; X^+ \sim R_\varphi \;\Leftrightarrow\; P_\varphi X^+ \sim R_1 \;\Leftrightarrow\; P_\varphi X^+ =_{\beta\eta} R_1$. This reduces DP_A to MP_A^+ and we may conclude:

$$\mathrm{DP}_A \text{ decidable} \;\Leftrightarrow\; \mathrm{MP}_A^+ \text{ decidable.}$$

[2] Note that Proposition 3 may in turn be easily derived from this apparently weaker theorem by considering the cartesian products $(\mathcal{F}_n)^k$. Indeed, the λ-definability of $\psi \in \mathcal{F}_n^{\mathbb{M}}$ is equivalent to $\exists X \in \Lambda^\varnothing\; [\![X]\!]\boldsymbol{\varphi}\boldsymbol{s} = \boldsymbol{\psi}\boldsymbol{\varphi}\boldsymbol{s}$ where $\boldsymbol{\psi} = (\psi, \dots, \psi) \in (\mathcal{F}_n^{\mathbb{M}})^k$, $\boldsymbol{\varphi} = (\varphi_1, \dots, \varphi_k)$, $\boldsymbol{s} = (s_1, \dots, s_k)$, $\{(\varphi_i, s_i); 1 \leqslant i \leqslant k\} = \mathcal{F}_n^3 \times \mathcal{F}_n^o$ and the latter is equivalent to the existence of $\varphi \in \mathcal{F}_{n^k}^{\mathbb{M}}$ such that $\exists X \in \Lambda^\varnothing\; [\![X]\!]\boldsymbol{\varphi}\boldsymbol{s} = \boldsymbol{\psi}\boldsymbol{\varphi}\boldsymbol{s}$ and $f(\varphi) = \boldsymbol{\varphi}$ where $f : \mathcal{F}_{n^k} \to (\mathcal{F}_n)^k$ is the partial surjective morphism defined in [Fri75].

In particular, Proposition 3 and Theorem 4 can also be restated as follows.

Proposition 5. *The following problem is undecidable.*

> *"Given $M^{3^+}, S^{o^+}, H^{o^+} \in \Lambda^{\varnothing}$ where $+$ is any type substitution $[o{:=}1_n]$, is there a closed λ-term $X^{\mathbb{M}}$ such that $X^+MS =_{\beta\eta} H$?"*

Proof. Let $\mathcal{F}_n^o = \{1, \ldots, n\}$. For any functional c of the 1-section of $\mathcal{F}_n$, say $c \in \mathcal{F}_n^{1_k}$, we define by induction on k a closed λ-term $|c|^{1_k^+}$ as follows: if $k = 0$ then $|c| = \lambda z_1^o \ldots z_n^o . z_c$ else $|c| = \lambda x^{o^+} \vec{a}. x(|c\,1|\vec{a}) \ldots (|c\,n|\vec{a})$ where $\vec{a} = x_2^{o^+} \ldots x_k^{o^+} z_1^o \ldots z_n^o$. Let us put $\mathcal{F}_n^1 = \{c_1, \ldots, c_r\}$ $(r = n^n)$ and let us define $|\varphi|^{3^+}$ for all $\varphi \in \mathcal{F}_n^3$ by $|\varphi|^{3^+} = \lambda f^{2^+}. |c|(f|c_1|) \ldots (f|c_r|)$, where c is any element of $\mathcal{F}_n^{1_r}$ such that $c(ac_1) \ldots (ac_r) = \varphi a$ for all $a \in \mathcal{F}_n^2$. From these definitions, we may check by induction on the normal λ-terms N^o with free variables among l^3, e^o, x_i^1 $(i \geqslant 0)$:

$$|[\![N]\!]_\rho| =_{\beta\eta} N^+[l{:=}|\rho(l)|,\ e{:=}|\rho(e)|,\ x_i{:=}|\rho(x_i)|\,]_{i \geqslant 0}.$$

It follows $|[\![X]\!]\,\varphi\, s| =_{\beta\eta} X^+|\varphi|\,|s|$ for all $X^{\mathbb{M}} \in \Lambda^{\varnothing}$, $\varphi \in \mathcal{F}_n^3$, $s \in \mathcal{F}_n^o$, hence any equation $[\![X]\!]\,\varphi\, s = h$ is equivalent to $X^+|\varphi|\,|s| =_{\beta\eta} |h|$ and we are done by Proposition 3. □

Recall that we may replace the equation in either Matching Problem (MP, MP$^+$, β-MP) with several ones of the same kind without generalizing the problem. Indeed, a system of finitely many equations $P_i X^{(+)} =_{\beta(\eta)} Q_i$ $(1 \leqslant i \leqslant n)$ is equivalent to the single equation $PX^{(+)} =_{\beta(\eta)} Q$ where $P = \lambda xc.c(P_1 x) \ldots (P_n x)$ and $Q = \lambda c.cQ_1 \ldots Q_n$. By virtue of this remark, Proposition 5 and the following lemma yield a fairly simple proof of the undecidability of β-MP.

Lemma 6. *For any type substitution $+$ and any $Y^{\mathbb{M}^+} \in \Lambda^{\varnothing}$,*

$$\exists X^{\mathbb{M}} \in \Lambda^{\varnothing}\ \ Y =_\beta X^+ \ \Leftrightarrow\ Y(\lambda f^{2^+}. f\mathbf{I}^{1^+}) =_\beta \mathbf{I}^{1^+}.$$

Proof. Let $B^3 = \lambda f^2. f\mathbf{I}^1$. If $Y =_\beta X^+$ $(X \in \Lambda^{\varnothing})$, then XB is a closed λ-term of type 1, hence $XB =_\beta \mathbf{I}^1$ and $YB^+ =_\beta (XB)^+ =_\beta \mathbf{I}^{1^+}$.

For the converse, we may assume w.l.o.g. that Y is β-normal and consider a particular reduction sequence $C \twoheadrightarrow_\beta \mathbf{I}^{1^+}$, where C is the contractum of the redex YB^+. Let us say that a β-redex is *nice* whenever it has one of the forms B^+M, $M^{2^+}\mathbf{I}^{1^+}$, $\mathbf{I}^{1^+}M$ or $M^T N$ where T is a subtype of o^+ (possibly $T = o^+$). We may easily check that if $P \rightarrow_\beta Q$ and if all the β-redexes of P are nice then so are the ones of Q and that if moreover Q is of the form X^+, then so is P. It follows that all the β-redexes of the terms of the sequence $C \twoheadrightarrow_\beta \mathbf{I}^{1^+}$ are nice and that C has the form X^+. By the definition of C, Y is then also of the form X^+. □

Theorem 7. (R. Loader, [Loa03]) *The problem β-MP is undecidable.*

Proof. Indeed, remark that for all $G^{o^+}, H^{o^+} \in \Lambda^{\varnothing}$ where $+ = [o{:=}1_n]$, we have $G =_{\beta\eta} H \Leftrightarrow G =_\beta H$. By Lemma 6, we then get for any given $M^{3^+}, S^{o^+}, H^{o^+} \in \Lambda^{\varnothing}$:

$$\exists X^{\mathbb{M}} \in \Lambda^{\varnothing}\ \ X^+MS =_{\beta\eta} H \ \Leftrightarrow\ \exists Y^{\mathbb{M}^+} \in \Lambda^{\varnothing}\ \left[\begin{array}{l} YMS =_\beta H \\ Y(\lambda f^{2^+}. f\mathbf{I}) =_\beta \mathbf{I}^{1^+}. \end{array}\right.$$

If β-MP were decidable, then we could decide the right hand side of this equivalence, but this contradicts Proposition 5. □

The absence of the η-rule is strongly used above in order to sift out the λ-terms not of the form X^+, and an equation like the one of Lemma 6 does not seem to exist in the case of the $\beta\eta$-equality, even if we allow an additional variable substitution[3] "$Y = FZ$", i.e. even if we look for F, P, Q such that:

$$\exists X^{\mathbb{M}} \in \Lambda^{\varnothing} \;\; Y =_{\beta\eta} X^+ \quad \Leftrightarrow \quad \exists Z^A \in \Lambda^{\varnothing} \;\; \begin{bmatrix} Y =_{\beta\eta} FZ \\ PZ =_{\beta\eta} Q. \end{bmatrix}$$

4 Back to the Definability Problem

Although this task of distinguishing the λ-terms $X^{\mathbb{M}^+}$ of the form X^+ does not seem possible by means of $\beta\eta$-equations of λ^{τ}, it is done by any FTS $\mathcal{F}_q$ such that $q \geqslant 3$ (see Lemma 11 below), so that another consequence of Proposition 5 is the undecidability of the Definability Problem for these FTS. For convenience, we adopt in the sequel the following conventions and notations.

Definition II. (i) We choose arbitrarily in every FTS $\mathcal{F}_n$ such that $n \geqslant 2$ two distinct elements of $\mathcal{F}_n^o$ that we denote $\top, \bot$.
(ii) Let $\Lambda_\bot$ (resp. $\Lambda_\bot^{\varnothing}$) be the set of the λ-terms (resp. the closed λ-terms) where an additional constant $\bot$ of type o may occur. This constant $\bot$ is interpreted in every FTS $\mathcal{F}_n$ $(n \geqslant 2)$ by $\bot \in \mathcal{F}_n^o$.
(iii) Let ι be the type substitution $[o := 1]$ and let $\mathbb{N} = 1^{\iota} = 1 \to 1$.
(iv) For any type substitution τ, let $\mathcal{S}_\tau = \{M \in \Lambda^{\varnothing} \;;\; \exists X \in \Lambda^{\varnothing} \;\; M =_{\beta\eta} X^{\tau}\}$.

Lemma 8. *For any $q \geqslant 3$, there is $\psi \in \mathcal{F}_q^{\mathbb{N}^{\iota} \to o}$ such that for all $M^{\mathbb{N}^{\iota}} \in \Lambda_\bot^{\varnothing}$,*

$$M \in \mathcal{S}_\iota \quad \Leftrightarrow \quad \psi\, [\![M]\!] = \top.$$

Proof. Let $a \in \mathcal{F}_q^o \smallsetminus \{\top, \bot\}$ and let ρ be a valuation into $\mathcal{F}_q$ such that for every variable z of type o, we have $\rho(z) = a$ and for some variables $e^1, x^{\mathbb{N}}$:

$$\begin{bmatrix} \rho(e^1)\, a = \top, \\ \rho(e^1)\, c = \bot \;\; \text{if } c \neq a, \end{bmatrix} \qquad \begin{bmatrix} \rho(x^{\mathbb{N}})\, \rho(e^1) = \rho(e^1), \\ \rho(x^{\mathbb{N}})\, \varphi = \lambda\!\!\lambda d.\bot \;\; \text{if } \varphi \neq \rho(e^1). \end{bmatrix}$$

We may check by a straightforward induction on all the long normal λ-terms $G^1 \in \Lambda_\bot \smallsetminus \{\mathbf{I}^1\}$ with free variables among $x^{\mathbb{N}}, e^1, z_i^o$ $(i \geqslant 0)$, that if G η-reduces to a λ-term of the form X^{ι}, then $[\![G]\!]_\rho = \rho(e)$ else $[\![G]\!]_\rho = \lambda\!\!\lambda d.c$ for some $c \in \mathcal{F}_q^o$. In particular we get for all $M^{\mathbb{N}^{\iota}} \in \Lambda_\bot^{\varnothing}$, $M \in \mathcal{S}_\iota \Leftrightarrow [\![M]\!]\, \rho(x)\, \rho(e) = \rho(e)$, so that we are done by taking any $\psi \in \mathcal{F}_q^{\mathbb{N}^{\iota} \to o}$ such that $\varphi\, \rho(x)\, \rho(e) = \rho(e) \Leftrightarrow \psi\, \varphi = \top$ for all $\varphi \in \mathcal{F}_q^{\mathbb{N}^{\iota}}$. □

---

[3] This technique is used in [Loa03] to get the undecidability of β-PM.

Definition III. Let us define the λ-terms:

- $\mathrm{R}^{\mathbb{M}\to\mathbb{N}} = \lambda r^{\mathbb{M}} x^1.r(\lambda f^2.x(fx)) \in \Lambda^{\varnothing}$,
- $\mathrm{P}_{n,k}^{1_n\to 1} = \lambda f^{1_n} z^o.f\bot\ldots\bot z\bot\ldots\bot$ where z is the kth argument of f,
- $\mathrm{R}_{n,k}^{\mathbb{N}^+\to\mathbb{N}^\iota} = \lambda r^{\mathbb{N}^+} x^{1^\iota} e^{o^\iota}.\mathrm{P}_{n,k}(rX_{n,k}E_{n,k}) \quad (1\leqslant k\leqslant n)$,

where $+ = [o{:=}1_n]$, $X_{n,k}^{1^+} = \lambda f^{1_n} z_1^o \ldots z_n^o.x^{1^\iota}(\mathrm{P}_k f)z_k$ and $E_{n,k}^{o^+} = \lambda z_1^o\ldots z_n^o.e^{o^\iota} z_k$.

Lemma 9. *Let* $+ = [o{:=}1_n]$ $(n\geqslant 1)$. *We have for any* $M^{\mathbb{N}^+}\in\Lambda^{\varnothing}$,

$$M \in \mathcal{S}_+ \quad\Leftrightarrow\quad \forall k\in\{1,\ldots,n\}\ \mathrm{R}_{n,k}M \in \mathcal{S}_\iota.$$

Proof. Let $\sigma_k = [x^{1^+}{:=}X_{n,k},\, e^{o^+}{:=}E_{n,k}]$ and let N^{o^+} be any $\beta\eta$-normal λ-term with free variables among x^{1^+}, e^{o^+}, z_i^o $(i \geqslant 0)$. Through an easy case study, we may check by induction on the size of N that N has the form X^+ iff for every $k\in\{1,\ldots,n\}$, $\mathrm{P}_k N^{\sigma_k}$ $\beta\eta$-reduces to a λ-term of the form X^ι. It follows for any λ-term $M^{\mathbb{N}^+}$ that $M^{\mathbb{N}^+}\in\mathcal{S}_+$ iff for all k, $\mathrm{P}_k(MX_{n,k}E_{n,k})\in\mathcal{S}_\iota$. □

Lemma 10. *For any type substitution* τ *and any* $M^{\mathbb{M}^\tau}\in\Lambda^{\varnothing}$, *we have*

$$M \in \mathcal{S}_\tau \quad\Leftrightarrow\quad \mathrm{R}^\tau M \in \mathcal{S}_\tau.$$

Proof. Clearly, if $M\in\mathcal{S}^\tau$, say $M =_{\beta\eta} X^\tau$ $(X\in\Lambda^{\varnothing})$, then $R^\tau M =_{\beta\eta} (RX)^\tau \in \mathcal{S}^\tau$. Conversely, let σ be the variable substitution $[l^{3^\tau}{:=}\lambda f^{2^\tau}.x^{1^\tau}(fx^{1^\tau}),\, x_i^{1^\tau}{:=}x^{1^\tau}]_{i\geqslant 0}$ and let N^{o^τ} be any $\beta\eta$-normal λ-term with free variables among l^{3^τ}, e^{o^τ}, $x_i^{1^\tau}$, z_i^o $(i \geqslant 0)$. We may check by a straightforward induction on the size of N that if N^σ $\beta\eta$-reduces to a λ-term of the form X^τ then N has the form X^τ. It follows that a closed $\beta\eta$-normal λ-term $M^{\mathbb{M}^\tau} = \lambda l^{3^\tau} e^{o^\tau}.N^{o^\tau}$ has the form X^τ whenever $\mathrm{R}^\tau M =_{\beta\eta} \lambda x^{1^\tau} e^{o^\tau}.N^\sigma$ $\beta\eta$-reduces to a λ-term of the form X^τ. □

Lemma 11. *Let* $+ = [o := 1_n]$ $(n \geqslant 1)$. *For every* $q \geqslant 3$, *there is a functional* $\phi \in \mathcal{F}_q^{\mathbb{M}^+\to o}$ *such that for all* $M^{\mathbb{M}^+}\in\Lambda^{\varnothing}$,

$$M \in \mathcal{S}_+ \quad\Leftrightarrow\quad \phi\,[\![M]\!] = \top.$$

Proof. Indeed, for any $M^{\mathbb{M}^+}\in\Lambda^{\varnothing}$ we have by Lemmata 8, 9 and 10, $M \in \mathcal{S}_+$ iff for all $k\in\{1,\ldots,n\}$ $\psi\,[\![\mathrm{R}^+(\mathrm{R}_{n,k}M)]\!] = \top$, where ψ is like in Lemma 8. Hence, we are done by choosing ϕ such that for all $\varphi \in \mathcal{F}_q^{\mathbb{M}^+}$, $\phi\,\varphi = \top$ iff $\psi([\![\mathrm{R}^+\circ\mathrm{R}_{n,k}]\!]\,\varphi) = \top$ for all $k\in\{1,\ldots,n\}$. □

Theorem 12. *For every* $q \geqslant 3$, $\mathrm{DP}_{\mathcal{F}_q}$ *is undecidable.*

Proof. Note that for all $G^{o^+}, H^{o^+} \in \Lambda^{\varnothing}$ where $+ = [o := 1_n]$ $(n \geqslant 1)$, we have $G =_{\beta\eta} H$ iff $[\![G]\!]^{\mathcal{F}_q} = [\![H]\!]^{\mathcal{F}_q}$. Hence, we get for all $M^{3^+}, S^{o^+}, H^{o^+} \in \Lambda^{\varnothing}$:

$$\exists X^{\mathbb{M}}\in\Lambda^{\varnothing}\ \ X^+MS =_{\beta\eta} H \quad\Leftrightarrow\quad \exists Y^{\mathbb{M}^+}\in\Lambda^{\varnothing}\ \ \begin{cases}[\![YMS]\!] = [\![H]\!]\\ \phi\,[\![Y]\!] = \top\end{cases}$$

where $\phi \in \mathcal{F}_q^{\mathbb{M}^+\to o}$ is like in the previous lemma. Now, the second member of this equivalence holds iff one of the finitely many $\varphi \in \mathcal{F}_q^{\mathbb{M}^+}$ such that $\varphi\,[\![M]\!]\,[\![S]\!] = [\![H]\!]$ and $\phi\,\varphi = \top$ is λ-definable. If $\mathrm{DP}_{\mathcal{F}_q}$ were decidable, then we could decide it, but this contradicts Proposition 5. □

5 Undecidability of $\mathrm{DP}_{\mathcal{F}_2}$

Unfortunately, Lemma 11 does not extend to the FTS $\mathcal{F}_2$. Indeed, every functional $\varphi \in \mathcal{F}_2^1$ satisfies $\varphi \circ \varphi \circ \varphi = \varphi$, so that e.g. $\mathcal{F}_2$ does not distinguish between the λ-terms (of type $\mathbb{M}^\iota$) $\lambda l^{3^\iota} e^1.e^1 \in \mathcal{S}_\iota$ and $\lambda l^{3^\iota} e^1.(e^1 \circ e^1 \circ e^1) \notin \mathcal{S}_\iota$. In order to derive the undecidability of $\mathrm{DP}_{\mathcal{F}_2}$ from Proposition 5, we will borrow a useful tool of [Loa03] that overcomes the lack of points in $\mathcal{F}_2$: surjective λ-terms $\mathsf{proj}_A^{A^\iota \to A}$ which allow us to replace artificially every domain $\mathcal{F}_2^A$ with the larger domain $\mathcal{F}_2^{A^\iota}$. We then may obtain in replacement of the functional of Lemma 11 a functional $\phi \in \mathcal{F}_2^{\mathbb{M}^{+\iota} \to o}$ such that:

$$Y^{\mathbb{M}^+} \in \mathcal{S}_+ \quad \Leftrightarrow \quad \exists X^{\mathbb{M}^{+\iota}} \in \Lambda^\varnothing \quad \begin{bmatrix} Y =_{\beta\eta} \mathsf{proj}_{\mathbb{M}^+} X \\ \phi[\![X]\!] = \top. \end{bmatrix}$$

Definition IV. (i) Let $\mathrm{N}_\perp^\iota$ be the set of the β-normal λ-terms $M \in \Lambda_\perp$ such that the type of every subterm (including M) has the form A^ι or is equal to o. (ii) Let $\pi : \mathrm{N}_\perp^\iota \to \Lambda_\perp$ be the map such that every $\pi(M^{A^\iota})$ has type A and every $\pi(M^o)$ has type o, defined by:

- $\pi(x^{A^\iota}) = x^A$,
- $\pi(P^{A^\iota \to B^\iota} Q^{A^\iota}) = \pi(P)\pi(Q)$,
- $\pi(\lambda x^{A^\iota}.P^{B^\iota}) = \lambda x^A.\pi(P)$,
- $\pi(x^o) = \perp$,
- $\pi(P^1 Q^o) = \pi(P)$,
- $\pi(\lambda x^o.P^o) = \pi(P)$.

(iii) For every type A of λ^τ, let $\mathsf{proj}_A^{A^\iota \to A}$ and $\mathsf{inj}_A^{A \to A^\iota}$ be the closed λ-terms mutually defined by:

- $\mathsf{proj}_o = \lambda x^1.x\perp$, $\mathsf{inj}_o = \lambda x^o d^o.x$,
- $\mathsf{proj}_{A \to B} = \lambda f^{A^\iota \to B^\iota} x^A.\mathsf{proj}_B(f(\mathsf{inj}_A x))$,
- $\mathsf{inj}_{A \to B} = \lambda f^{A \to B} x^{A^\iota}.\mathsf{inj}_B(f(\mathsf{proj}_A x))$.

The following lemma is essentially lemma 18 of [Loa03].

Lemma 13. *For any β-normal λ-term $M^{A^\iota} \in \Lambda_\perp^\varnothing$, we have* $\mathsf{proj}_A M =_{\beta\eta} \pi(M)$.

Proof. We obtain $\mathsf{proj}_A \circ \mathsf{inj}_A =_{\beta\eta} \mathbf{I}^{A \to A}$ from the definition of $\mathsf{proj}_A, \mathsf{inj}_A$ by a straightforward induction on A. Let σ be the substitution such that $\sigma(x^{A^\iota}) = \mathsf{inj}_A\, x^A$, $\sigma(y^o) = \perp$ for all the variables x^{A^ι}, y^o. We may check by a single induction on a λ-term $M^{A^\iota} \in \mathrm{N}_\perp^\iota$ that $\mathsf{proj}_A M^\sigma =_{\beta\eta} \pi(M)$ and moreover in the case where M is not an abstraction $M =_{\beta\eta} \mathsf{inj}_A \pi(M)$ (the latter relation implies clearly the previous one by virtue of the relation $\mathsf{proj}_A \circ \mathsf{inj}_A =_{\beta\eta} \mathbf{I}$). The lemma follows. □

Let us now introduce cousins[4] $\phi_A \in \mathcal{F}_2^{A \to o}$ of the λ-terms proj_A that distinguish the λI-terms.

[4] These functionals ϕ_A are actually interpretations of λ-terms θ_A and the λ-terms proj_A are instantiations of the same terms θ_A. The latters are themselves extracted from an intuitionistic proof of weak normalisation for λ^τ. Unsurprisingly, they compute

Definition V. (i) Let Λ_I be the set of the (possibly open) λI-terms where the constant $\bot$ does not occur.
(ii) Let $\mathrm{And} \in \mathcal{F}_2^{1_2}$, $\chi \in \mathcal{F}_2^2$ be the functionals such that: $\mathrm{And}\, a\, b = \top$ iff $a = b = \top$ and $\chi\, \varphi = \top$ iff $\varphi = \mathrm{Id}_{\mathcal{F}_2^o}$ and for any type A of λ^τ, let $\phi_A \in \mathcal{F}_2^{A\to o}$ and $\psi_A \in \mathcal{F}_2^{o\to A}$ be mutually defined by:

- $\phi_o = \psi_o = \mathrm{Id}_{\mathcal{F}_2^o}$,
- $\phi_{A\to B} = \lambda\!\!\lambda \varphi \in \mathcal{F}_2^{A\to B} . \chi(\phi_B \circ \varphi \circ \psi_A)$,
- $\psi_{A\to B} = \lambda\!\!\lambda a \in \mathcal{F}_2^o . \psi_B \circ (\mathrm{And}\, a) \circ \phi_A$.

(iii) Let ρ_I be the valuation into $\mathcal{F}_2$ such that $\rho_I(x^A) = \psi_A \top$ for all x^A.

Lemma 14. *For every β-normal λ-term $M^A \in \Lambda_\bot$,*

$$M^A \in \Lambda_I \quad \Leftrightarrow \quad \phi_A\, [\![M]\!]_{\rho_I} = \top\,.$$

Proof. We get at once $\phi_A \circ \psi_A = \mathrm{Id}_{\mathcal{F}_2^o}$ by induction on A. Let $\mathcal{V}$ be the set of the valuations ρ into $\mathcal{F}_2$ such that $\rho(x^A) = \psi_A \top$ or $\rho(x^A) = \psi_A \bot$ for all variables x^A. For any $\rho \in \mathcal{V}$, let us define $f_\rho : \Lambda_\bot \to \mathcal{F}_2^o$ by $f_\rho(M) = \top$ iff $M \in \Lambda_I$ and $\rho(x^A) = \psi_A \top$ for all the free variables x^A of M. We may check by a single induction on a β-normal λ-term $M^A \in \Lambda_\bot$ that we have for every $\rho \in \mathcal{V}$, $\phi_A\, [\![M]\!]_\rho = f_\rho(M)$ and moreover in the case where M is not an abstraction $[\![M]\!]_\rho = \psi_A(f_\rho(M))$ (note that the latter relation implies the previous one by virtue of the relation $\phi_A \circ \psi_A = \mathrm{Id}_{\mathcal{F}_2^o}$). The lemma follows from the particular case where $\rho = \rho_I$. □

Lemma 15. *There is a functional $\psi \in \mathcal{F}_2^{\mathbb{N}^{\iota\iota}\to o}$ such that for every $M^{\mathbb{N}^{\iota\iota}} \in \Lambda_\bot^\varnothing$,*

$$M \in \mathcal{S}_{\iota\iota} \quad \Rightarrow \quad \psi\, [\![M]\!] = \top \quad \Rightarrow \quad \mathsf{proj}_{\mathbb{N}^\iota} M \in \mathcal{S}_\iota\,.$$

Proof. Let $\psi_k \in \mathcal{F}_2^{\mathbb{N}^{\iota\iota}\to o}$ $(0 \leqslant k \leqslant 2)$ be the functionals defined by:

- $\psi_0\, \varphi = \phi_{\mathbb{N}^\iota}(\varphi(\rho_I(x^{\mathbb{N}^\iota})))$,
- $\psi_1\, \varphi = \phi_{\mathbb{N}}([\![\, \mathsf{proj}_{\mathbb{N}^\iota}]\!]\, \varphi\, (\rho_I(x^{\mathbb{N}})))$,
- $\psi_2\, \varphi = \phi_1([\![\, \mathsf{proj}_{\mathbb{N}} \circ \mathsf{proj}_{\mathbb{N}^\iota}]\!]\, \varphi\, (\rho_I(x^1)))$,

From Lemmata 13 and 14, we get for any β-normal λ-term $M^{\mathbb{N}^{\iota\iota}} = \lambda x.N \in \Lambda_\bot^\varnothing$:

$$\psi_k\, [\![M]\!] = \top \quad \Leftrightarrow \quad \pi^k(N) \in \Lambda_I \qquad (0 \leqslant k \leqslant 2).$$

---

for every $M^A \in \Lambda^\varnothing$ some code $\underline{N}$ of its long normal form N^A: $\theta_A M^A \twoheadrightarrow_\beta \underline{N}$ (see [Jol00], Chapter 1 for details). Their interest lies in that the code $\underline{N}$ produced allows to interpret N in all kinds of λ-models and to make various syntactical treatments on it. E.g. the above functionals ϕ_A are obtained from a 1-element partial model of the untyped λ-calculus, as defined in [Sel96], in which the λ-terms interpreted are exactly the λI-terms: ϕ_A is just the interpretation of θ_A in a model environment which evaluates in this finite partial model the terms N through their code $\underline{N}$. Other interpretation and instantiation of the very same λ-terms θ_A are the functionals of [BS91] and the λ-terms $\lambda z^A.\tau_z^o$ of [Jol01a].

Let ρ be a valuation into $\mathcal{F}_2$ such that:

$$\begin{bmatrix} \rho(x^{\mathbb{N}})\,\lambda\!\!\lambda d.\top = \lambda\!\!\lambda d.\top, \\ \rho(x^{\mathbb{N}})\,\varphi = \lambda\!\!\lambda d.\bot \ \text{ if } \varphi \neq \lambda\!\!\lambda d.\top, \end{bmatrix} \qquad \begin{bmatrix} \rho(e^{\mathbb{N}})\,\mathrm{Id}_{\mathcal{F}_2^o} = \lambda\!\!\lambda d.\top, \\ \rho(e^{\mathbb{N}})\,\varphi = \lambda\!\!\lambda d.\bot \ \text{ if } \varphi \neq \mathrm{Id}_{\mathcal{F}_2^o}. \end{bmatrix}$$

At last, let us define $\psi_3 \in \mathcal{F}_2^{\mathbb{N}^{\iota\iota}\to o}$ by $\psi_3\,\varphi = \varphi\,[\![\lambda f^{\mathbb{N}}.\,x^{\mathbb{N}} \circ f]\!]_\rho\,\rho(e^{\mathbb{N}})\,\mathrm{Id}_{\mathcal{F}_2^o}\top$ and $\psi \in \mathcal{F}_2^{\mathbb{N}^{\iota\iota}\to o}$ by $\psi\,\varphi = \top$ iff $\psi_k\,\varphi = \top$ for all $k \leqslant 3$.

We may check by a straightforward induction on the long normal λ-terms $G^1 \in \Lambda_I \setminus \{\mathbf{I}^1\}$ with free variables among $x^{\mathbb{N}}, e^{\mathbb{N}}, a_i^o$ $(i \geqslant 0)$, that if G has the form $\lambda v_n^o.x(\ldots(\lambda v_1^o.x(\lambda v_0^o.e\,\mathbf{I}^1 H_0^o)H_1^o)\ldots)H_n^o$, then $[\![G]\!]_\rho = \lambda\!\!\lambda d.\top$ else $[\![G]\!]_\rho = \lambda\!\!\lambda d.\bot$.

For every Church numeral $M^{\mathbb{N}^{\iota\iota}} = \lambda x^{\mathbb{N}^\iota} e^{\mathbb{N}}.x^n e$, we have $\pi^k(M) = \lambda xe.x^n e$ hence $\psi_k\,[\![M]\!] = \top$ for all $k \leqslant 2$. Moreover, $M(\lambda f^{\mathbb{N}}.\,x^{\mathbb{N}}\circ f)e^{\mathbb{N}}\mathbf{I}^1 \twoheadrightarrow x^n(e\,\mathbf{I}^1)$, hence $\psi_3\,[\![M]\!] = [\![M(\lambda f^{\mathbb{N}}.\,x^{\mathbb{N}}\circ f)e^{\mathbb{N}}\mathbf{I}^1]\!]_\rho\top = [\![x^n(e\,\mathbf{I}^1)]\!]_\rho\top = \top$ and therefore $\psi\,[\![M]\!] = \top$. This proves the first implication of the proposition.

Now, let $M^{\mathbb{N}^{\iota\iota}} = \lambda x.N \in \Lambda_\bot^\varnothing$ be any long normal λ-term such that $\psi\,[\![M]\!] = \top$. It follows $\psi_k\,[\![M]\!] = \top$ and then $\pi^k(N) \in \Lambda_I$ for all $k \leqslant 2$. We get in particular $\bot \notin \pi^2(N)$, hence $\pi^2(M)$ is a Church numeral, say $\pi^2(M) = \lambda x^1 e^o.x^n e$, and M has the form:

$$M = \lambda x^{\mathbb{N}^\iota} e^{\mathbb{N}} z_n^1 a_n^o.x(\ldots x(\lambda z_1^1 a_1^o.x(\lambda z_0^1 a_0^o.eP_0Q_0)P_1Q_1)P_2Q_2\ldots)P_nQ_n$$

for some λ-terms P_i^1, Q_i^o. Suppose for a contradiction that for some i the head variable of P_i is not z_i^1 and let k be the smallest such index i. The head variable of P_k could not be of type o, otherwise $\bot$ would occur in $\pi(N) \in \Lambda_I$. It could not be another variable z_i^1, otherwise $\pi(N)$ would not be a λI-term. Therefore, this head variable would be $x^{\mathbb{N}^\iota}$ or $e^{\mathbb{N}}$ and $M(\lambda f^{\mathbb{N}}.\,x^{\mathbb{N}} \circ f)e^{\mathbb{N}}\mathbf{I}^1$ would reduce into a long normal λ-term $G^1 = \lambda a_n^o.x(\ldots(\lambda a_1^o.x(\lambda a_0^o.eF^1 H_0^o)H_1^o)\ldots)H_n^o$ where the head variable of F is $x^{\mathbb{N}}$ or $e^{\mathbb{N}}$. Moreover, we would have $G \in \Lambda_I$ since $N = \pi^0(N) \in \Lambda_I$ and $(N[x^{\mathbb{N}^\iota} := \lambda f^{\mathbb{N}}.\,x^{\mathbb{N}} \circ f])e^{\mathbb{N}}\mathbf{I}^1 \twoheadrightarrow G$. By the remark above, this would yield $[\![G]\!]_\rho = \lambda\!\!\lambda d.\bot$ and then $\psi_3\,[\![M]\!] = [\![G]\!]_\rho\top = \bot$, contradicting the assumption $\psi\,[\![M]\!] = \top$. It follows $\pi(P_i) = z_i^o$ for all i and by Lemma 13: $\mathsf{proj}_{\mathbb{N}^\iota} M =_{\beta\eta} \pi(M) = \lambda x^{\mathbb{N}} e^1 z_n^o.x(\ldots(\lambda z_1^o.x(\lambda z_0^o.e\,z_0)z_1)\ldots)z_n \twoheadrightarrow_\eta \lambda x^{\mathbb{N}} e^1.x^n e$. □

Lemma 16. *Let* $\iota_\bot = [o{:=}1,\ \bot{:=}\lambda d^o.\bot]$. *We have:*

$$\begin{aligned} \mathsf{proj}_{o^\iota} \circ \mathrm{P}_{n,k}^{\iota_\bot} &=_{\beta\eta} \mathrm{P}_{n,k} \circ \mathsf{proj}_{o^+}, \\ \mathsf{proj}_{\mathbb{N}^\iota} \circ \mathrm{R}_{n,k}^{\iota_\bot} &=_{\beta\eta} \mathrm{R}_{n,k} \circ \mathsf{proj}_{\mathbb{N}^+}, \\ \mathsf{proj}_{\mathbb{N}^+} \circ \mathrm{R}^{+\iota} &=_{\beta\eta} \mathrm{R}^+ \circ \mathsf{proj}_{\mathbb{M}^+}. \end{aligned}$$

Proof. These $\beta\eta$-equalities are easily checked by reducing their members. □

Theorem 17. $\mathrm{DP}_{\mathcal{F}_2}$ *is undecidable.*

Proof. Let $\psi \in \mathcal{F}_2^{\mathbb{N}^{\iota\iota}\to o}$ be like in Lemma 15 and let us prove for any closed λ-terms $M^{3^+}, S^{o^+}, H^{o^+}$:

$$\exists X^{\mathbb{M}} \in \Lambda^\varnothing\ \ X^+ MS =_{\beta\eta} H \ \Leftrightarrow\ \exists Y^{\mathbb{M}^{+\iota}} \in \Lambda^\varnothing \begin{bmatrix} \mathsf{proj}_{\mathbb{M}^+} YMS =_{\beta\eta} H \\ \psi\,[\![\mathrm{R}_{n,k}^{\iota_\bot}(\mathrm{R}^{+\iota}Y)]\!] = \top \ \ (1 \leqslant k \leqslant n) \end{bmatrix}$$

If some $X^{\mathbb{M}} \in \Lambda^{\varnothing}$ is such that $X^{+}MS =_{\beta\eta} H$ then $Y = X^{+\iota}$ satisfies the right hand side. Indeed, by Lemma 13 we then have $\mathsf{proj}_{\mathbb{M}^+} YMS =_{\beta\eta} H$. Moreover, for every k there is $Z \in \Lambda^{\varnothing}$ such that $\mathrm{R}_{n,k}(\mathrm{R}X)^{+} =_{\beta\eta} Z^{\iota}$ by Lemma 9, hence $\mathrm{R}^{\iota_\perp}_{n,k}(\mathrm{R}^{+\iota}Y) = (\mathrm{R}_{n,k}(\mathrm{R}X)^{+})^{\iota_\perp} =_{\beta\eta} Z^{\iota\iota}$ and by Lemma 15 $\psi\, [\![\mathrm{R}^{\iota_\perp}_{n,k}(\mathrm{R}^{+\iota}Y)]\!] = \top$. Conversely, suppose that $Y^{\mathbb{M}^{+\iota}} \in \Lambda^{\varnothing}$ satisfies the right hand side. By Lemma 16, every λ-term $\mathrm{R}_{n,k}(\mathrm{R}^{+}(\mathsf{proj}_{\mathbb{M}^+} Y))$ converts to $\mathsf{proj}_{\mathbb{N}^\iota}(\mathrm{R}^{\iota_\perp}_{n,k}(\mathrm{R}^{+\iota}Y))$ and has therefore the form Z^{ι} up to $\beta\eta$-equality by Lemma 15; hence by Lemmata 9 and 10 there is a closed λ-term $X^{\mathbb{M}}$ such that $\mathsf{proj}_{\mathbb{M}^+} Y = X^{+}$ and then $X^{+}MS =_{\beta\eta} H$.

Now, we have $\mathsf{proj}_{\mathbb{M}^+} YMS =_{\beta\eta} H$ iff $\mathcal{F}_2 \vDash \mathsf{proj}_{\mathbb{M}^+} YMS = H$ because the members of these relations have the type 1_n. Therefore, the right hand side of the equivalence above just states the existence of a λ-definable functional $\varphi \in \mathcal{F}_2^{\mathbb{M}^{+\iota} \to o}$ such that $[\![\mathsf{proj}_{\mathbb{M}^+}]\!]\, \varphi\, [\![M]\!]\, [\![S]\!] = [\![H]\!]$ and $(\psi \circ [\![\mathrm{R}^{\iota_\perp}_{k} \circ \mathrm{R}^{+\iota}]\!])\varphi = \top$. It follows that if $\mathrm{DP}_{\mathcal{F}_2}$ were decidable, then we could decide whether there is $X^{\mathbb{M}} \in \Lambda^{\varnothing}$ such that $X^{+}MS =_{\beta\eta} H$ for any given $M^{3^+}, S^{o^+}, H^{o^+} \in \Lambda^{\varnothing}$, but this contradicts Proposition 5. □

Acknowledgement. I would like to thank Henk Barendregt, who offered me a long and peaceful time to achieve the present study. I am also greatly indebted to Ralph Loader, whose last work ([Loa03]) gave me the idea of Sections 3, 4 and 5.

References

[BS91] Ulrich Berger and Helmut Schwichtenberg. An inverse of the evaluation functional for typed λ-calculus. *6th Annual IEEE Symposium on Logic in Computer Science*, 203–211, 1991.

[Fri75] Harvey Friedman. Equality between functionals. *LNM*, 453:22–37, 1975.

[Hue75] Gérard Huet. A unification algorithm for typed λ-calculus. *TCS*, 1(1):27–57, 1975.

[Jol00] Thierry Joly. *Codages, séparabilité et représentation de fonctions en λ-calcul simplement typé et dans d'autres systèmes de types.* Thèse de Doctorat, Université Paris VII, January 2000.

[Jol01a] Thierry Joly. Constant time parallel computations in λ-calculus. *TCS*, 266: 975–985, 2001.

[Jol01b] Thierry Joly. The finitely generated types of the λ-calculus. *LNCS*, 2044:240–252, 2001. (proc. TLCA'01).

[Loa01] Ralph Loader. The Undecidability of λ-Definability. In *Logic, Meaning and Computation: Essays in Memory of Alonzo Church*, 331–342, C.A. Anderson and M. Zelëny editors, Kluwer Academic Publishers, Dordrecht, 2001.

[Loa03] Ralph Loader. Higher β matching is undecidable. *Logic Journal of the IGPL*, Vol. 11, No. 1:51–68, 2003.

[Sel96] Peter Selinger. Order-Incompleteness and Finite Lambda Models. *11th Annual IEEE Symposium on Logic in Computer Science*, 432–439, 1996.

[Sta82] Richard Statman. Completeness, invariance and λ-definability. *JSL*, 47:17–26, 1982.

[Sta85] Richard Statman. Logical Relations and the Typed λ-calculus. *IC*, 65:85–97, 1985.

Well-Going Programs Can Be Typed

Stefan Kahrs

University of Kent at Canterbury
Computing Laboratory
Canterbury CT2 7NF
United Kingdom
smk@ukc.ac.uk

Abstract. Our idiomatically objectionable title is a pun on Milner's "well-typed programs do not go wrong" — because we provide a completeness result for type-checking rather than a soundness result.
We show that the well-behaved functions of untyped PCF are already expressible in typed PCF: any equivalence class of the partial logical equivalence generated from the flat natural numbers in the model given by PCF's operational semantics is inhabited by a well-typed term.

1 Introduction

The functional programming language PCF [13,10,5] can be regarded as the kernel of all typed functional programming languages. It is expressive enough to closely mirror the operational or denotational semantics of the full languages and small enough to make a serious study of these semantics feasible.

The subject of this paper is the operational semantics of PCF and the question whether the type system of PCF prevents us from writing meaningful PCF programs. Of course, there are programs which syntactically fail to type-check and behave in exactly the same way as well-typed programs (hint: think about dead code), but it is not *a priori* clear whether there are others that also fail to type-check but exceed the expressive power of well-typed programs.

However, we do not want to give up soundness as a price for occasional wizardry. In other words, such programs should never exhibit elementary soundness violations such as using a number as a function or a function as a number (in an evaluation context). The absence of such additional functions would mean that PCF is *complete*, in the sense of [6,7].

This notion of completeness is indeed a dual to soundness — based on the partial logical equivalence we get by relating every value[1] of the flat natural numbers to itself. PCF's type system is sound since every well-typed term at type σ inhabits one of the equivalence classes, and we show that it is complete by demonstrating that every equivalence class is inhabited by a well-typed term. One consequence of this is that quotienting this term model by the partial equivalence gives us a fully abstract model of typed PCF.

[1] This is a slight simplification as we are actually dealing with a term model and have to close the relation under reduction.

M. Hofmann (Ed.): TLCA 2003, LNCS 2701, pp. 167–179, 2003.

We show completeness by writing an interpreter for untyped PCF in typed PCF — or rather, one interpreter for each PCF type. There is a technical difficulty here: when we interpret an untyped term at *a function type* $\alpha \rightarrow \beta$ then the interpretation has to incorporate values of type α; moreover, if α is itself a function type then it becomes impossible to syntactify these values — function types are abstract data types in a functional programming language. The solution to this problem is a combination of symbolic evaluation and evaluation in Dana Scott's model D_∞ [1]. We cannot implement full D_∞ in a PCF program, because that would involve infinitely many types. We can implement a finite fragment though, and we can statically determine (from the type into which we want to interpret) which fragment suffices for this purpose.

The aim of this paper is to show that every untyped PCF program p that *behaves well* at type σ is equivalent to well-typed program p' at that type. Technically, p "behaving well" means that $p \overset{\sigma}{\approx} p$, and p being "equivalent" to p' means $p \overset{\sigma}{\approx} p'$, where $\overset{\sigma}{\approx}$ is the partial equivalence generated from the flat natural numbers in the term model. Notice that this is a rather strong notion of well-behaviour: p not only fails to exhibit soundness violations, it also fails to break extensionality.

2 Implementing PCF

We assume familiarity with functional programming and the simply typed λ-calculus. PCF itself is a very simple typed functional programming language, which has natural numbers as its base type, supports general recursion and the simply typed λ-calculus.

As concrete syntax for writing PCF programs we choose Haskell [4] and allow ourselves the liberty to exploit its much richer syntax, for example we use general (and mutual) recursion instead of fixpoint operators and its class system. The source code of the implementation can be found at
`http://www.cs.ukc.ac.uk/~smk/PCF/Untyped_PCF.html`.

There are several reasons for not using PCF itself. First, there is the pragmatic one that too many things would have to be encoded, making the programs much longer and more difficult to read. Moreover, the structuring mechanism of Haskell (we exploit classes) can be used to emphasize general ideas. Most importantly, it also means that we can write a single schematic (and thus finite!) implementation instead of a separate one for each type — the techniques used to achieve that are similar to the ones describe by McBride in [9]; this necessarily involves the use of some compiler-supported language extensions such as *multi-parameter type classes* and *functional dependencies*, because we need to make some computations at type level.

However, these computations at type-level are static and serve to stress that the implementation is schematic. The types we use remain pure PCF types.

2.1 Symbolic Evaluation

Part of the implementation is entirely syntactic, i.e. it operates directly on the abstract syntax and does not involve imported values.

```
data Exp = App Exp Exp | Lam Exp | Var Int |
           Zero | Succ Exp | Pred Exp | If0 Exp Exp Exp |
           Fix Exp
```

The above defines (the abstract syntax of) open PCF expressions as a Haskell datatype; we represent variable binding via de Bruijn indices [2]. For our target language of typed PCF the type Exp is no more than syntactic sugar for ι.

The first line of the type definition shows the λ-calculus fragment of PCF, the second all the operations dealing with natural numbers; the final is a fixpoint operator — which the *untyped* language could actually do without, because its λ-calculus fragment is powerful enough to express it.

We need a notion of substitution for these terms, but it is completely routine — see appendix A.

The goal of the implementation is to stay syntactic as much as possible, and only resort to type casting once we use a variable in an evaluation context that demands its value. Therefore, we first define a type for normal forms (another piece of syntactic sugar for ι) and a normalizing function that reduces expressions to their normal forms. Although this works in an untyped setting, certain elementary soundness violations can occur — this is the reason why the result type of normalisation is Maybe NForm: a Nothing result indicates the occurrences of such an elementary soundness violation during expression normalisation.

```
data NForm =
    Hnf Int [Exp] | Abst Exp |
    Con Int | Pr NForm | Su NForm | Test NForm Exp Exp

normalize :: Exp -> Maybe NForm
normalize e = norm [] e

norm :: [Exp] -> Exp -> Maybe NForm
norm xs (Var k) = return (Hnf k xs)
norm [] Zero = return (Con 0)
norm _ Zero = Nothing
norm xs (Succ t) = norm xs t >>= su
norm xs (Pred t) = norm xs t >>= pr
norm xs (Fix t) = norm xs (subst0 (Fix t) t)
norm xs (App f a) = norm (a:xs) f
norm (x:xs) (Lam t) = norm xs (subst0 x t)
norm [] (Lam t) = return(Abst t)
norm (x:xs) (If0 c t e) =
    norm xs (If0 c (App t x)(App e x))
norm [] (If0 c t e) =
```

```
        normalize c >>= (\nf->
            case nf of
                Con 0 -> norm [] t
                Con k -> norm [] e
                Abst _ -> Nothing
                z -> return (Test z t e))
```

For non-Haskellists it should be explained that `>>=` is function application in the Kleisli category of a monad, in this case the lifting monad. For example, the function `su` (see appendix B) attempts to add 1 to a normal form, but we only use it here if the evaluation of its meant-to-be argument gave a normal form and not a type violation.

2.2 Interpreting Normal Forms at a Type

We are going to evaluate normal forms in an environment with bound variables of certain type. We will use only certain types for this purpose. More specifically, these types need to be instances of the following classes:

```
class IntEmbed a where
    int2a :: Int -> a
    a2int :: a -> Int

class SelfEmbed a where
    apply :: a -> a -> a
    lambd :: (a->a) -> a
```

Implicitly[2] we also require that these two maps form in each case retractions. We come to that later.

Concerning the relationship between this code and its PCF equivalent, it should be clear that the use of Haskell's class mechanism improves the presentation, but it is actually unnecessary: the four components of these two classes can always be passed around as additional parameters. Also, the following functions have as one of their parameters a list with component type `a` — we do not have lists in PCF but we can easily model them, e.g. as an integer (the length) plus a function of type $\iota \to$ `a` for the list indexing.

This list argument incidentally serves as a stack, giving meaning to the free variables of the expressions (or normal forms). The de Bruijn indices are relative addresses into this stack.

```
evalNF :: (SelfEmbed a,IntEmbed a) => [a] -> NForm -> a
evalNF xs (Hnf n es) = applyargs (xs!!n) [evalExp xs e | e<-es]
evalNF xs (Abst e) = lambd (\a -> evalExp (a:xs) e)
evalNF xs (Con n) = int2a n
evalNF xs (Pr e) = liftInt prednat (eval xs e)
```

[2] This is not directly enforceable in Haskell, but we achieve it by only providing instances that match or preserve the properties.

```
evalNF xs (Su e) = liftInt (+1) (eval xs e)
evalNF xs (Test c t e) =
    case a2int (evalNF xs c) of
        0 -> evalExp xs t
        _ -> evalExp xs e

applyargs :: SelfEmbed a => a -> [a] -> a
applyargs v [] = v
applyargs v (e:es) = applyargs (apply v e) es

evalExp :: (SelfEmbed a,IntEmbed a) => [a] -> Exp -> a
evalExp xs e = evalNF xs (unwrap(normalize e))

unwrap :: Maybe a -> a
unwrap (Just x) = x
unwrap Nothing = error "type violation"
```

The most interesting part of this is how we deal with abstractions and applications, because these are the two cases which involve the self-embedding. The first case deals with the application of a variable to a sequence of expressions. We fetch the value of the variable from the stack and subsequently apply it to the evaluations results of the expressions.

Notice that although we pass *evaluated* expressions to the application this does not mean that the application mechanism is using applicative evaluation order. If our implementation language uses normal evaluation order then these evaluations remain delayed until their values are requested. However, to achieve that preservation of evaluation order we had to employ the above `unwrap` function instead of evaluating in the lifting monad. The code of some auxiliary functions (e.g. `liftInt`) can be found in appendix B.

2.3 Interpreting at a Type

Our goal is to interpret (untyped) expressions at an arbitrary type. This effectively means to have one function $\texttt{interpret}_\sigma$ of type $\texttt{Exp} \to \sigma$ for every PCF type σ. In Haskell we can express this using a class:

```
class Run t where interpret :: Exp -> t
```

Clearly, we can instantiate this class at `Int`.

```
instance Run Int where interpret = evalExp []
```

However, things get a bit more complicated when we want to interpret function types. As we shall see, this requires certain computations on types, and Haskell 98 is too restrictive to aid in that task. Still, the Glasgow Haskell Compiler `GHC` supports several language extensions which *are* sufficient for our purposes. Be aware that we only need type computations in order to provide a *single, schematic* implementation of `interpret`; for the completeness goal in PCF it suffices to provide an implementation for `interpret`$_\sigma$ which can be obtained from instantiating the schema with σ.

The first of the language extensions we employ are multi-parameter classes [12].

```
instance (LUB a b c,Full c d) => Run(a->b) where
    interpret e a = (p1.p2)(evalExp [(e2.e1) a] (expand e))
                    where e1 = embed :: a->c
                          e2 = embed :: c->d
                          p2 = project :: d->c
                          p1 = project :: c->b
```

This may look rather bewildering at first. The essential idea is the following:

- we try to interpret an expression as a function of type $\alpha \to \beta$,
- we do this by waiting for the argument, pushing it on the stack and running our interpreter,
- except that the interpreter does not work with arbitrary types, only with the approximants of D_∞,
- so we need to find a sufficiently large one of those (which is `d`),
- lift the argument `a` via an embedding into `d` and project the result to the expected type `b`.

The function `expand` is defined in appendix A. Essentially, it is like an η-expansion without the λ, where the meaning of the free variable 0 is now provided by the stack. The indicated computations on types appear in the premise of the class instance.

The class `LUB` is provided in section 3; it exploits another extension feature, functional dependencies, which in this case guarantee that the first two parameters uniquely determine the third. Without that guarantee the use of the free type variable `c` in the premise of the class instance would not be permitted. A similar exploitation is made for class `Full`. The type annotation inside the right-hand side are necessary to disambiguate which embedding we use. `GHC`[3] keeps the type variables from an instance declaration free in the context in which its members are type-checked; this means that we are referring to the same type `d` here.

In general, it will not be the case that the type `c` we get from the least upper bound computation is suitable for evaluation. For example, the type $(\iota \to \iota \to \iota) \to \iota$ would not be. Therefore we have to find a type which is, and this is the purpose of the class constraint `Full c d`. Class `Full` is defined in such a way that `d` will be the smallest of the D_i domains of D_∞ into which both `a` and `b` can be faithfully embedded.

[3] The Haskell interpreter `Hugs` which generally supports very similar language extensions to `GHC` does not do that, e.g. `d` would be bound more locally — which is not what we want here. As a result, it is awkward if not impossible to replicate this implementation in Hugs.

This provides us now with an interpretation of expressions at *any* PCF type, provided we can find for any PCF type `a` a (unique) type `b` for which we have `Full a b`.

3 Retractions

Just as the D_∞ model, our implementation makes extensive use of retractions. We give retractions an extra section, partly because of this extensive use and partly because they play a central role when we want to reason about the implementation. However, most of this is fairly routine and follows the typical patterns one can find e.g. in chapters 11 and 12 of [3].

Definition 1. *A* retraction *between two objects A and B in a category is a pair of morphisms $\phi : A \to B$ and $\psi : B \to A$ such that $\psi \circ \phi = id_A$. We call ϕ the* embedding *and ψ the* projection.

We replicate that definition (without the condition) in Haskell, keeping in mind that all instances of the class should satisfy the condition. This requires multi-parameter classes. Of course, the PCF implementation cannot use such a schematic presentation and we have to provide these functions by hand; as a result, the PCF implementation increases in size as we move up the type hierarchy.

```
class Retraction a b where
    embed :: a->b
    project :: b->a
```

There is an added bit of security in Haskell which we get for free: although there are typically many possible retractions between two types (provided there is one at all), the compiler enforces that we only have one instance of the class for the given types.

We can give retractions between base type ι and any PCF type. For technical reasons (to pacify the `GHC` compiler), the code has to single them out.

```
class IntEmbed a where
    a2int :: a -> Int
    int2a :: Int -> a

instance IntEmbed Int where
    a2int = id
    int2a = id

instance (IntEmbed a,IntEmbed b) => IntEmbed (a->b) where
    a2int f = a2int (f (error "function is not a number"))
    int2a n _ = int2a n
```

```
instance IntEmbed a => Retraction Int a where
    embed = int2a
    project = a2int
```

The call to error is rather convenient here in Haskell, since it is triggered only if the function from which we retract touches its argument. In PCF we would have to resort to an ordinary non-termination instead.

In the categorical setting it should be obvious that functors preserve retractions: $F(\psi)\circ F(\phi) = F(\psi\circ\phi) = F(id) = id$; if F is contravariant then embedding and projection swap places. We use this to lift retractions to function types, e.g. defining the function space functor.

```
(-->) :: (a->b) -> (c->d) -> (b->c) -> (a->d)
f --> g = \ h -> g . h . f

instance (Retraction a c,Retraction b d) =>
    Retraction (a->b) (c->d) where
        embed = project --> embed
        project = embed --> project
```

This already sets up all the retractions between PCF types. Retractions generally define a preorder between the objects in a category, where $a \leq b$ iff there is a retraction from a to b. As previously indicated, we are interested in the least upper bounds of this preorder.

```
class (Retraction a c, Retraction b c) => LUB a b c | a b -> c
```

The bit at the end is a "functional dependency" which declares that types a and b uniquely determine c. This is enforced by the compiler.

Notice that the class has no members of its own. It is solely a device to convince the system that certain retractions exist.

```
instance LUB Int Int Int
instance (IntEmbed b,IntEmbed a,Retraction a a,Retraction b b) =>
    LUB Int (a->b) (a->b)
instance (IntEmbed a,IntEmbed b,Retraction a a,Retraction b b) =>
    LUB (a->b) Int (a->b)
instance (LUB a1 b1 c1, LUB a2 b2 c2) =>
    LUB (a1->a2) (b1->b2) (c1->c2)
```

Lemma 1. *All PCF types have least upper bounds in the retraction preorder.*

The proof is a trivial induction on the type structure, using exactly the pattern from the class instances.

We also need to single out types which are suitable for self-embeddings.

```
class SelfEmbed a
    where apply :: a -> a -> a
          lambd :: (a->a) -> a
```

For this purpose, we single out the finite approximations of D_∞:

```
instance SelfEmbed Int
    where apply n _ = n
          lambd f = f (error "use of function as number")

instance SelfEmbed a => SelfEmbed (a->a)
    where apply = lambd --> apply
          lambd = apply --> lambd
```

Lemma 2. *Every PCF type can be embedded into a PCF type that supports self-embedding.*

To prove the lemma, we need instances `Full a b` for every type `a`, for the following class `Full`, again a class without members.

```
class (Retraction a b,IntEmbed b,SelfEmbed b) => Full a b | a -> b
```

These instances are generated recursively as follows:

```
instance Full Int Int
instance (Full a x, Full b y, LUB x y c,
    Retraction a c, Retraction b c, IntEmbed c, SelfEmbed c) =>
        Full (a->b) (c->c)
```

The instance for the function type has a few more conditions than one might expect. The reason is that it is (i) impossible to tell `GHC` that the retraction relation is transitive and (ii) we do not know (in general) that the least upper bound of two self-embedding types has the required properties. For our purposes this is true, because the least upper bound of D_k and D_p is simply $D_{\max(k,p)}$.

4 The Term Model

The implementation gives us a function to interpret expressions at some type. But is its behaviour really the same as that of the original expression? In other words, we are trying to compare a PCF expression e at type σ with the PCF expression $\texttt{interpret}_\sigma(\ulcorner e \urcorner)$, where $\ulcorner e \urcorner$ is the Gödel number of e.

We need a bit of machinery to turn this into a meaningful question. First, recall from [11] that a *typed applicative structure* is a tuple

$$\langle \{A^\sigma\}, \{\mathbf{App}^{\sigma,\tau}\} \rangle$$

of families of sets and application maps indexed by type expressions. Compared to [11] I omitted the constants, i.e. this is viewed w.r.t. the empty signature.

Our untyped expressions provide us with a very simple typed applicative structure: A^σ is `Exp`, regardless of type, and **App** is just the `App` operation on `Exp`.

From this model we want to carve out the well-behaved expressions at the various types. We do this by defining a *partial equivalence relation*, i.e. a symmetric and transitive relation. Moreover, we define a family of such relations $\stackrel{\sigma}{\approx}$, one for each type σ — a *logical* relation. At base type we define:

$$a \stackrel{\iota}{\approx} b \iff \begin{cases} \exists k.\,\texttt{normalize}\ a = \texttt{Just(Con}\ k\texttt{)} = \texttt{normalize}\ b, \text{or} \\ (\texttt{normalize}\ a)\uparrow \wedge (\texttt{normalize}\ b)\uparrow \end{cases}$$

Here, $p\uparrow$ means that the evaluation of the expression p fails to terminate. In other words, at base type we relate all expressions to other expressions that compute the same number, and we also relate expressions that fail to terminate to each other. Not covered by this relation are the expressions which either evaluate to `Nothing` (a run-time type violation) or which evaluate to a normal form which is not a number.

As usual, this is lifted in a canonical way to function types:

$$a \stackrel{\sigma\to\tau}{\approx} b \iff \forall d, e.\, d \stackrel{\sigma}{\approx} e \Rightarrow \texttt{App}\ a\ d \stackrel{\tau}{\approx} \texttt{App}\ b\ e$$

An expression e is now viewed to be well-behaved at type σ iff it is related to itself at that type: $e \stackrel{\sigma}{\approx} e$.

Theorem 1. *For all types σ and all PCF expressions e:*

$$e \stackrel{\sigma}{\approx} e \Rightarrow e \stackrel{\sigma}{\approx} \texttt{interpret}_\sigma(\ulcorner e \urcorner)$$

where $\ulcorner e \urcorner$ is the Gödel-number of e, i.e. the corresponding value at type ***Exp***.

This implies that every equivalence class of the partial equivalence is inhabited by a well-typed term, simply because $\texttt{interpret}_\sigma(\ulcorner e \urcorner)$ is well-typed by construction.

The proof works by induction on the type structure, here is a sketch:

- At base type we get this very easily: every integer is expressible in PCF, and we do have well-typed non-terminating programs at this type as well.
- Suppose we have $t \stackrel{\sigma\to\tau}{\approx} t$ and arbitrary $u \stackrel{\sigma}{\approx} u'$. We need to show $t\ u \stackrel{\tau}{\approx} \texttt{interpret}_{\sigma\to\tau}(\ulcorner t \urcorner) u'$. Exploiting (i) the induction hypotheses on both σ and τ, (ii) the PER properties, and (iii) the fact that t and $\texttt{interpret}_{\sigma\to\tau}(\ulcorner t \urcorner)$ both inhabit $\stackrel{\sigma\to\tau}{\approx}$ we can reduce this problem to showing the following:

  $$\texttt{interp}_{\sigma\to\tau}(\ulcorner t \urcorner)\,(\texttt{interp}_\sigma(\ulcorner u \urcorner)) \stackrel{\tau}{\approx} \texttt{interp}_\tau(\ulcorner t\ \texttt{interp}_\sigma(\ulcorner u \urcorner) \urcorner).$$

  However, the only difference between interpretation on the left and the right is merely that the left will keep the term $\texttt{interp}_\sigma(\ulcorner u \urcorner)$ in the environment while the right has it in-lined.

The equivalence between evaluation in an environment and evaluation after substitution is normally explained by the substitution lemma for PCF. This does not quite apply here since the term t may not itself be well-typed, i.e. we may not have $\vdash t : \sigma \to \tau$. However, syntactic well-typedness can easily be replaced (for the purposes of a substitution lemma) by a semantic property, here that t inhabits $\stackrel{\sigma\to\tau}{\approx}$.

5 Related Work

In [7] we showed the *absence* of completeness for the programming language Standard ML. Although it would be possible to mirror the approach from this paper and write an interpreter for untyped SML within typed SML, it is not always possible to import/export values between these worlds and it is this fact that makes all the difference.

Marz and others [8] also linked typed PCF to an untyped language they called $\mathcal{L}$, but they are interpreting typed PCF via $\mathcal{L}$ while we are doing essentially the opposite. Their paper is lacking (but conjecturing) a "universality" result, and this is essentially the completeness of our paper. There is one major difference between $\mathcal{L}$ and untyped PCF: $\mathcal{L}$ has a unary operator "`where`" that enables $\mathcal{L}$ programs to distinguish between numbers and functions, and that goes beyond the capabilities of untyped PCF. Is typed PCF complete w.r.t. $\mathcal{L}$? This is not as clear-cut as one might think, and the reason is the following: the problematic `where` operator allows one to distinguish at a function type (say $\iota \to \iota$) between $\perp$ and $\lambda x.\perp$ — where $\perp$ is some suitable non-terminating expression. Typed PCF cannot make this distinction since these two terms are logically related, and thus we cannot implement this operator in its full glory in typed PCF.

6 Conclusion

We have shown how all "sound" values of PCF types in the untyped term model can be expressed by typed PCF expressions. It also follows from this result that if we quotient the term model of untyped PCF by the indicated partial equivalence we get fully abstract model of typed PCF: it is a model because PCF typing is sound, and it is fully abstract because every equivalence class is inhabited by a typed term.

The reason a self-implementation in this style does not work for Standard ML with "nested" recursive datatypes [7] is that the argument type of a constructor for nested datatypes can be "larger" than its result type and as a result we would not get away with having a *finite* number of class instances. A more interesting question is whether such a self-implementation would be possible in Haskell.

References

1. Hendrik P. Barendregt. *The Lambda-Calculus, its Syntax and Semantics*. North-Holland, 1984.
2. N. G. de Bruijn. Lambda calculus notation with nameless dummies, a tool for automatic formula manipulation. *Indagationes Mathematicae*, 34:381–392, 1972.
3. J. R. Hindley and J. P. Seldin. *Introduction to Combinators and λ-Calculus*. Cambridge University Press, 1986.
4. P. Hudak, S. Peyton Jones, and P. Wadler. Report on the Programming Language Haskell, a Non-strict, Purely Functional Language version 1.3. Technical report, University of Glasgow, 1996.

5. Achim Jung and Allen Stoughton. Studying the fully abstract model of PCF within its continuous function model. In *Typed Lambda Calculi and Applications*, 1993. LNCS 664.
6. Stefan Kahrs. About the completeness of type systems. In *Proceedings ESSLLI Workshop on Observational Equivalence and Logical Equivalence*, 1996.
7. Stefan Kahrs. Limits of ML-definability. In *Proceedings of PLILP'96*, pages 17–31, 1996. LNCS 1140.
8. Michael Marz, Alexander Rohr, and Thomas Streicher. Full abstraction and universality via realisability. In *LICS'99*, pages 174–182, 1999.
9. Conor McBride. Faking it. *Journa of Functional Programming*, 12(4,5):375–392, July 2002.
10. Robin Milner. Fully abstract models of types λ-calculi. *Theoretical Computer Science*, 4:1–22, 1977.
11. J.C. Mitchell. Type systems for programming languages. In Jan van Leeuwen, editor, *Handbook of Theoretical Computer Science*, volume B, chapter 8, pages 365–458. Elsevier, 1990.
12. Simon Peyton Jones, Mark Jones, and Erik Meijer. Type classes: an exploration of the design space. In *Haskell Workshop 1997*, pages 1–16, 1997.
13. Gordon Plotkin. LCF considered as a programming language. *Theoretical Computer Science*, 5:223–256, 1977.

A Substitution Code

```
subst :: Exp -> (Int->Exp) -> Exp
subst (Var n) f = f n
subst (App t u) f = App (subst t f) (subst u f)
subst (Lam t) f = Lam (subst t (Var 0 <: f))
subst (Fix t) f = Fix (subst t (Var 0 <: f))
subst Zero f = Zero
subst (Succ t) f = Succ (subst t f)
subst (Pred t) f = Pred (subst t f)
subst (If0 c t e) f =
    If0 (subst c f) (subst t f) (subst e f)

(<:) :: Exp -> (Int->Exp) -> Int -> Exp
t <: e = aux
         where aux 0 = t
               aux n = lift (e(n-1))

lift :: Exp -> Exp
lift t = subst t (Var . (+1))

subst0 :: Exp -> Exp -> Exp
subst0 t u = subst u (t <: (Var . (subtract 1)))

expand :: Exp -> Exp
expand e = App(lift e)(Var 0)
```

Remark: the described functions `subst`, `lift`, and `<:` are mutually recursive but it is clear that the mutual recursion could be resolved.

B Auxiliary Functions on Normal Forms

```
su :: NForm -> Maybe NForm
su (Con x) = return (Con (x+1))
su (Abst _) = Nothing
su x = return (Su x)

pr :: NForm -> Maybe NForm
pr (Con x) = return (Con (prednat x))
pr (Abst _) = Nothing
pr x = return (Pr x)

prednat :: Int -> Int
prednat n = if n==0 then 0 else (n-1)

liftInt :: IntEmbed a => (Int->Int) -> (a->a)
liftInt = a2int --> int2a
```

Parameterizations and Fixed-Point Operators on Control Categories

Yoshihiko Kakutani[1] and Masahito Hasegawa[1 2]

[1] Research Institute for Mathematical Sciences, Kyoto University
{kakutani,hassei}@kurims.kyoto-u.ac.jp
[2] PRESTO21, Japan Science and Technology Corporation

Abstract. The $\lambda\mu$-calculus features both variables and names, together with their binding mechanisms. This means that constructions on open terms are necessarily parameterized in two different ways for both variables and names. Semantically, such a construction must be modeled by a bi-parameterized family of operators. In this paper, we study these bi-parameterized operators on Selinger's categorical models of the $\lambda\mu$-calculus called control categories. The overall development is analogous to that of Lambek's functional completeness of cartesian closed categories via polynomial categories. As a particular and important case, we study parameterizations of uniform fixed-point operators on control categories, and show bijective correspondences between parameterized fixed-point operators and non-parameterized ones under uniformity conditions.

1 Introduction

1.1 Parameterization on Models of $\lambda\mu$-Calculus

The simply typed $\lambda\mu$-calculus introduced by Parigot [9] is an extension of the simply typed λ-calculus with first-class continuations. In the $\lambda\mu$-calculus, every judgment has two kinds of type declarations: one is for variables and the other is for continuation variables, which are often called names. So, it is natural that an operator $(-)^\dagger$ on the $\lambda\mu$-calculus takes the following form:

$$\frac{\Gamma \vdash M : A \mid \Delta}{\Gamma \vdash M^\dagger : B \mid \Delta}$$

The typed call-by-name $\lambda\mu$-calculus (with classical disjunctions) has a sound and complete class of models called control categories [11]. An interpretation of a judgment $x_1 : B_1, \dots, x_n : B_n \vdash M : A \mid \alpha_1 : A_1, \dots, \alpha_n : A_n$ in a control category $\mathcal{P}$ is a morphism $f \in \mathcal{P}([\![B_1]\!] \times \cdots \times [\![B_n]\!], [\![A]\!] \parr [\![A_1]\!] \parr \cdots \parr [\![A_n]\!])$. So, the above syntactic construction is modeled in a control category $\mathcal{P}$ like the following:

$$\frac{f \in \mathcal{P}(X, [\![A]\!] \parr Y)}{f^\dagger \in \mathcal{P}(X, [\![B]\!] \parr Y)}$$

M. Hofmann (Ed.): TLCA 2003, LNCS 2701, pp. 180–194, 2003.

Therefore, from the semantic point of view, operators should be understood as parameterized construction with two parameters X and Y. We call such a parameterization *bi-parameterization*. The study of this sort of parameterization on cartesian categories was initiated by Lambek [7], and its significance in modeling parameterized constructs associated to algebraic data-types has been studied by Jacobs [5]. The aim of this work is to derive analogous results for bi-parameterization on control categories.

1.2 Fixed-Point Operators and Parameterizations

Our motivation to study parameterization on control categories comes from our previous work about fixed-point operators on the $\lambda\mu$-calculi in [4] and [6]. The equational theories of fixed-point operators in call-by-name λ-calculi have been studied extensively, and now there are some canonical axiomatizations including iteration theories [1] and Conway theories, equivalently traced cartesian categories [3] (see [12] for recent results). Because the $\lambda\mu$-calculus is an extension of the simply typed call-by-name λ-calculus, it looks straightforward to consider fixed-point operators in the $\lambda\mu$-calculus and indeed we have considered an appropriate model of the $\lambda\mu$-calculus with a fixed-point operator. In a control category, however, the possible forms of parameterized fixed-point operators are various and their relation is sensitive. To understand our problem, let us recall a folklore construction.

In a cartesian closed category, it is possible to derive a parameterized fixed-point operator from a non-parameterized one:

$$\dfrac{\dfrac{\dfrac{\dfrac{X \times A \xrightarrow{f} A}{A \xrightarrow{\mathrm{cur}(f)} A^X}\ \text{Currying}}{A^X \xrightarrow{(\cdot)^X} (A^X)^X \xrightarrow{A^{\Delta}} A^X}}{1 \xrightarrow{(\cdot)^*} A^X}\ \text{Fixed point}}{X \xrightarrow{f^{\dagger}} A}\ \text{Uncurrying}$$

If we use the simply typed λ-calculus as an internal language of cartesian closed categories, this construction amounts to taking the fixed-point of $k : X \to A \vdash \lambda x^X . f(x, kx) : X \to A$ for $x : X, a : A \vdash f(x, a) : A$. So, by letting f be $\lambda(f^{A \to A}, x^A).fx$, we obtain a fixed-point combinator $\mathtt{fix}_A : (A \to A) \to A$. It is routine to see that this $\mathtt{fix}_A$ is indeed a fixed-point combinator. In a cartesian closed category, to give a parameterized fixed-point operator is to give a fixed-point combinator. Thus, we have a parameterized fixed-point operator from a non-parameterized one.

In this paper, we investigate such a construction of fixed-point operators on control categories. Since we have to consider parameters for free names in the $\lambda\mu$-calculus, a control category has three patterns of parameterizations: one is

Object constructors:

$$1 \quad \bot \quad A \times B \quad B^A \quad A \parr B$$

Morphism constructors:

$$\begin{array}{ll}
\mathrm{id}_A & : A \to A \\
\diamond_A & : A \to 1 \\
\pi_1 & : A \times B \to A \\
\pi_2 & : A \times B \to B \\
\varepsilon_{A,B} & : B^A \times A \to B \\
\mathrm{a}_{A,B,C} & : (A \parr B) \parr C \to A \parr (B \parr C) \\
\mathrm{l}_A & : A \to A \parr \bot \\
\mathrm{c}_{A,B} & : A \parr B \to B \parr A \\
\mathrm{i}_A & : \bot \to A \\
\nabla_A & : A \parr A \to A \\
\mathrm{d}_{A,B,C} & : (A \parr C) \times (B \parr C) \to (A \times B) \parr C \\
\mathrm{s}^{-1}_{A,B,C} & : (B \parr C)^A \to B^A \parr C
\end{array}$$

$$\frac{f : A \to B \quad g : B \to C}{g \circ f : A \to C} \qquad \frac{f : A \to B \quad g : A \to C}{\langle f, g \rangle : A \to B \times C}$$

$$\frac{f : A \times B \to C}{\mathrm{cur}(f) : A \to C^B} \qquad \frac{f : A \to B}{f \parr C : A \parr C \to B \parr C}$$

Fig. 1. Signature of Control Categories

standard *parameterization* and another is a parameterization for names called *co-parameterization*, and the last one is *bi-parameterization* which has both parameterization and co-parameterization. In a control category, a bi-parameterized fixed-point operator can be derived from a co-parameterized one in analogous way of the cartesian closed case. Moreover, a bi-parameterized fixed-point operator can be derived from a non-parameterized one under suitable uniformity principles. An interesting and important observation is that these correspondences are indeed bijective. This result simplifies semantic structure needed in [6].

1.3 Construction of This Paper

Section 2 is a reminder of control categories and the typed call-by-name $\lambda\mu$-calculus. In Section 3, we introduce polynomial categories with respect to control categories. In Section 4, we consider generic parameterized operators on control categories. The rest of this paper gives observations of parameterized fixed-point operators on control categories and their uniformity principles. Uniformity conditions enable us to prove bijective correspondence between uniform bi-parameterized fixed-point operators and uniform non-parameterized ones.

2 Preliminaries

2.1 Control Categories

Control categories introduced by Selinger [11] are sound and complete models of the typed call-by-name $\lambda\mu$-calculus. A control category is a cartesian closed

$$\frac{x:A \in \Gamma}{\Gamma \vdash x:A \mid \Delta} \qquad \frac{}{\Gamma \vdash *:\top \mid \Delta}$$

$$\frac{\Gamma, x:A \vdash M:B \mid \Delta}{\Gamma \vdash \lambda x^A.\, M:A \to B \mid \Delta} \qquad \frac{\Gamma \vdash M:A \to B \mid \Delta \quad \Gamma \vdash N:A \mid \Delta}{\Gamma \vdash MN:B \mid \Delta}$$

$$\frac{\Gamma \vdash M:A \mid \Delta \quad \Gamma \vdash N:B \mid \Delta}{\Gamma \vdash \langle M,N\rangle : A \wedge B \mid \Delta} \qquad \frac{\Gamma \vdash M:A_1 \wedge A_2 \mid \Delta}{\Gamma \vdash \pi_i M:A_i \mid \Delta}$$

$$\frac{\Gamma \vdash M:\bot \mid \alpha:A, \Delta}{\Gamma \vdash \mu\alpha^A.\, M:A \mid \Delta} \qquad \frac{\Gamma \vdash M:A \mid \Delta \quad \alpha:A \in \Delta}{\Gamma \vdash [\alpha]\, M:\bot \mid \Delta}$$

$$\frac{\Gamma \vdash M:\bot \mid \beta:B, \alpha:A, \Delta}{\Gamma \vdash \mu(\alpha^A, \beta^B).\, M:A \vee B \mid \Delta} \qquad \frac{\Gamma \vdash M:A \vee B \mid \Delta \quad \alpha:A, \beta:B \in \Delta}{\Gamma \vdash [\alpha,\beta]\, M:\bot \mid \Delta}$$

Fig. 2. Deduction Rules of $\lambda\mu$-Calculus

category together with a premonoidal structure [10]. In this section, we recall some definitions about control categories but may omit the detail, which are found in [11].

Definition 1. *A morphism $f:A \to B$ in a premonoidal category $\mathcal{P}$ is* ***central*** *if for every morphism $g \in \mathcal{P}(C,D)$, $(B \parr g) \circ (f \parr C) = (f \parr D) \circ (A \parr g)$ and $(g \parr B) \circ (C \parr f) = (D \parr f) \circ (g \parr A)$.*

Definition 2. *A morphism $f:A \to B$ in a symmetric premonoidal category with codiagonals $\mathcal{P}$ is* ***focal*** *if f is central, discardable and copyable [11]. The subcategory formed by the focal morphisms of $\mathcal{P}$ is called the* ***focus*** *of $\mathcal{P}$ and denoted by $\mathcal{P}^\bullet$.*

Remark 1. In general, the focus of a symmetric premonoidal category is not the same as its center. (For example, detailed analysis are found in [2].) However, in a control category, the center and the focus always coincide [11].

Definition 3. *Suppose $\mathcal{P}$ is a symmetric premonoidal category with codiagonals and also suppose $\mathcal{P}$ has finite products. We say that $\mathcal{P}$ is* ***distributive*** *if the projections of products are focal and the functor $(-) \parr C$ preserves finite products for all objects C.*

Definition 4. *Suppose $\mathcal{P}$ is a symmetric premonoidal category with codiagonals and also suppose $\mathcal{P}$ is cartesian closed. $\mathcal{P}$ is a* ***control category*** *if the canonical morphism $\mathrm{s}_{A,B,C} \in \mathcal{P}(B^A \parr C, (B \parr C)^A)$ is a natural isomorphism in A, B and C, satisfying certain coherence conditions.*

Definition 5. *A* ***(strict) functor of control categories*** *is a functor that preserves all the structures of a control category on the nose.*

The structure of control categories is equational in the sense of Lambek and Scott [8]. The object and morphism constructors of control categories are shown in Figure 1. Some other canonical morphisms that are not shown here

$$
\begin{array}{lll}
(\beta_{\rightarrow})\ (\lambda x^A.\,M)N = M\,[N\,/x] & :B & \\
(\eta_{\rightarrow})\ \lambda x^A.\,Mx = M & :B & x \notin \mathrm{FV}(M) \\
(\beta_{\wedge})\ \pi_i\langle M_1, M_2\rangle = M_i & :A_i & \\
(\eta_{\wedge})\ \langle \pi_1 M, \pi_2 M\rangle = M & :A \wedge B & \\
(\eta_{\top})\ * = M & :\top & \\
(\beta_{\mu})\ [\beta]\mu\alpha^A.\,M = M\,[\beta\,/\alpha] & :\bot & \\
(\eta_{\mu})\ \mu\alpha^A.[\alpha]\,M = M & :A & \alpha \notin \mathrm{FN}(M) \\
(\beta_{\vee})\ [\gamma,\delta]\mu(\alpha^A,\beta^B).\,M = M\,[\gamma\,/\alpha, \delta\,/\beta] & :\bot & \\
(\eta_{\vee})\ \mu(\alpha^A,\beta^B).[\alpha,\beta]\,M = M & :A \vee B & \alpha,\beta \notin \mathrm{FN}(M) \\
(\beta_{\bot})\ [\beta]\,M = M & :\bot & \\
(\zeta_{\rightarrow})\ (\mu\alpha^{A\rightarrow B}.\,M)N = \mu\beta^B.\,M\,[[\beta](-)N\,/[\alpha](-)] & :B & \\
(\zeta_{\wedge})\ \pi_i(\mu\alpha^{A_1\wedge A_2}.\,M) = \mu\beta^{A_i}.\,M\,[[\beta]\pi_i(-)\,/[\alpha](-)] & :A_i & \\
(\zeta_{\vee})\ [\gamma,\delta]\mu\alpha^{A\vee B}.\,M = M\,[[\gamma,\delta](-)\,/[\alpha](-)] & :\bot &
\end{array}
$$

Fig. 3. Axioms of $\lambda\mu$-Calculus

but definable from the constructors, $f \times A$, $\mathrm{w_r} = \mathrm{i}_A \parr B \circ \mathrm{c}_{B,\bot} \circ \mathrm{l}_B : B \rightarrow A \parr B$, $\mathrm{w_l} = \mathrm{c}_{B,A} \circ \mathrm{w_r} : A \rightarrow A \parr B$ and so on are also used in this paper. Since coherence theorems for premonoidal categories have been shown by Power and Robinson in [10], we may elide not only cartesian structural isomorphisms but also premonoidal ones.

2.2 $\lambda\mu$-Calculus

According to Selinger, the typed call-by-name $\lambda\mu$-calculus can be considered as an internal language of control categories [11]. The types and the terms of our $\lambda\mu$-calculus are defined as follows:

$$
\begin{aligned}
\text{Types } A, B &::= \sigma \mid A \rightarrow B \mid \top \mid A \wedge B \mid \bot \mid A \vee B, \\
\text{Terms } M, N &::= x \mid * \mid \lambda x^A.\,M \mid MN \mid \langle M, N\rangle \mid \pi_1 M \mid \pi_2 M \\
&\quad\ \mid \mu\alpha^A.\,M \mid [\alpha]\,M \mid \mu(\alpha^A,\beta^B).\,M \mid [\alpha,\beta]\,M,
\end{aligned}
$$

where σ, x and α (β) range over base types, variables and names respectively. Every judgment takes the form $\Gamma \vdash M : A \mid \Delta$, where Γ denotes a sequence of pairs $x : A$, and Δ denotes a sequence of pairs $\alpha : A$. The typing rules and the axioms are given by Figure 2 and 3.

3 Parameterizations on Control Categories

3.1 Polynomial Control Categories

Since parameterization is nicely modeled by polynomial categories like the case of cartesian closed categories (see [7]), we introduce polynomial control categories and show their functional completeness a la Lambek. Because we have to deal with not only variables but also names on control categories unlike on cartesian closed categories, an additional parameter for free names is required.

We construct a new category $\mathcal{P}_Y^X$ from a control category $\mathcal{P}$ by

$$\mathcal{P}_Y^X(A,B) = \mathcal{P}(X \times A, B \parr Y).$$

The identity arrow of $\mathcal{P}_Y^X$ is the projection from $X \times A$ to A followed by the weakening arrow to $A \parr Y$ in $\mathcal{P}$. We define $g \circ_Y^X f$, the composite of $f \in \mathcal{P}_Y^X(A,B)$ and $g \in \mathcal{P}_Y^X(B,C)$, by

$$\begin{array}{l} X \times A \xrightarrow{\Delta \times A} X \times X \times A \xrightarrow{X \times f} X \times (B \parr Y) \\ \xrightarrow{\mathrm{w_l} \times (B \parr Y)} (X \parr Y) \times (B \parr Y) \xrightarrow{\mathrm{d}_{X,B,Y}} (X \times B) \parr Y \\ \xrightarrow{g \parr Y} C \parr Y \parr Y \xrightarrow{C \parr \nabla} C \parr Y. \end{array}$$

Hereafter, we write $marhrmd'_{A,B,C}$ for $\mathrm{d}_{A,B,C} \circ \mathrm{w_l} \times (B \parr C) \in \mathcal{P}(A \times (B \parr C), (A \times B) \parr C)$.

Proposition 1. *$\mathcal{P}_Y^X$ is a control category.*

We can regard $\mathcal{P}_Y^X$ as the polynomial control category obtained from $\mathcal{P}$ by adjoining an indeterminate of X and a name of Y. We call $\mathrm{w_l} : X \to X \parr Y$ in $\mathcal{P}$ the indeterminate variable of $\mathcal{P}_Y^X$ and $\pi_2 : X \times Y \to Y$ in $\mathcal{P}$ the indeterminate name of $\mathcal{P}_Y^X$. $\mathcal{P}$ can be embedded into $\mathcal{P}_Y^X$ through the weakening functor $\mathrm{I}_Y^X : \mathcal{P} \to \mathcal{P}_Y^X$ defined by

$$\mathrm{I}_Y^X(f) = X \times A \xrightarrow{\pi_2} A \xrightarrow{f} B \xrightarrow{\mathrm{w_l}} B \parr Y.$$

The following theorem justifies to regard $\mathcal{P}_Y^X$ as a polynomial category with respect to control categories.

Theorem 1. *Let $\mathcal{P}$ be a control category. Given a control category $\mathcal{O}$ and a functor of control categories $F : \mathcal{P} \to \mathcal{O}$ with morphisms $a \in \mathcal{O}(1, FX)$ and $k \in \mathcal{O}^\bullet(FY, \bot)$, there exists a unique functor of control categories $F' : \mathcal{P}_Y^X \to \mathcal{O}$ such that it sends the indeterminate variable and the indeterminate name of $\mathcal{P}_Y^X$ to a and k respectively and the following diagram commutes:*

$$\begin{array}{ccc} \mathcal{P} & \xrightarrow{\mathrm{I}_Y^X} & \mathcal{P}_Y^X \\ & {\scriptstyle F} \searrow & \downarrow {\scriptstyle F'} \\ & & \mathcal{O} \end{array}$$

Proof. (Outline) F' is constructed by

$F'f =$

$$FA \cong 1 \times FA \xrightarrow{a \times FA} FX \times FA \xrightarrow{Ff} FB \parr FY \xrightarrow{FB \parr k} FB \parr \bot \cong FB$$

for $f \in \mathcal{P}(X \times A, B \parr Y)$. □

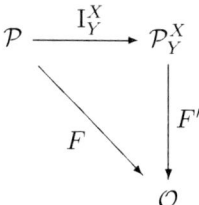

Remark 2. If we introduce a polynomial control category "$\mathcal{P}[x\!:\!X \mid \alpha\!:\!Y]$" syntactically (as in [8]), it is characterized by the universal property above. Hence we have $\mathcal{P}_Y^X \cong \mathcal{P}[x\!:\!X \mid \alpha\!:\!Y]$ and the functional completeness of control categories: for any "polynomial" $\phi(x,\alpha) \in \mathcal{P}[x\!:\!X \mid \alpha\!:\!Y](A,B)$, there exists a unique morphism $f \in \mathcal{P}(X \times A, B \parr Y)$ such that $\phi(x,\alpha) = (B \parr \alpha) \circ f \circ (x \times A)$.

By trivializing the parameter Y, we obtain a control category $\mathcal{P}^X$ with $\mathcal{P}^X(A,B) = \mathcal{P}(X \times A, B)$, and by trivializing the parameter X, we obtain a control category $\mathcal{P}_Y$ with $\mathcal{P}_Y(A,B) = \mathcal{P}(A, B \parr Y)$.

It can be seen easily that $\mathcal{P}_Y^X$, $(\mathcal{P}_Y)^X$ and $(\mathcal{P}^X)_Y$ are the same control category.

It is also useful to see that the mapping $(X,Y) \mapsto \mathcal{P}_Y^X$ gives rise to an indexed category $\mathcal{P}^{\mathrm{op}} \times \mathcal{P}^\bullet \to \mathbf{ContCat}$, where **ContCat** is the category of small control categories and functors of control categories. Indeed, $g \in \mathcal{P}(X, X')$ and $h \in \mathcal{P}^\bullet(Y', Y)$ determine the re-indexing functor from $\mathcal{P}_{Y'}^{X'}$ to $\mathcal{P}_Y^X$ which sends $f\!: X' \times A \to B \parr Y'$ to $(B \parr h) \circ f \circ (g \times A)\!: X \times A \to B \parr Y$.

Lemma 1. *The focus of $\mathcal{P}_Y$ agrees with the focus of $\mathcal{P}$,* i.e.,

$$(\mathcal{P}_Y)^\bullet(A,B) = \mathcal{P}^\bullet(A, B \parr Y).$$

For the focus of $\mathcal{P}^X$, see the next subsection.

3.2 Currying

The theorem below tells us how an indeterminate variable can be eliminated with a name, hence gives us a way to reduce the parameterized constructs on control categories to a simpler form: from the bi-parameterized form to the co-parameterized form.

Theorem 2. *The isomorphisms*

$$\mathcal{P}(X \times A, B \parr Y) \underset{\lfloor - \rfloor}{\overset{\lceil - \rceil}{\rightleftarrows}} \mathcal{P}(A, B \parr Y^X)$$

give rise to isomorphisms of control categories between $\mathcal{P}_Y^X$ and $\mathcal{P}_{Y^X}$.

Since a direct proof is very lengthy, we find it much easier to use the $\lambda\mu$-calculus as an internal language of control categories.

$$\frac{f = x\!:\!X, a\!:\!A \vdash M\!:\!B \mid \gamma\!:\!Y}{\lceil f \rceil = a\!:\!A \vdash \mu\beta^B.[\delta](\lambda x^X.\mu\gamma^Y.[\beta]M)\!:\!B \mid \delta\!:\!Y^X}$$

$$\frac{g = a\!:\!A \vdash N\!:\!B \mid \delta\!:\!Y^X}{\lfloor g \rfloor = x\!:\!X, a\!:\!A \vdash \mu\beta^B.[\gamma]((\mu\delta^{Y^X}.[\beta]N)x)\!:\!B \mid \gamma\!:\!Y}$$

It is routine to verify $\lceil - \rceil$ and $\lfloor - \rfloor$ preserve all the structures of control categories.

Remark 3. The above theorem suggests that any reasonable parameterized construct on a control category $\mathcal{P}$ must be compatible with $\lceil - \rceil$. Furthermore, the indexed categorical view mentioned before requires such a construct must be natural in parameter X in $\mathcal{P}$ and parameter Y in $\mathcal{P}^\bullet$. This consideration leads us to introduce the axiomatization of bi-parameterized fixed-point operators shortly.

The following results are part of this theorem, but we shall state them separately for future reference.

Lemma 2. *$\lceil - \rceil$ and $\lfloor - \rfloor$ preserve composites:*

$$\lceil g \circ_Y^X f \rceil = \lceil g \rceil \circ_{Y^X} \lceil f \rceil$$
$$\lfloor g \circ_{Y^X} f \rfloor = \lfloor g \rfloor \circ_Y^X \lfloor f \rfloor$$

Lemma 3. *$\lceil - \rceil$ and $\lfloor - \rfloor$ preserve focuses:*

$$f \in (\mathcal{P}_Y^X)^\bullet(A, B) \quad \textit{iff} \quad \lceil f \rceil \in (\mathcal{P}_{Y^X})^\bullet(A, B)$$
$$f \in (\mathcal{P}_{Y^X})^\bullet(A, B) \quad \textit{iff} \quad \lfloor f \rfloor \in (\mathcal{P}_Y^X)^\bullet(A, B)$$

In particular, from Lemma 1 and 3, we have $\mathcal{P}^{X^\bullet}(A, B) \cong \mathcal{P}^\bullet(A, B^X)$.

4 Parameterized Operators on Control Categories

In this section, we introduce three parameterization patterns based on our observation of polynomial control categories. One is standard parameterization in cartesian categories, and another parameterization is co-parameterization, which has a parameter for free names. The last one is bi-parameterization, which combines both parameterization and co-parameterization. Interaction between co-(bi-)parameterization and focuses of control categories is crucial.

Definition 6. *A **parameterized operator** of type $(A_1, B_1) \times \cdots \times (A_n, B_n) \to (A, B)$ on a control category $\mathcal{P}$ is a family of functions of the form*

$$\alpha^X : \mathcal{P}^X(A_1, B_1) \times \ldots \times \mathcal{P}^X(A_n, B_n) \to \mathcal{P}^X(A, B)$$

indexed by X, such that natural in X in $\mathcal{P}$.

Since a control category is a cartesian closed category, the following proposition holds.

Proposition 2. *Parameterized operators of type $(A_1, B_1) \times \cdots \times (A_n, B_n) \to (A, B)$ on a control category $\mathcal{P}$ are in bijective correspondence with arrows of $\mathcal{P}(B_1^{A_1} \times \ldots \times B_n^{A_n}, B^A)$.*

Definition 7. *A **co-parameterized operator** of type $(A_1, B_1) \times \cdots \times (A_n, B_n) \to (A, B)$ on a control category $\mathcal{P}$ is a family of functions of the form*

$$\alpha_Y : \mathcal{P}_Y(A_1, B_1) \times \ldots \times \mathcal{P}_Y(A_n, B_n) \to \mathcal{P}_Y(A, B)$$

indexed by Y, such that natural in Y in $\mathcal{P}^\bullet$.

Proposition 3. *Co-parameterized operators of type* $(A_1, B_1) \times \cdots \times (A_n, B_n) \to (A, B)$ *on a control category* $\mathcal{P}$ *are in bijective correspondence with arrows of* $\mathcal{P}(B_1^{A_1} \times \ldots \times B_n^{A_n}, B^A)$.

Proof. It follows from the isomorphisms $\mathcal{P}_Y(A, B) \cong \mathcal{P}^\bullet(\bot^{B^A}, Y)$ and $\mathcal{P}^\bullet(\bot^A, \bot^{A'}) \cong \mathcal{P}(A', A)$. □

Corollary 1. *Parameterized operators and co-parameterized operators of the same type are in bijective correspondence.*

The following bi-parameterization is important for control categories as semantic models of the $\lambda\mu$-calculus.

Definition 8. *A* ***bi-parameterized operator*** *of type* $(A_1, B_1) \times \cdots \times (A_n, B_n) \to (A, B)$ *on a control category* $\mathcal{P}$ *is a family of functions of the form*

$$\alpha_Y^X : \mathcal{P}_Y^X(A_1, B_1) \times \ldots \times \mathcal{P}_Y^X(A_n, B_n) \to \mathcal{P}_Y^X(A, B)$$

indexed by X *and* Y*, such that natural in* X *in* $\mathcal{P}$ *and natural in* Y *in* $\mathcal{P}^\bullet$*. A bi-parameterized operator is* ***strongly bi-parameterized*** *if it is compatible with currying:*

$$\lceil \alpha_Y^X(f_1, \ldots, f_n) \rceil = \alpha_{Y^X}^1(\lceil f_1 \rceil, \ldots, \lceil f_n \rceil).$$

The following lemma is immediate from the compatibility.

Lemma 4. *Co-parameterized operators and strongly bi-parameterized operators of the same type are in bijective correspondence.*

5 Parameterized Fixed-Point Operators on Control Categories

5.1 Uniform Co-parameterized Fixed-Point Operators

In this section, general approach to parameterizations in the previous section is specialized to fixed-point operators on control categories.

First, we define uniform non-parameterized fixed-point operators and uniform co-parameterized ones, and investigate their bijective correspondence.

Definition 9. *A* ***fixed-point operator*** *on a control category* $\mathcal{P}$ *is a family of functions* $(-)^* : \mathcal{P}(A, A) \to \mathcal{P}(1, A)$ *such that* $f^* = f \circ f^*$ *hold. A fixed-point operator on* $\mathcal{P}$ *is* ***uniform*** *if* $h \circ f = g \circ h$ *implies* $h \circ f^* = g^*$ *for any morphisms* $f \in \mathcal{P}(A, A)$*,* $g \in \mathcal{P}(B, B)$ *and* $h \in \mathcal{P}^\bullet(A, B)$*.*

Definition 10. *A* ***co-parameterized fixed-point operator*** *on a control category* $\mathcal{P}$ *is a family of functions* $(-)^\dagger : \mathcal{P}(A, A \parr Y) \to \mathcal{P}(1, A \parr Y)$ *such that the following conditions hold:*

1. *(naturality)*
   $(A \parr h) \circ f^\dagger = ((A \parr h) \circ f)^\dagger$ *for any* $f \in \mathcal{P}(A, A \parr Y')$ *and* $h \in \mathcal{P}^\bullet(Y', Y)$

2. *(fixed-point property)*
   $f^\dagger = f \circ_Y f^\dagger$ *for any* $f \in \mathcal{P}(A, A \parr Y)$, *that is,*

$$f^\dagger = 1 \xrightarrow{f^\dagger} A \parr Y \xrightarrow{f \parr Y} A \parr Y \parr Y \xrightarrow{A \parr \nabla} A \parr Y$$

It is ***uniform*** *if* $h \circ_Y f = g \circ_Y h$ *implies* $h \circ_Y f^\dagger = g^\dagger$ *for any morphisms* $f \in \mathcal{P}_Y(A, A)$, $g \in \mathcal{P}_Y(B, B)$ *and* $h \in (\mathcal{P}_Y)^\bullet(A, B)$.

$$\begin{array}{ccccc}
A & \xrightarrow{f} & A \parr Y & \xrightarrow{h \parr Y} & B \parr Y \parr Y \\
{\scriptstyle h}\downarrow & & & & \downarrow{\scriptstyle B \parr \nabla} \\
B \parr Y & \xrightarrow{g \parr Y} & B \parr Y \parr Y & \xrightarrow{B \parr \nabla} & B \parr Y
\end{array}$$

$\Rightarrow (B \parr \nabla) \circ (h \parr Y) \circ f^\dagger = g^\dagger$

Remark 4. In other words, a (uniform) co-parameterized fixed-point operator on $\mathcal{P}$ is a family of (uniform) fixed-point operators on $\mathcal{P}_Y$ that are preserved by re-indexing functors.

Remark 5. The word 'operator' in this section has not the same meaning as that of the previous section. In Section 4, a co-parameterized operator is a family of functions $(-)^\dagger : \mathcal{P}(A, A \parr Y) \to \mathcal{P}(1, A \parr Y)$ with fixed A, only indexed by Y. In this section, however, a co-parameterized fixed-point operator is a family of functions $(-)^\dagger : \mathcal{P}(A, A \parr Y) \to \mathcal{P}(1, A \parr Y)$ indexed by both A and Y.

Proposition 4. *On a control category, uniform co-parameterized fixed-point operators are in bijective correspondence with uniform fixed-point operators.*

Proof. Given a uniform fixed-point operator $(-)^*$, we define a uniform co-parameterized operator $(-)^\dagger$ by $f^\dagger = ((A \parr \nabla) \circ (f \parr Y))^*$ for $f \in \mathcal{P}(A, A \parr Y)$. Though we can directly check the naturality, the fixed-point property and the uniformity of $(-)^\dagger$ through chasing many diagrams, we will give a simpler proof via adjunctions later.

Conversely, from a uniform co-parameterized fixed-point operator $(-)^\dagger$, we obtain an operator $(-)^*$ just by trivializing the parameter. In this case, it is obvious that $(-)^*$ is a uniform fixed-point operator.

It is sufficient for a bijective correspondence to show that

$$((A \parr \nabla) \circ (f \parr Y))^\dagger = f^\dagger : 1 \to A \parr Y$$

holds for a uniform co-parameterized fixed-point operator $(-)^\dagger$. Since $(f \parr Y) \circ_Y^X (\mathrm{w_l} \circ \mathrm{w_l}) = (\mathrm{w_l} \circ \mathrm{w_l}) \circ_Y^X f$ holds, by the uniformity we have

$$(f \parr Y)^\dagger = (\mathrm{w_l} \circ \mathrm{w_l}) \circ_Y^X f^\dagger : 1 \to A \parr Y \parr Y.$$

Applying $A \parr \nabla$ to the both sides of the equation, we get $((A \parr \nabla) \circ (f \parr Y))^\dagger = f^\dagger$. $\square$

For the rest of the proof, we consider the weakening functor and the following results.

Proposition 5. *The weakening functor* $\mathrm{I}_Y : \mathcal{P} \to \mathcal{P}_Y$ *has a right adjoint* U_Y *given by* $\mathrm{U}_Y(A) = A ⅋ Y$ *and*

$$\mathrm{U}_Y(f) = A ⅋ Y \xrightarrow{f ⅋ Y} B ⅋ Y ⅋ Y \xrightarrow{B ⅋ \nabla} B ⅋ Y.$$

Moreover U_Y *preserves the focus.*

Corollary 2. *The weakening functor* $\mathrm{I}^X_Y : \mathcal{P} \to \mathcal{P}^X_Y$ *has a right adjoint* U^X_Y *given by* $\mathrm{U}^X_Y(A) = A ⅋ Y^X$ *and* $\mathrm{U}^X_Y(f) = \mathrm{U}_{Y^X}(\lfloor f \rfloor)$. *Moreover* U^X_Y *preserves the focus.*

This adjunction gives us a simpler proof of the construction and bijectivity between uniform fixed-point operators and co-parameterized ones.

As before, we define a uniform co-parameterized operator $(-)^\dagger$ from a uniform fixed-point operator $(-)^*$ by $f^\dagger = ((A ⅋ \nabla) \circ (f ⅋ Y))^*$. This $(-)^\dagger$ is just the same as $(\mathrm{U}_Y(f))^*$. We show that $(-)^\dagger$ is indeed a uniform co-parameterized fixed-point operator. Now we note that $\mathrm{U}_Y(g) \circ f = g \circ_Y f$.

- Naturality:
  For $f \in \mathcal{P}_Y(A, A')$ and $h \in \mathcal{P}^\bullet(Y', Y)$, $(A ⅋ h) \circ \mathrm{U}_Y(f) = \mathrm{U}_Y((A ⅋ Y) \circ f) \circ (A ⅋ h)$ holds. So, the uniformity of $(-)^*$ gives us the equation $(A ⅋ h) \circ f^\dagger = ((A ⅋ h) \circ f)^\dagger$.
- Fixed-point property:
  The co-parameterized fixed-point property trivially follows from the fixed-point property of $(-)^*$:

$$f^\dagger = (\mathrm{U}_Y(f))^* = \mathrm{U}_Y(f) \circ (\mathrm{U}_Y(f))^* = f \circ_Y f^\dagger$$

- Uniformity:
  We assume $h \circ_Y f = g \circ_Y h$ holds and h is focal. It follows that $\mathrm{U}_Y(h) \circ \mathrm{U}_Y(f) = \mathrm{U}_Y(g) \circ \mathrm{U}_Y(h)$ holds and $\mathrm{U}_Y(h)$ is focal. Therefore the uniformity of $(-)^*$ induces $\mathrm{U}_Y(h) \circ f^\dagger = g^\dagger$.

Similar result about the weakening functors and their adjunctions also help us to understand the relation between uniform fixed-point operators and uniform parameterized ones such as sketched in the introduction.

Definition 11. *A* ***parameterized fixed-point operator*** *on a control category* $\mathcal{P}$ *is a family of functions* $(-)^\# : \mathcal{P}(X \times A, A) \to \mathcal{P}(X, A)$ *such that the following conditions hold:*

1. *(naturality)*
   $f^\# \circ g = (f \circ (g \times A))^\#$ *for any* $f \in \mathcal{P}(X' \times A, A)$ *and* $g \in \mathcal{P}^\bullet(X, X')$
2. *(fixed-point property)*
   $f^\# = f \circ^X f^\#$ *for any* $f \in \mathcal{P}(X \times A, A)$.

It is ***uniform*** *if* $h \circ^X f = g \circ^X h$ *implies* $h \circ^X f^\# = g^\#$ *for any morphisms* $f \in \mathcal{P}^X(A, A)$, $g \in \mathcal{P}^X(B, B)$ *and* $h \in (\mathcal{P}^X)^\bullet(A, B)$.

Proposition 6. *The weakening functor* $I^X : \mathcal{P} \to \mathcal{P}^X$ *has a right adjoint* U^X *given by* $U^X(A) = A^X$ *and*

$$U^X(f) = A^X \xrightarrow{\mathrm{cur}(f)^X} (B^X)^X \cong B^{X \times X} \xrightarrow{B^\Delta} B^X.$$

Moreover U^X *preserves the focus.*

We construct a uniform parameterized fixed-point operator $(-)^\#$ from a uniform fixed-point operator $(-)^*$ by $f^\# = \varepsilon \circ^X I^X(U^X(f)^*)$, where $\varepsilon \in \mathcal{P}(X \times A^X, A)$ is the counit of the adjunction. Bijectivity follows from the equation $f^\# = \varepsilon \circ^X I^X((U^X(f))^\#)$, which is derived from the uniformity of $(-)^\#$ from the focality of ε and $\varepsilon \circ^X I^X(U^X(f)) = f \circ^X \varepsilon$.

This correspondence is generalized to a relation between uniform co-parameterized fixed-point operators and uniform bi-parameterized ones in the next subsection.

5.2   Uniform Bi-parameterized Fixed-Point Operators

Our goal is to show a bijective correspondence between uniform non-parameterized fixed-point operators and uniform bi-parameterized fixed-point operators introduced below.

Definition 12. *A (strongly) bi-parameterized fixed-point operator on a control category* $\mathcal{P}$ *is a family of functions* $(-)^\ddagger : \mathcal{P}(X \times A, A \,\mathregular{⅋}\, Y) \to \mathcal{P}(X, A \,\mathregular{⅋}\, Y)$ *such that the following conditions hold:*

1. *(naturality)*
   $(A \,\mathregular{⅋}\, h) \circ f^\ddagger \circ g = ((A \,\mathregular{⅋}\, h) \circ f \circ (g \times A))^\ddagger$ *for any* $f \in \mathcal{P}(X' \times A, A \,\mathregular{⅋}\, Y')$, $g \in \mathcal{P}(X, X')$ *and* $h \in \mathcal{P}^\bullet(Y', Y)$.
2. *(compatibility with currying)*
   $\lceil f^\ddagger \rceil = \lceil f \rceil^\ddagger$ *for any* $f \in \mathcal{P}(X \times A, A \,\mathregular{⅋}\, Y)$.
3. *(fixed-point property)*
   $f^\ddagger = f \circ^X_Y f^\ddagger$ *for any* $f \in \mathcal{P}(X \times A, A \,\mathregular{⅋}\, Y)$, *that is,*

$$f^\ddagger = X \xrightarrow{\Delta;\, X \times f^\ddagger} X \times (A \,\mathregular{⅋}\, Y) \xrightarrow{d'} (X \times A) \,\mathregular{⅋}\, Y \xrightarrow{f \,\mathregular{⅋}\, Y;\, A \,\mathregular{⅋}\, \nabla} A \,\mathregular{⅋}\, Y$$

*It is **uniform** if* $h \circ^X_Y f = g \circ^X_Y h$ *implies* $h \circ^X_Y f^\ddagger = g^\ddagger$ *for any morphisms* $f \in \mathcal{P}^X_Y(A, A)$, $g \in \mathcal{P}^X_Y(B, B)$ *and* $h \in (\mathcal{P}^X_Y)^\bullet(A, B)$.

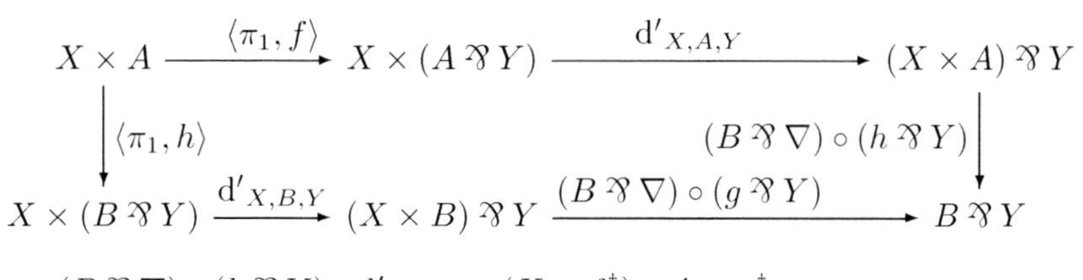

$$
\begin{array}{ccccc}
X \times A & \xrightarrow{\langle \pi_1, f \rangle} & X \times (A \,\mathregular{⅋}\, Y) & \xrightarrow{\quad d'_{X,A,Y} \quad} & (X \times A) \,\mathregular{⅋}\, Y \\
\Big\downarrow{\scriptstyle \langle \pi_1, h \rangle} & & & & \Big\downarrow{\scriptstyle (B \,\mathregular{⅋}\, \nabla) \circ (h \,\mathregular{⅋}\, Y)} \\
X \times (B \,\mathregular{⅋}\, Y) & \xrightarrow{d'_{X,B,Y}} & (X \times B) \,\mathregular{⅋}\, Y & \xrightarrow{(B \,\mathregular{⅋}\, \nabla) \circ (g \,\mathregular{⅋}\, Y)} & B \,\mathregular{⅋}\, Y
\end{array}
$$

$$\Rightarrow (B \,\mathregular{⅋}\, \nabla) \circ (h \,\mathregular{⅋}\, Y) \circ d'_{X,B,Y} \circ (X \times f^\ddagger) \circ \Delta = g^\ddagger$$

Remark 6. In other words, a (uniform) bi-parameterized fixed-point operator on $\mathcal{P}$ is a family of (uniform) fixed-point operators on $\mathcal{P}_Y^X$ that are preserved by re-indexing functors and compatible with currying.

Proposition 7. *On a control category, uniform bi-parameterized fixed-point operators are in bijective correspondence with uniform co-parameterized fixed-point operators.*

Proof. Given a uniform co-parameterized fixed-point operator$(-)^\dagger$, we define a uniform bi-parameterized fixed-point operator $(-)^\ddagger$ by $f^\ddagger = \lfloor \lceil f \rceil^\dagger \rfloor$.

$(-)^\ddagger$ satisfies the naturality, the compatibility with $\lfloor - \rfloor$, the bi-parameterized fixed-point property and the uniformity.

- Naturality:

$$\begin{array}{ll} ((B \parr h) \circ f \circ (g \times A))^\ddagger & \\ = \lfloor \lceil (B \parr h) \circ f \circ (g \times A) \rceil^\dagger \rfloor & \text{definition of } (-)^\ddagger \\ = \lfloor ((B \parr h^g) \circ \lceil f \rceil)^\dagger \rfloor & \text{naturality of } \lceil - \rceil \\ = \lfloor (B \parr h^g) \circ \lceil f \rceil^\dagger \rfloor & \text{naturality of } (-)^\dagger \\ = (B \parr h) \circ \lfloor \lceil f \rceil^\dagger \rfloor \circ g & \text{naturality of } \lfloor - \rfloor \\ = (B \parr h) \circ f^\ddagger \circ g & \text{definition of } (-)^\ddagger \end{array}$$

- Compatibility with $\lceil - \rceil$:

$$\lceil f^\ddagger \rceil = \lceil \lfloor \lceil f \rceil^\dagger \rfloor \rceil = \lceil f \rceil^\dagger = \lceil f \rceil^\ddagger.$$

- Fixed-point property:

$$f^\ddagger = \lfloor \lceil f \rceil^\dagger \rfloor = \lfloor \lceil f \rceil \circ_{Y^X} \lceil f \rceil^\dagger \rfloor = \lfloor \lceil f \rceil \rfloor \circ_Y^X \lfloor \lceil f \rceil^\dagger \rfloor = f \circ_Y^X f^\ddagger$$

  follows from Lemma 2.
- Uniformity:
  $g \circ_Y^X h = h \circ_Y^X f$ implies $\lceil g \rceil \circ_{Y^X} \lceil h \rceil = \lceil h \rceil \circ_{Y^X} \lceil f \rceil$ by Lemma 2. Lemma 3 means that if h is focal, so is $\lceil h \rceil$. Hence by the uniformity of $(-)^\dagger$, $\lceil g \rceil^\dagger = \lceil h \rceil \circ_{Y^X} \lceil f \rceil^\dagger$ for $f \in \mathcal{P}_Y^X(A, A)$, $g \in \mathcal{P}_Y^X(B, B)$ and $h \in (\mathcal{P}_Y^X)^\bullet(A, B)$ such that $g \circ_Y^X h = h \circ_Y^X f$. $\lceil g \rceil^\dagger = \lceil h \rceil \circ_{Y^X} \lceil f \rceil^\dagger$ implies $g^\ddagger = h \circ_Y^X f^\ddagger$.

For bijectivity, it is sufficient to show

$$\lfloor \lceil f \rceil^\ddagger \rfloor = f^\ddagger : X \to A \parr Y$$

for a uniform bi-parameterized fixed-point operator $(-)^\ddagger$, but that is equivalent to $(-)^\ddagger$'s compatibility with $\lceil - \rceil$. □

The following theorem is deduced from Proposition 4 and Proposition 7.

Theorem 3. *On a control category, uniform bi-parameterized fixed-point operators are in bijective correspondence with uniform fixed-point operators.*

6 Fixed-Point Operator in $\lambda\mu$-Calculus

In the previous work [6], we have extended the call-by-name $\lambda\mu$-calculus with a uniform fixed-point operator. In this section, we recall its definition (including the syntactic notion of focality which is used for determining the uniformity) and refine the completeness theorem of [6] using the results from Section 5.

Definition 13. *In a call-by-name $\lambda\mu$-theory [11], $\Gamma \vdash H : A \to B \mid \Delta$ is **focal** if*

$$\Gamma, k : \neg\neg A \vdash H(\mu\alpha^A.\, k(\lambda x^A.[\alpha]\, x)) = \mu\beta^B.\, k(\lambda x^A.[\beta]\, Hx) : B \mid \Delta$$

holds.

This syntactic notion of focus precisely corresponds to the semantic focality.

Proposition 8. *Given a control category $\mathcal{P}$, $h \in \mathcal{P}_Y^X(A, B)$ is focal if and only if the term $x : X \vdash \lambda a^A.\mu\beta^B.[\beta, \gamma]h\langle x, a\rangle : A \to B \mid \gamma : Y$ in the internal language is focal in the sense of Definition 13.*

Definition 14. *A type-indexed family of closed terms $\{\texttt{fix}^A : (A \to A) \to A\}$ in a $\lambda\mu$-theory is called a **uniform fixed-point operator** if the following conditions hold:*

1. *(fixed-point property)*
   $\texttt{fix}^A\, F = F(\texttt{fix}^A\, F)$ *holds for any term* $F : A \to A$.
2. *(uniformity)*
   For any terms $F : A \to A$, $G : B \to B$ and focal $H : A \to B$, $H \circ F = G \circ H$ implies $H(\texttt{fix}^A\, F) = \texttt{fix}^B\, G$.

Indeed, this axiomatization is sound and complete for control categories with uniform bi-parameterized fixed-point operators [6]. However, since we know that uniform bi-parameterized fixed-point operators are reducible to non-parameterized ones, we have the following theorem, which strengthens and simplifies the completeness result under the uniformity conditions.

Theorem 4. *Control categories with uniform fixed-point operators provide a sound and complete class of models of the $\lambda\mu$-calculus extended with a uniform fixed-point operator.*

7 Conclusion

In this paper, we have introduced polynomial categories for control categories, which are required to deal with not only free variables but also free names, and shown their functional completeness a la Lambek [7]. Based on those consideration, we defined strongly bi-parameterized operators. Bi-parameterized operators have more complicated forms than standard parameterized ones since they have both parameterization and co-parameterization. Co-parameterization is for free names while usual parameterization is for free variables. Our strongly bi-parameterized operators can be reduced to co-parameterized operators by the compatibility with currying.

General approach to parameterizations is specialized to parameterizations on fixed-point operators. In this paper, we introduced uniform co-parameterized fixed-point operators and bi-parameterized ones. Our bi-parameterized fixed-point operators are in bijective correspondence with co-parameterized ones. As we have shown in the paper, the uniformity conditions imply the bijective correspondence between bi-parameterized fixed-point operators and non-parameterized ones.

The technical novelty of this approach is that we closely look at co-parameterization and its interaction with the focus of a control category. We believe our observations are useful not merely for fixed-point operators but also for other parameterized constructs on control categories and the $\lambda\mu$-calculus, in particular, for modeling parameterized data-type constructions as in [5].

References

1. S. Bloom and Z. Ésik. *Iteration Theories.* EATCS Monographs on Theoretical Computer Science, Springer-Verlag, 1993.
2. C. Führmann. Varieties of effects. In *Foundations of Software Science and Computation Structures*, volume 2303 of LNCS, pages 144–158, Springer-Verlag, 2002.
3. M. Hasegawa. *Models of Sharing Graphs: A Categorical Semantics of `let` and `letrec`.* PhD thesis, University of Edinburgh, 1997.
4. M. Hasegawa and Y. Kakutani. Axioms for recursion in call-by-value. *Higher-Order and Symbolic Computation*, 15(2):235–264, 2002.
5. B. Jacobs. Parameters and parameterization in specification, using distributive categories. *Fundamenta Informaticae*, 24(3):209–250, 1995.
6. Y. Kakutani. Duality between call-by-name recursion and call-by-value iteration. In *Computer Science Logic*, volume 2471 of LNCS, pages 506–521, Springer-Verlag, 2002.
7. J. Lambek. Functional completeness of cartesian categories. *Annals of Mathematical Logic*, 6:259–292, 1970.
8. J. Lambek and P.J. Scott. *Introduction to Higher-Order Categorical Logic.* Cambridge Studies in Advanced Mathematics, Cambridge University Press, 1986.
9. M. Parigot. $\lambda\mu$-calculus: an algorithmic interpretation of classical natural deduction. In *Logic Programming and Automated Reasoning*, volume 624 of LNCS, pages 190–201, Springer-Verlag, 1992.
10. A.J. Power and E.P. Robinson. Premonoidal categories and notions of computation. *Mathematical Structures in Computer Science*, 7(5):453–468, 1997.
11. P. Selinger. Control categories and duality: on the categorical semantics of the lambda-mu calculus. *Mathematical Structures in Computer Science*, 11(2):207–260, 2001.
12. A. Simpson and G. Plotkin. Complete axioms for categorical fixed-point operators. In *Proceedings of 15th Annual Symposium on Logic in Computer Science*, pages 30–41, 2000.

Functional In-Place Update with Layered Datatype Sharing

Michal Konečný

LFCS Edinburgh, Mayfield Rd, Edinburgh EH9 3JZ, UK
mkonecny@inf.ed.ac.uk,
//homepages.inf.ed.ac.uk/mkonecny

Abstract. Hofmann's LFPL is a functional language with constructs which can be interpreted as referring to heap locations. In this view, the language is suitable for expressing and verifying in-place update algorithms. Correctness of this semantics is achieved by a linear typing. We introduce a non-linear typing of first-order LFPL programs which is more permissive than the recent effect-based typing of Aspinall and Hofmann. The system efficiently infers separation assertions as well as destruction and re-use effects for individual layers of recursive-type values. Thus it is suitable for in-place update algorithms with complicated data aliasing.

1 Introduction

First-order LFPL (Linear Functional Programming Language) [1] is a functional programming language which has a straightforward compositional denotational semantics and at the same time can be evaluated imperatively without dynamic memory allocation. The higher-order version of this language is interesting among other reasons because it captures non-size increasing polynomial time/space computation with primitive/full recursion [2]. We focus on the first order version of LFPL in this paper.

For example, in LFPL we can write the following program to append two lists of elements of type A:

$$\begin{aligned} append_A(x,y) = {} & \mathsf{match}\ x\ \mathsf{with}\ \mathsf{nil} \Rightarrow y \\ & \mid \mathsf{cons}(h,t)@d \Rightarrow \mathsf{cons}(h, append_A(t,y))@d \end{aligned}$$

In the denotational semantics of the program, the attachment $@d$ of cons is (virtually) ignored in both cases and thus the semantics is list concatenation as expected. Nevertheless, in the operational semantics $@d$ indicates that the cons-cell is (or will be) at location d on the heap. Thus the above program appends the lists in-place, rewriting the first list's end with a reference to the second list. This amounts to changing its last cons-cell if the first list is not empty. The other cons-cells of the first list are overwritten with the same content.

This evaluation strategy can go wrong easily, for example for $append_A(x,x)$ with a non-empty list x. Such terms might fail to evaluate or evaluate with incorrect results. We will call a term *sound* if it evaluates in harmony with its

M. Hofmann (Ed.): TLCA 2003, LNCS 2701, pp. 195–210, 2003.

denotational semantics independently of the values of its free variables and how they are represented on the heap.

We also need a finer notion of soundness relative to some *extra condition* on the representation of the arguments on the heap. For example, we would like to declare $append_A(x, y)$ correct under certain condition on how lists x and y are represented. A necessary and sufficient condition for $append_A(x, y)$ to evaluate correctly is that the heap region occupied by the cons-cells of x should be separated (disjoint) from the whole region of y.

Apart from stating conditions for correctness, we need to mark certain *guarantees* about the heap representation of the result and the change of heap during the evaluation. For example, $append_A(x, y)$ preserves the value of y and this might be crucial for showing the correctness of a bigger term of which this is a subterm.

LFPL uses linearity to achieve soundness of its terms. This means that LFPL maintains the "single pointer property". This implies the precondition that different variables always refer to disjoint regions on the heap.

It is impossible to achieve a full characterisation of semantical correctness of the first-order LFPL evaluation by a simple typing system. Nevertheless, one can improve LFPL by relaxing its linearity in order to recognise more of the correct algorithms. This means losing the single pointer property and considering sharing. One such system, which we call UAPL, is provided by Aspinall and Hofmann [3]. It assigns a *usage aspect* to each variable in the typing context.

UAPL still rejects many correct programs, e.g. those with $append_A(x, y)$ where the whole regions of x and y are not guaranteed to be disjoint. UAPL needs this stronger precondition because it cannot distinguish between the cons-cell (shallow) and data (deep) levels of a list. This paper develops a typing for the underlying language of LFPL which improves over UAPL in that it can distinguish between these two levels (and more than that) for arbitrary recursive types (with no restriction on nesting).

More precisely, we note each place in a type term which may involve a heap location in the operational representation of the values of the type. For example, there is one such place within each list type constructor. We give names (ranged over by ζ) to all such places and mark them in the type (e.g. $\mathsf{L}^{[\mathsf{a}]}(\_)$). To each such place we logically associate a portion of the heap region taken by a value of the type (e.g. the locations of the cons-cells of a list).

We formulate preconditions and rely-guarantees using assertions about the portion names given in the types within a typing judgement. The most straightforward atomic assertion is $\zeta_1 \otimes \zeta_2$ which stands for the separation (disjointness) of two portions ζ_1, ζ_2 on the heap[1]. We need other atomic separation assertions related to the unfolding of a recursive type. For example, we need to express that a list has its elements separated from each other or that a tree's skeleton is laid out without overlapping.

For this reason we give names also to each occurrence of a recursive type constructor within the types of a typing judgement. For a list or tree constructor

[1] A set of such assertions corresponds to a *may-alias relation* in alias analysis [4]

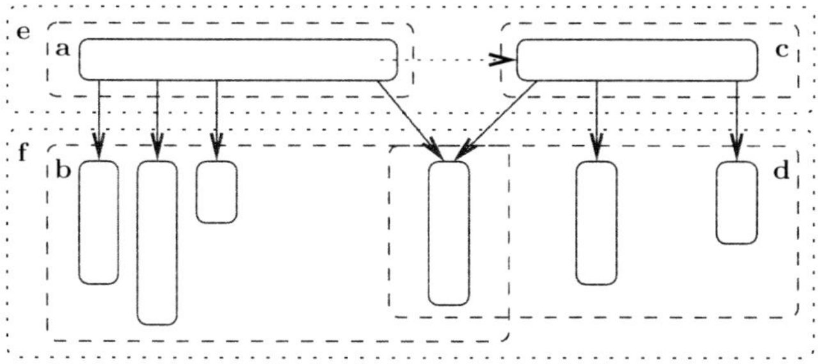

Fig. 1. Heap portions in the *append* program

we will simply reuse the name of the only portion associated with it. Given a recursive type occurrence name ζ_R and a portion name ζ which occurs within the scope of ζ_R, we will define two assertions $\otimes\zeta/\zeta_R$ and $\otimes\zeta\curlywedge\zeta_R$. The former one states that the sub-portions of ζ which correspond to the unfoldings of the type ζ_R are pairwise disjoint. The latter one is the same but requires disjointness only for pairs of unfoldings in which one is not a sub-unfolding of the other (see Def. 4 for details).

Now we can express that the elements of a list of the type $\mathsf{L}^{[\mathbf{a}]}(\mathsf{L}^{[\mathbf{b}]}(\mathsf{Bool}))$ do not share by $\otimes\mathbf{b}/\mathbf{a}$ and that the skeleton of a tree of the type $\mathsf{T}^{[\mathbf{c}]}_{\mathsf{b}}(\mathsf{Bool})$ has no confluences by $\otimes\mathbf{c}\curlywedge\mathbf{c}$.

The typing for $append_A$ in our new system would be:

$$x : \mathsf{L}^{[\mathbf{a}]}(A^{[\mathbf{b}]}), y : \mathsf{L}^{[\mathbf{c}]}(A^{[\mathbf{d}]}); \{\mathbf{a}\otimes\mathbf{b}, \mathbf{a}\otimes\mathbf{c}, \mathbf{a}\otimes\mathbf{d}\} \vdash$$
$$append_A(x, y) : \mathsf{L}^{[\mathbf{e}\subseteq\{\mathbf{a},\mathbf{c}\}]}(A^{[\mathbf{f}\subseteq\{\mathbf{b},\mathbf{d}\}]}); \{\mathbf{a}\}; \otimes\mathbf{f}/\mathbf{e} \Longleftarrow \{\mathbf{b}\otimes\mathbf{d}, \otimes\mathbf{b}/\mathbf{a}, \otimes\mathbf{d}/\mathbf{c}\}$$

The participating heap portions are illustrated in Fig. 1. Just before $\vdash$ there is a precondition demanding that three pairs of argument portions should be separated. Whenever this condition holds, the term will evaluate in harmony with the obvious denotational semantics. The result type has also two portions, named $\mathbf{e}$ and $\mathbf{f}$. The notation with $\subseteq$ indicates the guarantee that the corresponding result portion is contained within the union of the given argument portions. Behind the result type is the set of possibly destroyed portions, i.e. $\{\mathbf{a}\}$ in this case. The last guarantee is a list of implications giving a set of separation preconditions necessary for each basic separation guarantee about the result. The expression $\otimes\mathbf{f}/\mathbf{e}$ indicates internal separation between individual elements of the resulting list which depends on the same property for both argument lists and the separation of $\mathbf{b}$ from $\mathbf{d}$.

The containment and destruction-limitation guarantees in our system play a role analogous to that of usage aspects in UAPL.

Our motivating example of an algorithm which makes use of the distinction between shallow and deep levels is an in-place update binary tree reversal program $paths_A$ converting an A-labelled binary tree to the list containing for each leaf the list of node labels as read along the path from the leaf to the root. The program can be found in the report version of this paper [5] (as well as many other details omitted here). Automatically inferred annotation for $paths_{\mathsf{L(Bool)}}$ can be found in Sect. 6.

In the operational point of view, the program $paths_A(t)$ simply reverses all the pointers in the structure of the tree t and turns its leaves into cons-cells of the resulting list. It is important here that leaves are not heap-free and their locations can be reused. Consequently, any labels on leaves are ignored.

The crucial point in the algorithm is when after recursive calls for the two branches of a tree, two intermediate result lists are appended. These are lists whose spines are disjoint but whose data are lists that share among each other extensively.

2 Underlying Language

First, we define the types, terms, typing judgements and denotational semantics of a language which is a version of LFPL without linearity and with arbitrary heap-aware recursive types. The types include (nested) recursive types, sums and products, unit and LFPL-like memory-resource types:

$$A ::= \diamond \mid {}^{\diamond}A \mid \mathsf{Unit} \mid A_1 \times A_2 \mid A_1 + A_2 \mid X \mid \mu X.A$$

where X ranges over a set of type variables. The type ${}^{\diamond}A$ has the same values as A but in the operational semantics a value would be represented by a pointer to a heap location which contains its ordinary type A representation[2].

The pre-terms feature function calls which provide for general recursion as well as pointer manipulation operators:

$$\begin{aligned} e ::= {} & x \mid f(x_1, \ldots, x_n) \mid \mathsf{let}\ x = e_1\ \mathsf{in}\ e_2 \\ \mid {} & \mathsf{unit} \mid (x_1, x_2) \mid \mathsf{match}\ x\ \mathsf{with}\ (x_1, x_2) \Rightarrow e \\ \mid {} & \mathsf{inL}\,(x) \mid \mathsf{inR}\,(x) \mid \mathsf{case}\ x\ \mathsf{of}\ \mathsf{inL}\,(x_L) \Rightarrow e_L | \mathsf{inR}\,(x_R) \Rightarrow e_R \\ \mid {} & \mathsf{fold}_{\mu X.A}\,(x) \mid \mathsf{unfold}_{\mu X.A}\,(x) \\ \mid {} & x@d \mid \mathsf{loc}\,(x) \mid \mathsf{get}\,(x) \end{aligned}$$

where x, d range over a set of variables and f over a set of function symbols.

The plain typing defines *typing judgements* of the form $\Gamma \vdash e : A$ and is standard apart from the axiom rules for the new heap-aware term constructors:

[PUT] $x : A, d : \diamond \vdash x@d : {}^{\diamond}A$ [GET] $x : {}^{\diamond}A \vdash \mathsf{get}\,(x) : A$ [LOC] $x : {}^{\diamond}A \vdash \mathsf{loc}\,(x) : \diamond$

All types in a typing judgement are closed (without free type variables).

[2] Operationally viewed, the construction ${}^{\diamond}A$ is analogous to Standard ML reference type A ref as well as to the pointer type construction `(A *)` in C. Unlike in SML, our constructor is transparent to the denotational semantics.

$$\begin{array}{ll}
\mathsf{Bool} = \mathsf{Unit} + \mathsf{Unit} & \mathsf{tt} = \mathsf{inL}\,(\mathsf{unit})\,, \mathsf{ff} = \mathsf{inR}\,(\mathsf{unit}) \\
\mathsf{L}(A) = & \mathsf{nil}_A = \mathsf{fold}_{\mathsf{L}(A)}\,(\mathsf{inL}\,(\mathsf{unit})) \\
\quad \mu X.\mathsf{Unit} + {}^{\diamond}(A \times X) & \mathsf{cons}(h,t)@d = \mathsf{fold}_{\mathsf{L}(A)}\,(\mathsf{inR}\,((h,t)@d)) \\
\mathsf{T_b}(A) = & \mathsf{leaf_b}(a)@d = \mathsf{fold}_{\mathsf{T_b}(A)}\,(\mathsf{inL}\,(a@d)) \\
\quad \mu X.{}^{\diamond}A + {}^{\diamond}(A \times (X \times X)) & \mathsf{node_b}(a,l,r)@d = \mathsf{fold}_{\mathsf{T_b}(A)}\,(\mathsf{inR}\,((a,(l,r))@d))
\end{array}$$

Fig. 2. Shortcut notation for booleans, lists and trees.

A *program* P is a finite domain partial function from function symbols to typing judgements (i.e. $P(f) = \Gamma_f \vdash e_f : A_f$) which capture the signature Γ_f, A_f and the definition e_f of each symbol f.

We consider the straightforward least fixpoint denotational semantics of types as sets in which ${}^{\diamond}(\_)$ is ignored and $\diamond$ does not play any significant role either: $[\![\diamond]\!] = \{\diamond\}$, $[\![{}^{\diamond}A]\!] = [\![A]\!]$. Denotation of programs $[\![P]\!]$ and terms $[\![e]\!]_{\eta,[\![P]\!]}$ with valuation η of free variables in e is defined as usual apart from the pointer-related constructs which are virtually ignored:

- $[\![\mathsf{loc}\,(x)]\!]_{\eta,[\![P]\!]} = \diamond$
- $[\![x@d]\!]_{\eta,[\![P]\!]} = [\![\mathsf{get}\,(x)]\!]_{\eta,[\![P]\!]} = [\![x]\!]_{\eta,[\![P]\!]} = \eta(x)$

In examples, we will use a straightforward *shortcut notation* for booleans, lists and labelled binary trees as shown in Fig. 2 (elimination terms are omitted).

3 Operational Semantics

In order to gain a good intuition for the annotated typing which will follow, let us first consider the details of the operational semantics to which the annotations will refer.

Data representation. The occurrences of ${}^{\diamond}(\_)$ in types correspond to the intended heap layout of their values. For example, compare the given type of binary trees with $\mu X.{}^{\diamond}(A + A \times (X \times X))$

Any binary tree would take the same amount of heap locations according to both of the types. A representation of a value of the above type is a pointer to a location with a value of the sum type while a value of the type in Fig. 2 is a sum value which contains a pointer to either a leaf or to a node. This difference is irrelevant at this point. It could make a difference later when a heap portion is assigned to each diamond in the type. The chosen type then allows one to divide the heap portion taken by a binary tree skeleton into two parts, one with the leaves and one with the nodes.

Let Loc be a set of *locations* which model memory addresses on a heap. We use ℓ to range over elements of Loc. Let Val be the set of (operational) *values* defined as follows:

$$v ::= \mathrm{unit} \mid \ell \mid (v_1, v_2) \mid \mathrm{inl}\,(v) \mid \mathrm{inr}\,(v)$$

An *environment* is a partial mapping S Var $\rightharpoonup$ Val from variables to operational values. A *heap* σ Loc $\rightharpoonup$ Val is a partial mapping from heap locations to operational values.

The way in which a denotational value a of type A is represented by an operational value v on a heap σ is formalised using a 4-ary relation $v \Vdash^{\sigma}_{A} a$ defined in the obvious way (see [5]), e.g.

$$\ell \Vdash^{\sigma}_{\diamond} \diamond \qquad v \Vdash^{\sigma}_{\mu X.A} a \text{ if } v \Vdash^{\sigma}_{A[\mu X.A/X]} a \qquad \ell \Vdash^{\sigma}_{^{\diamond}A} a \text{ if } \sigma(\ell) \Vdash^{\sigma}_{A} a$$

Any heap location can be used for values of any type and thus there is no general bound on the size of a heap location. For a particular program it is desirable that such a bound could be derived statically. A simple sufficient condition for this is that all recursive types mentioned in P are bounded. We call a recursive type $\mu X.A$ *bounded* if all occurrences of X in A are within some $^{\diamond}(\_)$.

Term evaluation. Using the heap representation of values, define the operational semantics by a big-step evaluation relation $S, \sigma \vdash e \rightsquigarrow v, \sigma'$ which is defined as usual (see [5]). Folding and unfolding has no operational effect.

The only heap-modifying operation is $x@d$ which overwrites a previously referenced location:

$$S, \sigma \vdash x@d \rightsquigarrow S(d), \sigma[S(d) \mapsto S(x)]$$

Heap regions. Whenever we have $v \Vdash^{\sigma}_{A} a$, we denote by $R_A(v, \sigma) \subseteq \mathrm{Dom}(\sigma)$ the straightforwardly defined complete heap region taken by v on the heap.

As we said in the introduction, we will view all recursive types as container types, i.e. we will view the region taken by a representation of a value of a recursive type $\mu X.A$ on the heap as consisting of two parts. Firstly, it is the "skeleton" layer corresponding to the locations associated with those $\diamond$'s within A which are not within any deeper recursive type. The data layer then consist of the locations associated with all the other $\diamond$'s within A. For example, in a value of the list type

$$\mathsf{L}(\mathsf{L}(\mathsf{Bool})) = \mu X.\mathsf{Unit} + {}^{\diamond}((\mu Y.\mathsf{Unit} + {}^{\diamond}(\mathsf{Bool} \times Y)) \times X)$$

there are locations holding cons-cells of the top level list corresponding to the outer-level $^{\diamond}(\_)$ and cons-cells of the element lists of the top-level list that correspond to the deeper $^{\diamond}(\_)$.

Type addresses. In order to be able to make this distinction, we need to define the *portions* of a region $R_A(v, \sigma)$ corresponding to each $\diamond$ in A. Thus we need to give formal addresses to the occurrences of $\diamond$ in a type A, as well as the occurrences of recursive types and type variables within them. These addresses are sequences of the following symbols: L and R selecting sub-terms of sum types, 1 and 2 for product types, $\overline{\mu}_X$ for descending into a recursive type binding variable X and $\overline{\diamond}$ for descending into a boxed type. The address must end with $\overline{\diamond}$, $\overline{\mu}_X$ or $\overline{X}$

denoting an occurrence of a diamond or a box subtype, recursive type binder and a type variable, respectively. A formal definition can be found in [5].

Let $\mathrm{Addr}(A)$ be the set of all such addresses in A and let $\mathrm{AddrD}(A)$, $\mathrm{AddrR}(A)$ and $\mathrm{AddrX}_X(A)$ be its subsets consisting of the addresses which end with $\overline{\diamond}$, $\overline{\mu}_X$ and $\overline{X}$, respectively, where X is a given type variable.

For $\xi \in \mathrm{Addr}(A)$ let A_ξ be the subterm of A corresponding to the address ξ. For example, using the type $A = \diamond \times \left(\mu Y.\mathsf{Unit} + {}^{\diamond}Y\right)$ we get $A_{1\overline{\diamond}} = \diamond$, $A_{2\overline{\mu}_Y} = \mu Y.\mathsf{Unit} + {}^{\diamond}Y$ and $A_{2\overline{\mu}_Y \mathsf{R}\overline{Y}} = Y$.

Unfolded types. When it holds $v \Vdash^{\sigma}_{A} a$, we may unfold the recursive types within A arbitrarily and still yielding a valid type A' for which it holds $v \Vdash^{\sigma}_{A'} a$. Moreover, by such unfoldings we can "cover" any possible shape that a value might take on the heap. Thus we will define addresses into the infinite full unfolding of A to help us with the definition of heap portions later (see [5] for a completely formal definition).

For a closed type A, let $\mathrm{Uddr}(A) = \mathrm{Addr}(A^*)$ be the set of *unfolded addresses* of A where A^* is the infinite term obtained from A by co-inductively replacing each type variable by the recursive type which binds it. Analogously, we define $\mathrm{UddrD}(A)$ and $\mathrm{UddrR}(A)$ (there are no type variables in A^*).

Moreover, we associate to every unfolded address $\xi \in \mathrm{Uddr}(A)$ its corresponding non-unfolded address $\xi^{\mathrm{F},A} \in \mathrm{Addr}(A)$ which is a certain postfix of ξ (see [5] for details).

Assuming $v \Vdash^{\sigma}_{A} a$, for every unfolded address $\xi \in \mathrm{UddrD}(A)$ we define $v_\xi(A, v, \sigma) \in \mathrm{Dom}(\sigma)$ to be the unique location, if it exists, which we get by intuitively following the address. If that location does not exist, we let $v_\xi(A, v, \sigma)$ be undefined.

For example, consider a value of the list type $\mu X.\mathsf{Unit} + {}^{\diamond}(A \times X)$. All the unfolded addresses $\overline{\mu}_X \mathsf{R}\,(\overline{\diamond}2\overline{\mu}_X)^n\,\overline{\diamond}$ for $n \in \mathbb{N}$ correspond to one and the same diamond in the type. If such a list has length m, then for every $n < m$ the address above can be associated with the heap location that contains the n-th cons-cell of this list. The rest of the addresses (i.e. for $n \geq m$) cannot be associated with any heap location.

Portions and subportions. Now we are ready to "collect" all the locations that correspond to a given occurrence of a diamond in a type. We will also define a bit more subtle sub-collection of these locations—those which are limited to a certain sub-value. The sub-value is given by an unfolded address of a recursive type which is then forbidden to unfold any more.

Definition 1. *Whenever $v \Vdash^{\sigma}_{A} a$ and $\xi \in \mathrm{AddrD}(A)$, we define the* portion *of the region $R_A(v, \sigma)$ corresponding to the unfolded type address ξ as the set*

$$P^{\xi}_{A}(v, \sigma) = \left\{ v_{\xi'}(A, v, \sigma) \,\middle|\, \xi' \in \mathrm{UddrD}(A),\ (\xi')^{\mathrm{F},A} = \xi \right\}$$

For an unfolded address $\xi_R\xi \in \mathrm{UddrD}(A)$ *with* $A_{\xi_R} = \mu X.A'$, *its* subportion along ξ_R *is the set*

$$P_A^{\xi_R\xi/\xi_R}(v,\sigma) = \left\{ v_{\xi_R\xi'}(A,v,\sigma) \,\middle|\, \xi_R\xi' \in \mathrm{UddrD}(A),\ \xi' \textit{ has no } \overline{\mu}_X,\ (\xi_R\xi')^{\mathrm{F},A} = (\xi_R\xi)^{\mathrm{F},A} \right\}$$

where A' *is assumed not to contain a* bound *type variable named* X.

Different heap portions of a value may overlap with each other in some cases.

4 Annotations

We will add several kinds of annotations to every typing judgement $\Gamma \vdash e : A$ as shown on an example in the Introduction. We will use ζ to range over names (e.g. $\mathbf{a}, \mathbf{b}, \dots$) from a set Nm.

Types. We first need to formalise a type labelled at its diamond and rec-type addresses. We define two syntactical constructions of annotated types, one suitable for generic treatment of the whole type and one suitable for accessing specific labels in a specific annotated type.

An annotated type is an expression $A\left[\frac{f_D}{f_R}\right]$ where A is a type and

$$f_D\ \mathrm{AddrD}(A) \to T, f_R\ \mathrm{AddrR}(A) \to T$$

are functions. Another syntax for the same annotated type involves placing all the images of f_D and f_R at appropriate places withing the type A (see [5]).

For example, annotated versions of the list and tree types are listed below together with their shortcut notation incorporating a simplification enforcing the use of the same names at certain addresses:

$$\mathsf{L}^{[\zeta]}(A) = \mu X^{(\zeta)}.\mathsf{Unit} + {}^{\diamond[\zeta]}(A \times X)$$
$$\mathsf{T}_{\mathsf{b}}^{[\zeta]}(A) = \mu X^{(\zeta)}.{}^{\diamond[\zeta]}A + {}^{\diamond[\zeta]}(A \times X \times X)$$

where A stands for an annotated type in this case.

Let $\alpha(\_)$ be the obvious polymorphic forgetful mapping from annotated to plain types. All the notions defined for plain types, in particular $\mathrm{Addr}(A)$ and A_ξ, extend by analogy to annotated types.

Definition 2 (Types annotated with portion names and containment). *Let* BType *and* EType *be the sets of all annotated types of the form* $A[\frac{\delta}{\rho}]$ *and* $A\left[\frac{(\delta,\gamma)}{\rho}\right]$, *respectively where*

$$\delta\ \mathrm{AddrD}(A) \to \mathrm{Nm},\ \ \rho\ \mathrm{AddrR}(A) \to \mathrm{Nm},\ \ \gamma\ \mathrm{AddrD}(A) \to \wp(\mathrm{Nm}).$$

In the other syntax for $A\left[\frac{(\delta,\gamma)}{\rho}\right]$, *a pair* $(\zeta, \{\zeta_1, \zeta_2, \dots\})$ *from the image of* (δ,γ) *will be usually written as* $[\zeta \subseteq \{\zeta_1, \zeta_2, \dots\}]$. *For* $E = A\left[\frac{(\delta,\gamma)}{\rho}\right] \in$ EType *let*

$$\mathrm{N_D}(E) = \mathrm{Ran}(\delta),\ \mathrm{N_R}(E) = \mathrm{Ran}(\rho),\ \mathrm{N_C}(E) = \bigcup_{\xi \in \mathrm{AddrD}(A)} \gamma(\xi)$$

and also $\mathrm{N}(E) = \mathrm{N_D}(E) \cup \mathrm{N_R}(E)$. *All these but* $\mathrm{N_C}()$ *apply to* $B \in$ BType *too.*

From now on, whenever using the symbol B or E (with sub- or super-scripts) let us implicitly assume that it stands for a type from BType or EType, respectively. In plain words, $\mathrm{N_D}(B)$ and $\mathrm{N_R}(B)$ are the sets of the names of all the diamonds and recursive type constructors, respectively, within B. For any $\xi \in \mathrm{AddrD}(B) \cup \mathrm{AddrR}(B)$, let ζ_ξ^B denote the name from Nm at the top of B_ξ (i.e. the image of δ or ρ). We will often leave the superscript B out from ζ_ξ^B. Analogously define ζ_ξ^E.

Definition 3 (Syntax of separation assertions).
For any type $B \in$ BType let $\mathrm{BasicSep}(B)$ be the least set containing the following basic separation assertions*:*

$$\frac{}{\bot \in \mathrm{BasicSep}(B)} \qquad \frac{\xi_1, \xi_2 \in \mathrm{AddrD}(B)}{\zeta_{\xi_1} \otimes \zeta_{\xi_2} \in \mathrm{BasicSep}(B)}$$

$$\frac{\xi_R \in \mathrm{AddrR}(B), \xi_D \in \mathrm{AddrD}(B),\ \xi_R \sqsubseteq \xi_D}{\otimes\zeta_{\xi_D} \curlywedge \zeta_{\xi_R} \in \mathrm{BasicSep}(B),\ \otimes\zeta_{\xi_D}/\zeta_{\xi_R} \in \mathrm{BasicSep}(B)}$$

Let $\otimes\zeta$ be a shortcut for $\otimes\zeta \curlywedge \zeta$.
Separation assertions $C \in \mathrm{Sep}(B)$ *are sets of basic separation assertions. Sets* $\mathrm{BasicSep}(E)$ *and* $\mathrm{Sep}(E)$ *are defined analogously.*

Definition 4 (Meaning of separation assertions).
We say that a valid representation $v \Vdash_A^\sigma a$ satisfies an assertion $C \in \mathrm{Sep}(B)$ (written as $v \Vdash_{B;C}^\sigma a$) where $\alpha(B) = A$, if:

- $\bot \notin C$
- *for every $\zeta_{\xi_1} \otimes \zeta_{\xi_2} \in C$ it holds $P_A^{\xi_1}(v, \sigma) \cap P_A^{\xi_2}(v, \sigma) = \emptyset$*
- *for every $\otimes\zeta_\xi \curlywedge \zeta_{\xi_R} \in C$ with $A_{\xi_R} = \mu X.A'$ and $\xi = \xi_R\xi'$ (implying $A_\xi = A'_{\xi'}$) it holds*

$$\forall \xi_1, \xi_2 \in \mathrm{UddrR}(A),\ \xi_1^{\mathrm{F},A} = \xi_2^{\mathrm{F},A} = \xi_R,\ \xi_1 \not\sqsubseteq \xi_2,\ \xi_2 \not\sqsubseteq \xi_1$$
$$\Longrightarrow P_A^{\xi_1\xi'/\xi_R}(v, \sigma) \cap P_A^{\xi_2\xi'/\xi_R}(v, \sigma) = \emptyset$$

- *for every $\otimes\zeta_\xi/\zeta_{\xi_R} \in C$ with $A_{\xi_R} = \mu X.A'$ and $\xi = \xi_R\xi'$ it holds*

$$\forall \xi_1, \xi_2 \in \mathrm{UddrR}(A),\ \xi_1^{\mathrm{F},A} = \xi_2^{\mathrm{F},A} = \xi_R,\ \xi_1 \neq \xi_2$$
$$\Longrightarrow P_A^{\xi_1\xi'/\xi_R}(v, \sigma) \cap P_A^{\xi_2\xi'/\xi_R}(v, \sigma) = \emptyset$$

The definition follows the informal description of separation assertions in the Introduction. The assertions $\otimes\zeta'/\zeta$ and $\otimes\zeta' \curlywedge \zeta$ correspond to internal separation *along* and *across* a recursive type (see Fig. 3).

For a list $\mathsf{L}^{[\zeta]}(A)$, the assertions $\otimes\zeta' \curlywedge \zeta$ (and $\otimes\zeta$ in particular) are void (because among any two unfolded addresses of cons-cells one is a prefix of the other) and thus trivially true in any context. We will therefore ignore them from now on. Assertions $\zeta \otimes \zeta'$ for the type $\mathsf{L}^{[\zeta]}(A)$ are not always tautological and will be kept explicitly in the $\mathrm{Sep}(B)$ sets.

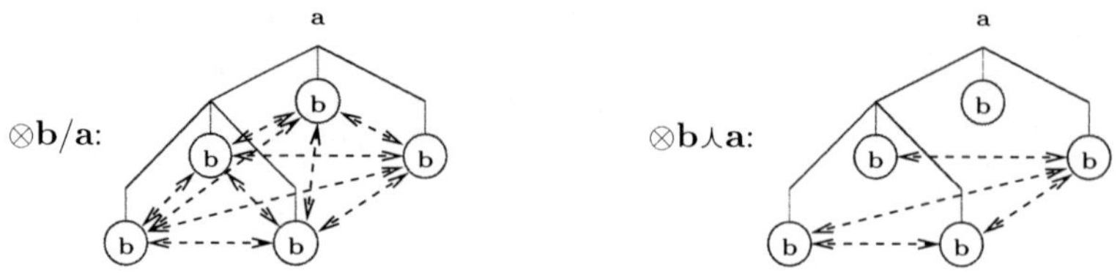

Fig. 3. Separation along and across a recursive type

Typing judgements. Before tackling the annotation of typing judgements, we need to extend some type-related notions to (annotated) typing contexts viewed as tuples of (annotated) types. We put $\mathrm{Addr}(\Gamma) = \{x\xi \mid \xi \in \mathrm{Addr}(\Gamma(x))\}$ and also define regions, name sets, separation assertions and a heap representation of value tuples and their portions and subportions related to typing contexts in a straightforward manner.

Definition 5 (Syntax of typing judgements).
The set of annotated typing judgements ZJudg *is defined as follows:*

$$\frac{\Gamma = x_1 : B_1, \dots, x_n : B_n \qquad C \in \mathrm{Sep}(\Gamma) \\ E \in \mathrm{EType},\ \mathrm{N_C}(E) \subseteq \mathrm{N_D}(\Gamma) \qquad D \subseteq \mathrm{N_D}(\Gamma) \qquad G \in \mathrm{BasicSep}(E) \to \mathrm{Sep}(\Gamma)}{(\Gamma; C \vdash e : E; D; G) \in \mathrm{ZJudg}}$$

We will represent a concrete guarantee function G by a series of statements $s \Longleftarrow G(s)$ one for each $s \in \mathrm{Dom}(G) \setminus \{\bot\}$ with $G(s) \neq \emptyset$. We will always assume that a guarantee function G maps $\bot$ (i.e. false) to itself.

Definition 6 (Meaning of typing judgements).
A typing judgement $\Gamma; C \vdash e : E; D; G$ is sound *if whenever it holds $S \Vdash^{\sigma}_{\Gamma;C} \eta$ (separation preconditions) and $S, \sigma \vdash e \rightsquigarrow v, \sigma'$ then it holds*

- *(correctness and internal separation guarantees) $v \Vdash^{\sigma'}_{E;C_G} [\![e]\!]_{\eta,[\![P]\!]}$ where $C_G = \{s \in \mathrm{BasicSep}(E) \,|\, S \Vdash^{\sigma}_{\Gamma;G(s)} \eta\}$*
- *(containment guarantees) for each $\xi \in \mathrm{AddrD}(E)$ with $[\zeta_\xi \subseteq \gamma(\xi)]$ $P^{\xi}_{E}(v, \sigma') \subseteq \bigcup_{\zeta_{x\xi'} \in \gamma(\xi)} P^{x\xi'}_{\Gamma}(S, \sigma)$*
- *(preservation guarantees) for each $x\xi \in \mathrm{AddrD}(\Gamma)$ it holds $\sigma|_P = \sigma'|_P$ where $P = P^{x\xi}_{\Gamma}(S, \sigma) \setminus \bigcup_{\zeta_{x'\xi'} \in D} P^{x'\xi'}_{\Gamma}(S, \sigma)$*

Consider the typing judgement for a predictably defined function which negates the first boolean (if existent) of every element in a list of lists of booleans:

$$\begin{aligned} & x : \mathsf{L}^{[\mathbf{a}]}(\mathsf{L}^{[\mathbf{b}]}(\mathsf{Bool})); \{\mathbf{a}\otimes\mathbf{b}, \otimes\mathbf{b}/\mathbf{a}\} \vdash \\ & \quad \mathit{negheadlist}(x) : \mathsf{L}^{[\mathbf{c}\subseteq\{\mathbf{a}\}]}(\mathsf{L}^{[\mathbf{d}\subseteq\{\mathbf{b}\}]}(\mathsf{Bool})); \{\mathbf{a}, \mathbf{b}\};\ \ \mathbf{c}\otimes\mathbf{d} \Longleftarrow \{\mathbf{a}\otimes\mathbf{b}\} \\ & \qquad\qquad\qquad\qquad\qquad\qquad\qquad\qquad\quad \otimes\mathbf{d}/\mathbf{c} \Longleftarrow \{\otimes\mathbf{b}/\mathbf{a}\} \end{aligned}$$

The precondition means that the elements of the list cannot overlap on the heap with each other. With our system, this is the only way to guarantee the correctness, i.e. that no boolean on the heap would be negated more than once.

One might argue that, in fact, only portion **b** gets modified and portion **a** does not change at all. This is true and follows from the fact that lists in portion **b** are modified in-place and moreover their resulting cons-cells occupy exactly the same positions as the original ones. Nevertheless, we have no means of expressing such a condition in our assertion language and when deriving the annotation in a purely compositional manner we have to assume that the lists could have been modified in a way which would change their reference location and thus modify portion **a** too.

5 Typing Rules

For the complete set of annotated typing rules see [5]. In this paper we show only some representative cases which are listed in Fig. 4. For illustration, consider the following instantiation of the fairly complex typing rule [FOLD] for binary trees labelled with diamonds:

$$\begin{array}{c} x : {}^{\diamond[\mathbf{a}]}(\diamond^{[\mathbf{b}]}) + {}^{\diamond[\mathbf{c}]}(\diamond^{[\mathbf{d}]} \times \mathsf{T}_{\mathsf{b}}^{[\mathbf{e}]}(\diamond^{[\mathbf{f}]}) \times \mathsf{T}_{\mathsf{b}}^{[\mathbf{g}]}(\diamond^{[\mathbf{h}]})); \emptyset \vdash \\ \qquad \mathsf{fold}_{\mathsf{T}_{\mathsf{b}}(\diamond)}(x) : \mathsf{T}_{\mathsf{b}}^{[\mathbf{i}\subseteq\{\mathbf{a},\mathbf{c},\mathbf{e},\mathbf{g}\}]}(\diamond^{[\mathbf{j}\subseteq\{\mathbf{b},\mathbf{d},\mathbf{f},\mathbf{h}\}]}); \emptyset; \end{array}$$

$$\begin{array}{rlrl} \mathbf{i}\otimes\mathbf{j} \Longleftarrow & \{\mathbf{a},\mathbf{c},\mathbf{e},\mathbf{g}\}\otimes\{\mathbf{b},\mathbf{d},\mathbf{f},\mathbf{h}\} & \otimes\mathbf{j}\curlywedge\mathbf{i} \Longleftarrow & \{\mathbf{f}\otimes\mathbf{h}\} \\ \otimes\mathbf{j}/\mathbf{i} \Longleftarrow & \{\mathbf{f}\otimes\mathbf{h}\} \cup \{\mathbf{b},\mathbf{d}\}\otimes\{\mathbf{f},\mathbf{h}\} & \otimes\mathbf{i} \Longleftarrow & \{\mathbf{e}\otimes\mathbf{g}\} \end{array}$$

In the [FUNC] rule we assume an annotated program P with $P(f) = \Gamma_f; C_f \vdash e_f : B_f; D_f; G_f$ where Γ_f contains exactly the variables $y_1, \ldots, y_{n_f}$ in this order.

Given a result type E, we define the minimal default guarantee $\mathrm{G_M}(E)$ for E as $[\zeta_1\otimes\zeta_2 \Longleftarrow Z_1\otimes_{\mathrm{D}}Z_2]_{\zeta_1\otimes\zeta_2\in\mathrm{BasicSep}(E)}$ where $[\zeta_1\subseteq Z_1]$ and $[\zeta_2\subseteq Z_2]$ are annotations in E and $\otimes_{\mathrm{D}}$ is one of the following two similar operations:

$$\begin{aligned} Z_1\otimes Z_2 &= \{\zeta_1\otimes\zeta_2 \mid \zeta_1 \in Z_1, \zeta_2 \in Z_2\} \\ Z_1\otimes_{\mathrm{D}} Z_2 &= \{\zeta_1\otimes\zeta_2 \mid \zeta_1 \in Z_1, \zeta_2 \in Z_2, \zeta_1 \neq \zeta_2\} \end{aligned}$$

The difference is significant only in the [UNFOLD] rule where every portion is being split into several pieces some pairs of which might be disjoint depending on the internal separation preconditions.

Several rules use a "copied" type $E \in \mathrm{E_Y}(B)$ for (part of) the result type which declares the one-to-one containment of E portions within corresponding B portions. With it go its default guarantees $\mathrm{G_Y}(E, B)$ which state that every basic separation assertion of E relies on the corresponding assertion for B.

In [FOLD], we construct the result type by a point-wise union of types which differ only in their containment sets. In several places also various rely-guarantee functions are merged via a point-wise union of images.

The notation $\Gamma_1 \wedge \Gamma_2$ stands for the merge of the context Γ_1, Γ_2. The use of this notation contains the implicit condition that every variable that appears in both contexts has the same type and portion name annotations in both of them. To be able to apply rules that contain $\Gamma_1 \wedge \Gamma_2$ as intended, we might need to apply the [RENAME] rule first.

[FUNC]

$$\frac{\Gamma = \Gamma_f[x_1/y_1, \ldots, x_{n_f}/y_{n_f}]}{\Gamma; C_f \vdash f(x_1, \ldots, x_{n_f}) : E_f; D_f; G_f}$$

[RENAME]

$$\frac{\Gamma; C \vdash e : E; D; G \quad \tau\ \mathrm{N}(\Gamma) \to \mathrm{Nm}}{\Gamma[\tau]; C[\tau] \vdash e : E[\tau]; D[\tau]; G[\tau]}$$

[LET]

$$\frac{\Gamma_1; C_1 \vdash e_1 : A\left[\tfrac{\delta,\gamma}{\rho}\right]; D_1; G_1 \quad \Gamma_2, x : A\left[\tfrac{\delta}{\rho}\right]; C_2 \vdash e_2 : E; D_2; G_2}{\Gamma_1 \wedge \Gamma_2; C_1 \cup T(C_2) \cup \big(D_1 \otimes \mathrm{N_D}(\Gamma_2)\big) \vdash \mathsf{let}\ x = e_1\ \mathsf{in}\ e_2 : E[\gamma/\delta]; D_1 \cup D_2[\gamma/\delta]; T \circ G_2}$$

where $T = C \mapsto \big((C \setminus \mathrm{BSx})[\gamma/\delta]\big) \cup G_1\big(C \cap \mathrm{BSx}\big)$ and $\mathrm{BSx} = \mathrm{BasicSep}(A\left[\tfrac{\delta}{\rho}\right])$.

[PUT]

$$\frac{E \in \mathrm{E_Y}(^{\diamond[\zeta]}B)}{x : B, d : \diamond^{[\zeta]}; \{\zeta\}\otimes\mathrm{N_D}(B) \vdash x@d : E; \{\zeta\}; \mathrm{G_Y}(E, {}^{\diamond[\zeta]}B)}$$

[GET]

$$\frac{E \in \mathrm{E_Y}(B)}{x : {}^{\diamond[\zeta]}B; \emptyset \vdash \mathsf{get}\,(x) : E; \emptyset; \mathrm{G_Y}(E, B)}$$

[LOC]

$$\frac{}{x : {}^{\diamond[\zeta]}B; \emptyset \vdash \mathsf{loc}\,(x) : \diamond^{[\zeta' \subseteq \{\zeta\}]}; \emptyset; \emptyset}$$

[PAIR]

$$\frac{E \in \mathrm{E_Y}(B_1 \times B_2)}{x_1 : B_1, x_2 : B_2; \emptyset \vdash (x_1, x_2) : E; \emptyset; \mathrm{G_Y}(E, B_1 \times B_2)}$$

[PAIR-ELIM]

$$\frac{\Gamma, x_1 : B_1, x_2 : B_2; C \vdash e : E; D; G}{\Gamma, x : B_1 \times B_2; C \vdash \mathsf{match}\ x\ \mathsf{with}\ (x_1, x_2) \Rightarrow e : E; D; G}$$

[FOLD]

$$\frac{B = (A[\mu X.A/X])\left[\tfrac{\delta}{\rho}\right] \quad B' = A\left[\tfrac{\delta|_{\mathrm{AddrD}(A)}}{\rho|_{\mathrm{AddrR}(A)}}\right] \qquad E = E' \cup \bigcup_{\xi\overline{X} \in \mathrm{AddrX}_X(A)} E_{\xi\overline{\mu}_X} \quad E' \in \mathrm{E_Y}(B') \quad E_\xi \in \mathrm{E_Y}(B_\xi)}{x : B; \emptyset \vdash \mathsf{fold}_{\mu X.A}\,(x) : E; \emptyset; \mathrm{G_M}(E) \cup G \cup G' \cup G''}$$

where
$G = \mathrm{G_Y}(E', B') \cup \bigcup_{\xi\overline{X} \in \mathrm{AddrX}_X(A)} \mathrm{G_Y}(E_{\xi\overline{\mu}_X}, B_{\xi\overline{\mu}_X})$
$G' = \left[\otimes\zeta_\xi \curlywedge \zeta_{\overline{\mu}_X} \Longleftarrow \left\{\zeta_{x\xi'\xi}\otimes\zeta_{x\xi''\xi} \,\middle|\, \xi'\overline{X}, \xi''\overline{X} \in \mathrm{AddrX}_X(A)\right\}\right]_{\xi\in\mathrm{AddrD}(A)}$ and
$G'' = \left[\otimes\zeta_\xi/\zeta_{\overline{\mu}_X} \Longleftarrow \mathrm{ditto} \cup \left\{\zeta_{x\xi'\xi}\otimes\zeta_{x\xi} \,\middle|\, \xi'\overline{X} \in \mathrm{AddrX}_X(A)\right\}\right]_{\xi\in\mathrm{AddrD}(A)}$

[UNFOLD]

$$\frac{E \in \mathrm{E_Y}(B[\mu X^{(\zeta)}.B/X])}{x : \mu X^{(\zeta)}.B; \emptyset \vdash \mathsf{unfold}_{\mu X.\alpha(B)}\,(x) : E; \emptyset; \mathrm{G_Y}(E, B[\mu X^{(\zeta)}.B/X]) \cup G' \cup G''}$$

where
$G' = [\zeta_{x\xi'\xi}\otimes\zeta_{x\xi''\xi} \Longleftarrow \{\otimes\zeta_\xi \curlywedge \zeta_{\overline{\mu}_X}\}]_{\xi\in\mathrm{AddrD}(A), \xi'\overline{X}, \xi''\overline{X}\in\mathrm{AddrX}_X(A)}$ and
$G'' = \left[\zeta_{x\xi'\overline{\mu}_\xi}\otimes\zeta_{x\xi} \Longleftarrow \{\otimes\zeta_\xi/\zeta_{\overline{\mu}_X}\}\right]_{\xi\in\mathrm{AddrD}(A), \xi'\overline{X}\in\mathrm{AddrX}_X(A)}$

Fig. 4. Example annotated typing rules

When δ AddrD$(A) \rightarrow$ Nm and γ AddrD$(A) \rightarrow \wp(\mathrm{Nm})$, then we derive a set-valued substitution $\gamma/\delta\ \mathrm{Nm} \rightharpoonup \wp(\mathrm{Nm})$ as $\zeta \mapsto \gamma(\delta^{-1}(\zeta))$. This is used in the [LET] rule.

Notice that, unlike in other extensions of LFPL, there is no side condition in [LET]. Instead, the illegal cases yield an inconsistent assertion set C, i.e. one containing $\bot$ or $\zeta\otimes\zeta$ for some portion name ζ. Similarly, when an additional separation guarantee s is unachievable, its rely-assertion $G(s)$ would contain an unsatisfiable basic separation assertion.

6 Correctness and Inference

A typing rule is sound if whenever all the premises of an instance of the rule are sound, then so is the conclusion. Assuming that all the typing judgements $P(f)$ are sound, it is possible to prove that all the typing rules in our system as listed in [5] are correct. The proofs are conceptually simple but technical due to the complexity of the annotations. See the report for a sample proof of correctness for [LET].

Given a well-typed and well-annotated program P, for every term e over P we can perform the inference of its unannotated type deterministically as usual but we can also infer its annotation at the same time. Moreover, we can derive an annotation which does not reuse any name for two different portions or recursive type constructors. Such a derivable annotation of e is unique modulo renaming of portions and recursive type constructors.

Given an unannotated well-typed program P, we can infer its strongest derivable annotation by iteratively inferring weaker and weaker annotations starting from the strongest annotation: $C_f = \emptyset$, $E_f = A_f[\frac{-,\xi\mapsto\emptyset}{-}]$, $D_f = \emptyset$ and $G_f = [s \Longleftarrow \emptyset]$ for all f in the program.

We have studied this process in a more abstract setting in [6] and showed that it will always terminate with the strongest derivable annotation for P provided that all the typing rules are monotone. In [5] we have outlined the proof that our rules are monotone.

Example. The following is the strongest correct annotated judgement for the binary tree reversal algorithm mentioned in the Introduction:

$$x : \mathsf{T}_{\mathsf{b}}^{[\mathbf{a}]}(\mathsf{L}^{[\mathbf{b}]}(\mathsf{Bool})); \{\mathbf{a}\otimes\mathbf{b}, \otimes\mathbf{a}\} \vdash$$
$$\mathit{paths}_{\mathsf{L(Bool)}}(x) : \mathsf{L}^{[\mathbf{c}\subseteq\{\mathbf{a}\}]}(\mathsf{L}^{[\mathbf{d}\subseteq\{\mathbf{a}\}]}(\mathsf{L}^{[\mathbf{e}\subseteq\{\mathbf{b}\}]}(\mathsf{Bool}))); \{\mathbf{a}\};$$
$$\begin{array}{llllll} \mathbf{c}\otimes\mathbf{d} \Longleftarrow & \{\otimes\mathbf{a}\} & \mathbf{c}\otimes\mathbf{e} \Longleftarrow & \{\mathbf{a}\otimes\mathbf{b}\} & \mathbf{d}\otimes\mathbf{e} \Longleftarrow & \{\mathbf{a}\otimes\mathbf{b}\} \\ \otimes\mathbf{d}/\mathbf{c} \Longleftarrow & \{\bot, \otimes\mathbf{a}\} & \otimes\mathbf{e}/\mathbf{c} \Longleftarrow & \{\bot, \otimes\mathbf{b}/\mathbf{a}\} & \otimes\mathbf{e}/\mathbf{d} \Longleftarrow & \{\otimes\mathbf{b}/\mathbf{a}\} \end{array}$$

Due to the lack of parametric polymorphism, the example cannot involve types with an unbound variable. Therefore we have substituted a simple non-heap-free type $\mathsf{L(Bool)}$ in place of the usual type meta-variable A.

The annotation of *paths* tells us about the function that in order to work correctly, it needs the input tree to have non-overlapping skeleton ($\otimes\mathbf{a}$) disjoint from the labels ($\mathbf{a}\otimes\mathbf{b}$). The cons-cells of the result list ($\mathbf{c}$) as well as the cons-cells of the lists in it ($\mathbf{d}$) are constructed solely from the skeleton of the argument tree ($\mathbf{c}\subseteq\{\mathbf{a}\}$, $\mathbf{d}\subseteq\{\mathbf{a}\}$) which is also the only argument portion being destroyed. Thus the labels are preserved.

Notice that the elements of the result list cannot be guaranteed to be separated from each other ($\otimes\mathbf{d}/\mathbf{c}$) under any preconditions (indicated by $\perp$). Consequently, the same holds for the labels when viewed from the result lists' cons-cells ($\otimes\mathbf{e}/\mathbf{c}$). Labels viewed from the individual elements' cons-cells can be guaranteed to be separated from each other ($\otimes\mathbf{e}/\mathbf{d}$) provided that the labels of the argument tree are separated from each other ($\otimes\mathbf{b}/\mathbf{a}$).

7 Conclusion

We have enhanced the typing system of [3] following the natural idea that data and skeleton should be treated separately. This lead to a, perhaps surprisingly, complicated system of annotations. Despite using complex notation, the system provides efficient annotation inference.

A possible direct application of various extensions of LFPL including ours is to generate complex efficient imperative algorithms automatically from easy-to-verify functional programs. At the same time, a formal proof of correctness or permissible resource consumption could be generated fairly easily. For some problem areas the present language might not be sufficiently expressive. It should be possible, though, to adapt the annotation system in this paper to other language features including higher order functions and other kinds of resource-aware typings than that of LFPL.

For example, it could be used for functional languages without the $\diamond$ types that have a fairly straightforward heap-based evaluation strategy. Recently Hofmann, Jost [7,8] and Kirlı developed a statical analysis of memory consumption in a $\diamond$-free version of LFPL with implicit memory allocation and deallocation. Combining the present system with the above is a subject of current research.

There is an implementation of the present typing system adapted to a language with ML-style datatypes as an add-on to an LFPL compiler by Robert Atkey. The compiler can be tested through a web interface [9]. It was used to generate the examples in this paper. To make the typing more practicable, a faily sophisticted system for generating helpful error messages is under development.

Related work. The theory of *Shapely types* by Jay [10] treats certain types as containers explicitly. Thus the idea to treat differently the skeleton (i.e. shape) and data layers of data types is common to both this and Jay's work. Nevertheless, the tools and the purpose of shape theory are very different from ours and do not seem to be applicable to the present system. Shape theory is formulated in categorical terms and studies mainly shape polymorphism and statical shape analysis on the semantical level. In contrast, we develop a statical analysis of non-interference in a particular operational semantics with in-place update.

The idea to indicate the level of boxing for representing values of recursive types using a "boxing" type constructor ($^{\diamond}(\_)$ in our case) has been also used by

Shao [11] using the notation `Boxed(_)`. This idea appears often in the context of the strongly-typed intermediate language for compilation of functional languages *FLINT* [12].

The use of pre- and postconditions for certifying in-place update with sharing is not new. Recent work includes *Alias types* [13,14] which have been designed to express when heap is manipulated type-safely in a typed assembly language. Alias types express properties about heap layout and could be considered as representations of our assertions. Unfortunately, they cannot express that two heap location variables *may* have the same value. In an alias type, two different location variables always take different values.

Separation Logic [15] may serve a similar purpose as alias types but is also applicable to higher-level languages. The logic is in Hoare style where postconditions cannot refer to the original state and cannot therefore capture effects. If the logic is adapted so that it can refer to the original state, we believe that it is expressive enough to encode our assertions. To find a natural way of doing this is a promising ongoing research.

Our system bears some resemblance to the *Region inference* of Tofte and Talpin [16]. They also associate regions with types and infer certain effects made on these regions. One major difference is that they also infer where deallocation of regions should take place. This means that they derive special annotation in *terms* which influences their evaluation. An analogy to this in the context of LFPL might be the inference of diamond typed arguments ($@d$) within programs mentioned above. Another major difference is that the regions of Tofte and Talpin are mutually disjoint and cannot be explicitly overwritten.

The usage aspects of UAPL which are subsumed in the present system are similar to *use types* in linear logic [17] and also to *passivity* [18] within syntactic control of interference [19,20]. These aspects can be also viewed as *effects* (in the sense of [21]) somewhat similar to get and put of [16]. Our system can be viewed as inferring a combination of effects on finely specified regions and separation assertions in the style of Reynolds.

Acknowledgements. This research has been supported by the EPSRC grant GR/N28436/01 and by the EC Fifth Framework Programme project Mobile Resource Guarantees. The author is grateful to David Aspinall, Robert Atkey and Martin Hofmann for discussion and comments on this work.

References

1. Hofmann, M.: A type system for bounded space and functional in-place update. Nordic Journal of Computing **7** (2000) 258–289
2. Hofmann, M.: The strength of non size-increasing computation. In: 29th ACM SIGPLAN-SIGACT Symposium on Principles of Programming Languages (POPL'02). (2002) 260–269
3. Aspinall, D., Hofmann, M.: Another type system for in-place update. In Métayer, D.L., ed.: Programming Languages and Systems, Proceedings of 11th European Symposium on Programming, Springer-Verlag (2002) 36–52 Lecture Notes in Computer Science 2305.

4. Appel, A.W.: Modern Compiler Implementation in Java. Cambridge University Press (1998)
5. Konečný, M.: LFPL with types for deep sharing. Technical Report EDI-INF-RR-157, LFCS, Division of Informatics, University of Edinburgh (2002)
6. Konečný, M.: Typing with conditions and guarantees for functional in-place update. In: TYPES 2002 Workshop, Nijmegen, Proceedings. (2003) to appear.
7. Hofmann, M., Jost, S.: Static prediction of heap space usage for first-order functional programs. In: 30th ACM SIGPLAN-SIGACT Symposium on Principles of Programming Languages (POPL '03). (2003) 185–197
8. Jost, S.: Static prediction of dynamic space usage of linear functional programs. Master's thesis, Technische Universität Darmstadt, Fachbereich Mathematik (2002)
9. Atkey, R., Konečný, M.: A prototype LFPL compiler (frontend DEEL). An interface available at: `http://www.dcs.ed.ac.uk/home/resbnd/prototypes/by_Robert_Atkey/deel/` (2003)
10. Jay, C.B.: A semantics for shape. Science of Computer Programming **25** (1995) 251–283
11. Shao, Z.: Flexible representation analysis. In: Proc. 1997 ACM SIGPLAN International Conference on Functional Programming (ICFP'97), Amsterdam, The Netherlands (1997) 85–98
12. League, C., Shao, Z.: Formal semantics of the FLINT intermediate language. Technical Report Yale-CS-TR-1171, Department of Computer Science, Yale University (1998)
13. Smith, F., Walker, D., Morrisett, G.: Alias types. In Smolka, G., ed.: 9th European Symposium on Programming (ESOP'2000), Springer-Verlag (2000) 366–381 Lecture Notes in Computer Science 1782.
14. Walker, D., Morrisett, G.: Alias types for recursive data structures. In: Types in Compilation 2000. (2001) 177–206 Lecture Notes in Computer Science 2071.
15. Reynolds, J.C.: Separation logic: A logic for shared mutable data structures. In: Proceedings of 17th Annual IEEE Symposium on Logic in Computer Science (LICS'02), Copenhagen, Denmark (2002) 55–74
16. Tofte, M., Talpin, J.P.: Region-based memory management. Information and Computation **132** (1997) 109–176
17. Guzmán, J.C., Hudak, P.: Single-threaded polymorphic lambda calculus. In: Proceedings of the Fifth Annual IEEE Symposium on Logic in Computer Science. (1990) 333–343
18. O'Hearn, P.W., Power, A.J., Takeyama, M., Tennent, R.D.: Syntactic control of interference revisited. Theoretical Computer Science **228** (1999) 211–252
19. Reynolds, J.C.: Syntactic control of interference. In: Proceedings of the 5th ACM SIGACT-SIGPLAN Symposium on Principles of Programming Languages (POPL), ACM Press (1978) 39–46
20. Reynolds, J.C.: Syntactic control of interference, part 2. In Ausiello, G., Dezani-Ciancaglini, M., Rocca, S.R.D., eds.: Automata, Languages and Programming, 16th International Colloquium, Springer-Verlag (1989) 704–722 Lecture Notes in Computer Science 372.
21. Lucassen, J.M., Gifford, D.K.: Polymorphic effect systems. In: Proceedings of the 15th ACM SIGPLAN-SIGACT symposium on Principles of programming languages, ACM Press (1988) 47–57

A Fully Abstract Bidomain Model of Unary FPC

Jim Laird

COGS, University of Sussex, UK
jiml@cogs.susx.ac.uk

Abstract. We present a fully abstract and effectively presentable model of unary FPC (a version of FPC with lifting rather than lifted sums) in a category of *bicpos and continuous and stable functions*. We show universality for the corresponding model of unary PCF, and then show that this implies full abstraction for unary FPC. We use a translation into this metalanguage to show that the "canonical" bidomain model of the lazy λ-calculus (with seqential convergence testing) is fully abstract.

1 Introduction

Over the past twenty-five years a great deal of effort has been invested in the search for fully abstract models of sequential functional languages such as PCF, and more generally, for a semantic characterisation of sequentiality. One of the earliest candidates is Berry's notion of *stability* [6,7], which is particularly attractive in that it is a simple, order-theoretic concept, fitting in well with Scott semantics and many of the subsequent developments in domain theory, via the notion of *bidomain*, which combines both orders on a single structure. It is well known that the bidomain model of PCF is not fully abstract, nor even sequential. Moreover, by showing that observational equivalence in finitary PCF is undecidable, Loader [19] has demonstrated that it is impossible to give a fully abstract model of PCF or its recursively typed extension FPC which is *effectively presentable*. However, we will present a fully abstract and effectively presentable bidomain model of a recursively typed metalanguage which is sufficiently expressive to give fully abstract interpretations of untyped sequential languages such as the lazy λ-calculus [1], showing that a significant amount of sequential functional computation can be characterised precisely using stability.

Fully abstract models of PCF and FPC have been described using game semantics [5,16,21] and Kripke logical relations [22,27]. However, their failure to be effectively presentable is a serious limitation. For instance, the games models give an elegant characterisation of *definability* in PCF but tell us little about observational equivalence, since they are quotiented by an *intrinsic equivalence* which is undecidable even for finitary elements. This seems to be a problem specific to purely functional languages; extensions of PCF with non-functional features such as control and state have been given fully abstract and effectively presentable models using sequential algorithms [10] and games [3].

Loader's result thus raises the question: which sequential functional languages can be modelled fully abstractly and effectively? PCF certainly does not provide

M. Hofmann (Ed.): TLCA 2003, LNCS 2701, pp. 211–225, 2003.

the only reasonable notion of sequentiality; the strongly stable [8] or "sequentially realisable" functionals [20] are an effectively representable, sequential and semantically natural alternative, although they do not appear to correspond (in terms of a full abstraction result) to a natural programming language. In another development, Loader has proved that observational equivalence in unary PCF (which has a single ground type containing a single value) is decidable [18]. However, this language seems very inexpressive and lacks a corresponding effectively presented, *syntax-independent* and fully abstract model.

This paper will show that the bidomain model of unary PCF is *universal* — i.e. every element is the denotation of a term — using a listing algorithm devised by Schmidt-Schauß [28]. In fact, we shall show that every order-extensional and sequential model of unary PCF is universal, and that up to isomorphism there are precisely two order-extensional models — the Scott model and the bidomain model. We will then use these results to prove full abstraction for a unary version of FPC — that is, FPC with lifting rather than a general sum constructor. The addition of recursive types represents a great increase in expressive power. For example, the *lazy λ-calculus* [1] has a natural interpretation in the category of bidomains via (fully abstract) translation into unary FPC [21]. The former has been proposed as a prototypical *untyped* lazy functional programming language, much as PCF is the prototypical typed functional language. The existence of non-sequential functions in the continuous function model of the lazy λ-calculus causes a failure of full abstraction [4] just as `parallel_or` does for the corresponding model of PCF, whilst fully abstract models of the calculus have been constructed using similar techniques to those used for PCF — e.g. by collapsing a category of games under its intrinsic preorder, without determining whether this order can be characterised effectively [2,14]. But we are able to use our full abstraction result for unary FPC to show that, unlike PCF, the fully abstract model of the lazy λ-calculus (with sequential convergence testing) is effectively presentable.

2 Unary PCF

Our starting point is the language unary PCF. This is a simply-typed λ-calculus over a single ground type unit containing two constants $\top, \bot$, with a "sequential composition" operation — from M : unit and N : unit, form $M \wedge N$: unit. The equational theory of unary PCF is generated by $\beta\eta$-equality together with the equations $\bot \wedge M = M \wedge \bot = \bot$ and $\top \wedge M = M \wedge \top = M$ (from which it is straightforward to derive a normal-order operational semantics).

We may think of unary PCF as the node $\Lambda_\bot^\top(\wedge)$ in the hierarchy of simply-typed λ-calculi with $\top$ and $\bot$ which is depicted in Fig. 1. Each language in the hierarchy contains some combination of $\wedge$ and and its dual $\vee$ ("parallel composition") where $M \vee \top = \top \vee M = \top$ and $\bot \vee M = M \vee \bot = M$. As is the case for `parallel_or` in PCF, $\vee$ is not definable in unary PCF, and it can break equivalences which hold there — e.g. $F_1 = \lambda f.f\,(f\,\bot\top)\,(f\,\top\bot)$ is equivalent to

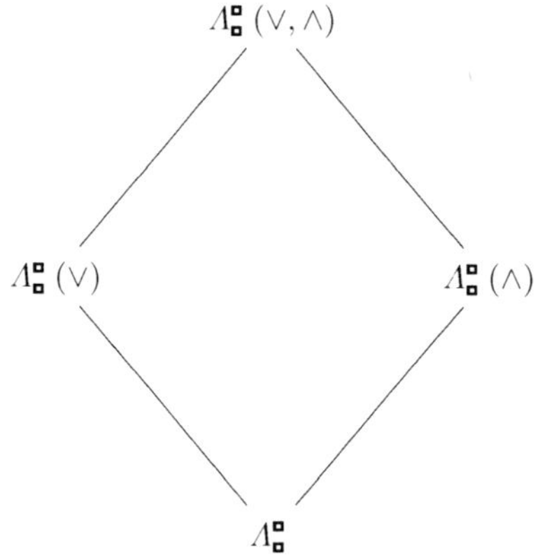

Fig. 1. A hierarchy of simply-typed λ-calculi

$F_2 = \lambda f.(f\ \bot\bot)$ in $\Lambda^{\top}_{\bot}(\wedge)$, but not in $\Lambda^{\top}_{\bot}(\vee,\wedge)$, since $F_1\ \lambda xy.x \vee y = \top$ and $F_2\ \lambda xy.x \wedge y = \bot$.

We shall assume that a "standard" model of unary PCF consists of a cartesian closed category in which unit is interpreted as an object Σ containing two "points" $\top, \bot : \mathbf{1} \to \Sigma$, and sequential composition as a morphism $\wedge : \Sigma \times \Sigma \to \Sigma$, such that the equational theory is satisfied.

Definition 1. *For any standard model $\mathcal{M}$ we can define a relation $\preceq_\tau$ on the points $e : \mathbf{1} \to [\![\tau]\!]_{\mathcal{M}}$ for each τ:*
$e \preceq_{\mathsf{unit}} e'$ if $e = \bot$ or $e' = \top$,
$f \preceq_{\sigma \Rightarrow \tau} g$ if $\forall e \preceq_\sigma e'.f\ e \preceq_\tau g\ e'$.
Then $\mathcal{M}$ is order-extensional *if $\preceq$ is a partial order at each type. An order-extensional model has greatest and least elements $\top$ and $\bot$ at each type.*

The simplest standard model of unary PCF is the "Scott Model" in which unit is interpreted as the partial order $\bot \sqsubseteq \top$ and function-types as sets of monotone functions (ordered extensionally). Because this contains an interpretation of $\vee$ it is also a model of $\Lambda^{\top}_{\bot}(\vee,\wedge)$, and hence neither universal nor fully abstract for unary PCF. The critical condition for definability, which fails in the Scott model, is Milner-Vuillemin *sequentiality.*

Definition 2. *Let $\mathcal{M}$ be a standard, order-extensional model of unary PCF. An element $f \in [\![\tau_1 \Rightarrow \ldots \tau_n \Rightarrow \mathsf{unit}]\!]_{\mathcal{M}}$ is* i-strict *if for every sequence of elements $e_1 \in [\![\tau_1]\!]_{\mathcal{M}}, \ldots, e_{i-1} \in [\![\tau_{i-1}]\!]_{\mathcal{M}}, e_{i+1} \in [\![\tau_{i+1}]\!]_{\mathcal{M}}, \ldots e_n \in [\![\tau_n]\!]_{\mathcal{M}}$, we have $f\ e_1 \ldots e_{i-1} \perp e_{i+1} \ldots e_n = \bot$.*

Definition 3. *A standard model of unary PCF is* sequential *if for all types τ, every $f \in [\![\tau]\!]$ is either constant or i-strict for some i.*

It is not the case that sequentiality for models of boolean PCF implies universality. However, this property is sufficient to establish universality for *any* standard and order-extensional model of unary PCF. As shown in Appendix 1, we can prove this by recasting Schmidt-Schauß's algorithm for listing representatives of observational equivalence classes of unary PCF [28] as a proof of universality for sequential models.

Theorem 1. *Any standard model of unary PCF which is order-extensional and sequential is universal.*

Theorem 1 also entails that the order-extensional collapses of many other models of unary PCF — including hypercoherences and strongly stable functions [8], sequential algorithms [11] and games models [16,5] — are also fully abstract. In fact, we can use the theorem to show that there are precisely two order-extensional models of unary PCF.

Lemma 1. *Suppose $\mathcal{M}$ is an order-extensional standard model of $\Lambda_\bot^\top(\wedge)$ which is not sequential. Then it is a model of $\Lambda_\bot^\top(\vee,\wedge)$.*

Proof. If $\mathcal{M}$ is not sequential, then there exists a type $\tau = \tau_1 \Rightarrow \ldots \Rightarrow \tau_n \Rightarrow \mathsf{unit}$ and an element $e \in [\![\tau]\!]_{\mathrm{n}}$ which is neither a constant nor i-strict for some i. We show by induction on n that $\vee$ is definable in unary PCF extended with a term M_e denoting e, and all of the equations which are sound with respect to interpretation in $\mathcal{M}$. Given a term $M : \mathsf{unit}$, we write $\mathsf{c}_\tau(M)$ for $\lambda x : \tau_1 \ldots x_n : \tau_n.M : \tau$, where $x_1, \ldots, x_n \notin FV(M)$.

The case $n = 1$ is trivial, since every monotone function of one argument is either strict or constantly $\top$. For the inductive case there are two possibilities. If $(e\,\top)\,\bot\ldots\bot = \top$ then $\lambda xy.x \vee y$ is definable as $\lambda xy.(M_e\;\mathsf{c}_{\tau_1}(x)\;(\mathsf{c}_{\tau_2}(y))\ldots(\mathsf{c}_{\tau_n}(y)))$, since by definition e is not strict in its first argument. Otherwise $(e\top)\,\bot\ldots\bot = \bot$. Then $e\,\top$ is not constant, and nor is it strict in any argument. Hence $\vee$ is definable from $M_e\,\top$ and hence from M_e.

Lemma 2. *The Scott model of $\Lambda_\bot^\top(\vee,\wedge)$ is universal.*

Proof. This is a simplified version of the proof that the finite elements of the Scott model of PCF plus `parallel_or` are all definable [24].

Corollary 1. *Every order-extensional model of unary PCF is isomorphic either to the bidomain model, or to the Scott model.*

Proof. Every order-extensional model is either sequential or not. In the former case it is universal and therefore isomorphic to the bidomain model. In the latter case it contains parallel composition by Lemma 1, and by Lemma 2 is a universal model for $\Lambda_\bot^\top(\vee,\wedge)$.

3 Bicpos

We shall first briefly recall the definitions of the cartesian closed category of bicpos from [6,12] (bidomains are a full subcategory of this CCC, but the difference is not important here). An alternative would be to construct the corresponding *bistructure* model — bistructures of a certain form give a precise and concrete representation of bidomains [12], allowing domain equations to be solved up to equality, and giving a decomposition into a model of classical linear logic.

Definition 4. *A bicpo [6,12] is a triple* $\langle D, \sqsubseteq, \leq \rangle$ *consisting of a set* D *with two partial orders* $\sqsubseteq, \leq$ *on it such that:*

- $(D, \sqsubseteq)$ *(the extensional order) is a cpo with least element and a continuous binary greatest lower bound operator* $\sqcap$.
- $(D, \leq)$ *(the stable order) is a cpo with the same least element as* $(D, \sqsubseteq)$*, and such that the identity from* $(D, \leq)$ *to* $(D, \sqsubseteq)$ *is continuous.*
- *The operator* $\sqcap$ *is continuous with respect to* $\leq$*, from which it follows that if* x, y *are bounded above in* $(D, \leq)$*, then* $x \sqcap y$ *is a greatest lower bound for* x, y *in* $(D, \leq)$.
- *For all* $\sqsubseteq$*-directed subsets* $S, S' \subseteq D$*, if for all* $s \in S$ *and* $s' \in S'$ *there exist* $t \in S$ *and* $t' \in S'$ *such that* $s \sqsubseteq t$ *and* $s' \sqsubseteq t'$ *and* $t \leq t'$ *then* $\bigsqcup S \leq \bigsqcup S'$.

Definition 5. *Let* $\mathcal{B}$ *be the category of bicpos and functions which are continuous with respect to* $\sqsubseteq$ *and continuous and stable with respect to* $\leq$*, where a function* f *from* $\langle D, \sqsubseteq, \leq \rangle$ *to* $\langle D', \sqsubseteq', \leq' \rangle$ *is* stable *if it preserves binary glbs, i.e. if* $x \uparrow y$ *then* $f(x \sqcap y) = f(x) \sqcap f(y)$.

Proposition 1. $\mathcal{B}$ *is cartesian closed.*

Proof. The terminal object in $\mathcal{B}$ is the one element bicpo. Cartesian product of bicpos is defined by taking the pairwise orderings on the product of the underlying sets. For bicpos $(D, \sqsubseteq_D, \leq_D)$ and $(D', \sqsubseteq_{D'}, \leq_{D'})$ the function-space is the bicpo over the set of continuous and stable functions from D to D', with $f \sqsubseteq_{D \Rightarrow D} g$ if $f\,x \sqsubseteq_{D'} g\,x$ for all $x \in D$. The *stable order* $\leq_{D \Rightarrow D'}$ is defined:

$$f \leq_{D \Rightarrow D'} g \text{ if for all } x, y \in D, x \leq_D y \Longrightarrow f(x) = f(y) \wedge g(x).$$

Thus the category of bicpos — with unit interpreted as the two-element "Sierpinski space" bicpo, and sequential composition as $\pi_1 \sqcap \pi_2 : \Sigma \times \Sigma \Rightarrow \Sigma$ — is easily checked to be an instance of a a standard and order-extensional model of unary PCF. Although the bicpo model of full PCF is not sequential [6], unary PCF is different.

Lemma 3. *The bicpo model of unary PCF is sequential.*

Proof. Suppose $f : [\![\tau_1]\!]_{\mathcal{B}} \times \ldots \times [\![\tau_n]\!]_{\mathcal{B}} \rightarrow \Sigma$ is stable and monotone but not i-strict for any i.
Then (by monotonicity) for each i, $f\langle \top_1, \ldots \top_{i-1}, \bot_i, \top_{i+1}, \ldots, \top_n \rangle = \top$. So $f\langle \bot, \ldots, \bot \rangle = f \bigsqcap_{i \leq n} \langle \top_1, \ldots, \top_{i-1}, \bot, \top_{i+1}, \ldots, \top_n \rangle$
$= \top$. Hence by monotonicity, $f\,\langle e_1, \ldots, e_n \rangle = \top$ for all $e_1 \ldots e_n$.

Corollary 2. *The bicpo model of unary PCF is universal and fully abstract.*

4 Unary FPC

Our meta-language — unary FPC, or UFPC — for solving recursive domain equations is just a call-by-name version of Plotkin's Fixed Point Calculus [26] in which the sum has been restricted to the corresponding unary operation — *lifting*. We assume a set of type variables, and generate the *types* of UFPC as follows:
$\tau ::= X \mid \tau_\bot \mid \tau \times \tau \mid \tau \Rightarrow \tau \mid \mu X.\tau$
The terms of UFPC are derived according to the following grammar:
$M ::= x \mid \mathsf{up}(M) \mid \mathsf{let}\ M = \mathsf{lift}(x)\ \mathsf{in}\ M \mid$
$\quad \langle M, M\rangle \mid \pi_1(M) \mid \pi_2(M) \mid$
$\quad \lambda x.M \mid M\ M \mid \mathsf{intro}_{\mu X.\tau}(M) \mid \mathsf{elim}(M).$

Table 1. Typing judgements for terms of UFPC

$$\frac{}{\Theta;\Gamma,x{:}\tau \vdash x{:}\tau}\ FV(\tau) \subseteq \Theta$$

$$\frac{\Theta;\Gamma,x{:}\sigma \vdash M{:}\tau}{\Theta;\Gamma \vdash \lambda x.M{:}\sigma \Rightarrow \tau} \qquad \frac{\Theta;\Gamma \vdash M{:}\sigma \Rightarrow \tau \quad \Theta;\Gamma \vdash N{:}\sigma}{\Theta;\Gamma \vdash M\ N{:}\tau}$$

$$\frac{\Theta;\Gamma \vdash M_1{:}\tau_1 \quad \Theta;\Gamma \vdash M_2{:}\tau_2}{\Theta;\Gamma \vdash \langle M_1,M_2\rangle{:}\tau_1 \times \tau_2} \qquad \frac{\Theta;\Gamma \vdash M{:}\tau_1 \times \tau_2}{\Theta;\Gamma \vdash \pi_i(M){:}\tau_i}\ i = 1,2$$

$$\frac{\Theta;\Gamma \vdash M{:}\tau}{\Theta;\Gamma \vdash \mathsf{up}(M){:}\tau_\bot} \qquad \frac{\Theta;\Gamma,x{:}\sigma \vdash M{:}\tau \quad \Theta;\Gamma \vdash N{:}\sigma_\bot}{\Theta;\Gamma \vdash \mathsf{let}\ N = \mathsf{lift}(x)\ \mathsf{in}\ M{:}\tau}$$

$$\frac{\Theta;\Gamma \vdash M{:}\tau[X \mapsto \mu X.\tau]}{\Theta;\Gamma \vdash \mathsf{intro}_{\mu X.\tau}(M){:}\mu X.\tau} \qquad \frac{\Theta;\Gamma \vdash M{:}\mu X.\tau}{\Theta;\Gamma \vdash \mathsf{elim}(M){:}\tau[X \mapsto \mu X.\tau]}$$

The well-typed terms are derived in a dual context $\Theta;\Gamma$, according to Table 1. A "big-step" call-by-name operational semantics of UFPC is given in Table 2.

We write $M \Downarrow$ if there exists M' such that $M \Downarrow M'$. For any UFPC type τ there is a divergent term $\Omega_\tau : \tau = \lambda x : S.(\mathsf{elim}(x)\ x)\ \mathsf{intro}_S(\lambda x : S.(\mathsf{elim}(x)\ x))$, where $S = \mu X.(X \Rightarrow \tau)$. We define a standard notion of contextual approximation on UFPC terms based on observation at lifted types.

Definition 6. *For terms $M, N : \tau$, $M \lesssim N$ if for all closing contexts $C[\cdot] : \sigma_\bot$, $C[M] \Downarrow$ implies $C[N] \Downarrow$.*
A semantics of UFPC *is inequationally* fully abstract *if for all closed terms M, N, $M \lesssim N$ if and only if $[\![M]\!] \sqsubseteq [\![N]\!]$.*

The bicpo model of UFPC is obtained via the biordered analogues of the standard constructions from cpos [29], so all that is required is to check that these yield suitable bicpos.

In $\mathcal{B}$, lifting is modelled in standard fashion by adjoining a new element which is least with respect to both orders.

Table 2. Operational semantics of UFPC

$$\frac{}{\lambda x.M \Downarrow \lambda x.M} \qquad \frac{M \Downarrow \lambda x.P \quad P[N/x] \Downarrow P'}{M\ N \Downarrow P'}$$

$$\frac{}{\mathsf{up}(M) \Downarrow \mathsf{up}(M)} \qquad \frac{M \Downarrow \mathsf{up}(M') \quad N[M'/x] \Downarrow P}{\mathsf{let}\ M = \mathsf{lift}(x)\ \mathsf{in}\ N \Downarrow P}$$

$$\frac{}{\langle M_1, M_2\rangle \Downarrow \langle M_1, M_2\rangle} \qquad \frac{M \Downarrow \langle M_1, M_2\rangle \quad M_i \Downarrow N}{\pi_i(M) \Downarrow N}\ i = 1, 2$$

$$\frac{}{\mathsf{intro}_{\mu X.\tau}(M) \Downarrow \mathsf{intro}_{\mu X.\tau}(M)} \qquad \frac{M \Downarrow \mathsf{intro}_{\mu X.\tau}(N) \quad N \Downarrow P}{\mathsf{elim}(M) \Downarrow P}$$

Definition 7. *Given a bicpo $(D, \sqsubseteq, \leq)$ the* lift *of D is the bicpo $(D_\perp, \sqsubseteq_\perp, \leq_\perp)$, where:*

- $D_\perp = (D \times \{0\}) \cup \{\perp\}$,
- $\forall e \in D, \perp \sqsubseteq_\perp \langle e, 0\rangle$ *and* $\forall e, e' \in D$, $\langle e, 0\rangle \sqsubseteq_\perp \langle e', 0\rangle$ *iff* $e \sqsubseteq e'$,
- $\forall e \in D, \perp \leq_\perp \langle e, 0\rangle$ *and* $\forall e, e' \in D$, $\langle e, 0\rangle \leq_\perp \langle e', 0\rangle$ *iff* $e \leq e'$.

For each A we have a function $\mathtt{up}_A : A \to A_\perp$, defined $\mathtt{up}_A(x) = \langle(x, 0)\rangle$, and for each function $f : A \to B$, we have $f^* : A_\perp \to B$, where $f^*(\perp) = \perp$, and $f^*(\langle e, 0\rangle) = f(e)$.
These two operations suffice to define the action of lifting as a strong monad, and to interpret the typing rules in UFPC.

To interpret recursive types of UFPC we can solve domain equations for bicpos by straightforward extension of the standard methodology for solving them for cpos.

Definition 8. *Let A, B be bicpos, then $A \trianglelefteq B$ if there exist morphisms $\mathtt{inj} : A \to B, \mathtt{proj} : B \to A$ such that $\mathtt{inj};\mathtt{proj} = \mathtt{id}_A$, $\mathtt{proj};\mathtt{inj} \leq \mathtt{id}_B$, and $\mathtt{proj};\mathtt{inj} \sqsubseteq \mathtt{id}_B$.*

The one-element bicpo I is least with respect to $\trianglelefteq$.

Proposition 2. *Every bounded ω-chain in $\mathcal{B}$ has a least upper bound in $\mathcal{B}$.*

Proof. Given a chain of bicpos, $\langle D_1, \sqsubseteq_1, \leq_1\rangle \trianglelefteq \langle D_2, \sqsubseteq_2, \leq_2\rangle \trianglelefteq \ldots$ with injections and projections $\mathtt{inj}_i : D_i \to D_{i+1}, \mathtt{proj}_i : D_{i+1} \to D_i$, define $\bigsqcup_{i\in\omega} D_i = (D, \sqsubseteq, \leq)$, where $D = \{x \in \Pi_{i\in\omega} D_i \mid \forall i \in \omega.\mathtt{proj}_i(\pi_{i+1}(x)) = \pi_i(x)\}$
and $x \sqsubseteq y$ if for all i, $\pi_i(x) \sqsubseteq_i \pi_i(y)$, and $x \leq y$ if for all i, $\pi_i(x) \leq_i \pi_i(y)$.

Definition 9. *Given a continuous functor, $F : \mathcal{B}^{OP} \times \mathcal{B} \to \mathcal{B}$, let $F^\Delta = \bigsqcup_{i\in\omega} F_i$, where $F_0 = I, F_{i+1} = F(F_i, F_i)$.*

The limit/co-limit coincidence theorem holds in $\mathcal{B}$, with a proof which follows that for cpos (see [25,23]). Thus a type $\tau(X_1, \ldots, X_n)$ of UFPC can be interpreted as a functor from $(\mathcal{B}^{OP} \times \mathcal{B})_1 \times \ldots \times (\mathcal{B}^{OP} \times \mathcal{B})_n$ to $\mathcal{B}$ using the function-space, product, lifting and fixed point constructions (see [15,21]).

We use the method of *formal-approximation relations* as described in [23] to prove that the bicpo model of UFPC is *computationally adequate.*

Proposition 3. *For all programs $M : \tau_\bot$ of* UFPC*, $M \Downarrow$ iff $[\![M]\!]_\mathcal{B} \neq \bot$.*

Proof. The proof is detailed but follows closely the examples set in [23,21].

We can deduce that inequational soundness holds in our model by the standard argument from computational adequacy.

Proposition 4. *For closed M, N: $[\![M]\!]_\mathcal{B} \sqsubseteq [\![N]\!]_\mathcal{B}$ implies $M \lesssim N$.*

Our model is also *effectively presentable*; we shall not give details here as, again, the definitions are essentially the same as those for domains [25]. So it remains to prove inequational completeness — if $M \lesssim N$ then $[\![M]\!]_\mathcal{B} \sqsubseteq [\![N]\!]_\mathcal{B}$.

5 Full Abstraction for UFPC

We shall prove that universality holds for certain *finite* types of UFPC, using our result for unary PCF, and then show that these provide enough approximants to prove full abstraction. We first identify a typed fragment which corresponds to unary PCF via a simple translation.

Definition 10. *Let $*$ be the "terminal type" $\mu X.X$. Define the translation of unary PCF types:*
$(\![\mathsf{unit}]\!) = *_\bot$, *and* $(\![\sigma \Rightarrow \tau]\!) = (\![\sigma]\!) \Rightarrow (\![\tau]\!)$,
and of unary PCF terms:

- $(\![\bot]\!) = \Omega_{*_\bot}$, $(\![\top]\!) = \mathsf{up}(\Omega_*)$,
- $(\![\lambda x.M]\!) = \lambda x.(\![M]\!)$, $(\![M\ N]\!) = (\![M]\!)\ (\![M]\!)$,
- $(\![M \wedge N]\!) = \mathsf{let}\ (\![M]\!)\ =\ \mathsf{lift}(x)\ \mathsf{in}\ (\![N]\!)$ *$(x \notin FV(N))$.*

Lemma 4. *For any type τ of unary PCF, we have $[\![(\![\tau]\!)]\!]_\mathcal{B} = [\![\tau]\!]_\mathcal{B}$ and for any $M : \tau$, we have $[\![M]\!]_\mathcal{B} = [\![(\![M]\!)]\!]_\mathcal{B}$.*

The PCF types of UFPC are those of the form $(\![\tau]\!)$ for some unary PCF type τ.

Corollary 3. *Universality holds at all PCF types of* UFPC.

It is straightforward to use currying and preservation of products by $\Rightarrow$ to show that universality is preserved when the unary types are extended with products (and the terminal type).

Lemma 5. *Universality holds at all PCF types with products. That is, all types generated by the following grammar:*
$\tau ::= * \mid *_\bot \mid \tau \Rightarrow \tau \mid \tau \times \tau$.

Proof. For any (non-empty) PCF type with products, τ, we can find PCF types $\sigma_1, \ldots \sigma_n$ and a term $\mathsf{IN} : \tau \Rightarrow \sigma_1 \times \ldots \times \sigma_n$ of UFPC which denotes an isomorphism. So given $e \in [\![\tau]\!]_\mathcal{B}$, we can find terms $M_1 : \sigma_1, \ldots M_n : \sigma_n$ such that $e = [\![\mathsf{IN}(\langle M_1, \ldots, M_n \rangle)]\!]_\mathcal{B}$.

We shall now show that universality holds at all the finite closed types of UFPC.

Definition 11. *Let the* finite types *of* UFPC *be the types generated from* $*$ *and the type-variables by the lifting, product and function-space constructions: i.e.* $\tau ::= * \mid X \mid \tau_\bot \mid \tau \Rightarrow \tau \mid \tau \times \tau$.
The finite closed *types are those finite types without free variables, i.e. those types generated from* $*$ *by the lifting, product and function-space constructions.*

Definition 12. *Let* σ, τ *be* UFPC *types. A definable retraction from* σ *to* τ *is a pair of (closed)* UFPC *terms* $\mathsf{Inj} : \sigma \Rightarrow \tau$ *and* $\mathsf{Proj} : \tau \Rightarrow \sigma$ *such that* $[\![\lambda x.\mathsf{Proj}\ (\mathsf{Inj}\ x)]\!]_{\mathcal{B}} = [\![\lambda x.x]\!]_{\mathcal{B}}$ *and* $[\![\lambda x.\mathsf{Inj}\ (\mathsf{Proj}\ x)]\!]_{\mathcal{B}} \sqsubseteq [\![\lambda x.x]\!]_{\mathcal{B}}$. *We shall write* $\mathsf{Inj} : \sigma \trianglelefteq \tau : \mathsf{Proj}$ *(or just* $\sigma \trianglelefteq \tau$*) if there is a definable retraction from* σ *to* τ.

Lemma 6. *For any finite type* $\tau(X)$*, and definable retraction* $\mathsf{Inj} : \sigma_1 \trianglelefteq \sigma_2 : \mathsf{Proj}$ *there is a definable retraction* $\tau(\mathsf{Inj}) : \tau(\sigma_1) \trianglelefteq \tau(\sigma_2) : \tau(\mathsf{Proj})$.

Proof. We define $\tau(\mathsf{Inj}), \tau(\mathsf{Proj})$ by a straightforward structural induction on τ: e.g. $(\sigma \Rightarrow \tau)(\mathsf{Inj}) = \lambda f x.\tau(\mathsf{Inj})\,(f\,(\sigma(\mathsf{Proj})\,x))$, $(\sigma \Rightarrow \tau)(\mathsf{Proj}) = \lambda f x.\tau(\mathsf{Proj})\,(f\,(\sigma(\mathsf{Inj})\,x))$.

To reduce universality at finite types to universality at the unary types with products we make use of the following observation.
Lemma 7. *For any type* τ *there is a definable retraction* $\tau_\bot \trianglelefteq \tau \times *_\bot$.

Proof. Define $\mathtt{in}_A : A_\bot \Rightarrow A \times \Sigma$:
$\mathtt{in}_A(\bot) = \langle \bot, \bot \rangle$, $\mathtt{in}_A(\langle e, 0 \rangle) = \langle e, \top \rangle$.
Then $\mathtt{in}_{[\![\tau]\!]} = [\![\lambda x.\mathsf{let}\ x = \mathsf{lift}(y)\ \mathsf{in}\ \langle y, \mathsf{up}(\Omega_*)\rangle]\!]_{\mathcal{B}}$.

Define $\mathtt{out}_A : A \times \Sigma \Rightarrow A_\bot$:
$\mathtt{out}_A(\langle e, \bot \rangle) = \bot$, $\mathtt{out}_A(\langle e, \top \rangle) = \langle e, 0 \rangle$.
Then $\mathtt{out}_\tau = [\![\lambda x.\mathsf{let}\ \pi_2(x) = \mathsf{lift}(y)\ \mathsf{in}\ \mathsf{up}(\pi_1(x))]\!]_{\mathcal{B}}$.

For a finite type τ, let $\tau^\diamondsuit$ be the unary PCF type with products obtained by replacing each lifting in τ with $\_ \times *_\bot$ — i.e.

- $*^\diamondsuit = *$,
- $(\sigma \Rightarrow \tau)^\diamondsuit = \sigma^\diamondsuit \Rightarrow \tau^\diamondsuit$,
- $(\sigma \times \tau)^\diamondsuit = \sigma^\diamondsuit \times \tau^\diamondsuit$,
- $(\tau_\bot)^\diamondsuit = \tau^\diamondsuit \times *_\bot$.

Proposition 5. *If* τ *is finite there is a definable retraction* $\mathsf{Inj}_\tau : \tau \trianglelefteq \tau^\diamondsuit : \mathsf{Proj}_\tau$.

Proof. This is by repeated application of Lemmas 6 and 7.

Corollary 4. *Universality holds at all finite types.*

Proof. Let τ be a finite type and e an element of $[\![\tau]\!]_{\mathcal{B}}$. By Lemma 5, there exists $M : \tau^\diamondsuit$ such that $\langle [\![\mathsf{Inj}_\tau]\!]_{\mathcal{B}}, e \rangle; \mathrm{App} = [\![M]\!]_{\mathcal{B}}$.
Therefore $[\![\mathsf{Proj}_\tau\ M]\!]_{\mathcal{B}} = \langle [\![\lambda x.(\mathsf{Proj}_\tau\ (\mathsf{Inj}_\tau\ x)]\!]_{\mathcal{B}}, e \rangle; \mathrm{App} = e$ as required.

We shall now show that universality at finite types implies full abstraction for UFPC.

Lemma 8. *Full abstraction holds at all finite types — i.e. for all closed $M, N : \sigma$ where σ is closed and finite, $[\![M]\!]_{\mathcal{B}} \sqsubseteq [\![N]\!]_{\mathcal{B}}$ if and only if $M \lesssim N$.*

Proof. Given $M, N : \tau$ such that $[\![M]\!]_{\mathcal{B}} \not\sqsubseteq [\![N]\!]_{\mathcal{B}}$, the map $\alpha_M \in [\![\tau \Rightarrow *_\perp]\!]_{\mathcal{B}}$ (defined $\alpha_M(x) = \top \iff [\![M]\!]_{\mathcal{B}} \sqsubseteq x$) is continuous and stable, and hence definable as a term $L : \tau \Rightarrow *_\perp$ such that $[\![L\ M]\!]_{\mathcal{B}} = \top$ and $[\![L\ N]\!]_{\mathcal{B}} = \perp$. Hence by computational adequacy, $L\ M \Downarrow$ and $L\ N \not\Downarrow$ and $M \not\lesssim N$ as required.

We can now define, for each UFPC type τ, a chain of finite types with definable retractions $\tau_1 \trianglelefteq \tau_2 \trianglelefteq \ldots$ which approximates τ. Given a type $\tau(X)$, define $\tau^i(X)$ for $i \geq 1$ by $\tau^1(X) = \tau(X)$, $\tau^{i+1}(X) = \tau(\tau^i(X))$.

Definition 13. *We define τ_i as follows:*
$X_i = X$, $(\tau_\perp)_i = (\tau_i)_\perp$, $(\sigma \Rightarrow \tau)_i = \sigma_i \Rightarrow \tau_i$, $(\sigma \times \tau)_i = \sigma_i \times \tau_i$, $(\mu X.\tau(X))_0 = *$, $(\mu X.\tau(X))_{i+1} = (\tau_i)^i(*)$,

Lemma 9. *For each $i \in \omega$, $\tau_i \trianglelefteq \tau$.*

Proof. We define embedding/projection pairs $\mathsf{Inj}_i(\tau) : \tau_i \trianglelefteq \tau : \mathsf{Proj}_i(\tau)$ for each τ by induction on type-structure.

Lemma 10. *For each type τ, $\bigsqcup_{i\in\omega}[\![\tau_i]\!]_{\mathcal{B}} \cong [\![\tau]\!]_{\mathcal{B}}$,*

Proof. By induction on type-structure, e.g. : $[\![\mu X.\tau]\!]_{\mathcal{B}} = \bigsqcup_{i\in\omega}[\![\tau^i(\mu X.X)]\!]_{\mathcal{B}} = \bigsqcup_{i\in\omega}(\bigsqcup_{j\in\omega}[\![(\tau^i(\mu X.X))_j]\!]_{\mathcal{B}}) = \bigsqcup_{i\in\omega}[\![(\tau^i(\mu X.X))_i]\!]_{\mathcal{B}} = \bigsqcup_{i\in\omega}[\![(\tau_i)^i(\mu X.X)]\!]_{\mathcal{B}} = \bigsqcup_{i\in\omega}[\![(\mu X.\tau)_i]\!]_{\mathcal{B}}$.

Theorem 2. *The bicpo model of* UFPC *is fully abstract.*

Proof. By Proposition 4 it is sufficient to show that if $M, N : \tau$ are closed terms such that $[\![M]\!]_{\mathcal{B}} \not\sqsubseteq [\![N]\!]_{\mathcal{B}}$, then $M \not\lesssim N$. By Lemma 10 $[\![\tau]\!]_{\mathcal{B}} \cong \bigsqcup_{i\in\omega}[\![\tau_i]\!]_{\mathcal{B}}$, and hence $[\![M]\!]_{\mathcal{B}} = \bigsqcup_{i\in\omega}[\![\mathsf{Inj}_i(\tau)\ (\mathsf{Proj}_i(\tau)\ M)]\!]_{\mathcal{B}} \not\sqsubseteq \bigsqcup_{i\in\omega}[\![\mathsf{Inj}_i(\tau)\ (\mathsf{Proj}_i(\tau)\ N)]\!]_{\mathcal{B}} = [\![N]\!]_{\mathcal{B}}$.
Hence there exists i such that $[\![\mathsf{Inj}_i(\tau)\ (\mathsf{Proj}_i(\tau)\ M)]\!]_{\mathcal{B}} \not\sqsubseteq [\![\mathsf{Inj}_i(\tau)\ (\mathsf{Proj}_i(\tau)\ N)]\!]_{\mathcal{B}}$, and so $[\![\mathsf{Proj}_i(\tau)\ M]\!]_{\mathcal{B}} \not\sqsubseteq [\![(\mathsf{Proj}_i(\tau)\ M]\!]_{\mathcal{B}}$. By Lemma 8, this implies that $(\mathsf{Proj}_i(\tau)\ M) \not\lesssim (\mathsf{Proj}_i(\tau)\ N)$ and hence $M \not\lesssim N$ as required.

5.1 The Lazy λ-Calculus

As a corollary of our full abstraction result for UFPC, we can now show that the canonical model of the lazy λ-calculus with "sequential convergence testing" (λl_{C}) obtained by solving $D \cong (D \Rightarrow D)_\perp$ in $\mathcal{B}$, is fully abstract. A fully abstract model of λl_{C} based on a quotiented category of games was described in [2]. (More recently, di Gianantonio [14] has refined the the games model to give a full abstraction result for the lazy λ-calculus without C, but, again, for a quotiented model.)

The terms of λl_{C} are given by the grammar:
$M ::= x \mid M\ M \mid \lambda x.M \mid \mathsf{C}M$.

The "big-step" operational semantics is defined:

$$\frac{}{\lambda x.P \Downarrow \lambda x.P} \qquad \frac{M \Downarrow \lambda x.P \quad P[N/x] \Downarrow V}{M\ N \Downarrow V} \qquad \frac{M \Downarrow \lambda x.P}{\mathsf{C}M \Downarrow \lambda x.x}$$

Let D be the least solution of $D \cong (D \Rightarrow D)_\bot$ in $\mathcal{B}$. Then $D = [\![T]\!]_{\mathcal{B}}$, where T is the UFPC type $\mu X.(X \Rightarrow X)_\bot$, and we can give the standard interpretation of λl_{C} in D by translation into UFPC: for a λl_{C} term M let $[\![x_1, \ldots, x_n \vdash M]\!]_{\mathcal{B}} = [\![\ x_1 : T, \ldots x_n : T \vdash (\![M]\!)]\!]_{\mathcal{B}}$, where:

- $(\![x]\!) = x$,
- $(\![\lambda x.M]\!) = \mathsf{intro}_D(\mathsf{up}(\lambda x.(\![M]\!)))$,
- $(\![M\ N]\!) = \mathsf{let\ elim}(\![M]\!)\ =\ \mathsf{lift}(x)\ \mathsf{in}\ (x\ (\![N]\!))\quad (x \notin FV(N))$,
- $(\![\mathsf{C}M]\!) = \mathsf{let\ elim}((\![M]\!))\ =\ \mathsf{lift}(x)\ \mathsf{in}\ \lambda y.y$.

Full abstraction of this translation has been proved by McCusker [21] (using a fully abstract games model).

Proposition 6 (McCusker [21]). *The translation* $(\![\_]\!)$ *from* λl_{C} *into* UFPC *is* fully abstract — *i.e. for every pair of* λl_{C}*-terms* M, N, $M \lesssim N$ *if and only if* $(\![M]\!) \lesssim (\![N]\!)$.

Corollary 5. *The bicpo model of* λl_{C} *is fully abstract.*

6 Conclusions

We have shown that bidomains do capture sequentiality in unary PCF, unary FPC and the lazy λ-calculus, and that this is sufficient to prove full abstraction for all three languages. We have also shown that there are only two distinct order-extensional interpretations of unary PCF.

Which other sequential languages can be given fully abstract and effectively presentable models using bidomains ? One obvious example is the untyped call-by-value λ-calculus which, like λl_{C}, can be fully abstractly translated into UFPC [21]. We can also give a fully abstract "pre-bidomain" semantics of call-by-value UFPC itself. However, the crucial feature of these models as far as full abstraction and effective presentability are concerned appears not to be the restriction to "unary" sums *per se*, but to domains which are bounded above in the extensional ordering.

The results described here are part of a wider ranging attempt to characterise a hierarchy of effective and fully abstract models of functional languages based on extensional models with additional orderings such as stability (as in bidomains), co-stability (the order-theoretic dual of stability) and bistability [17]. We can give universal biordered models of the hierarchy of languages described in Figure 1. Biorders can also be used to model a variety of richer languages, including SPCF [9], and its extension with separated sums (with additional $\top$-element) and recursive types, as well as effects such as non-determinism and first-class continuations. Full abstraction results for these languages can be reduced to our fundamental universality results.

Acknowledgement. This research was supported by the UK EPSRC. I would like to thank Guy McCusker, Thomas Streicher and Pierre-Louis Curien for useful discussions of these topics.

References

1. S. Abramsky. The lazy λ-calculus. In D. Turner, editor, *Research Topics in Functional Programming*, pages 65–117. Addison Wesley, 1990.
2. S. Abramsky and G. McCusker. Games and full abstraction for the lazy λ-calculus. In *Proceedings of the Tenth Annual IEEE Symposium on Logic in Computer Science*, pages 234–243. IEEE Computer Society Press, 1995.
3. S. Abramsky and G. McCusker. Linearity, Sharing and State: a fully abstract game semantics for Idealized Algol with active expressions. In P.W. O'Hearn and R. Tennent, editors, *Algol-like languages*. Birkhauser, 1997.
4. S. Abramsky and C.-H. L. Ong. Full abstraction in the lazy λ-calculus. *Information and Computation*, 105:159–267, 1993.
5. S. Abramsky, R. Jagadeesan and P. Malacaria. Full abstraction for PCF. *Information and Computation*, 163:409–470, 2000.
6. G. Berry. Stable models of typed λ-calculi. In *Proceedings of the 5th International Colloquium on Automata, Languages and Programming*, number 62 in LNCS, pages 72–89. Springer, 1978.
7. G. Berry. *Modèles complètement adéquats et stables des lambda-calculs typés.* PhD thesis, Université Paris 7, 1979.
8. A. Bucciarelli and T. Ehrhard. A theory of sequentiality. *Theoretical Computer Science*, 113:273–292, 1993.
9. R. Cartwright and M. Felleisen. Observable sequentiality and full abstraction. In *Proceedings of POPL '92*, 1992.
10. R. Cartwright, P.-L. Curien and M. Felleisen. Fully abstract semantics for observably sequential languages. *Information and Computation*, 1994.
11. P.-L. Curien. *Categorical combinators, sequential algorithms and functional programming.* Progress in Theoretical Computer Science series. Birkhauser, 1993.
12. P.-L. Curien, G. Winskell, and G. Plotkin. Bistructures, bidomains and linear logic. In *Milner Festschrift.* MIT Press, 1997.
13. N. Dershowitz and Z. Manna. Proving termination with multiset orderings. *Communications of the ACM*, 22:465–476, 1979.
14. P. di Gianantonio. Games semantics for the pure lazy λ-calculus. In S. Abramsky, editor, *Proceedings of TLCA '01*, number 2044 in LNCS. Springer, 2001.
15. M. Fiore and G. Plotkin. An axiomatisation of compuationally adequate domain thoeretic models of FPC. In *Proceedings of LICS '94*, pages 92–102. IEEE Computer Society Press, 1994.
16. J. M. E. Hyland and C.-H. L. Ong. On full abstraction for PCF: I, II and III. *Information and Computation*, 163:285–408, 2000.
17. J. Laird. Bistability and bisequentiality. Available from the author's home page, 2002.
18. R. Loader. Unary PCF is decidable. *Theoretical Computer Science*, 206, 1998.
19. R. Loader. Finitary PCF is undecidable. *Annals of Pure and Applied Logic*, 2000.
20. J. Longley. The sequentially realizable functionals. Technical Report ECS-LFCS-98-402, LFCS, Univ. of Edinburgh, 1998.

21. G. McCusker. *Games and full abstraction for a functional metalanguage with recursive types.* PhD thesis, Imperial College London, 1996.
22. P.W. O'Hearn and R. Tennent. Kripke logical relations and PCF. *Information and Computation*, 120(1):107–116, 1995.
23. A. M. Pitts. Relational properties of domains. *Information and Computation*, 127:66–90, 1996.
24. G. Plotkin. LCF considered as a programming language. *Theoretical Computer Science*, 5:223–255, 1977.
25. G. Plotkin. Postgraduate lecture notes in advanced domain theory (incorporating the 'Pisa notes'). Available from `http://www.dcs.ed.ac.uk/home/gdp/publications/`, 1981.
26. G. Plotkin. Lectures on predomains and partial functions, 1985. Notes for a course given at the Center for the study of Language and Information, Stanford.
27. J. Riecke and A. Sandholm. A relational account of call-by-value sequentiality. In *Proceedings of the Twelfth Annual Symposium on Logic in Computer Science, LICS '97.* IEEE Computer Society Press, 1997.
28. M. Schmidt-Schauß. Decidability of behavioural equivalence in unary PCF. *Theoretical Computer Science*, 216:363–373, 1999.
29. M. Smyth and G. Plotkin. The category-theoretic solution of recursive domain equations. *SIAM Journal on Computing*, 11(4):761–783, 1982.

Appendix 1: Universality for Unary PCF

We shall now describe the algorithm given in [28] for deriving representatives of operational equivalence classes of unary PCF, and use it to show that any sequential and order-extensional standard model of unary PCF is universal. The algorithm is based on giving combinators which allow definability of the i-strict functions at each type to be reduced to universality at types which are strictly smaller with respect to a nested multi-set ordering.

For a unary PCF type $\tau = \tau_1 \Rightarrow \tau_2 \Rightarrow \ldots \Rightarrow \tau_n \Rightarrow \mathsf{unit}$, define the *arity* of τ, $\mathrm{ar}(\tau)$, to be n. For τ with $\mathrm{ar}(\tau) > 0$ we shall write τ_i for the type of the ith argument to τ and $\tau_{i,j}$ for the type of the jth argument to τ_i.

A *well-founded* order on unary PCF types, based on the *nested multiset ordering* [13] can be defined as follows.

Definition 1 *Let* $\delta(\mathsf{unit}) = \{\mathsf{unit}\}$, *and* $\delta(\tau) = \{\delta(\tau_i) \mid i \leq \mathrm{ar}(\tau)\}$, *if* $\mathrm{ar}(\tau) > 0$. *Then* $\sigma \ll \tau$ *if* $\delta(\sigma) \ll_{ms} \delta(\tau)$, *where* $\ll_{ms}$ *is the nested multi-set ordering [13] over the single element* unit. *(Recall that this can be be defined as the least relation satisfying:*
If $M \neq \mathsf{unit}$ *then* $\mathsf{unit} \ll_{ms} M$.
If $M \neq M'$ *and* $\forall x \in M - M'.\exists y \in M' - M.x \ll_{ms} y$ *then* $M \ll_{ms} M'$.*)*

The following is then an instance of the general result in [13].

Proposition 1 (Dershowitz and Manna) $\ll$ *is well-founded.*

From each type τ we derive families of subtypes $\{\pi(\tau)^{i,j} \mid i \leq \mathrm{ar}(\tau), j \leq \mathrm{ar}(\tau_i)\}$ which are strictly less than τ with resepect to $\ll$.

Definition 2 *Given* $\tau = \tau_1 \Rightarrow \ldots \Rightarrow \tau_n \Rightarrow \mathsf{unit}$, *define:*
$\pi(\tau)^{i,0} = \tau_1 \Rightarrow \ldots \tau_{i-1} \Rightarrow \tau_{i+1} \Rightarrow \ldots \Rightarrow \tau_n \Rightarrow \mathsf{unit}$,
$\pi(\tau)^{i,j} = \pi(\tau_i)^{j,0} \Rightarrow \ldots \Rightarrow \pi(\tau_i)^{j,\mathrm{ar}(\tau_{i,j})} \Rightarrow \pi(\tau)^{i,0}$ *for* $j > 0$

Lemma 1 *For* $1 \leq i \leq \mathrm{ar}(\tau)$, *and* $0 \leq j \leq \mathrm{ar}(\tau_i)$, $\pi(\tau)^{i,j} \ll \tau$.

Proof. This is immediate for $j = 0$. For $0 < j \leq \mathrm{ar}(\tau_i)$, we use induction on type depth to get that for all k, $\delta(\pi(\tau_i)^{j,k} \ll \delta(\tau_i)$ and hence $\delta(\pi(\tau)^{i,j}) = \{\delta(\pi(\tau_i)^{j,0}, \ldots, \delta(\pi(\tau_i)^{j,\mathrm{ar}(\tau_{i,j})})\} \cup \{\delta(\tau_1), \ldots, \delta(\tau_{i-1}), \delta(\tau_{i+1}), \ldots, \delta(\tau_n)\} \ll \delta(\tau) = \{\delta(\tau_1), \ldots, \delta(\tau_n)\}$.

Given an element $e \in \Sigma$, define $\mathtt{con}_\tau(e) \in [\![\tau]\!]_{\mathcal{M}}$ by $\mathtt{con}_\tau(e) = \top_{[\![\tau]\!]}$ if $e = \top$ and $\mathtt{con}_\tau(e) = \bot_{[\![\tau]\!]}$ if $e = \bot$.

Two families of combinators will now be defined by mutual recursion on the multiset order.

For each type τ, and $1 \leq i \leq \mathrm{ar}(\tau)$:
$\mathsf{E}(\tau)_i : \pi(\tau)^{i,0} \Rightarrow \ldots \Rightarrow \pi(\tau)^{i,\mathrm{ar}(\tau_i)} \Rightarrow \tau$,
and for each $1 \leq i \leq \mathrm{ar}(\tau)$, and $0 \leq j \leq \mathrm{ar}(\tau_i)$, we give $\mathsf{P}(\tau)_{i,j} : \tau \Rightarrow \pi(\tau)^{i,j}$.

Definition 3 *If* $\tau_i = \mathsf{unit}$, *then* $\mathsf{E}(\tau)_i =$
$\lambda y : \pi(\tau)^{i,0}.x_1 : \tau_1 \ldots x_n : \tau_n.$
$\quad x_i \wedge (y\, x_1 \ldots x_{i-1}\, x_{i+1} \ldots x_n)$.
If $\tau_i \neq \mathsf{unit}$, *then* $\mathsf{E}(\tau)_i =$
$\lambda y_0 : \pi(\tau)^{i,0}, \ldots y_{\mathrm{ar}(\tau_i)} : \pi(\tau)^{i,\mathrm{ar}(\tau_i)}, x_1 : \tau_1, \ldots x_n : \tau_n.$
$\quad (x_i\ \mathsf{c}_{\tau_{i,1}}(t_1\ x_1 \ldots x_{i-1}\, x_{i+1} \ldots x_n)\ \top_2 \ldots \top_{\mathrm{ar}(\tau_i)}) \wedge$
$\quad \vdots$
$\quad (x_i\ \top_1 \ldots \top_{\mathrm{ar}(\tau_i)-1}\ \mathsf{c}_{\tau_{i,\mathrm{ar}(\tau_i)}}(t_{\mathrm{ar}(\tau_i)}\ x_1 \ldots x_{i-1}\, x_{i+1} \ldots x_n)) \wedge$
$\quad (y_0\, x_1 \ldots x_{i-1}\, x_{i+1} \ldots x_n)$,
where $t_j = y_j\ (\mathsf{P}(\tau)_{j,0}\, x_i) \ldots (\mathsf{P}(\tau_i)_{j,\mathrm{ar}(\tau_{i,j})}\, x_i)$.
$\mathsf{P}(\tau)_{i,0} =$
$\lambda f : \tau.x_1 : \tau_1 \ldots x_{i-1} : \tau_{i-1}.x_{i+1} : \tau_{i+1} \ldots x_n : \tau_n.$
$\quad (f\ x_1 \ldots x_{i-1}\ \top_{\tau_i}\ x_{i+1} \ldots x_n)$
For $1 \leq j \leq \mathrm{ar}(\tau_i)$, $\mathsf{P}(\tau)_{i,j} =$
$\lambda f : \tau.\lambda y_0 : \pi(\tau_i)^{j,0} \ldots y_{\mathrm{ar}(\tau_{i,j})} : \pi(\tau_i)^{j,\mathrm{ar}(\tau_{i,j})}.$
$\lambda x_1 : \tau_1 \ldots x_{i-1} : \tau_{i-1}.x_{i+1} : \tau_{i+1} \ldots x_n : \tau_n.$
$(f\ x_1 \ldots x_{i-1}\ (\mathsf{E}(\tau_i)_j\ y_0 \ldots y_{\mathrm{ar}(\tau_{i,j})})\ x_{i+1} \ldots x_n)$.
Let $\mathtt{E}(\tau)_i = [\![\mathsf{E}(\tau)_i]\!]_{\mathcal{M}}$ *and* $\mathtt{P}(\tau)_{i,j} = [\![\mathsf{P}(\tau)_{i,j}]\!]_{\mathcal{M}}$.

For $f \in [\![\tau]\!]_{\mathcal{M}}$, define:

$$\widehat{f^i} = \mathtt{E}(\tau)_i(\mathtt{P}(\tau)_{i,0}\, f) \ldots (\mathtt{P}(\tau)_{i,\mathrm{ar}(\tau_i)}\, f).$$

Proposition 2 *If* $f \in [\![\tau]\!]_{\mathcal{M}}$ *is i-strict then* $\widehat{f^i} = f$

Proof. It is necessary to show that for any elements $d_1 \in [\![\tau_1]\!], \ldots, d_n \in [\![\tau_n]\!]_{\mathcal{M}}$, $f\, d_1 \ldots d_n = \widehat{f^i}\, d_1 \ldots d_n$. The proof proceeds by induction on type depth.

- Suppose $\tau_i = \mathsf{unit}$. Then $\widehat{f^i}\, d_1 \ldots d_n = d_i \wedge (\mathsf{P}(\tau)_{i,0})\, d_1 \ldots d_{i-1}\, d_{i+1} \ldots d_n) = d_i \wedge f\, d_1 \ldots d_{i-1} \top_{\tau_i} d_{i+1} \ldots d_n$.
  If $d_i = \bot$, then $f\, d_1 \ldots d_n = \bot$ by i-strictness, and $\widehat{f^i}\, d_1 \ldots d_n = \bot \wedge f\, d_1 \ldots d_{i-1} \top_{\tau_i} d_{i+1} \ldots d_n = \bot$.
  If $d_i = \top$ then $\widehat{f^i}\, d_1 \ldots d_n = f\, d_1 \ldots d_{i-1} \top_{\tau_i} d_{i+1} \ldots d_n = f\, d_1 \ldots d_n$.
- Suppose $\mathrm{ar}(\tau_i) > 0$. Then for each $1 \leq j \leq \mathrm{ar}(\tau_i)$, let $e_j \in \Sigma = (\mathsf{P}(\tau)_{\tau,j} f)\, (\mathsf{P}(\tau_i)_{j,0}\, d_i) \ldots (\mathsf{P}(\tau_i)_{j,\mathrm{ar}(\tau_{i,j})}\, d_i)\, d_1 \ldots d_{i-1} d_{i+1} \ldots d_n = f\, d_1 \ldots d_{i-1} \widehat{(d_i)}^j d_{i-1} \ldots d_n$, since $\widehat{(d_i)}^j = \mathsf{E}(\tau_i)_j\, (\mathsf{P}(\tau_i)_{j,0}\, d_i) \ldots (\mathsf{P}(\tau_i)_{j,\mathrm{ar}(\tau_{i,j})}\, d_i)$. So by the induction hypothesis, if d_i is j-strict then $e_j = f\, d_1 \ldots d_n$.
  By simplifying definitions we have: $\widehat{f^i}\, d_1 \ldots d_n =$
  $(d_i\, \mathtt{con}_{\tau_{i,1}}(e_1) \top_{\tau_{i,2}} \ldots \top_{\tau_{i,\mathrm{ar}(\tau_i)}}) \wedge \ldots \wedge (d_i \top_{\tau_{i,1}} \ldots \top_{\tau_{i,\mathrm{ar}(\tau_i)-1}} \mathtt{con}_{\tau_{i,\mathrm{ar}(\tau_i)}}(e_{\mathrm{ar}(\tau_i)}))$
  $\wedge\, (f\, d_1 \ldots d_{i-1} \top_{\tau_i} d_{i+1} \ldots d_n)$.
  Now suppose for a contradiction that $\widehat{f^i}\, d_1 \ldots d_n \neq f\, d_1 \ldots d_n$. Then:
  Either $\widehat{f^i}\, d_1 \ldots d_n = \bot$ and $f\, d_1 \ldots d_n = \top$. Then there is some j such that $d_i \top_{\tau_{i,1}} \ldots \top_{\tau_{i,j-1}} e_j^{\tau_{i,j}} \top_{\tau_{i,j+1}} \ldots \top_{\tau_{i,\mathrm{ar}(\tau_i)}} = \bot$ — otherwise $\widehat{f^i}\, d_1 \ldots d_n = f\, d_1 \ldots d_{i-1} \top_{\tau_i} d_{i+1} \ldots d_n = f\, d_1 \ldots d_n$ by monotonicity. But then d_i is j-strict and hence $e_j = f\, d_1 \ldots d_n = \top$, and so $d_i \top_{\tau_{i,1}} \ldots \top_{\tau_{i,\mathrm{ar}(\tau_i)}} = \bot$, i.e. $d_i = \bot_{\tau_i}$, contradicting the assumption that f is i-strict.
  Or $\widehat{f^i}\, d_1 \ldots d_n = \top$ and $f\, d_1 \ldots d_n = \bot$. If $d_i = \top_{\tau_i}$, then by definition, $\widehat{f^i}\, d_1 \ldots d_{i-1} \top d_{i+1} \ldots d_n = \top$ which is a contradiction. So d_i is j-strict for some j, and hence $e_j = f\, t_1 \ldots t_n = \bot$, and moreover
  $d_i \top_{\tau_{i,1}} \ldots \top_{i,j-1} \mathtt{con}_{\tau_{i,j}}(e_j) \top_{\tau_{i,j+1}} \ldots \top_{\tau_{i,\mathrm{ar}(\tau_i)}}$
  $= \bot$, so by definition, $\widehat{f^i}\, d_1 \ldots d_n = \bot$, which is a contradiction.

Corollary 1 (Theorem 1) *Let $\mathcal{M}$ be an order-extensional and sequential standard model of unary PCF. Then for any type-object $[\![\tau]\!]_{\mathcal{M}}$, all elements of τ are definable.*

Proof. Proof is by induction on the nested multiset ordering of types.
Given $f \in [\![\tau_1 \Rightarrow \ldots \tau_n \Rightarrow \mathsf{unit}]\!]_{\mathcal{M}}$, by the definition of sequentiality, f is either a constant function (and hence definable) or else f is i-strict for some $i \leq n$. Then by Proposition 2, $f = \mathsf{E}(\tau)_i (\mathsf{P}(\tau)_{i,0}\, f) \ldots (\mathsf{P}(\tau)_{i,\mathrm{ar}(\tau_i)}\, f)$. By Lemma 1, $\pi(\tau^{i,j}) \ll \tau$ for each $j \leq \mathrm{ar}(\tau_i)$ and hence by hypothesis there exist $M_0, \ldots M_{\mathrm{ar}(\tau)_i}$ such that $[\![M_j]\!]_{\mathcal{M}} = (\mathsf{P}(\tau)_{i,\mathrm{ar}(\tau_i)}\, f)$. Hence $f = [\![\mathsf{E}(\tau)_i\, M_0 \ldots M_{\mathrm{ar}(\tau_i)}]\!]_{\mathcal{M}}$ as required.

On a Semantic Definition of Data Independence*

Ranko Lazić[1]** and David Nowak[2]

[1] Department of Computer Science, University of Warwick, UK
Ranko.Lazic@dcs.warwick.ac.uk
[2] LSV, CNRS & ENS Cachan, France
David.Nowak@lsv.ens-cachan.fr

Abstract. A variety of results which enable model checking of important classes of infinite-state systems are based on exploiting the property of data independence. The literature contains a number of definitions of variants of data independence, which are given by syntactic restrictions in particular formalisms. More recently, data independence was defined for labelled transition systems using logical relations, enabling results about data independent systems to be proved without reference to a particular syntax. In this paper, we show that the semantic definition is sufficiently strong for this purpose. More precisely, it was known that any syntactically data independent symbolic LTS denotes a semantically data independent family of LTSs, but here we show that the converse also holds.

Keywords: Data independence, definability, logical relations, nondeterminism

1 Introduction

Informally, a system is data independent with respect to a data type X when, apart from input, output and storage, the only operation that is performed on values of type X is testing a pair of them for equality. The stronger variant of data independence where equality on X is not available is also studied in the literature, as are weaker variants such as allowing constants of type X and unary predicates on X.

A variety of results which enable model checking [5] of important classes of infinite-state systems are based on exploiting data independence (e.g. [23,12,10, 20,6,14,22,19]). Although their proofs are in terms of semantics, most of these results are based on definitions of data independence which are by means of syntactic restrictions in particular formalisms.

* We acknowledge support from the EPSRC Standard Research Grant 'Exploiting Data Independence', GR/M32900. A part of this research was done at the Oxford University Computing Laboratory.

** Also affiliated to the Mathematical Institute, Belgrade. This author was supported in part by a grant from the Intel Corporation, a Junior Research Fellowship from Christ Church, Oxford, and previously by a scholarship from Hajrija & Boris Vukobrat and Copechim France SA.

M. Hofmann (Ed.): TLCA 2003, LNCS 2701, pp. 226–240, 2003.

The first semantic definition of data independence which accommodates the variants with or without equality, constants and predicates was given in [15]. The semantic entities used are families of labelled transition systems (LTSs), which consist of an LTS per instantiation of a signature. A signature is a set of type variables and a term context. Logical relations [21] are used to define when a family of LTSs is parametric, and the definition of data independence is the special case when the term context of the signature consists of only equality predicates, constants and unary predicates. It is shown in [15] that the semantics of any syntactically data independent UNITY program [4] is a data independent family of LTSs. The same paper also proves a theorem based on the semantic definition which enables the problem of model checking a data independent system for all instantiations of X to be reduced to model checking for a finite number of finite instantiations. Since it is proved from the semantic definition, the theorem applies to any formalism which can be given semantics by LTSs in which transition labels record values of transition parameters.

Although the definition in [15] was sufficiently restrictive to prove the particular reduction theorem, it was not known whether that was an accident. In other words, it was not known whether the collection of all data independent families of LTSs was equal to or strictly larger than the collection of those which arise as semantics of syntactically data independent systems. In this paper, we show that it is equal.

More precisely, we show that, in the absence of inductive types, any parametric family of LTSs whose signature consists of equality predicates, uninterpreted predicates of arbitrary arity, and uninterpreted constants, and whose types of states and transition labels do not contain functions, is definable by a symbolic LTS with the same signature. Data independence as in [15] is the special case when the uninterpreted predicates are only unary. Inductive types are not considered for simplicity, because they are orthogonal to definability of families constrained by logical relations, which is the topic of the paper. Functions within states or transition labels are also excluded in the reduction theorems in [15]. Symbolic LTSs are a basic formalism which combines simply typed λ-calculus and nondeterminism. They can be seen as a graphical variant of UNITY, and are a generalisation of first-order Kripke structures [2].

Compared with the literature on definability in models based on logical relations (e.g. [18,13,1,7]), we consider only first-order computation which can use equality testing and predicates of arbitrary arity, but the novelty in our result is that it applies to nondeterministic computation.

Section 2 is devoted to preliminaries. In Section 3, we define families of values, sets, and LTSs, and specify what it means for such families to be parametric. In Section 4, we define symbolic LTSs, and observe that their semantics is parametric. Section 5 is devoted to the definability result. In Section 6, we conclude and remark about future work.

Due to the space limit, proofs can be found in [16].

2 Preliminaries

We fix notation for the syntax and set-theoretic semantics of the simply typed λ-calculus with product and sum types, for binary logical relations, and for LTSs. Terms of the λ-calculus will be used to form symbolic LTSs. In addition to typing the terms, types will be used to structure the semantic entities we shall consider, such as families of LTSs. Logical relations will serve to constrain families indexed by signature instantiations, such as in the semantic definition of data independence.

λ-calculus syntax. We assume $TypeVars$ is an infinite set of names for type variables.

The syntax of types T is as follows:

$$T ::= X \in TypeVars \mid T_1 \times \cdots \times T_n \mid T_1 + \cdots + T_n \mid T_1 \to T_2$$

where n ranges over nonnegative integers.

For any type T, we write $Free(T)$ for the set of free type variables of T.

We assume $TermVars$ is an infinite set of names for term variables.

A *type context* Γ is a sequence of the form

$$\langle x_1 : T_1, \ldots, x_n : T_n \rangle$$

where the x_i are distinct term variables.

We write $\Gamma(x)$ for the type associated to the term variable x in Γ. The type context $\Gamma \backslash x$ is obtained by removing $x : \Gamma(x)$ from Γ if $x \in Dom(\Gamma)$, otherwise it is Γ. The type context $\Gamma\Gamma'$ is the concatenation of Γ and Γ', provided their domains are disjoint.

A *signature* is an ordered pair (Υ, Γ), where

- Υ is a finite subset of $TypeVars$, and
- Γ is a type context such that $Free(\Gamma(x)) \subseteq \Upsilon$ for each $x \in Dom(\Gamma)$.

Terms-in-context $\Upsilon, \Gamma \vdash t : T$ are built in the standard well-typed fashion from term variables, tuple formation, projection, injection, matching, λ-abstraction, and application.

$$\Upsilon, \Gamma \vdash x : T \text{ if } \Gamma(x) = T$$

$$\frac{\Upsilon, \Gamma \vdash t_i : T_i \text{ for } i = 1, \ldots, n}{\Upsilon, \Gamma \vdash \langle t_1, \ldots, t_n \rangle : T_1 \times \cdots \times T_n}$$

$$\frac{\Upsilon, \Gamma \vdash t : T_1 \times \cdots \times T_n}{\Upsilon, \Gamma \vdash \pi_i(t) : T_i} \text{ for } i = 1, \ldots, n$$

$$\frac{\Upsilon, \Gamma \vdash t : T_i}{\Upsilon, \Gamma \vdash in_i^{T_1 + \cdots + T_n}(t) : T_1 + \cdots + T_n} \text{ for } i = 1, \ldots, n$$

$$\frac{\Upsilon, \Gamma \vdash t : T_1 + \cdots + T_n \qquad \Upsilon, (\Gamma \setminus x)\langle x : T_i \rangle \vdash t_i : T \text{ for } i = 1, \ldots, n}{\Upsilon, \Gamma \vdash \mathit{match}\ t\ \mathit{with}\ \mathit{in}_1(x) \Rightarrow t_1 \ [\!]\ \cdots\ [\!]\ \mathit{in}_n(x) \Rightarrow t_n : T}$$

$$\frac{\Upsilon, (\Gamma \setminus x)\langle x : T_1 \rangle \vdash t : T_2}{\Upsilon, \Gamma \vdash \lambda x : T_1.t : T_1 \to T_2}$$

$$\frac{\Upsilon, \Gamma \vdash t_1 : T_1 \to T_2 \quad \Upsilon, \Gamma \vdash t_2 : T_1}{\Upsilon, \Gamma \vdash t_1\ t_2 : T_2}$$

We write $Free(t)$ for the set of free term variables of t.

λ-calculus semantics. A *set map* is a partial map from $TypeVars$ to sets, whose domain is finite.

For any type T, and any set map δ such that $Free(T) \subseteq Dom(\delta)$, we write $[\![T]\!]_\delta$ for the denotational semantics of T with respect to δ:

$$\begin{aligned}
[\![X]\!]_\delta &= \delta[\![X]\!] \\
[\![T_1 \times \cdots \times T_n]\!]_\delta &= [\![T_1]\!]_\delta \times \cdots \times [\![T_n]\!]_\delta \\
[\![T_1 + \cdots + T_n]\!]_\delta &= \{1\} \times [\![T_1]\!]_\delta \cup \cdots \cup \{n\} \times [\![T_n]\!]_\delta \\
[\![T_1 \to T_2]\!]_\delta &= [\![T_1]\!]_\delta \longrightarrow [\![T_2]\!]_\delta
\end{aligned}$$

A *value map* is a map whose domain is a finite subset of $TermVars$.

We write $\eta[x \mapsto v]$ for the value map whose domain is $Dom(\eta) \cup \{x\}$, and which is defined by $\eta[x \mapsto v][\![x]\!] = v$ and $\eta[x \mapsto v][\![y]\!] = \eta[\![y]\!]$ if $y \neq x$.

Given a type context Γ and a set map δ such that $Free(\Gamma(x)) \subseteq Dom(\delta)$ for each $x \in Dom(\Gamma)$, we say that a value map η is *with respect to Γ and δ* iff

- $Dom(\eta) = Dom(\Gamma)$, and
- $\eta[\![x]\!] \in [\![\Gamma(x)]\!]_\delta$ for each $x \in Dom(\Gamma)$.

An *instantiation* of a signature (Υ, Γ) is an ordered pair (δ, η) such that

- δ is a set map whose domain is Υ, and
- η is a value map with respect to Γ and δ.

For any term $\Upsilon, \Gamma \vdash t : T$ and any instantiation (δ, η), we write $[\![\Upsilon, \Gamma \vdash t : T]\!]_{(\delta,\eta)}$ for the denotational semantics of $\Upsilon, \Gamma \vdash t : T$ with respect to (δ, η).

Some abbreviations. We define some standard types:

$$\begin{aligned}
Unit &= \textit{the empty product type} \\
Bool &= Unit + Unit \\
Enum_k &= \underbrace{Unit + \cdots + Unit}_{k}
\end{aligned}$$

We also define terms and values:

$$\mathit{false} = in_1^{Bool}(\langle\rangle) \qquad \mathit{true} = in_2^{Bool}(\langle\rangle)$$
$$\overline{\mathit{false}} = (1, ()) \qquad \overline{\mathit{true}} = (2, ())$$

so that we have:

$$[\![Bool]\!]_{\{\}} = \{\overline{\mathit{false}}, \overline{\mathit{true}}\} \qquad [\![Enum_k]\!]_{\{\}} = \{(1, ()), \ldots, (k, ())\}$$

Logical relations. A *relation map* is a triple (ρ, δ, δ') such that

- ρ is a partial map from $TypeVars$ to relations, whose domain is finite,
- δ and δ' are set maps with the same domain as ρ, and
- $\rho[\![X]\!]$ is a relation between $\delta[\![X]\!]$ and $\delta'[\![X]\!]$, for each $X \in Dom(\rho)$.

A relation map (ρ, δ, δ') determines a logical relation [21] indexed by the types T such that $Free(T) \subseteq Dom(\rho)$. For any such type T, we write $[\![T]\!]_{(\rho,\delta,\delta')}$ for the component at T of the logical relation. This is a relation between the sets $[\![T]\!]_\delta$ and $[\![T]\!]_{\delta'}$, and is defined by

$$[\![X]\!]_{(\rho,\delta,\delta')} = \rho[\![X]\!]$$
$$[\![T_1 \times \cdots \times T_n]\!]_{(\rho,\delta,\delta')} = \{(\underline{a}, \underline{a'}) \mid \forall i \in \{1, \ldots, n\}.\pi_i(\underline{a})[\![T_i]\!]_{(\rho,\delta,\delta')}\pi_i(\underline{a'})\}$$
$$[\![T_1 + \cdots + T_n]\!]_{(\rho,\delta,\delta')} = \{((i, a), (i', a')) \mid i = i' \wedge a[\![T_i]\!]_{(\rho,\delta,\delta')}a'\}$$
$$[\![T_1 \to T_2]\!]_{(\rho,\delta,\delta')} = \{(f, f') \mid \forall a, a'.a[\![T_1]\!]_{(\rho,\delta,\delta')}a' \Rightarrow f(a)[\![T_2]\!]_{(\rho,\delta,\delta')}f'(a')\}$$

Labelled transition systems. An *LTS* is a tuple $S = (A, B, I, \longrightarrow)$ such that A and B are sets, $I \subseteq A$ and $\longrightarrow \subseteq A \times B \times A$.

We say that A is the *set of states*, B is the *set of transition labels*, I is the *set of initial states*, and $\longrightarrow$ is the *transition relation*. We write $a_1 \xrightarrow{b} a_2$ for $(a_1, b, a_2) \in \longrightarrow$.

3 Parametric Families

If (Υ, Γ) is a signature, then the semantics of a term or program which uses (Υ, Γ), with respect to a class $\mathcal{I}$ of instantiations of (Υ, Γ), can be seen as a family of semantic elements which is indexed by $\mathcal{I}$. We now define three kinds of such families, namely those of values, sets and LTSs. In the first two cases, there is a type T such that the family member whose index is (δ, η) is an element/subset of $[\![T]\!]_\delta$. In the case of LTSs, there are two types T and U which determine the sets of states and transition labels of family members.

Definition 1 (families). *A* family *of values, sets, or LTSs (respectively) is of the form* $(\Upsilon, \Gamma, T, \mathcal{I}, \underline{v})$, $(\Upsilon, \Gamma, T, \mathcal{I}, \underline{N})$, *or* $(\Upsilon, \Gamma, T, U, \mathcal{I}, \underline{S})$.

(Υ, Γ) *is a signature,* T *and* U *are types such that* $Free(T) \subseteq \Upsilon$ *and* $Free(U) \subseteq \Upsilon$, *and* $\mathcal{I}$ *is a class of instantiations of* (Υ, Γ).

The vectors $\underline{v}$, $\underline{N}$ *and* $\underline{S}$ *are indexed by elements of* $\mathcal{I}$. *For each* $(\delta, \eta) \in \mathcal{I}$, *we have* $v_{(\delta,\eta)} \in [\![T]\!]_\delta$, $N_{(\delta,\eta)} \subseteq [\![T]\!]_\delta$, *and* $S_{(\delta,\eta)}$ *is an LTS with set of states* $[\![T]\!]_\delta$ *and set of transition labels* $[\![U]\!]_\delta$. □

Logical relations can be used as follows to define when a family is parametric. We shall see below that families arising as semantics of λ-calculus terms or of symbolic LTSs have this property.

The details are as in [15], except that here we treat families of values and sets explicitly, because they will be used later in the paper. The definitions of when two sets/LTSs are related can be seen as liftings of logical relations to powerset/LTS types, although we do not give such types first-class status. A more general treatment of such liftings of logical relations can be found in [8].

Definition 2 (related sets). *Suppose:*

- P *is a relation between* N *and* N'*;*
- $M \subseteq N$ *and* $M' \subseteq N'$.

We say that P *relates* M *and* M' *iff*

$$\begin{array}{l} \forall x \in M \;\cdot \exists x' \in M' \cdot (x, x') \in P \\ \forall x' \in M' \cdot \;\exists x \in M \;\cdot (x, x') \in P \quad \square \end{array}$$

Definition 3 (universal partial R-bisimulation). *Suppose:*

- P *is a relation between* A *and* A'*;*
- R *is a relation between* B *and* B'*;*
- $S = (A, B, I, \longrightarrow)$ *and* $S' = (A', B', I', \longrightarrow')$ *are LTSs.*

We say that P *is a* universal partial R-bisimulation *between* S *and* S' *iff*

(i) whenever aPa' *then* $a \in I$ *iff* $a' \in I'$*, and*
(ii) whenever $a_1Pa'_1$ *and* bRb'*, then* P *relates* $\{a_2 \mid a_1 \xrightarrow{b} a_2\}$ *and* $\{a'_2 \mid a'_1 \xrightarrow{b'}{}' a'_2\}$. $\square$

Definition 4 (parametric families). *A family* $(\Upsilon, \Gamma, T, \mathcal{I}, \underline{v})$*,* $(\Upsilon, \Gamma, T, \mathcal{I}, \underline{N})$*, or* $(\Upsilon, \Gamma, T, U, \mathcal{I}, \underline{S})$ *(respectively) is* parametric *iff, for any* $(\delta, \eta), (\delta', \eta') \in \mathcal{I}$*, and any relation map* (ρ, δ, δ') *such that*

$$\forall x \in Dom(\Gamma) \cdot \eta[\![x]\!][\![\Gamma(x)]\!]_{(\rho,\delta,\delta')}\eta'[\![x]\!]$$

we have

- $v_{(\delta,\eta)}[\![T]\!]_{(\rho,\delta,\delta')}v_{(\delta',\eta')}$*,*
- $[\![T]\!]_{(\rho,\delta,\delta')}$ *relates* $N_{(\delta,\eta)}$ *and* $N_{(\delta',\eta')}$*, or*
- $[\![T]\!]_{(\rho,\delta,\delta')}$ *is a universal partial* $[\![U]\!]_{(\rho,\delta,\delta')}$*-bisimulation between* $S_{(\delta,\eta)}$ *and* $S_{(\delta',\eta')}$. $\square$

We can now state the Basic Lemma of logical relations [21] in the following way.

Proposition 1. *For any term* $\Upsilon, \Gamma \vdash t : T$ *and class* $\mathcal{I}$ *of instantiations of* (Υ, Γ), *the family* $(\Upsilon, \Gamma, T, \mathcal{I}, [\![t]\!])$ *is parametric.* □

Signatures which consist of equality predicates, uninterpreted predicates, and uninterpreted constants will be important later in the paper, as will types which are sums of products of type variables, and classes of instantiations whose only restriction is that equality predicates are interpreted as expected. These will determine the kind of parametric families of LTSs which our main result applies to.

Terminology 1. We say that a signature (Υ, Γ) is *EPC* iff Γ is of the form $\Gamma_E \Gamma_P \Gamma_C$ such that

- any $\Gamma_E(e)$ is of the form $(X \times X) \rightarrow Bool$,
- any $\Gamma_P(p)$ is of the form $T \rightarrow Enum_k$ where T is a product of type variables, and
- any $\Gamma_C(c)$ is a product of type variables.

A type T is *SP* iff it is a sum of products of type variables.

The *full* class of instantiations of an EPC-signature (Υ, Γ) consists of all (δ, η) such that $\eta[\![e]\!]$ is the equality predicate on $\delta[\![X]\!]$ for any $e : ((X \times X) \rightarrow Bool)$ in Γ_E. □

Data independence was defined semantically in [15] as parametricity of families of LTSs whose signatures are EPC with only unary uninterpreted predicates, and whose classes of instantiations are full.

Example 1. Consider a family of LTSs defined as follows. The signature

$$\Upsilon = \{X\}$$
$$\Gamma = \langle p : X \rightarrow Bool, q : (X \times X) \rightarrow Bool \rangle$$

consists of type variable X, unary uninterpreted predicate p on X, and binary uninterpreted predicate q on X.

The type of states $T = X + (X \times X)$ can be seen as two control states, the first with one data item of type X, the second with two data items of type X. The type of transition labels $U = X$ means that any transition has a parameter of type X.

$\mathcal{I}$ consists of all instantiations of (Υ, Γ), and any $S_{(\delta,\eta)}$ is such that

- a state is initial iff it is the first control state together with data u which satisfies p, and
- a transition is from the first control state to the second provided either the parameter w satisfies p and the two target data items v_1 and v_2 are set to w and the source data u, or (u, w) satisfies q and v_1, v_2 are set to u, w.

More formally:

$$I_{(\delta,\eta)} = \{(1,u) \mid \eta[\![p]\!](u)\}$$
$$\longrightarrow_{(\delta,\eta)} = \{((1,u), w, (2,(v_1,v_2))) \mid (\eta[\![p]\!](w) \wedge v_1 = w \wedge v_2 = u) \vee (\eta[\![q]\!](u,w) \wedge v_1 = u \wedge v_2 = w)\}$$

It is straightforward to check that this family is parametric. In fact, as we shall see in Example 2, it is the semantics of a symbolic LTS.

Let $\delta[\![X]\!] = \{\spadesuit, \clubsuit\}$ and $\delta'[\![X]\!] = \{\diamondsuit, \heartsuit, \triangle\}$. Let η and η' be such that $\eta[\![p]\!]$ holds only on $\spadesuit$, $\eta'[\![p]\!]$ holds on $\diamondsuit$ and $\heartsuit$, and $\eta[\![q]\!]$ and $\eta'[\![q]\!]$ hold on all pairs. Define $\rho[\![X]\!]$ by $\spadesuit\rho[\![X]\!]\diamondsuit$ and $\spadesuit\rho[\![X]\!]\heartsuit$. Then $[\![T]\!]_{(\rho,\delta,\delta')}$ is a universal partial $[\![U]\!]_{(\rho,\delta,\delta')}$ bisimulation between $S_{(\delta,\eta)}$ and $S_{(\delta,\eta)}$, and the following figure is a simple illustration.

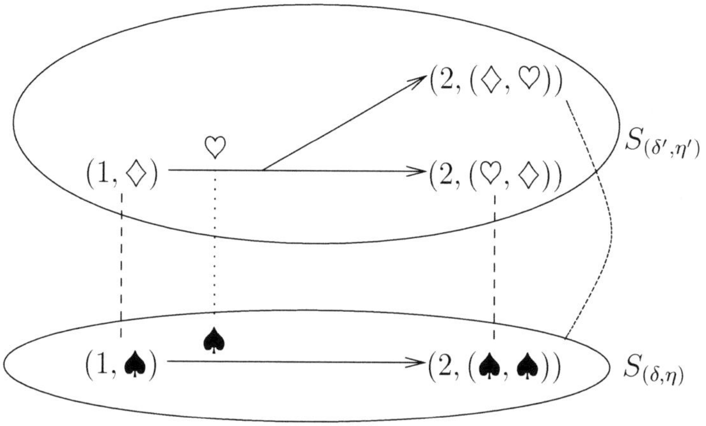

□

4 Symbolic Labelled Transition Systems

The notion of SLTSs we define below is a formalism for expressing nondeterministic reactive systems, which is based on the simply typed λ-calculus introduced in Section 2.

An SLTS $\mathbf{S}$ computes on types built from type variables from Υ, using operations from Γ, where (Υ, Γ) is a signature. $\mathbf{S}$ has a set $\mathbf{A}$ of symbolic states, and a set $\mathbf{B}$ of symbolic labels. The elements of $\mathbf{A}$ and $\mathbf{B}$ have associated type contexts $\Phi(\mathbf{a})$ and $\Psi(\mathbf{b})$. Symbolic states can be thought of as control states, where $\Phi(\mathbf{a})$ are data variables at $\mathbf{a}$. Similarly, symbolic labels can be thought of as kinds of transitions, so that $\Psi(\mathbf{b})$ are parameters for transitions of kind $\mathbf{b}$.

The initial states of $\mathbf{S}$ are given as a set of pairs $(\mathbf{a}, t)$, where t is a λ-calculus term of type *Bool* which specifies which data values associated with the symbolic state $\mathbf{a}$ form initial states.

The transitions of $\mathbf{S}$ are given by symbolic transitions. A symbolic transition has source and target symbolic states, a symbolic label, a guard, and an assignment. The guard is a λ-calculus term of type *Bool* which determines when the symbolic transition is enabled, in which case the action of the symbolic transition is to set each data variable at the target symbolic state according to the assignment. In particular, the lifetime of data variables and transition parameters is one transition. Nondeterminism is present when $\mathbf{S}$ has more than one symbolic transition with the same source symbolic state and the same symbolic label.

The fact that SLTSs allow different sets of data variables at different symbolic states can be used e.g. to model data which is local to a part of the system. Observe also that a portion of data in a system (such as data which is not treated in a data independent manner) can be modelled non-symbolically by regarding it as part of control.

SLTSs are similar to a number of formalisms in the literature, e.g. [9,2,3]. They can also be seen as a graphical variant of UNITY [4].

Definition 5 (SLTS). $\mathbf{S}$ *is an* SLTS *iff it is a tuple*

$$(\Upsilon, \Gamma, \mathbf{A}, \Phi, \mathbf{B}, \Psi, \mathbf{I}, \mathbf{R})$$

such that:

- (Υ, Γ) *is a signature.*
- $\mathbf{A}$ *and* $\mathbf{B}$ *are sets. We call elements of* $\mathbf{A}$ symbolic states, *and elements of* $\mathbf{B}$ symbolic labels.
- Φ *and* Ψ *are such that, for any* $\mathbf{a} \in \mathbf{A}$ *and* $\mathbf{b} \in \mathbf{B}$, *we have that* $(\Upsilon, \Phi(\mathbf{a}))$ *and* $(\Upsilon, \Psi(\mathbf{b}))$ *are signatures, and that* $Dom(\Phi(\mathbf{a}))$ *and* $Dom(\Psi(\mathbf{b}))$ *are disjoint from* $Dom(\Gamma)$.
- $\mathbf{I}$ *is a set of ordered pairs* $(\mathbf{a}, t)$, *where* $\mathbf{a} \in \mathbf{A}$ *and* $\Upsilon, \Gamma\Phi(\mathbf{a}) \vdash t : Bool$ *is a term. We say that* t *is an* initial condition, *and that elements of* $\mathbf{I}$ *are* symbolic initial states.
- $\mathbf{R}$ *is a set of tuples of the form* $(\mathbf{a}_1, \mathbf{b}, g, E, \mathbf{a}_2)$ *where* $\mathbf{a}_1, \mathbf{a}_2 \in \mathbf{A}$, $\mathbf{b} \in \mathbf{B}$, $Dom(\Phi(\mathbf{a}_1))$ *and* $Dom(\Psi(\mathbf{b}))$ *are disjoint,* $\Upsilon, \Gamma\Phi(\mathbf{a}_1)\Psi(\mathbf{b}) \vdash g : Bool$ *is a term, and* E *is such that, for any* $x \in Dom(\Phi(\mathbf{a}_2))$, $\Upsilon, \Gamma\Phi(\mathbf{a}_1)\Psi(\mathbf{b}) \vdash E(x) : \Phi(\mathbf{a}_2)(x)$ *is a term. We say that* $\mathbf{a}_1$ *is the* symbolic source state, $\mathbf{a}_2$ *is the* symbolic target state, g *is the* guard, E *is the* assignment, $\mathbf{R}$ *is the* symbolic transition relation *and its elements are* symbolic transitions. *We write* $\mathbf{a}_1 \, [\mathbf{b} : g \hookrightarrow E\rangle_{\mathbf{R}} \, \mathbf{a}_2$ *for* $(\mathbf{a}_1, \mathbf{b}, g, E, \mathbf{a}_2) \in \mathbf{R}$. □

Example 2. The following SLTS is illustrated in the figure.

- $\Upsilon = \{X\}$;
- $\Gamma = \langle p : X \to Bool, q : (X \times X) \to Bool\rangle$;
- $\mathbf{A} = \{\mathbf{a}_1, \mathbf{a}_2\}$;
- $\Phi(\mathbf{a}_1) = \langle x : X\rangle$, $\Phi(\mathbf{a}_2) = \langle y_1 : X, y_2 : X\rangle$;
- $\mathbf{B} = \{\mathbf{b}\}$;
- $\Psi(\mathbf{b}) = \langle z : X\rangle$;
- $\mathbf{I} = \{(\mathbf{a}_1, p\,x)\}$;
- $\mathbf{a}_1 \, [\mathbf{b} : p\,z \hookrightarrow \{y_1 \mapsto z, y_2 \mapsto x\}\rangle \, \mathbf{a}_2$ and $\mathbf{a}_1 \, [\mathbf{b} : q\,x\,z \hookrightarrow \{y_1 \mapsto x, y_2 \mapsto z\}\rangle \, \mathbf{a}_2$ are the symbolic transitions.

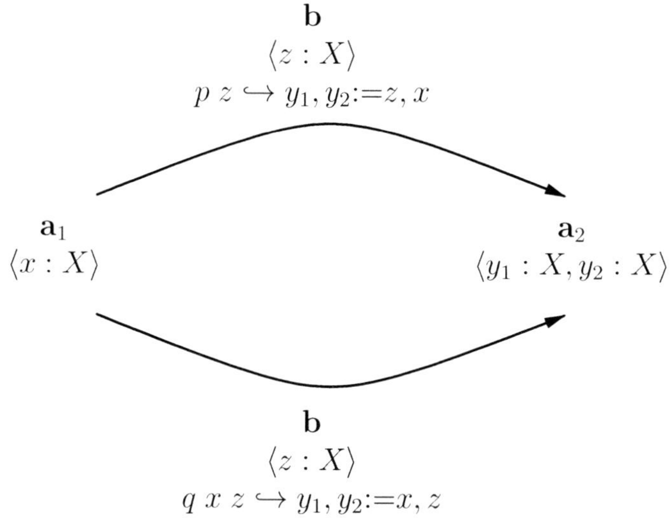

□

Given an SLTS $\mathbf{S}$ and an instantiation (δ, η) of its signature (Υ, Γ), we will define a concrete LTS $[\![\mathbf{S}]\!]_{(\delta,\eta)}$. Provided the sets of symbolic states and symbolic labels of $\mathbf{S}$ are finite, and given a class of instantiations of (Υ, Γ), the concrete LTSs $[\![\mathbf{S}]\!]_{(\delta,\eta)}$ will form a parametric family.

Notation 1. Given a signature (Υ, Δ), where $\Delta = \langle x_1 : T_1, \ldots, x_n : T_n \rangle$, and a set map δ such that $\Upsilon \subseteq \mathit{Dom}(\delta)$, let $[\![\Delta]\!]_\delta$ be defined by

$$[\![\Delta]\!]_\delta = \{(v_1, \ldots, v_n) \mid \forall i \cdot v_i \in [\![T_i]\!]_\delta\}$$

Given a type context $\Delta = \langle x_1 : T_1, \ldots, x_n : T_n \rangle$, a value map ζ such that $\mathit{Dom}(\zeta) \cap \mathit{Dom}(\Delta) = \{\}$, and a tuple $\underline{v} = (v_1, \ldots, v_n)$, let $\zeta \oplus_\Delta \underline{v}$ be the map ζ extended by $x_i \mapsto v_i$ for all $x_i \in \mathit{Dom}(\Delta)$. □

Definition 6 (semantics of SLTSs). *Given an SLTS*

$$\mathbf{S} = (\Upsilon, \Gamma, \mathbf{A}, \Phi, \mathbf{B}, \Psi, \mathbf{I}, \mathbf{R})$$

and an instantiation (δ, η) *of* (Υ, Γ), *let* $[\![\mathbf{S}]\!]_{(\delta,\eta)}$ *be the LTS* $(A, B, I, \longrightarrow)$ *defined as follows:*

- $A = \{(\mathbf{a}, \underline{v}) \mid \mathbf{a} \in \mathbf{A} \wedge \underline{v} \in [\![\Phi(\mathbf{a})]\!]_\delta\}$.
- $B = \{(\mathbf{b}, \underline{w}) \mid \mathbf{b} \in \mathbf{B} \wedge \underline{w} \in [\![\Psi(\mathbf{b})]\!]_\delta\}$.
- $I = \bigcup_{(\mathbf{a},t) \in \mathbf{I}} [\![(\mathbf{a}, t)]\!]_{(\delta,\eta)}$ *where*

$$[\![(\mathbf{a}, t)]\!]_{(\delta,\eta)} = \{(\mathbf{a}, \underline{v}) \in A \mid [\![\Upsilon, \Gamma\Phi(\mathbf{a}) \vdash t : \mathit{Bool}]\!]_{(\delta, \eta \oplus_{\Phi(\mathbf{a})} \underline{v})} = \overline{\mathit{true}}\}.$$

- *The transition relation* $\longrightarrow$ *is the set of triples*

$$((\mathbf{a}_1, \underline{v^1}), (\mathbf{b}, \underline{w}), (\mathbf{a}_2, \underline{v^2})) \in A \times B \times A$$

such that for some g *and* E *we have*

- $(\mathbf{a}_1, \mathbf{b}, g, E, \mathbf{a}_2) \in \mathbf{R}$,
- $[\![\Upsilon, \Gamma\Phi(\mathbf{a}_1)\Psi(\mathbf{b}) \vdash g : Bool]\!]_{(\delta,(\eta \oplus_{\Phi(\mathbf{a}_1)} \underline{v}^1) \oplus_{\Psi(\mathbf{b})} \underline{w})} = \overline{true}$, *and*
- *for all* $x_i \in Dom(\Phi(\mathbf{a}_2))$,

$$[\![\Upsilon, \Gamma\Phi(\mathbf{a}_1)\Psi(\mathbf{b}) \vdash E(x_i) : \Phi(\mathbf{a}_2)(x_i)]\!]_{(\delta,(\eta \oplus_{\Phi(\mathbf{a}_1)} \underline{v}^1) \oplus_{\Psi(\mathbf{b})} \underline{w})} = v_i^2$$

where x_i *is the ith component of* $Dom(\Phi(\mathbf{a}_2))$. □

Proposition 2. *Suppose* $\mathbf{S} = (\Upsilon, \Gamma, \mathbf{A}, \Phi, \mathbf{B}, \Psi, \mathbf{I}, \mathbf{R})$ *is an SLTS such that* $\mathbf{A} = \{1, \ldots, n\}$ *and* $\mathbf{B} = \{1, \ldots, m\}$. *Let*

$$T = \sum_{i=1}^{n} \prod_{x \in \Phi(\mathbf{a})} \Phi(\mathbf{a})(x)$$

$$U = \sum_{j=1}^{m} \prod_{y \in \Psi(\mathbf{b})} \Psi(\mathbf{b})(y)$$

For any class $\mathcal{I}$ *of instantiations of* (Υ, Γ), *we have that* $(\Upsilon, \Gamma, T, U, \mathcal{I}, [\![\mathbf{S}]\!])$ *is a parametric family of LTSs.* □

When restricted to EPC signatures with only unary uninterpreted predicates and to full classes of instantiations, Proposition 2 states that the semantics of any syntactically data independent SLTS is data independent according to the semantic definition in [15].

Example 3. The SLTS in Example 2 yields (up to isomorphism) the parametric family of LTSs in Example 1, for the class of all instantiations. □

5 Definability

This section contains the main result of the paper, namely that any parametric family of LTSs whose signature is EPC, whose types of states and transition labels are SP, and whose class of instantiations is full, is definable by an SLTS. In particular, this shows that the semantic definition of data independence in [15] is sufficiently strong. The SP assumption is equivalent to assuming absence of the function-type construct, which is done in the reduction theorems in [15].

Before the theorem, we present a proposition and a lemma which are used in its proof.

Proposition 3. *For any parametric family of values*

$$(\Upsilon, \Gamma, T, \mathcal{I}, \underline{v})$$

such that (Υ, Γ) *is EPC,* T *is SP, and* $\mathcal{I}$ *is full, there exists a term*

$$\Upsilon, \Gamma \vdash s : T$$

such that, for any $(\delta, \eta) \in \mathcal{I}$, $[\![s]\!]_{(\delta,\eta)} = v_{(\delta,\eta)}$, *and such that* s *is of the form*

$$\mathit{match}\ h\ \mathit{with}\ [\!]_{i=1}^{H}\ in_i(x) \Rightarrow in_{R_i}^{T}(r_i)$$

where $H \in \mathbb{N}$, $\Upsilon, \Gamma \vdash h : Enum_H$ *is a term,* $T = \sum_{i=1}^{n} T_i$, *and for each* i, $R_i \in \{1, \ldots, n\}$ *and* $\Upsilon, \Gamma_C \vdash r_i : T_{R_i}$ *is a term.*

Proof outline. The class $\mathcal{I}$ can be split into finitely many subclasses according to the results of all possible applications of the equality predicates and the uninterpreted predicates to the uninterpreted constants. The subclasses have the property that two instantiations can be related by a relation map iff they belong to the same subclass.

The term s can be defined by letting H be the number of subclasses. Each R_i and r_i are defined by considering an instantiation from the corresponding subclass which, for any $X \in \Upsilon$ without an equality predicate in Γ_E, instantiates any two uninterpreted constant components of type X by distinct values. □

Example 4. This example (due to Plotkin) shows that Proposition 3 cannot be extended straightforwardly to signatures which contain types such as $X \to X$. It is a parametric family of values which is not definable.

The signature consists of one type variable X, predicate $p : X \to Bool$, operation $s : X \to X$, and constant $z : X$. The type of the family is $Bool$, and for any instantiation (δ, η) of the signature, the member $v_{(\delta,\eta)}$ is defined to be $\overline{true}$ iff, for all $n \in \mathbb{N}$, the result of applying n times $\eta[\![s]\!]$ to $\eta[\![z]\!]$ satisfies $\eta[\![p]\!]$. □

Terminology 2. We say that a family of sets is *deterministic* iff each set of the family is either the empty set or a singleton. □

Notation 2. We write $(\Upsilon, \Gamma, T, \mathcal{I}, \underline{N}) \sqsubseteq (\Upsilon', \Gamma', T', \mathcal{I}', \underline{N'})$ iff we have $\Upsilon = \Upsilon'$, $\Gamma = \Gamma'$, $T = T'$, $\mathcal{I} = \mathcal{I}'$, and $N_{(\delta,\eta)} \subseteq N'_{(\delta,\eta)}$ for all $(\delta, \eta) \in \mathcal{I}$.

We write $(\Upsilon, \Gamma, T, \mathcal{I}, \underline{N}) \sqcup (\Upsilon, \Gamma, T, \mathcal{I}, \underline{N'})$ for the family of sets $(\Upsilon, \Gamma, T, \mathcal{I}, \underline{M})$ where, for any $(\delta, \eta) \in \mathcal{I}$, $M_{(\delta,\eta)} = N_{(\delta,\eta)} \cup N'_{(\delta,\eta)}$. □

Lemma 1. *Given any parametric family of sets*

$$\mathcal{N} = (\Upsilon, \Gamma, T, \mathcal{I}, \underline{N})$$

such that (Υ, Γ) is EPC, T is SP, and $\mathcal{I}$ is full, there are parametric families of sets $\mathcal{M}_1, \ldots, \mathcal{M}_m$ such that:

(i) $\mathcal{M}_i \sqsubseteq \mathcal{N}$ *for each* i;
(ii) $\mathcal{M}_i$ *is deterministic for each* i;
(iii) given any parametric family of sets $\mathcal{M}'$ *such that* $\mathcal{M}' \sqsubseteq \mathcal{N}$ *and* $\mathcal{M}'$ *is deterministic, it is equal to* $\mathcal{M}_i$ *for some* i;
(iv) $\bigsqcup_{i=1}^{m} \mathcal{M}_i = \mathcal{N}$. □

The proof of Lemma 1 has similar structure to the proof of Proposition 3, which means that the definability of families of sets can be shown somewhat more directly than by combining the two results. However, Lemma 1 is of wider interest, since it shows that definability of parametric nondeterministic families can be reduced to definability of finitely many parametric deterministic families.

Example 5. Recall the SLTS in Example 2 and its semantics (up to isomorphism) in Example 1.

Take $\mathcal{N}$ to be the family of sets corresponding to the nondeterministic computation at symbolic state $\mathbf{a}_1$ and symbolic label $\mathbf{b}$. Its signature is $\Upsilon = \{X\}$ and

$$\Gamma = \langle p : X \to Bool, q : (X \times X) \to Bool, x : X, z : X \rangle$$

The type of $\mathcal{N}$ is $X \times X$, and the class consists of all instantiations of (Υ, Γ). For any (δ, η), $N_{(\delta,\eta)}$ is the set of all outcomes of the two symbolic transitions when p, q, x and z have the values given by η. If neither symbolic transition is enabled, $N_{(\delta,\eta)} = \{\}$.

Similarly, we can let $\mathcal{M}_1$ and $\mathcal{M}_2$ be families of sets corresponding to the two symbolic transitions respectively.

Parametricity of these families of sets follows from parametricity of the family of LTSs. Also, $\mathcal{M}_1$ and $\mathcal{M}_2$ are deterministic, and $\mathcal{M}_1 \sqcup \mathcal{M}_2 = \mathcal{N}$. Therefore, $\mathcal{M}_1$ and $\mathcal{M}_2$ are two of the families corresponding to $\mathcal{N}$ in the statement of Lemma 1. There are others, e.g. the empty family. □

Theorem 1. *For any parametric family of LTSs $(\Upsilon, \Gamma, T, U, \mathcal{I}, \underline{S})$ such that (Υ, Γ) is EPC, T and U are SP, and $\mathcal{I}$ is full, there exists a finite SLTS $\mathbf{S}$ with the same signature and such that, for any $(\delta, \eta) \in \mathcal{I}$, $[\![\mathbf{S}]\!]_{(\delta,\eta)} = S_{(\delta,\eta)}$.*

Proof outline. Symbolic states and symbolic labels of $\mathbf{S}$ are defined to correspond to the sum components of T and U.

For each symbolic state $\mathbf{a}$, the sets of initial states of $S_{(\delta,\eta)}$ restricted to $\mathbf{a}$ form a parametric family of predicates, so that Proposition 3 can be applied to obtain the initial condition at $\mathbf{a}$.

For each symbolic state $\mathbf{a}$ and symbolic label $\mathbf{a}$, the transitions of the concrete LTSs $S_{(\delta,\eta)}$ form a parametric family of sets whose type is T. Lemma 1 can be applied to this family to yield a finite number of deterministic families. Symbolic transitions of $\mathbf{S}$ are then obtained by applying Proposition 3 to families of values which correspond to the deterministic families of sets. □

Example 6. Theorem 1 applies to the family of LTSs in Example 1. We already saw that this family is definable (up to isomorphism) by the SLTS in Example 2. □

6 Conclusions

This paper answers negatively the question of whether there are any data independent families of LTSs [15] which do not arise as semantics of any syntactically data independent system. Thus we confirm that the semantic definition of data independence is suitable for reasoning about data independent systems without being tied to a particular syntax.

More precisely, we showed that any parametric family of LTSs whose signature consists of equality predicates, uninterpreted predicates of arbitrary arity, and uninterpreted constants, and whose types of states and transition labels do not contain functions, is definable by a symbolic LTS. Data independence is the special case with only unary uninterpreted predicates.

From another point of view, we demonstrated that when binary logical relations are extended to nondeterministic computation by means of bisimulation, they can be used to ensure that any parametric family is definable.

Future work should investigate definability of parametric families in settings where powerset types have first-class status [17,11,8].

Acknowledgements. We are grateful to Samson Abramsky, Brian Dunphy, Andrew Pitts, Gordon Plotkin, Uday Reddy, Bill Roscoe and Alex Simpson for useful discussions, and to the anonymous referees for their helpful comments.

References

1. M. Alimohamed. A characterization of lambda definability in categorical models of implicit polymorphism. *Theoretical Computer Science*, 146:5–23, 1995.
2. J. Bohn, W. Damm, O. Grumberg, H. Hungar, and K. Laster. First-order-CTL model checking. In *Foundations of Software Technology and Theoretical Computer Science (FST&TCS'98)*, volume 1530 of *Lecture Notes in Computer Science*, pages 283–294. Springer-Verlag, 1998.
3. M. Calder and C. Shankland. A symbolic semantics and bisimulation for full LOTOS. In *International Conference on Formal Description Techniques for Networked and Distributed Systems (FORTE'01)*, pages 184–200. Kluwer Academic Publishers, 2001.
4. K. M. Chandy and J. Misra. *Parallel Program Design: A Foundation.* Addison-Wesley, 1988.
5. E. M. Clarke, O. Grumberg, and D. A. Peled. *Model Checking.* MIT Press, 1999.
6. D. Dill, R. Hojati, and R.K. Brayton. Verifying linear temporal properties of data intensive controllers using finite instantiations. In *Hardware Description Languages and their Applications (CHDL '97).* Chapman and Hall, 1997.
7. M. Fiore and A. Simpson. Lambda definability with sums via Grothendieck logical relations. In *Proceedings of the 4th International Conference on Typed Lambda Calculi and Applications (TLCA'99)*, volume 1581 of *Lecture Notes in Computer Science*, pages 147–161. Springer-Verlag, 1999.
8. J. Goubault-Larrecq, S. Lasota, and D. Nowak. Logical relations for monadic types. In *Proceedings of the 11th Annual Conference of the European Association for Computer Science Logic (CSL'02)*, volume 2471 of *Lecture Notes in Computer Science*, pages 553–568. Springer-Verlag, 2002.
9. M. Hennessy and H. Lin. Symbolic bisimulations. *Theoretical Computer Science*, 138(2):353–389, 1995.
10. C. N. Ip and D. L. Dill. Better verification through symmetry. *Formal Methods in System Design: An International Journal*, 9(1/2):41–75, 1996.
11. A. Jeffrey. A fully abstract semantics for a higher-order functional language with nondeterministic computation. *Theoretical Computer Science*, 228:105–150, 1999.
12. B. Jonsson and J. Parrow. Deciding bisimulation equivalences for a class of non-finite-state programs. *Information and Computation*, 107(2):272–302, 1993.

13. A. Jung and J. Tiuryn. A new characterization of lambda definability. In *Proceedings of the 1st International Conference on Typed Lambda Calculi and Applications (TLCA'93)*, volume 664 of *Lecture Notes in Computer Science*, pages 245–257. Springer-Verlag, 1993.
14. R. Lazić. *A Semantic Study of Data Independence with Applications to Model Checking*. DPhil thesis, Oxford University Computing Laboratory, 1999.
15. R. Lazić and D. Nowak. A unifying approach to data-independence. In *Proceedings of the 11th International Conference on Concurrency Theory (CONCUR 2000)*, volume 1877 of *Lecture Notes in Computer Science*, pages 581–595. Springer-Verlag, 2000.
16. R. Lazić and D. Nowak. On a semantic definition of data independence. Research report 392, Department of Computer Science, University of Warwick, 2003. http://www.dcs.warwick.ac.uk.
17. T. Nipkow. Non-deterministic data types: models and implementations. *Acta Informatica*, 22(6):629–661, 1986.
18. G. D. Plotkin. Lambda-definability in the full type hierarchy. In *To H. B. Curry: Essays on Combinatory Logic, Lambda Calculus and Formalism*, pages 363–373. Academic Press, 1980.
19. S. Qadeer. Verifying sequential consistency on shared-memory multiprocessors by model checking. Research Report 176, Compaq, 2001.
20. R. Hojati and R. K. Brayton. Automatic datapath abstraction in hardware systems. In *Proceedings of the 7th International Conference On Computer Aided Verification*, volume 939 of *Lecture Notes in Computer Science*, pages 98–113. Springer Verlag, 1995.
21. J. C. Reynolds. Types, abstraction and parametric polymorphism. In *Proceedings of the 9th IFIP World Computer Congress (IFIP'83)*, pages 513–523. North-Holland, 1983.
22. A. W. Roscoe and P. J. Broadfoot. Proving security protocols with model checkers by data independence techniques. *Journal of Computer Security, Special Issue on the 11th IEEE Computer Security Foundations Workshop (CSFW11)*, pages 147–190, 1999.
23. P. Wolper. Expressing interesting properties of programs in propositional temporal logic. In *Conference Record of the 13th Annual ACM Symposium on Principles of Programming Languages*, pages 184–193. ACM, 1986.

Nondeterministic Light Logics and NP-Time

François Maurel

Preuves, Programmes et Systèmes*
Francois.Maurel@pps.jussieu.fr

Abstract. This paper relates two distinct traditions: the one of complexity classes characterisations through light logics and models of nondeterminism. Light logics provide an implicit characterisation of P-Time algorithms through the Curry-Howard isomorphism: every derivation reduces to its normal form in polynomial time and every polynomial Turing machine can be simulated by a derivation. In this paper, we extend Intuitionistic Light Affine Logic, a logic with full weakening, with a simple rule for nondeterminism and get a completeness result for NP-Time algorithms which is, as far as we know, the first Curry-Howard characterisation of NP complexity. We conclude by a reformulation of the P $\neq$ NP conjecture.

Keywords. Light logics, implicit characterisations of complexity classes, NP complexity.

1 Introduction

There has been quite a huge work to characterise complexity classes through various machines [15] or programming languages [13,4,9]. Light logics have provided a new insight by giving Curry-Howard characterisations to P-Time. Girard introduced the first light logic in [7]. It is called Light Linear Logic. LLL proof nets represent P-Time in the sense that (1) every net in LLL can only perform polynomial cut-eliminations on its *size*, the polynomial bound depending on a graph theoretic value, its *depth*; (2) conversely, every polynomial Turing machine is encodable inside LLL. Light Affine Logic, defined by Asperti in [1] and its intuitionnistic variant ILAL lead to many simplifications. Some links between these traditions (lambda calculus, safe recursion and light logics) are studied by Roversi in [17] and Murawsky and Ong in [14].

In this paper, we first recall the syntax and reduction rules of ILAL. Then, we introduce nILAL, obtained by adding a Sum deduction rule to ILAL. This deduction rule is a logical counterpart to nondeterministic choice in process calculi. We show that nILAL represents NP in the same way that ILAL represents P. Finally, we conclude by reformulating the P $\neq$ NP conjecture in our setting.

* CNRS & Université Paris VII, Case 7014, 2 place Jussieu, 75251 Paris cedex 05.

M. Hofmann (Ed.): TLCA 2003, LNCS 2701, pp. 241–255, 2003.

1.1 NP-Time Problems

The notion of polynomial time problems can be defined in many models and, fortunately, these classes generally coincide. A deterministic Turing machine is an abstract machine with finitely many states and an unbounded finite memory but with a bounded local view on it. A configuration of the machine is the data of the state and the memory. A transition, i.e. a step between two configurations, is uniquely determined by the current state of the machine and the local access to the memory. A deterministic Turing machine representing a function is in P-Time if there exists a polynomial bound (on the size of the input) for the number of transitions before it stops. Note that we consider functions or relations from integers to booleans, as is customary, but also to integers.

The notion of nondeterministic Turing machines is a generalisation of deterministic Turing machines where many transitions can occur: at each step, the machine makes a nondeterministic choice between these transitions. The notion of calculation is usually enriched by an *accepting* state: the result of a computation is taken into account if and only if it ends in the accepting state. We then say that the machine answers. A NP-Turing machine is a machine for which there is a polynomial bound on the size of its input for the number of transitions it may perform before it stops. There is an asymmetry between *true* and *false*: a class of inputs is said to be in NP if there exists a NP-Turing machine which may answer *true* on a given input if and only if this input is in the considered class; a class is said to be in coNP if there exists a NP-Turing machine which may answer *false* on a given input if and only if this input is not in the considered class. A problem is said to be in P if it is solvable by a P-Time Turing machine. A problem is said to be in NP if it is solvable by a NP-Turing machine *i.e.* a NP-Turing machine which may answer *true* if and only if the answer is *true*. Symmetrically, a problem is said to be in coNP if it is co-solvable by a NP-Turing machine i.e. a NP-Turing machine which may answer *false* if and only if the answer is *false*.

1.2 Intuitionistic Light Affine Logic

Light Affine Logic [1,2] has been introduced by Asperti in 1998 as a simplification of LLL. It leads to a similar logical characterisation of P-Time algorithms.

Syntax of LAL. To impose a polynomial cut elimination, the syntax of LAL is very much constrained. This requires a new connective: the *paragraph*. The syntax of formulas, ranged over by $A, B \ldots$, is $\alpha \mid \alpha^\perp \mid A \parr A \mid A \otimes A \mid !A \mid ?A \mid \S A$. The negation $\perp$ is defined by de Morgan laws. Dual pairs of connectives or constants are : $(\otimes, \parr)$, $(!, ?)$, $(\S, \S)$ and $(\forall, \exists)$. The linear implication $\multimap$ is defined by $A \multimap B = A^\perp \parr B$.

The elimination of a cut between two weakened formulas (which can occur in LAL) is problematical; so, we simply restrict ourselves to the intuitionistic fragment of LAL (called ILAL), described in Fig. 1. Following [1,7,2], these rules may be explained by some remarks. In linear logic, the *of course* modality enables one to duplicate a proof net, whatever the number of formulas $?A$ in its context.

$$\frac{}{A \vdash A}\ \text{Ax} \qquad \frac{\Gamma \vdash A \quad A, \Delta \vdash B}{\Gamma, \Delta \vdash B}\ \text{Cut}$$

$$\frac{\Gamma, A, B, \Delta \vdash C}{\Gamma, B, A, \Delta \vdash C}\ \text{Perm.}$$

$$\frac{\Gamma, !A, !A \vdash B}{\Gamma, !A \vdash B}\ \text{Contr.} \qquad \frac{\Gamma \vdash B}{\Gamma, A \vdash B}\ \text{Weak.}$$

$$\frac{\Gamma, A, B \vdash C}{\Gamma, A \otimes B \vdash C}\ \otimes_l \qquad \frac{\Gamma \vdash A \quad \Delta \vdash B}{\Gamma, \Delta \vdash A \otimes B}\ \otimes_r$$

$$\frac{\Gamma \vdash A \quad B, \Delta \vdash C}{\Gamma, A \multimap B, \Delta \vdash C}\ \multimap_l \qquad \frac{\Gamma, A \vdash B}{\Gamma \vdash A \multimap B}\ \multimap_r$$

$$\frac{B \vdash A}{!B \vdash !A}\ \text{of course} \qquad \frac{\vdash A}{\vdash !A}\ \text{of course}$$

$$\frac{\Gamma, \Delta \vdash A}{!\Gamma, \S\Delta \vdash \S A}\ \text{paragraph}$$

$$\frac{\vdash \Gamma, A}{\vdash \Gamma, \forall\alpha.A}\ \text{for all}\ (\alpha \notin FV(\Gamma)) \qquad \frac{\Gamma, A[B/\alpha] \vdash C}{\Gamma, \forall\alpha.A \vdash C}\ \text{there is}$$

Fig. 1. Sequent calculus presentation of ILAL

In light logics, one needs to constrain the duplication in order to remain polytime. Hence, the *of course* rule admits only a single or empty context. Besides, the usual dereliction rule is replaced by the *paragraph* rule, which tags the conclusion with §.

Proof Nets of ILAL. Proof nets in ILAL are just like standard intuitionistic proof nets (plus a §-box). Since [7], it is customary to consider a stratified cut-elimination strategy on them. The intuitive idea is to reduce at depth 0 then 1 and so on. Every link has a measure (called size in this paper) and the total measure does not increase too fast during reduction. The enables a simple enough proof of P-Time. However, a recent work [18] shows that alternative ILAL *non stratified* strategies cut-eliminate in P-Time.

Every proof net has a unique output (conclusion) but many inputs (hypothesis). For example, an axiom introduces a net with one input linked to one output and a cut links the output of a net with an input of another net. Links $\lambda, \otimes_r$ and contraction have two outputs and a single input, @ and $\otimes_l$ have two inputs and a single output, $\forall$ has one input and one output, and the weakening has only one input and no output. There are boxes for $\forall$ (on the right), and exponential rules (*of course* and *paragraph*). A complete description of ILAL proof nets can be found in [2]. This cannot be included here for space reasons but Fig. 2 gives

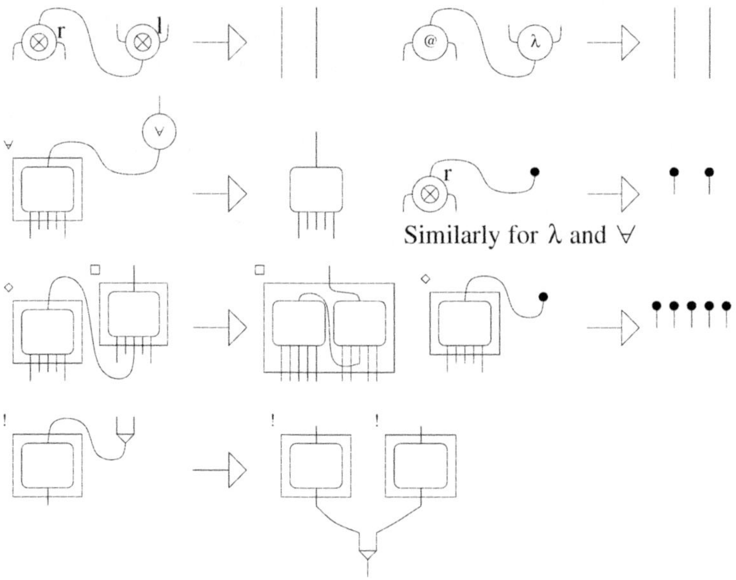

Fig. 2. Cut elimination for ILAL

the normalisation rules (where the diamond and box symbols can be either ∀, ! or § depending upon the types) and important points are recalled in Definition 2.1. Examples of derivations can be found in Fig. 3. They represent both nonequivalent kinds of integers (unary and binary ones) and the iterator which only works on unary integers.

Definition 1.1 (Depth). *The* depth *of a net is the maximal nesting of exponential (!/§)-boxes.*

A crucial property in ILAL is that the depth of a net cannot increase during a reduction and boxes cannot be opened since there is no dereliction. In particular, a cut between exponential boxes (either ! or §) is eliminated by merging the boxes. From that, Asperti adapts Girard's proof in [7] and proves

Proposition 1.1 (ILAL is in P-Time). *For every $d \in \mathbb{N}$, there exist polynomials S_d and R_d such that for every net of size s and depth d, the maximal size of its reducts is bounded by $S_d(s)$ and the maximal number of steps necessary to reach the normal form is bounded by $R_d(s)$.*

Conversely, Asperti and Roversi [2] adapt Girard's encoding of P-Time Turing machines inside proof nets, and prove that

Proposition 1.2 (P-Time is in ILAL). *Any P-Time function from binary integers to binary integers or booleans can be represented in ILAL.*

A Peano int: SS0=$\lambda fx.f(f\ x)$

$$\dfrac{\dfrac{\dfrac{\dfrac{\dfrac{\dfrac{\dfrac{}{\alpha \vdash \alpha} \quad \dfrac{\dfrac{}{\alpha \vdash \alpha} \quad \dfrac{}{\alpha \vdash \alpha}}{\alpha \multimap \alpha, \alpha \vdash \alpha}}{\alpha \multimap \alpha, \alpha \multimap \alpha, \alpha \vdash \alpha}}{\alpha \multimap \alpha, \alpha \multimap \alpha \vdash \alpha \multimap \alpha}}{!(\alpha \multimap \alpha), !(\alpha \multimap \alpha) \vdash \S(\alpha \multimap \alpha)}}{!(\alpha \multimap \alpha) \vdash \S(\alpha \multimap \alpha)}}{\vdash !(\alpha \multimap \alpha) \multimap \S(\alpha \multimap \alpha)}}{\vdash \forall\alpha.!(\alpha \multimap \alpha) \multimap \S(\alpha \multimap \alpha)}$$

A Binary int: $11_2 = \lambda s_1 s_0\ x.s_1(s_1\ x)$

$$\dfrac{\dfrac{\dfrac{\dfrac{\dfrac{\dfrac{\dfrac{\dfrac{}{\alpha \vdash \alpha} \quad \dfrac{\dfrac{}{\alpha \vdash \alpha} \quad \dfrac{}{\alpha \vdash \alpha}}{\alpha \multimap \alpha, \alpha \vdash \alpha}}{\alpha \multimap \alpha, \alpha \multimap \alpha, \alpha \vdash \alpha}}{\alpha \multimap \alpha, \alpha \multimap \alpha \vdash \alpha \multimap \alpha}}{!(\alpha \multimap \alpha), !(\alpha \multimap \alpha) \vdash \S(\alpha \multimap \alpha)}}{!(\alpha \multimap \alpha) \vdash \S(\alpha \multimap \alpha)}}{!(\alpha \multimap \alpha), !(\alpha \multimap \alpha) \vdash \S(\alpha \multimap \alpha)}}{\vdash !(\alpha \multimap \alpha) \multimap !(\alpha \multimap \alpha) \multimap \S(\alpha \multimap \alpha)}}{\vdash \forall\alpha.!(\alpha \multimap \alpha) \multimap !(\alpha \multimap \alpha) \multimap \S(\alpha \multimap \alpha)}$$

The iterator on Peano integers: $\lambda nfx.nfx$

$$\dfrac{\dfrac{\dfrac{\dfrac{}{!(A \multimap A) \vdash !(A \multimap A)} \quad \dfrac{\dfrac{\dfrac{}{A \vdash A} \quad \dfrac{}{A \vdash A}}{A \multimap A, A \vdash A}}{\S(A \multimap A), \S A \vdash \S A}}{!(A \multimap A) \multimap \S(A \multimap A), !(A \multimap A), \S A \vdash \S A}}{\forall\alpha.!(\alpha \multimap \alpha) \multimap \S(\alpha \multimap \alpha), !(A \multimap A), \S A \vdash \S A}}{\vdash (\forall\alpha.!(\alpha \multimap \alpha) \multimap \S(\alpha \multimap \alpha)) \multimap !(A \multimap A) \multimap \S A \multimap \S A}$$

Fig. 3. Integers and iterator

2 An Analysis of P-Time in ILAL

At this point, it is worth recalling the main steps leading to the proof of P-Time completeness for ILAL, as formulated in Propositions 1.1 and 1.2. These steps are mainly guidelines and we refer the reader to [2] for an explicit proof. We will see in Sect. 3 that the proof of NP-Time completeness for nILAL follows essentially the same steps.

2.1 Polynomial Bounds: ILAL in P-Time (Proposition 1.1)

Step 2.1 (Reduction Rules) *In ILAL, the rewriting rules over proof structures are of the form* $\overline{R} \mapsto \overline{S}$ *where:*

- ***Connectedness.*** $\overline{R}$ *is a connected proof structure extended with (meta) variables for proof structures, proof nets, formulas and links.*
- ***Locality.*** *All links in* $\overline{R}$ *are in the same depth (depth is defined in Definition 2.1).*

Remark 2.1. The locality property is not true for LLL because the reduction of the visible cut in Fig. 4 cannot be performed before some reduction steps occur inside the !-box. These "internal" reduction steps are necessary to remove every auxiliary door but one of the !-box (this is mandatory to ensure a polynomial

The reduction is not local as seen in this net: if the left auxiliary door is removed (by reducing the weakening) then the contraction disappears and the box may be duplicated.

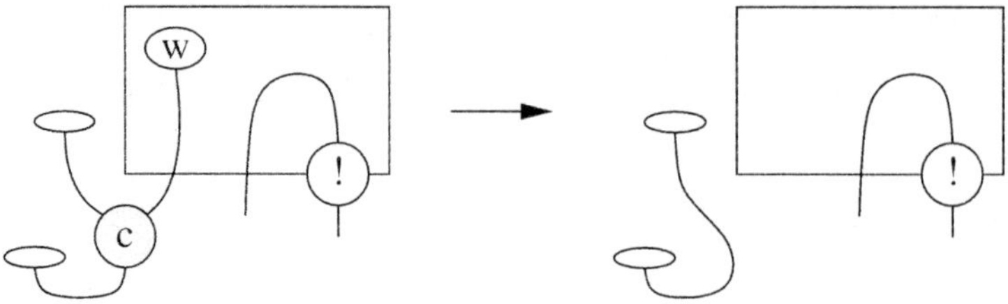

Fig. 4. Non locality of LLL

reduction). We will see next section that locality enables to extend the reduction strategies to nondeterministic SUM boxes; as for connectedness, that it enables us to *select* sub-nets and compute inside them.

Definition 2.1 (Boxes, depth and size). *The* depth *of a link in a ILAL net is the number of the exponential boxes which contain this link. The* size *of a net is the sum of the sizes of its links. The* partial size s_i *of a net is the sum of the sizes of the links at depth* i.

Step 2.2 (Size) *The size and partial sizes of a proof net are polynomially equivalent to the actual size and partial sizes of the proof net as a graph.*

Step 2.3 (Cuts) *There are two kinds of cuts: basic ones and exponential ones. At depth d, the reduction of:*

- *a basic cut only creates cuts at depth d and reduces the partial size* s_d,
- *an exponential cut creates only redexes at depths at least* $d+1$.

Step 2.4 (Normalisation) *A complete ILAL normalisation strategy of a net of depth d is:*

- *in a preliminary round, reduce basic cuts at depth 0,*
- *for* $i < d$*, in round* i*, reduce exponential cuts at depth* i *and then basic ones at depth* $i+1$.

Remark 2.2. The standard definitions of LLL or ILAL are more liberal, since they enable to interleave reduction of exponential cuts at depth i and of basic ones at depth $i+1$; whereas here, we impose that basic cuts are reduced after exponential ones. This is possible in the case of ILAL, thanks to locality. This stricter strategy is introduced here because it is needed for confluence in the case of nondeterminism.

	depth 0	*depth* 1	*depth* 2	...	*depth d*
preliminary	s_0	s_1	s_2	...	s_d
round 0	s_0	$(s_0)s_1$	$(s_0)s_2$	...	$(s_0)s_d$
round 1	s_0	$(s_0)s_1$	$(s_0 s_1)(s_0)s_2$	...	$(s_0 s_1)(s_0)s_d$
round 2	s_0	$(s_0)s_1$	$(s_0 s_1)(s_0)s_2$	...	$(s_0 s_1 s_0 s_2)(s_0 s_1)(s_0)s_d$
⋮	⋮	⋮	⋮	⋱	⋮
round $d-1$	s_0	$s_0 s_1$	$s_0^2 s_1 s_2$	...	$s_0^{2^d} \dots s_{d-1} s_d$

Fig. 5. Partial sizes

Step 2.5 (Partial sizes and normalisation) *If $s_i^{(k)}$ is the maximal partial size at depth i during round k and $S_i^{(k)}$ the maximal partial size during first part of round k, we have $\forall k \leq k'$, $S_{k'}^{(k+1)} \leq s_{k'}^{(k+1)} \leq s_k^{(k)}.s_{k'}^{(k)}$. Hence partial sizes are bounded by the values in Fig. 5.*

Proposition 2.1 (Time). *There are polynomials R_d and S_d such that for every proof net of size s and depth d, the maximal number of steps before reaching the normal form is bounded by $R_d(s)$ and the size of its reducts is bounded by $S_d(s)$.*

Proof. The preliminary round takes at most s_0 step since basic cuts reduce partial size. The exponential sub-round i takes at most $(s_0 \dots s_i) \dots (s_0 s_1)(s_0)s_i$ steps it reduces the number of cuts. The basic sub-round i is similar to the preliminary round. □

Step 2.6 (Polystep/polytime) *Every step is polynomially bounded by the size of the net.*

Therefore, we remark that the reduction is polytime and not only polystep.

2.2 Completeness: P-Time in ILAL (Proposition 1.2)

Step 2.7 (Iterator) *A net $iter(n, f, input)$ where f is a net of type $A \multimap A$, n a Peano integer, and input an arbitrary input of the type A, reduces to $P(f \circ \dots \circ f(input))$ where P is some context independent from f, n or input.*

Step 2.8 (Encoding) *Turing machines of type bint $\to \tau$ are encoded (and simulated) in some types that are in a set $\Theta_{TM} = \{!bint \multimap \S^k \tau\}$. When given the translation n of an integer, the encoding of a Turing machine is equivalent to a net $C(iter(pol(length(n)), T, config(n)))$ where:*

- *C is a context from configurations to a representation of τ (there might be some* paragraphs *which come from the polynomial pol),*
- *iter is the iterator,*
- *length is the function from bint to int equivalent to $\lambda nfx.nffx$,*
- *pol is a polynomial, i.e. a function from int to int,*
- *config is a function from binary integers to configurations,*
- *T is the encoding of a deterministic transition.*

3 Adding Nondeterminism

Remark 3.1. In the sequel of this paper, the representation of an output type needs to take into account the notion of accepting state. We represent an output type τ (which needs not be iterable), where $\tau = \mathit{bint}$ or $\tau = \mathit{bool}$, by $\mathit{bool} * \tau$ where the first element is *true* if the computation ends in the accepting state.

3.1 The Connective $\oplus$ and the Sum Box

Although ILAL has no additive connectives, the connective $\oplus$ led to our first intuition on nondeterminism. The usual understanding of polarities (in games or ludics [12,8]) expresses that, during an interaction, a positive connective commits the first move and a negative one answers. The duality between the connectives $\oplus$ and & can be explained as: the connective $\oplus$ chooses what to do, it always play either $\oplus_1$ or $\oplus_2$ and the connective & can answer to any of $\oplus_1$ and $\oplus_2$. Hence interaction is well determined between these connectives. A personal discussion with Paul-André Melliès led the author to the study of a strange connective interpreted in terms of nondeterminism: the connective $\oplus_3$ is positive, but chooses nondeterministically between $\oplus_1$ and $\oplus_2$. So, we add a new rule $\oplus_3$ for $\oplus$ which is the same as the rule for & and get three rules for this connective:

$$\frac{\vdash \Gamma, A}{\vdash \Gamma, A \oplus B}\,\oplus_1 \qquad \frac{\vdash \Gamma, B}{\vdash \Gamma, A \oplus B}\,\oplus_2 \qquad \frac{\vdash \Gamma, A \quad \vdash \Gamma, B}{\vdash \Gamma, A \oplus B}\,\oplus_3$$

Cut-elimination has to be performed in the following proof π:

$$\frac{\dfrac{\vdash \Gamma, A \quad \vdash \Gamma, B}{\vdash \Gamma, A \oplus B}\,\oplus_3 \qquad \dfrac{\vdash \Delta, A^\perp \quad \vdash \Delta, B^\perp}{\vdash \Delta, A^\perp \& B^\perp}\,\&}{\vdash \Gamma, \Delta}\,\mathit{cut}$$

Hence, since we are seeking for a confluent system, we need a rule for choice. We introduce a Sum rule, inspired by what Ehrhard and Regnier use in [5] for algebraic purposes:

$$\frac{\vdash \Gamma \quad \ldots \quad \vdash \Gamma}{\vdash \Gamma}\,\textsc{Sum}$$

The reduct of π is then:

$$\frac{\dfrac{\vdash \Gamma, A \quad \vdash \Delta, A^\perp}{\vdash \Gamma, \Delta}\,\mathit{cut} \qquad \dfrac{\vdash \Gamma, B \quad \vdash \Delta, B^\perp}{\vdash \Gamma, \Delta}\,\mathit{cut}}{\vdash \Gamma, \Delta}\,\textsc{Sum}$$

The rule $\oplus_3$ is then definable with the Sum rule.

Definition 3.1 (The logic nILAL). *The logic nILAL is formed by the rules of ILAL and a* Sum *rule:*

$$\frac{\Gamma \vdash A \quad \ldots \quad \Gamma \vdash A}{\Gamma \vdash A}\,\textsc{Sum}$$

Reduction rules for nILAL are given in Fig. 6 and 7. The strategy is described in Definition 3.4. One may remark that the DOWN rule is not local. This is not surprising since it changes the complexity (if $P \neq NP$).

Remark 3.2. We may proceed in the same way with a variant of LLL called Elementary Light Logic. In that case, however, SUM boxes are (intuitively) encodable inside the logic. Indeed, ELL corresponds to elementary time and elementary time is closed under exponential. It is easy to deduce from the sequel of the paper, that this encoding cannot be considered in ILAL for complexity reasons: it would impose an exponential cut elimination.

3.2 Proof Nets

Proof structures for nILAL are proof structures for ILAL plus SUM boxes of conclusion Γ with *one or more* sub-proof structures of conclusions Γ. The proof net associated to a proof ending by a SUM rule is the SUM box of the proof nets associated to every immediate sub-proofs of the considered proof.

Definition 3.2 (Summand). *In a given net, a* summand *is a proof structure obtained by erasing all but one sub-net in every* SUM *box.*

Proposition 3.1 (Correctness). *A nILAL proof structure is a proof net (i.e. is sequentialisable) if and only if all its summands are ILAL proof nets.*

Proof. By induction on the structure of proofs or proof nets. □

Definition 3.3 (Depth). *The* depth *of a net is the maximal depth (in the sense of ILAL) of its summands.*

3.3 Normalisation

In order to define a P-Time normalisation strategy, cut elimination for ILAL is stratified (i.e. executed in several rounds). Similarly, the strategy that we define for nILAL is stratified. This is necessary for the representation theorem: one shouldn't decide a nondeterministic choice too early. The first rules for nILAL are described in Fig. 6.

Definition 3.4 (Normalisation). *We define a normalisation strategy for a proof net R of depth d:*

- *Preliminary. In a first preliminary round, eliminate every sub-*SUM *boxes of* SUM *boxes using the* MERGE *rule. One can suppose that the whole net is included in a* SUM *box*[1]*. This outer* SUM *box is called* principal. *Its immediate sub-nets are also called* principal.

[1] By eventually plugging R in a singleton SUM box: SUM(R).

The MERGE **rule (associativity):**

A net $\textsc{Sum}(R_1, \ldots, \textsc{Sum}(S_1, \ldots, S_n), \ldots, R_k)$ may be reduced into the net $\textsc{Sum}(R_1, \ldots, S_1, \ldots, S_n, \ldots, R_k)$.

The DOWN **rule (linearity):**

A net of the form $\textsc{Sum}(\ldots, C[\textsc{Sum}(S_1, \ldots, S_k)], \ldots)$, where C is an arbitrary context, may be rewritten into $\textsc{Sum}(\ldots, \textsc{Sum}(C[S_1], \ldots, C[S_k]), \ldots)$.

The ILAL rules:

Reduce basic and exponential cuts as in ILAL.

Fig. 6. Cut elimination rules

The rest of the strategy is also described by rounds for nILAL but here, rounds only apply to principal nets: during a normalisation, one may encounter different rounds for different principal nets.

- *In a second preliminary round, apply* DOWN *rule,* MERGE *rule and basic reduction at depth* 0.
- *For* $i < d$, *in round* i, *eliminate exponential cuts at depth* i *and then apply* DOWN *rule,* MERGE *rule and basic reduction at depth* $i+1$.

Remark 3.3. SUM boxes rise to the top and gather in a single SUM box. One may also check that SUM boxes disappear: after round $i > 0$, there is no SUM box at depth i. This rule involves a lot of duplications. We add *selection* in Sect. 3.5 to avoid exponential complexity.

Let ψ be the function which takes a net and returns the set (and not the multi-set) of its principal sub-nets. A strategy is said to be confluent up to multiplicities when the strategy is confluent after applying ψ.

Proposition 3.2 (Confluence). *The normalisation strategy in nILAL is confluent up to multiplicities.*

Remark 3.4. The notion of confluence has to be relaxed because of the case of a net with a weakened $\forall$-box that contain a SUM box. There is a reduction path where the SUM box duplicates the context before being erased and another path where the SUM box is erased when the $\forall$-box is erased. Hence, the number of summands is not the same in both cases but the sets of (normal) summands are the same. This stratified strategy is also needed because of nondeterminism: a lax strategy where one can reduce SUM boxes before exponential ones is not confluent because one would do identical choices in a given box, and this is too uniform. If one needs a strictly confluent strategy, one may require that SUM boxes are reduced before basic cuts. A better solution is to slightly change the reduction rules of ILAL: the reduction of a weakened $\forall$-box would be the opening of the box with a weakening on top. The complete erasing of the box would wait for the reduction of inner SUM boxes.

3.4 An Encoding of NP-Turing Machines

Our encodings are still denoted by $[\![\ ]\!]$. To encode a nondeterministic transition, we simply use a SUM box of deterministic transitions, see [7,16,2]. To encode the Turing machine, we iterate this nondeterministic transition on an input with a polynomial bound (depending on the input) on steps. This net, when fed with an input (the encoding of the initial configuration of the machine), reduces to a SUM box of every possible results.

Lemma 3.1 (Iteration of Sum boxes).
A net $Iter(n, \text{SUM}((f_i)_{i\in[1..k]}), input)$ where the f_i are deterministic nets of type $A \multimap A$ reduces to $\text{SUM}(P(f_{i_1} \circ f_{i_2} \circ \ldots \circ f_{i_n}(input))_{i_1,\ldots,i_n\in[1..k]})$ where P is the same context as for Step 2.7.

Definition 3.5 (Representation). *A nondeterministic Turing machine f is* representable *in nILAL if and only if the net obtained by linking its representation in nILAL and the representation of an arbitrary input x reduces to a net which has a summand representing in the sense of ILAL any possible $f(x)$.*

Proposition 3.3 (Completeness). *Any NP-Time function from binary integers to binary integers or booleans can be represented in nILAL.*

The obvious reduction algorithm derived from this representation is exponential. However, next section proves that reduction can be done polynomially.

3.5 Time

When reducing a cut with a SUM box, one needs to be careful in order to avoid useless duplications. One can remark that two principal sub-nets cannot communicate, hence since we are only interested in summands, if we seek for a future summand of a given principal sub-net, we only need to reduce this sub-net. This given sub-net is called selected. In any computation, one may only have one selected sub-net, the one leading to the desired summand.

Definition 3.6 (Selected sub-net). *In a reduction, one may consider a particular principal sub-net and perform the next reduction step if possible inside this sub-net. This sub-net is said to be* selected.

For the simplicity of exposition, we change the DOWN rule into the DOWN' one. When applying the DOWN and MERGE rules,

$$\begin{aligned} &\text{SUM}(\ldots, C[\text{SUM}_i(S_i)], \ldots) \\ \longmapsto\ &\text{SUM}(\ldots, \text{SUM}_i(C[S_i]), \ldots) \\ \longmapsto\ &\text{SUM}(\ldots, C[S_1], \ldots, C[S_k], \ldots) \end{aligned}$$

one may exponentially increase the amount of duplications. To avoid this, we use selection: we write the net as a SUM box with at most two principal sub-nets: the selected one and a SUM box with other former sub-nets. We add a new rule in

The DOWN' **rule**:

Every net of the form $\text{SUM}(C[\text{SUM}_i(S_i)], D)$ can be rewritten into $\text{SUM}(C[S_{i_0}], \text{SUM}(D, C[\text{SUM}_{i \neq i_0}(S_i)]))$ where i_0 corresponds to the newly selected sub-net.

Fig. 7. The DOWN' rule

Fig. 7, the DOWN' rule, which preserve the property of having only two principal sub-nets. The DOWN' is more complex than the DOWN rule but its reduction is cheaper since it only imposes one duplication of the context C. However, we could have kept the DOWN rule but proofs would have been heavier. We adapt our normalising strategy described in Definition 3.4 by replacing DOWN rules by DOWN' rules. This is safe since DOWN and DOWN' lead to the same normal forms.

Definition 3.7 (Selective reduction). *A reduction is said to be* selective *if at every step, there is a selected principal sub-net, and at every non initial step, the selected principal sub-net is a reduct of the previous one.*

Remark 3.5. The MERGE rule does not cost much: one has at most one MERGE rule after every DOWN' rules. Therefore, in the sequel, we may suppose that no MERGE rule can be applied in the selected principal net whenever needed.

Proposition 3.4 (Selective reduction). *For every net R and any summand of its normal form S, there is a selective reduction to a net which has S as a summand.*

Proof. The connectedness hypothesis and the definition of the normalisation strategy enable one to reduce principal sub-nets independently. □

Definition 3.8 (Sizes). *The* size *of a net is the maximal size its summands and the* SUM size *of a net is the sum of the sizes of its links, where the size of a* SUM *box is* 1.

Definition 3.9 (Sum reduction step). *We call* SUM reduction steps *the ones using* MERGE *or* DOWN' *rules. Other reduction steps are called* non-SUM reduction steps.

Proposition 3.5 (Time). *As in ILAL, for a net of size s and depth d, for every summand S of its normal form, the maximal number of non-*SUM *reduction steps necessary to reach a net which has S as a summand is bounded by $R_d(s)$.*

Proof. Consider a selective reduction leading to S and use Step 2.4 (normalisation). Check that Step 2.5 (partial sizes) is still valid. The sequel of the proof is similar of the one of Proposition 2.1. □

Now that non-SUM reduction steps are controlled, it is time to go on with SUM ones. They are as well controlled. Since it is valid, we suppose that no MERGE rule is applicable in the considered selected net.

Proposition 3.6 (Sum size). *During a selective reduction, the* SUM *size of the selected principal sub-net is bounded by* $s.S_d(s)$.

Proof. One can prove a result analogous to step 2.5 which establishes that each partial SUM size is bounded by the initial SUM size of the net times the partial size given in step 2.5. □

The DOWN' rules strictly lower the maximal nesting number of SUM boxes in the selected principal sub-net which is bounded by $S_d(s)$. Hence their applications are bounded by $S_d(s)$.

Proposition 3.7 (Maximal Sum size). *For every* $d \in \mathbb{N}$, *there exists a polynomial* M_d *such that for every selective reduction of a net* R *of* SUM *size* s *and depth* d, *the maximal* SUM *size of its reducts is bounded by* $M_d(s)$.

Proof. The non selected part of the net is a comb tree. Hence its SUM size is bounded by $s.S_d(s) + 1$ times the number of SUM reductions plus 1. Therefore, we require $M_d(s) \geq (s.S_d(s) + 1).(R_d(s') + 1)$ where s' is the size of R. Since s' is smaller than s, we take $M_d(X) = (X.S_d(X) + 1)(R_d(X) + 1)$. □

Proposition 3.8 (Time). *For every* $d \in \mathbb{N}$, *there exists a polynomial* R_d *such that for every net* R *of size* s *and depth* d, *for every summand* S *of the normal form of* R, *the maximal number of cut reductions to reach a net which has* S *as a summand is bounded by* $R_d(s)$.

Proof. For a net of size s_0 and depth d, for every summand S of its normal form, the maximal number of non SUM reduction steps necessary to reach a net which has S as a summand is bounded by $R_d(s_0)$. We only need to prove that there are not too many SUM reduction steps and that they don't involve too many duplications. We need not be too careful, we can bound the number of DOWN' reductions by $S_d(s_0)$ and every duplication costs less than $M_d(s_0)$. So, the work to reduce the net is bounded by $((R_d + S_d)M_d)(s_0)$ and hence it is polynomial. □

Proposition 3.9 (Correctness). *For every net* $R :!bint \multimap \S^k\tau$ *there is a polynomial* P *such that for every input* x *of size* s *and depth* 1, *for every summand* S *of the normal form of the net* $f(x)$, *the maximal number of cut reductions to reach a net which has* S *as a summand is bounded by* $P(s)$.

Proof. Take an input of R of type *bint* and size s. Since a binary integer does not increase depth, the reduction toward a summand of $f(x)$ is polynomial in $s + s_0$, and hence polynomial in s. □

As for ILAL, we remark that the reduction is, here nondeterministically, polytime and not only polystep.

4 Reformulation of the P $\neq$ NP Conjecture

In this section, we reformulate the P $\neq$ NP conjecture. For simplicity, we only consider boolean problems. The following results are directly justified by the fact that we have a complete model of NP. However, we believe that they might have an interest thanks to the model.

If R is a net and b a boolean value, we write $R \rightsquigarrow b$ to express that the normal form of R has a summand representing b (in the accepting state 0).

Definition 4.1 (The relations $\cong_{\mathbf{NP}}$, $\cong_{\mathbf{coNP}}$ and ${}_{\mathbf{NP}}\leftrightharpoons_{\mathbf{coNP}}$). *We define three relations on nets of conclusions in Θ_{TM}:*

$R \cong_{NP} S \iff (\forall n \in \mathbb{N}, R([\![n]\!]) \rightsquigarrow true \Leftrightarrow S([\![n]\!]) \rightsquigarrow true)$

$R \cong_{coNP} S \iff (\forall n \in \mathbb{N}, R([\![n]\!]) \rightsquigarrow false \Leftrightarrow S([\![n]\!]) \rightsquigarrow false)$

$R \; {}_{NP}\leftrightharpoons_{coNP} S \iff (\forall n \in \mathbb{N}, R([\![n]\!]) \rightsquigarrow true \Leftrightarrow S([\![n]\!]) \not\rightsquigarrow false)$

Proposition 4.1 (Correctness/completeness). *Some natural results hold:*

1. *For every NP Turing machines M and N, M and N are NP-equivalent if and only if $[\![M]\!] \cong_{NP} [\![N]\!]$ and similarly for coNP.*
2. *For every proof net R of type in Θ_{TM}, there is a Turing machine M such that $[\![M]\!] \cong_{NP} R$ and similarly for coNP.*

Proof. This comes from the representation property and Proposition 3.9. □

Corollary 4.1 (Equivalence classes). *The translation $[\![\ ]\!]$ can be lifted in a bijection between NP and classes of $\cong_{NP}$ and between coNP and classes of $\cong_{coNP}$.*

Definition 4.2 (Uniformity and determinism). *A net is said to be* uniform *if $R \; {}_{NP}\leftrightharpoons_{coNP} R$. A net is said to be* deterministic *if R is uniform and every computation ends in the accepting state. The set $\mathfrak{Unif}$ is the set of uniform nets, and the set $\mathfrak{Det}$ is the set of deterministic nets.*

Recall from [15] that if NP = coNP, the whole polynomial hierarchy[2] collapses which implies P = NP. Therefore, P $\neq$ NP if and only if NP $\neq$ coNP.

Proposition 4.2 (Reformulation of the P $\neq$ NP conjecture). *We have:*

$P \neq NP \iff \exists R. \nexists S. R \; {}_{NP}\leftrightharpoons_{coNP} S \iff \exists S. \nexists R. R \; {}_{NP}\leftrightharpoons_{coNP} S$

$P \neq NP \iff \pi_1({}_{NP}\leftrightharpoons_{coNP}) \neq NP \iff \pi_2({}_{NP}\leftrightharpoons_{coNP}) \neq coNP$ *where π_i is the i-projection.*

We define three functions. The functions $/_{true}$ (*resp.* $/_{false}$) which takes a net of any type in Θ_{TM} and returns an NP-equivalent (*resp.* coNP-equivalent) net that never answers false (*resp.* true) and the function ϕ defined by:

$$\begin{array}{llcl} \phi : & \mathrm{NP} \cap \mathrm{coNP} & \longrightarrow & \mathfrak{Unif} \\ & (M : \mathrm{NP}, N : \mathrm{coNP}) & \longmapsto & \textsc{Sum}(M/_{true}, N/_{false}) \end{array}$$

Remark that ϕ is a bijection up to equivalence of uniform nets.

Proposition 4.3 (Uniformity and NP $\cap$ coNP). *We have:*

$P = NP \cap coNP \iff \forall R \in \mathfrak{Unif}. \exists S \in \mathfrak{Det}.\ R \cong_{NP} S \iff \forall R \in \mathfrak{Unif}. \exists S \in \mathfrak{Det}.\ R \cong_{coNP} S$

[2] The polynomial hierarchy is an ordered set of complexity classes, smaller than P-SPACE, containing P, NP and coNP.

5 Conclusions

We have presented a new model of NP which is, as far as we know, the first implicit characterisation of NP through the Curry-Howard isomorphism.

Concerning the possible applications of this work for the study of the class NP, we cannot see a clear guideline but some points should be interesting. Some logical languages (such as Λ_{LA} defined in [16]) seem sufficient for programming purposes and one might have intuitions on their nondeterministic versions. Another work might be to characterise NP-complete problems in our system. One may also try to find a geometrical notion as the nesting of SUM boxes to get some continuum of complexity classes. Some links with semantics works [3,10] may also be interesting. Finally, following the first motivation, one should also understand whether the model can be adapted to LLL, SLL or FALL [11,17].

Acknowledgements. Thanks to Patrick Baillot, Pierre-Louis Curien and Paul-André Melliès as well as to anonymous referees for their valuable comments and remarks.

References

1. Asperti A., *Light Affine Logic.* LICS98, 1998.
2. Asperti A., Roversi L., *Intuitionistic Light Affine Logic.* ACM Transactions on Computational Logic, 2002.
3. Baillot P., *Stratified coherent spaces: a denotational semantics for Light Linear Logic.* ICC00, 2000.
4. Bellantoni S., Cook S., *A new recursion-theoretic characterisation of the polytime functions.* Computational Complexity, 1992.
5. Ehrhard T., Regnier L., *The Differential Lambda-Calculus.* 2001.
6. Girard J.-Y., *Proof-nets: the parallel syntax for proof-theory.* Logic and Algebra. eds Ursini and Agliano, Marcel Dekker, New York 1996.
7. Girard J.-Y., *Light Linear Logic.* Information and Computation, 1998.
8. Girard J.-Y., *Locus Solum.* Mathematical Structures in Computer Science 11(3), 2001.
9. Hofmann M., *Linear types and non-size increasing polynomial time computation.* To appear in TCS, 1999.
10. Kanovich M. I., Okada M., Scedrov A., *Phase semantics for light linear logic.* Technical report, 1997.
11. Lafont Y., *Soft Linear Logic and Polynomial Time.* To appear in TCS.
12. Laurent O., *Polarized games.* Seventeenth annual IEEE symposium on Logic In Computer Science, IEEE Computer Society, 2002.
13. Leivant D., Marion J.Y., *Lambda calculus characterisations of poly-time.* TLCA 1993.
14. Murawski A. S., Ong C.-H. L., *Can safe recursion be interpreted in light logic?.* Second International Workshop on Implicit Computational Complexity, 2000.
15. Papadimitriou C., *Computational complexity.* Addison Wesley, 1994.
16. Roversi L., *A P-Time Completeness Proof for Light Logics.* CSL99, 1999.
17. Roversi L., *Flexible Affine Light Logic.* Submitted, 2001.
18. Terui K., *Light Affine Calculus and Polytime Strong Normalisation.* LICS01, 2001.

Polarized Proof Nets with Cycles and Fixpoints Semantics

Raphaël Montelatici

Équipe Preuves, Programmes et Systèmes*
Raphael.Montelatici@pps.jussieu.fr

Abstract. Starting from Laurent's work on *Polarized Linear Logic*, we define a new polarized linear deduction system which handles recursion. This is achieved by extending the *cut*-rule, in such a way that iteration unrolling is achieved by cut-elimination. The *proof nets* counterpart of this extension is obtained by allowing oriented cycles, which had no meaning in usual polarized linear logic. We also free proof nets from additional constraints, leading up to a correctness criterion as straightforward as possible (since almost all proof structures are correct). Our system has a sound semantics expressed in traced models.

1 Introduction

This paper investigates two different topics, proof nets and fixpoints semantics, in order to present a new syntax, *polarized proof nets with cycles*, as a way of encoding recursion within a logical-like framework.

Proof nets. Girard's Linear Logic [4] introduced new syntactic objects, proof nets, that represent proofs by means of graphs. They provide a much less sequential syntax than sequent calculus and yield a more convenient way to study cut-elimination (see for instance [3]). However, LL as a whole is pretty complex. A new system named Polarized Linear Logic (LLP), introduced by Laurent, arose from the study of a sub-system of LL ruled by polarity constraints. LLP enjoys simpler properties than LL, and is expressive enough to interpret classical systems such as Girard's LC or Parigot's $\lambda\mu$-calculus [14].
Laurent has given a fairly comprehensive description of LLP proof nets [9], which enjoy a good accommodation of additive connectives, a confluent and strongly normalizing procedure of cut-elimination and a simple correctness criterion.
Fixpoints semantics. Domain theory generally interprets recursive programs by means of least fixed-points of continuous functions. In categorical semantics, fixpoints are axiomatized as *fixpoint operators*. Simpson and Plotkin described in [16] various classes of fixpoint operators and provided axiomatic methods to construct them. Alternatively, *trace operators* [8] interpret recursive functions as well as cycling data structures [5].

* CNRS & Université Paris VII, Case 7014, 2 place Jussieu, 75251 Paris cedex 05.

M. Hofmann (Ed.): TLCA 2003, LNCS 2701, pp. 256–270, 2003.

The proof nets with cycles introduced in this paper are a new instance of such cyclic structures. As such, they have a natural interpretation in traced cartesian closed categories. The idea of adding a fixpoint primitive to LLP and the way of doing it is a real semantic backlash. In [13] we studied LLP semantics in *categories of continuations*, following Laurent's work on polarized games [10] as well as Melliès and Selinger's ideas [12]. We noticed that relaxing the *totality* constraint on strategies corresponds to adding strategies computed by fixpoints operations. The following step was to move from semantics to syntax, add a fixpoint deduction rule to a polarized logical system inspired by LLP, then study the associated proof nets and correctness criterion. Polarized proof nets with cycles appear as an elegant solution. The idea that cyclic nets could handle recursion has been suggested informally by Danos. Today the modern theory of polarized proof nets allows to formalize and study them.

2 Polarized Linear Logic

Here we briefly recall features of LLP that are relevant to our presentation. The reader should consult [9] for a complete account. In this paper, we only deal with MELLP, the *Multiplicative Exponential Polarized Linear Logic* fragment of LLP, though most of the properties stated below still hold for LLP.

2.1 MELLP Syntax

The formulas of MELLP are divided into positive and negative ones:

$$\begin{array}{rcccccc} P & ::= & 1 & \mid & P \otimes P & \mid & !N \\ N & ::= & \bot & \mid & N ⅋ N & \mid & ?P \end{array}$$

Notational conventions. We adopt the following conventions: P, Q denote positive formulas and N, M denote negative formulas. Γ and Δ stand for multisets of formulas (of any polarity) and $\mathcal{N}$ for a multiset of negative only formulas. The perpendicular symbol $\perp$ is used as a meta-operation negating formulas, according to the usual De Morgan rules of LL.

The deduction rules presented in figure 1 allow to prove sequents of shape $\vdash N_1, \ldots, N_k$ or $\vdash N_1, \ldots, N_k, P$ (*i.e.* with at most one positive formula).

2.2 Polarized Proof Nets

MELLP proof structures are oriented graphs inspired by LL proof structures. LLP introduces an orientation on edges which is determined by formulas polarities. Check figure 2 for the grammar of MELLP nodes.
As for LL proof structures, !-nodes are associated to !-boxes, such that edges crossing auxiliary doors must be labelled by negative formulas, see figure 3. The *level* of a node is defined to be the number of nested boxes which contain it.

$$\frac{}{\vdash N, N^{\perp}}ax \qquad \frac{\vdash \Gamma, N \quad \vdash \Delta, N^{\perp}}{\vdash \Gamma, \Delta}cut$$

$$\frac{\vdash \Gamma, P \quad \vdash \Delta, Q}{\vdash \Gamma, \Delta, P \otimes Q}\otimes \qquad \frac{\vdash \Gamma, N, M}{\vdash \Gamma, N \parr M}\parr$$

$$\frac{}{\vdash 1}1 \qquad \frac{\vdash \Gamma}{\vdash \Gamma, \perp}\perp$$

$$\frac{\vdash \mathcal{N}, N}{\vdash \mathcal{N}, !N}! \qquad \frac{\vdash \Gamma, P}{\vdash \Gamma, ?P}? \qquad \frac{\vdash \Gamma, N, N}{\vdash \Gamma, N}c \qquad \frac{\vdash \Gamma}{\vdash \Gamma, N}w$$

Fig. 1. MELLP deduction rules

The obvious transformation mapping MELLP deduction trees to proof structures is not onto and thus defines a strict subset of proof structures: *proof nets*. A so-called *correctness criterion* decides which proof structures are proof nets.

Definition 2.1 (Correctness graph). *The correctness graph of a proof structure is obtained by replacing* !*-boxes of the* 0*-level by generalized axiom nodes.*

Definition 2.2 (Correctness criterion). *Let* $C(\mathcal{R})$ *be the following property: The correctness graph of the proof structure* $\mathcal{R}$ *has no oriented cycle and has exactly one initial node*[1] *which is not a* w*-node or a* $\perp$*-node.*
A proof structure $\mathcal{R}$ *is a proof net if and only if it satisfies* $C(\mathcal{R})$ *and all the proof structures enclosed in any* !*-box inside* $\mathcal{R}$ *do so.*

This criterion is really simple, because it does not require to switch proof nets, like LL criterion does.

We recall the important notion of *positive trees*:

Definition 2.3 (Positive trees). *Positive trees are proof structures with one positive conclusion, inductively defined as:*

- *ax nodes,* 1*-nodes and* !*-boxes wrapping a proof structure are positive trees,*
- *two positive trees on top of a* $\otimes$*-node form a positive tree.*

Any positive edge within a proof structure is the root of a positive tree. Positive trees are well-delimited structures that allow copying and deleting: from this point of view, they generalize !-boxes. See figure 3 for their representation.

A set of *cut-elimination rules* makes proof nets a rewriting system, which task is to eliminate the *cut*-nodes. In the case of LLP and MELLP, cut-elimination is confluent and strongly normalizing. In figure 4, we give the rules for MELLP (note that polarized systems take advantage of polarities and use positive trees to express the cut-elimination rules).

[1] with respect to the orientation relation on nodes

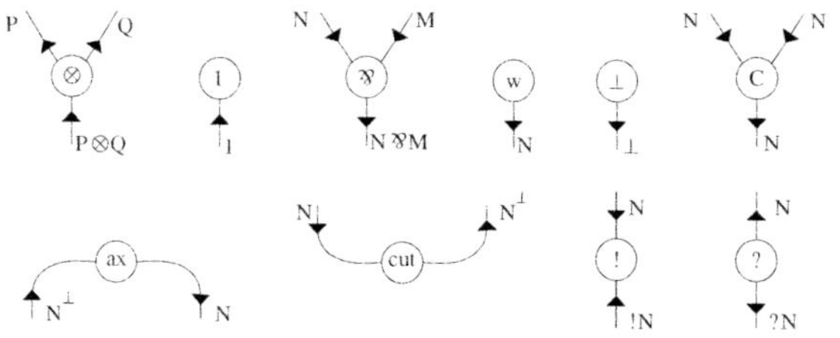

Fig. 2. Nodes

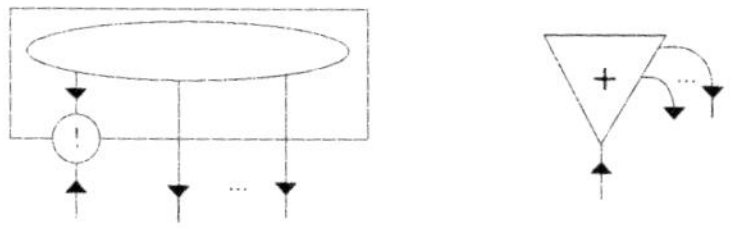

Fig. 3. Exponential box and positive tree representations

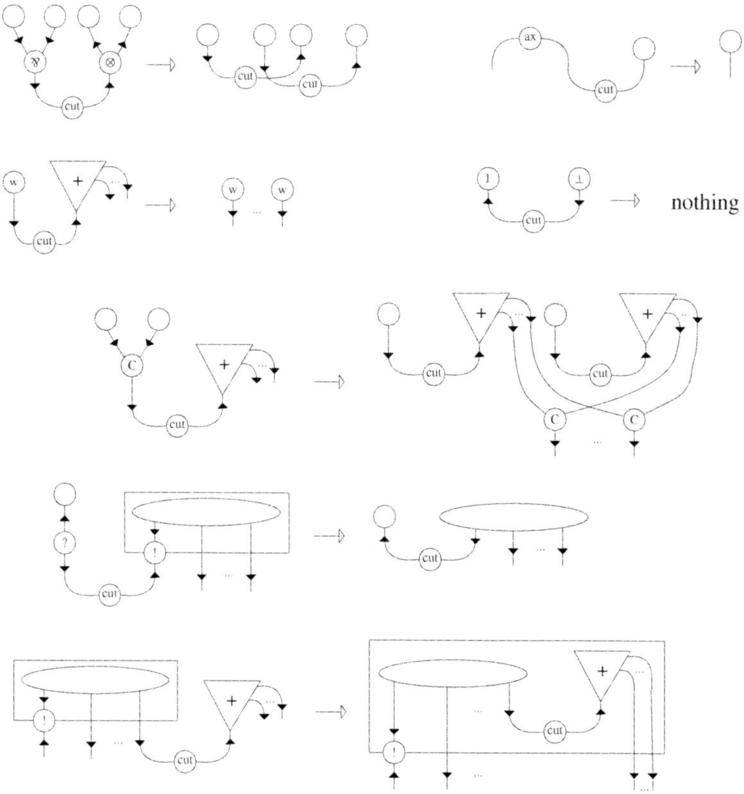

Fig. 4. Cut-elimination rules

3 The Semantic Pathway from LLP to MELLP with Cycles

This paper is about adding fixpoints to LLP. It is well-known that fixpoints can only be computed on a category of domains, not on a category of predomains. Logically, this means moving from LLP to MELLP, that is forgetting the additive connective $\oplus$.

Modulo translation, LLP and the extended $\lambda\mu$-calculus of Selinger are equivalent. Hence LLP can be interpreted in any model of the $\lambda\mu$-calculus, like *control categories* [15]. Selinger proves a representation theorem, stating that every control category is isomorphic (in the sense of control categories) to some category $R^{\mathcal{C}}$ where $\mathcal{C}$ is a *response category.* Categorically, moving from LLP to MELLP means relaxing the notion of response categories, by discarding coproducts as well as the so-called mono requirement [15]. This leads to *categories of continuations*, in the sense of Hofmann and Streicher [7]:

Definition 3.1. *A category of continuations $\mathcal{C}$ is a category:*

- *with a cartesian product $\otimes$ and a terminal object* 1,
- *with a distinguished object R and a counter-variant functor $R^{(-)} : \mathcal{C}^{op} \to \mathcal{C}$ such that there is a natural family of bijections $\Lambda_{A,B}$*

$$\frac{A \otimes B \to R}{A \to R^B}$$

Positive formulas of MELLP are interpreted according to the following rules:

$$[\![1]\!] = 1 \qquad [\![P \otimes Q]\!] = [\![P]\!] \otimes [\![Q]\!] \qquad [\![!N]\!] = R^{[\![N^\perp]\!]}$$

A deduction tree of conclusion sequent $\vdash N_1, \ldots, N_k, P$ is interpreted by a morphism $[\![N_1^\perp]\!] \otimes \ldots \otimes [\![N_k^\perp]\!] \to [\![P]\!]$ and a deduction tree of conclusion sequent $\vdash N_1, \ldots, N_k$ is interpreted by a morphism $[\![N_1^\perp]\!] \otimes \ldots \otimes [\![N_k^\perp]\!] \to R$.

Important examples of categories of continuations are cartesian closed categories. Let us recall the definition of a fixpoint operator in a cartesian category:

Definition 3.2 (Fixpoint operator). *Let $(\mathcal{D}, \times, 1)$ be a cartesian category with a terminal object. A parameterized fixpoint operator is a family of functions $(\_)^\nabla : \mathcal{D}(X \times A, A) \to \mathcal{D}(X, A)$ such that:*

1. *(Fixpoint property) For all $f : X \times A \to A$, $< id_X, f^\nabla >; f = f^\nabla$*
2. *(Naturality) For all $f : X \times A \to A$ and $g : Y \to X$, $(g \times A; f)^\nabla = g; f^\nabla$*

An interesting example: we noticed in [13] that the category of continuations associated to Hyland-Ong games and innocent strategies [10] coincides with the underlying control category; and thus admits the fixpoint operator associated to Y in the usual interpretation of PCF.
Adding a trace to the semantics may be understood from several perspectives.

Proof-theoretically, it means adding the following fixpoint deduction rule to MELLP sequent calculus:

$$\frac{\vdash \Gamma, P^{\perp}, P}{\vdash \Gamma, P} Y$$

Syntactically, it means extending polarized proof nets with so-called Y-boxes; and adding the reduction rule of figure 5 to usual proof net cut-elimination [13].

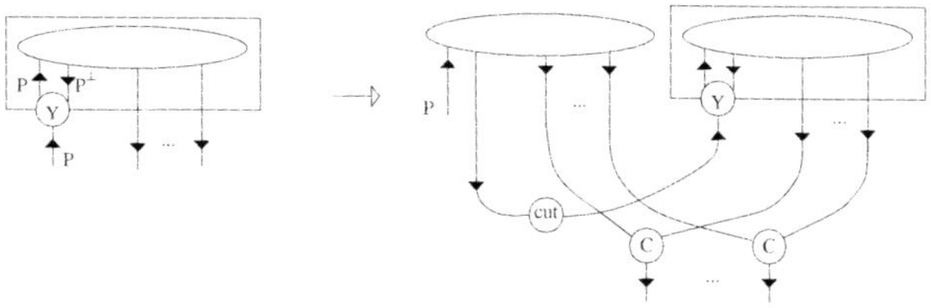

Fig. 5. Y-box and associated reduction rule

Here, we want to go one step further and set up a system that simulates the dynamical behavior of the Y-box by means of cut-elimination rules only. As duplication is typically performed by the elimination of a *cut*-node below a contraction node, it looks like the proof net of figure 6 is a nice candidate.

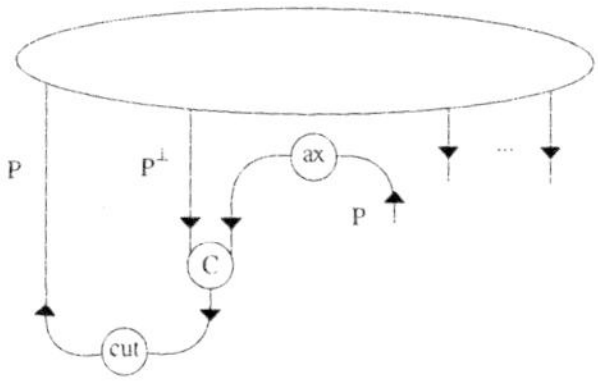

Fig. 6. An alternative for Y

This MELLP proof structure is generally not a MELLP proof net because it may contain oriented cycles. In order to see such cyclic structures as "proofs", we shall introduce a deduction system with (1) a *mix*-rule which allows putting several proofs side by side and (2) a relaxed *cut*-rule which applies to a single premise.

4 Multiplicative Exponential Polarized Linear Logic with Cycles

Our system is based on MELLP and is called MELLPC (MELLP *with cycles*). Since cut-elimination is not strongly normalizing in the presence of recursion, MELLPC is closer to a programming language, but still enjoys a flavor of logic.

4.1 Sequents-Style Syntax

We keep MELLP class of formulas:

$$\begin{array}{rcl} P & ::= & 1 \mid\ !N \mid P \otimes P \\ N & ::= & \bot \mid\ ?P \mid N \parr N \end{array}$$

The sequents are of the form $\{n\} \vdash \Gamma$ where Γ is a multiset of polarized formulas and n is a natural number. The deduction rules are presented in figure 7.

$$\frac{}{\{0\} \vdash \Gamma, \Gamma^{\perp}}ax \qquad \frac{\{n\} \vdash \Gamma \quad \{m\} \vdash \Gamma'}{\{n+m\} \vdash \Gamma, \Gamma'}mix \qquad \frac{\{n\} \vdash \Gamma, N, N^{\perp}}{\{n\} \vdash \Gamma}cut$$

$$\frac{\{n\} \vdash \Gamma, P, Q}{\{n\} \vdash \Gamma, P \otimes Q}\otimes \qquad \frac{\{n\} \vdash \Gamma, N, M}{\{n\} \vdash \Gamma, N \parr M}\parr$$

$$\frac{}{\{0\} \vdash 1}1 \qquad \frac{\{n\} \vdash \Gamma}{\{n\} \vdash \Gamma, \bot}\bot$$

$$\frac{\{1\} \vdash \mathcal{N}, N}{\{0\} \vdash \mathcal{N}, !N}! \qquad \frac{\{n\} \vdash \Gamma, P}{\{n+1\} \vdash \Gamma, ?P}? \qquad \frac{\{n\} \vdash \Gamma, N, N}{\{n\} \vdash \Gamma, N}c \qquad \frac{\{n\} \vdash \Gamma}{\{n\} \vdash \Gamma, N}w$$

Fig. 7. MELLPC deduction rules

Remark 4.1. The natural numbers labelling the sequents hold a counter of the number of ?-rules used in the sub-tree above each sequent, which are not above any !-rule in the same sub-tree. We could have discarded these numbers since they are determined by the deduction rules, but we keep them anyway, because they provide a convenient way to express the new !-rule and also because they directly take part in the semantics of proofs (see next subsection).

Remark 4.2. Since multisets can be empty, the axiom rule may be instantiated in order to prove the empty sequent $\{0\} \vdash$ in one step. Even if we disallow such a use of the axiom rule, our system would be able to prove the empty sequent anyway, by applying the *cut*-rule to the sequent $\{0\} \vdash N, N^{\perp}$ obtained after an axiom rule on the formula N. From a programming point of view, these proofs correspond to non-terminating recursions, *i.e.* diverging fixpoint applications.

Proposition 4.3. *By labelling* MELLP *sequents (*$\vdash N_1, \dots, N_k$ *becoming* $\{1\} \vdash N_1, \dots, N_k$ *and* $\vdash N_1, \dots, N_k, P$ *becoming* $\{0\} \vdash N_1, \dots, N_k, P$*),* MELLP *can be embedded in* MELLPC*, making every* MELLP *deduction rule admissible.*

Proof. By induction. For instance, the usual *cut*-rule is simulated by:

$$\frac{\dfrac{\{0\} \vdash \Gamma, N \qquad \{i\} \vdash \Delta, N^{\perp}}{\{i\} \vdash \Gamma, \Delta, N, N^{\perp}}mix}{\{i\} \vdash \Gamma, \Delta}cut \qquad \text{where } i = 0 \text{ or } 1$$

4.2 Semantics in Traced Cartesian Closed Categories

We designed MELLPC in the light of both semantics and proof net technology. On one hand we want a correctness criterion as simple as possible, and on the other hand, we want to stick to a straightforward semantics. In order to allow more correct proof structures, the MELLPC dereliction ?-rule handles a context $\Gamma^\perp$ of negative formulas and Δ of positive formulas. Semantically, this implies to interpret:

$$\frac{\Gamma \to \Delta \otimes A}{\Gamma \otimes R^A \to \Delta \otimes R}$$

This can be achieved in cartesian closed categories but not in categories of continuations.

We recall the definition of *traced symmetric monoidal categories*, as found in [5]. This definition is a slight adaptation from the original one given in [8].

Definition 4.4. *A symmetric monoidal category* $(\mathcal{C}, \otimes, 1)$ *is said to be* traced *if it is equipped with a natural family of functions, called a* trace,

$$Tr^A_{\Gamma,\Delta} : \mathcal{C}(\Gamma \otimes A, \Delta \otimes A) \longrightarrow \mathcal{C}(\Gamma, \Delta)$$

which satisfies the following properties:

1. ***Vanishing**:* $Tr^1_{A,B}(f) = f$*, for all* $f : A \longrightarrow B$
   and $Tr^{X\otimes Y}_{A,B}(f) = Tr^X_{A,B}(Tr^Y_{A\otimes X,B\otimes X}(f))$ *for all* $f : A\otimes X\otimes Y \longrightarrow B\otimes X\otimes Y$
2. ***Superposing**:* $Tr^X_{C\otimes A,C\otimes B}(id_c \otimes f) = id_C \otimes Tr^X_{A,B}(f)$
   for all $f : A \otimes X \longrightarrow B \otimes X$
3. ***Yanking**:* $Tr^X_{X,X}(s_{X,X}) = id_X$
   (where $s_{X,X} : X \otimes X \longrightarrow X \otimes X$ *is the symmetry morphism)*

Note that the naturality of $Tr^A_{\Gamma,\Delta}$ in Γ, Δ and A is usually called *left tightening*, *right tightening* and *sliding*, respectively.

On the semantical side, trace operators perfectly match the extended *cut*-rule of MELLPC. It is worth noting that in the case of cartesian closed categories, trace operators are equivalent to *Conway operators*, which are fixpoints operators with some additional properties [5].

We interpret MELLPC in *traced cartesian closed categories* with a distinguished object R, as follows:

- Positive formulas are interpreted according to the following rules:

$$[\![1]\!] = 1 \qquad [\![P \otimes Q]\!] = [\![P]\!] \otimes [\![Q]\!] \qquad [\![!N]\!] = [\![N^\perp]\!] \Rightarrow R$$

- A deduction tree of root $\{n\} \vdash N_1, \ldots, N_k, P_1, \ldots, P_l$ is interpreted by a morphism:

$$[\![N_1^\perp]\!] \otimes \ldots \otimes [\![N_k^\perp]\!] \longrightarrow [\![P_1]\!] \otimes \ldots \otimes [\![P_l]\!] \otimes R \otimes \ldots \otimes R$$

where the number of R's in the right-hand side product is equal to n. The semantics of an empty product is equal to the terminal object 1, as usual.

The interpretation morphism is constructed by induction on the deduction tree according to the rules of figure 8. Although these rules build interpretation morphisms that depend on an assumed ordering of the multisets composing the sequents, the axiomatic properties of trace operators allow to interpret the sequents up to commutations of formulas.

Rule	Premise morphism(s)	Conclusion morphism
ax		$\Gamma \xrightarrow{id_\Gamma} \Gamma$
1		$1 \xrightarrow{id_1} 1$
mix	$\Gamma \xrightarrow{f} \Delta$ and $\Gamma' \xrightarrow{g} \Delta'$	$\Gamma \otimes \Gamma' \xrightarrow{f \otimes g} \Delta \otimes \Delta'$
cut	$\Gamma \otimes A \xrightarrow{f} \Delta \otimes A$	$\Gamma \xrightarrow{Tr(f)} \Delta$
c	$\Gamma \otimes A \otimes A \xrightarrow{f} \Delta$	$\Gamma \otimes A \xrightarrow{(id_\Gamma \otimes d_A);f} \Delta$
w	$\Gamma \xrightarrow{f} \Delta$	$\Gamma \otimes A \xrightarrow{(id_\Gamma \otimes e_A);f} \Delta$
!	$\Gamma \otimes A \xrightarrow{f} R$	$\Gamma \xrightarrow{\Lambda(f)} A \Rightarrow R$
?	$\Gamma \xrightarrow{f} \Delta \otimes A$	$\Gamma \otimes (A \Rightarrow R) \xrightarrow{\Lambda^{-1}(f;\Delta \otimes \nu_A; m_{A,\Delta})} \Delta \otimes R$
$\otimes$, $\parr$, $\bot$	nothing to do	

where $A \xrightarrow{d_A} A \otimes A$ and $A \xrightarrow{e_A} 1$ are respectively the diagonal and the erasing morphism, Λ denotes curryfication, $\nu_A : A \longrightarrow (A \Rightarrow R) \Rightarrow R$ is a canonical morphism and $m_{A,\Delta}$ is computed as follows (here we need cartesian closeness):

$$\cfrac{\cfrac{\cfrac{(A \Rightarrow R) \Rightarrow R \longrightarrow (A \Rightarrow R) \Rightarrow R}{((A \Rightarrow R) \Rightarrow R) \otimes (A \Rightarrow R) \longrightarrow R}}{\Delta \otimes ((A \Rightarrow R) \Rightarrow R) \otimes (A \Rightarrow R) \longrightarrow \Delta \otimes R}}{m_{A,\Delta} : \Delta \otimes ((A \Rightarrow R) \Rightarrow R) \longrightarrow (A \Rightarrow R) \Rightarrow (\Delta \otimes R)}$$

Fig. 8. Interpretation of rules in traced cartesian closed categories

5 Polarized Proof Nets with Cycles

5.1 Definition

Definition 5.1. *The proof structures of* MELLPC *are exactly those of* MELLP, *except that we also allow the empty proof structure.*

Definition 5.2 (Proof nets). *We define* $\mathfrak{N}$ *to be the straightforward transformation which maps* MELLPC *deduction trees to proof structures. The set of proof nets is the image of* $\mathfrak{N}$.

Proposition 5.3. *The translation* $\mathfrak{N}$ *maps a* MELLPC *deduction tree of conclusion* $\{n\} \vdash \Gamma$ *to a proof net of conclusions* Γ.

The conclusions of the proof net $\mathfrak{N}(\pi)$ associated to a proof π do not keep track of the natural numbers labelling the sequents. However the label of the conclusion sequent of π is equal to $r(\mathfrak{N}(\pi))$, where r is defined as:

Definition 5.4. *For any proof structure* $\mathcal{R}$, *let* $r(\mathcal{R})$ *be the number of ?-nodes lying at the level* 0 *of* $\mathcal{R}$.

5.2 Correctness Criterion

Proposition 5.5. *Each connected component of a proof structure contains at most one node that falls in either of the following categories:*

- *nodes having a positive conclusion that is a conclusion of the proof structure,*
- *?-nodes of the* 0*-level.*

This has to be compared with MELLP's correctness criterion (definition 1): initial nodes in the correctness graph fall in either of the above categories. Allowing juxtaposition of proof nets by means of the *mix*-rule discards the requirement of having only one such node. The acyclicity constraint is also discarded by our system. This yields a much simpler correctness criterion:

Definition 5.6 (Correctness criterion). *A proof structure* $\mathcal{R}$ *satisfies the correctness criterion (i.e. is* correct*) when:*
all proof structures $\mathcal{R}_0$ *enclosed in a !-box of any level of* $\mathcal{R}$ *satisfy* $r(\mathcal{R}_0) = 1$.

The criterion merely makes sure that proof net boxing is performed consistently with the restriction enforced on the !-rule of MELLPC.

Theorem 5.7 (Proof nets characterization). *A proof structure is a proof net if and only if it satisfies the correctness criterion.*

It is easy to check that proof nets satisfy the correctness criterion. The converse is a consequence of the following sequentialization property.

Proposition 5.8 (Sequentialization). *From any correct proof structure* $\mathcal{R}$, *we can compute a proof* π *of* MELLPC *such that* $\mathfrak{N}(\pi) = \mathcal{R}$, *by means of the sequentialization algorithm described below. If* $\mathcal{R}$ *has conclusions* $N_1, \ldots, N_k, P_1, \ldots, P_l$ *then* π *has conclusion* $\{r(\mathcal{R})\} \vdash N_1, \ldots, N_k, P_1, \ldots, P_l$.

Proof. The sequentialization algorithm S builds the deduction tree π from root up to leaves. It picks a node out of the proof structure $\mathcal{R}$ that corresponds to the final deduction rule of π and goes on building the deduction tree by a recursive call on the remaining sub-structure(s). More precisely, if $\mathcal{R}$ has a terminal node[2] N which is different from an *ax*-node, 1-node or !-node, then S

[2] Here by terminal node, we mean a node which does not have any conclusion (such as the *cut*-node), or a node whose conclusions are all conclusions of the proof structure.

picks out one of these nodes. If $\mathcal{R}$ does not have such terminal nodes, then it is a mere juxtaposition of *ax*-nodes, 1-nodes and !-boxes. Then S is recursively called on each sub-structure enclosed in the !-boxes, and computes the desired deduction tree using *mix*-rules. At each step, the deduction rule added to π is guaranteed to be consistently applied because the decomposition performed by S preserves correctness.

Remark 5.9. This algorithm is not deterministic. Anyway for any output $S(\mathcal{R})$ produced by a particular run of S on input $\mathcal{R}$, we have $\mathfrak{N}(S(\mathcal{R})) = \mathcal{R}$.

5.3 Proof Nets Semantics

The proof nets syntax is closer to semantics than the sequent calculus syntax:

Theorem 5.10. *Let π and π' be two proofs of* MELLPC. *We have*

$$\mathfrak{N}(\pi) = \mathfrak{N}(\pi') \Longrightarrow [\![\pi]\!] = [\![\pi']\!]$$

This result allows to define the interpretation of polarized proof nets with cycles in traced cartesian closed categories:

Definition 5.11. *Let $\mathcal{R}$ be a polarized proof net with cycles. We define its interpretation $[\![\mathcal{R}]\!]$ to be any $[\![\pi]\!]$ where π is a proof of* MELLPC *such that $\mathfrak{N}(\pi) = \mathcal{R}$ (the sequentialization proposition ensures that such a π always exists).*

6 Cut Elimination

We now describe the dynamic behavior of MELLPC by explaining the cut-elimination process. We keep the standard MELLP cut-elimination rules of figure 4 unchanged. One need to be careful however. The MELLP rewriting rules involve premise nodes belonging to the contextual net and conclusion edges that may be linked to nodes of the contextual net. When dealing with cycles, some of those premise nodes and conclusion nodes might be equal. For instance, keeping the contraction rule unchanged enables to simulate Y by the MELLPC proof net of figure 6. The cut-elimination patterns of figure 4 lead to corresponding reduction rules for MELLPC in all cases but two pathological cases: the so-called axiom loop and exponential loop. We give two solutions to reduce these patterns.

6.1 MELLPC with β_0-Reduction

The axiom and exponential loops can be replaced by unambiguous patterns by expansing an edge into a sequence of an axiom node and a cut node. This trick yields the rewriting relation β_0, which handles axiom and exponential loops as displayed on figure 9. As a special case of the exponential loop reduction, a !-box whose !-node is cut against one of its own auxiliary doors is reduced to itself.

Proposition 6.1. *(MELLPC,β_0) satisfies the following properties: (1) it preserves proof nets correctness, (2) it is sound with respect to traced cartesian closed categories, (3) it is not locally confluent.*

Proof. Preservation of proof correctness is easy to check, thanks to the simplicity of the correctness criterion. The soundness result follows from the axiomatic properties of trace operators. Figure 10 shows a counterexample of local confluence, where *tt* and *ff* are two distinct normal nets of the same negative type.

6.2 MELLPC with β-Reduction

The axiom and exponential loops are some kinds of "ghost nets" which only possible interaction is to swallow through their auxiliary doors some positive trees. The net enclosed in the !-box cannot be probed by its context (this is confirmed by both syntax and semantics). This observation leads up to the cut-elimination rules β which policy is to delete ghost nets, see figure 11.

Proposition 6.2. *(MELLPC,β) satisfies the following properties: (1) it preserves proof nets correctness, (2) it is sound with respect to traced cartesian closed categories, (3) it is locally confluent, (4) it is not confluent.*

Proof. The local confluence follows from an overview of critical pairs: each diagram can be closed by means of a few rewriting steps. See figure 12 for a counterexample of confluence, where *tt*+ and *ff*+ are two distinct normal proof nets of the same positive type. This should be compared with the counterexample in cyclic λ-calculus observed by Ariola and Klop in [1].

6.3 A Confluent Subsystem: MELLPY

In order to achieve confluence, we may define Y-cycles as oriented cycles that step over exactly one contraction node and one !-box on their way.

Definition 6.3. *Let* MELLPY *be the subset of* MELLPC *proof nets such that every cycle is a Y-cycle.*

The axiom and exponential loops are not part of MELLPY; and MELLPY is closed under cut-elimination. Thus β_0 and β agree on MELLPY.

Proposition 6.4. *(MELLPY,β) satisfies the following properties: (1) it preserves proof nets correctness, (2) it is sound with respect to traced cartesian closed categories, (3) it is confluent.*

Proof. The confluence follows from a direct and tedious proof, based on the decreasing diagram method and a sharp analysis of commuting reduction steps.

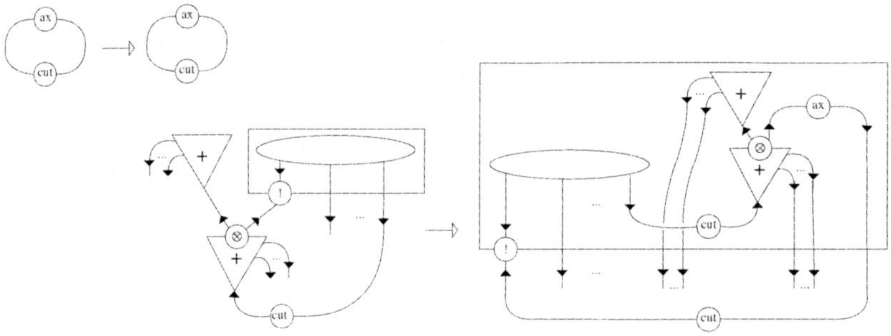

Fig. 9. β_0-reduction on axiom and exponential loops

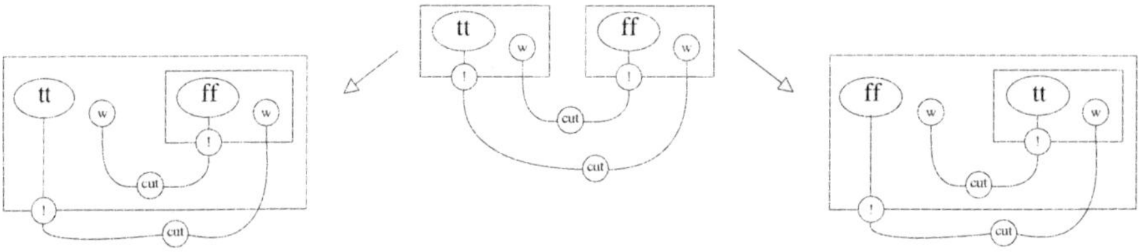

Fig. 10. Counterexample: β_0 is not locally confluent

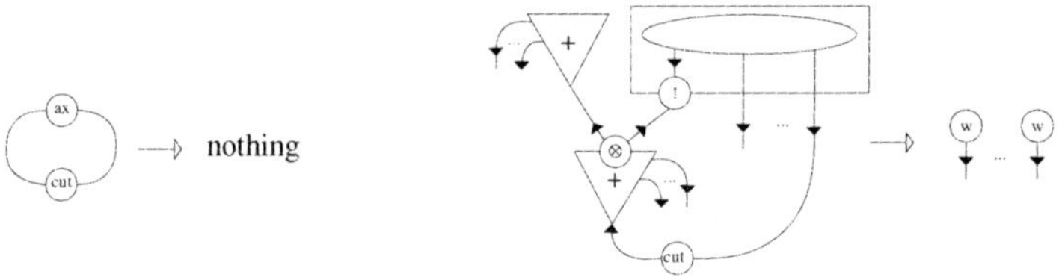

Fig. 11. β-reduction on axiom and exponential loops

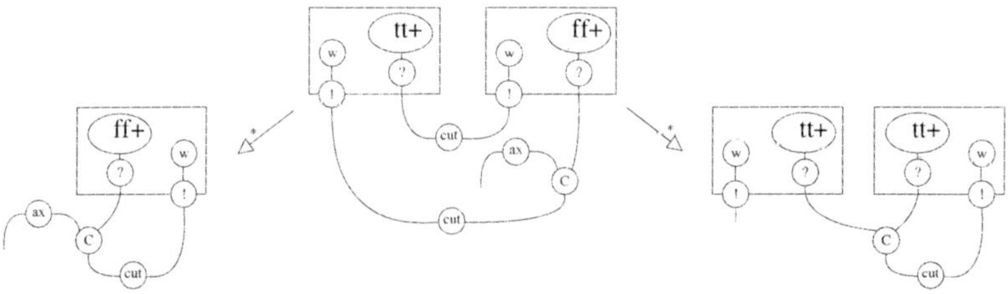

Fig. 12. Counterexample: β is not confluent

7 Interpretation of $\lambda\mu$-Calculus with Fixpoint

Laurent showed [11,9] that MELLP is a suitable target for interpreting both call-by-name and call-by-value $\lambda\mu$-calculus.

The call-by-name interpretation $(.)^-$ uses negative formulas for interpreting types: $(A \to B)^- = ?(A^-)^\perp \parr B^-$ and translates a typing derivation of conclusion $\Gamma \vdash t : A \mid \Delta$ into a proof net of conclusions $?(\Gamma^-)^\perp$, Δ^- and A^-.
On the other hand, the call-by-value interpretation $(.)^+$ uses positive formulas for interpreting types: $(A \to B)^+ = !((A^+)^\perp \parr ?(B^+))$ and translates a typing derivation of conclusion $\Gamma \vdash t : A \mid \Delta$ into a proof net of conclusions $(\Gamma^+)^\perp$, $?(\Delta^+)$ and $?(A^+)$.

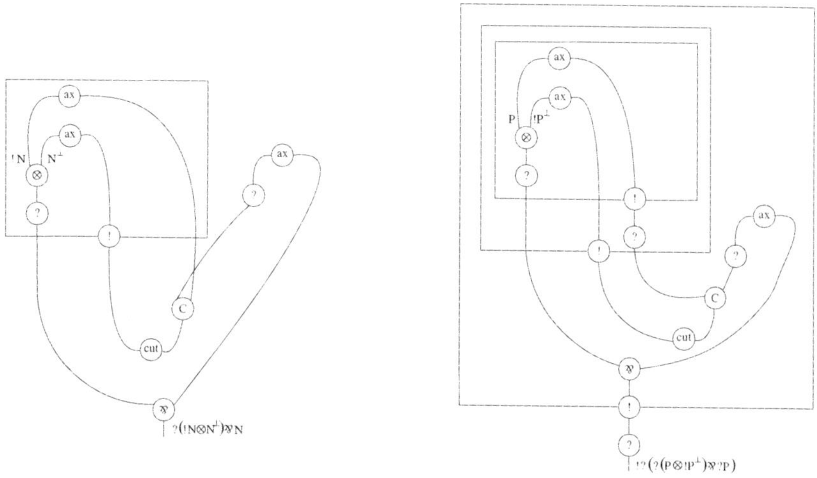

Fig. 13. The Y proof nets: call-by-name (left) and call-by-value (right)

It is possible to extend the $\lambda\mu$-calculus with fixpoint terms Y_A of type $(A \to A) \to A$ and interpret the resulting call-by-name and call-by-value calculi in MELLPC, using the call-by-name and call-by-value variants of figure 6, see figure 13. It is easy to check that these interpretations actually target MELLPY. These translations do not yield exact simulation properties, but the following equational properties do hold:

Proposition 7.1. *In the call-by-value setting:* $(Y(\lambda x.t))^+ =_\beta ((\lambda x.t)(Y(\lambda x.t)))^+$.

Proposition 7.2. *In the call-by-name setting:* $(Yf)^- \simeq (f(Yf))^-$.

Where $\simeq$ is $=_\beta$ together with additional structural rules, generated by a η-rule and a generalized auxiliary exponential cut elimination rule, which are sound in traced cartesian closed categories (details omitted here).

8 Conclusion

We extend polarized linear logic with a fixpoint operator, and obtain a deduction system with (1) a sound denotational semantics in traced cartesian closed categories, (2) a theory of cyclic proof nets, (3) a correctness criterion in the spirit of Laurent's criterion. One main observation is that one needs to move from response categories to traced cartesian closed categories in order to reach that nice balance (see subsection 4.2). Polarized linear logic provides a logical account of the (non-commutative) continuation monad. This should be related to alternative axiomatics of commutative and non-commutative monads by Benton and Hyland [2], and also by Hasegawa and Kakutani [6]. It should be informative to translate various existing calculi with recursion into proof nets with cycles, and study the resulting equational properties.

References

1. Z. M. Ariola and J.W. Klop, Cyclic Lambda Graph Rewriting. In *Proc. of the Eight IEEE Symposium on Logic in Computer Science*, Paris, July 1994.
2. N. Benton and M. Hyland, Traced Premonoidal Categories (Extended Abstract). Workshop on Fixed Points in Computer Science (FICS 2002). July 2002.
3. V. Danos, Une Application de la Logique Linéaire à l'Étude des Processus de Normalisation (principalement du λ-calcul). Thèse de Doctorat, Université Paris 7, 1990
4. J-Y. Girard, Linear Logic. *Theoretical Computer Science* 50 :1–102, 1987.
5. M. Hasegawa, Recursion from Cyclic Sharing: Traced Monoidal Categories and Models of Cyclic Lambda Calculi. In *Proc. 3rd International Conference on Typed Lambda Calculi and Applications*, volume LNCS1210. Springer, 1997.
6. M. Hasegawa and Y. Kakutani, Axioms for recursion in call-by-value. Higher-Order and Symbolic Computation 15(2/3):235–264, September 2002
7. M. Hofmann and T. Streicher, Continuation models are universal for lambda-mu-calculus. In *Proc. LICS '97*, Warsaw, IEEE, pp. 387–397.
8. A. Joyal, R. Street and D. Verity, Traced monoidal categories. *Mathematical Proceedings of the Cambridge Philosophical Society*, 119(3), pages 447–468, 1996.
9. O. Laurent, *Etude de la polarisation en logique.* PHD thesis, Université Aix-Marseille II, March 2002.
10. O. Laurent, Polarized games (extended abstract). *Seventeenth annual IEEE symposium on Logic In Computer Science*, p. 265–274. IEEE Computer Society, 2002.
11. O. Laurent, Polarized proof nets and λμ-calculus. *Theoretical Computer Science*, 290(1):161–188, Dec. 2002.
12. P-A. Melliès and P. Selinger, Polarized categories and polarized games. Forthcoming paper, 2003.
13. R. Montelatici, Présentation axiomatique de théorèmes de complétude forte en sémantique des jeux et en logique classique. Mémoire de DEA, Univ. Paris 7, 2001.
14. M. Parigot, λμ-calculus: an algorithmic interpretation of classical natural deduction. In *Proc. of International Conference on Logic Programming and Automated Deduction*, volume LNCS624. Springer, 1992.
15. P. Selinger, Control Categories and Duality: on the Categorical Semantics of the Lambda-Mu Calculus. *Mathematical Structures in Computer Science*, 2001.
16. A. K. Simpson and G. D. Plotkin, Complete Axioms for Categorical Fixed-Point Operators. In *Logic in Computer Science*, pages 30–41, 2000.

Observational Equivalence and Program Extraction in the Coq Proof Assistant

Nicolas Oury

Laboratoire de Recherche en Informatique,
CNRS UMR 8623,
Université Paris-Sud,
Bâtiment 490,
91405 Orsay Cedex, France
Nicolas.Oury@lri.fr, http://www.lri.fr/ noury

Abstract. The *Coq* proof assistant allows one to specify and certify programs. Then, code can be extracted from proofs to different programming languages. The goal of this article is to substitute, at extraction time, some complex and fast data structures for the structures used for specification and proof. This is made under two principal constraints: (1) this substitution must be *correct*: the optimized data structures in the extracted program must have the same properties as the original ones, (2) on the proof side, the structure must keep a *computable nature*. If the framework described here is general, we focus on the case of functional arrays. This work leads us to formalize the notion of *observational equivalence* in the *Coq* system. We conclude with benchmarks.

Coq [1] is a proof assistant. It allows to define properties and to conduct formal proofs, while ensuring the logical soundness. It is based on the *Calculus of Inductive Constructions* [2,7,8], a constructive logic. From a proof of P, it is possible to get a program realizing the specification expressed by P. This derivation is called *program extraction* and can be used to produce certified programs.

In order for these programs to be useful, we want the extracted code to be optimized. Though, we want to have data structures easy to understand and to work with during the certification process. This simplifies proofs of the algorithms. Moreover, it is better to have a data structure defined, instead of being axiomatized, in order to be able to make proofs by computation: since the *Coq* system can do automatic reductions, it allows a shorter and easier certification process. Moreover, *Coq* can do reductions and computations during typing, so it is important to have data structures defined—instead of being axiomatized—in order to type some terms.

This paper presents one solution to this problem. We have simple data structures in the *Coq* world that are extracted to optimized data structures. These structures are defined *Coq* terms, and it is possible to compute with them. The substitution is safe: the optimized program is *observationally equivalent* to the original one.

M. Hofmann (Ed.): TLCA 2003, LNCS 2701, pp. 271–285, 2003.

This method makes intensive use of *objects with axioms*. These objects—introduced later—are used to model data structures while hiding their representation. They are objects wrapping together a state, an implementation and a proof that the latter verifies some axioms of the data structure. In order to formalize the substitution, we define the notion of *observational equivalence* in the *Coq* system.

1 Presentation

1.1 The Coq Proof System and Program Extraction

Coq is a proof assistant based on the *Calculus of Inductive Constructions*. The user can specify and prove theorems and algorithms. We do not describe precisely here this software. One can look at [9] and [10] for more precise introduction and description of *Coq* and *Calculus of Inductive Constructions*. We just mention some important points. Then, we present program extraction in the *Coq* system.

The *Coq* proof system. *Coq* is based on *the calculus of inductive constructions*, a multisorted logic. An important distinction is made between sorts *Set* and *Prop*. Both are included in $Type_1$, yet they are distinct: *Set* contains types of computable terms and *Prop* contains types of logical terms. This distinction is only relevant for extraction.

The user can define *inductive* data types and predicates that are similar to concrete types in ML. The user can then use inductive reasoning on these types.

It is usually difficult to write proof terms directly, so the user can define them interactively, using tactics.

Program extraction. *Coq*'s logic—without additional axioms—is *constructive*. Existential proofs contain a witness, *i.e.* a term that satisfies the property. This leads to a remarkable fact: it is possible to get programs from proofs. For example, if one has proved $\forall n : \mathbb{Z}, \exists m : \mathbb{Z}, m > n$, the system can extract a function that, provided an integer n, gives back an integer greater than n. This function can be, for now, produced to both *Objective Caml* [5] and *Haskell* [4] syntax. But the concept could work with other programming languages.

The main difference between sorts *Set* and *Prop* appears in extraction. Terms in *Set* are seen as computations and kept in the extracted code, whereas terms in *Prop* are seen as logic and erased.

As axioms has no defined value, the user has to specify an extracted value in order to extract them.

```
Axiom f : Z -> Z.
Extract Constant f => "my_f_in_ML".
```

This can also be done for non-axioms. For example, the user can specify an optimized function to use. The same apply for *inductive types*:

```
Extract Inductive Z => "myZ" ["MyNeg" "MyZero" "MyPos"].
```

indicates to extract Z to the *Ocaml* type myZ, whose constructors are MyNeg, MyZero and MyPos.

These extraction values are not checked. The user should take care of not breaking correctness of extracted programs.

This paper takes place in the context of extraction. We want to produce extracted code using complex and optimized data structure; while keeping the representation of this structures in *Coq* simple and easy to use. We also want to keep the correctness of the extraction: we want the extracted program to verify the properties of the proved algorithm.

The notion of persistent data structures is central in the remaining of this paper, and we present it in the next section.

1.2 Functional Data Structures versus Persistent Data Structures

Coq is a system made for proving and reasoning. It has to be purely functional: it has no meaning to use a function with side effects in a specification. How could we do mathematics with a language where $2.f(x) \neq f(x) + f(x)$? This means that *referential transparency* is needed for *Coq* logic.

However, there is a family of imperative structures that are *observationally* functional: *persistent* data structures.

Definition 1 (persistence). *A data structure—that may or not be functional—is persistent if it has a functional behavior,* i.e. *if it keeps referential transparency.*

The notion of functional behavior can be found in [6].

When it is imperative, such a data structure cannot be written in *Coq*. But since it is persistent, its behavior can be described in *Coq*. As this allows access to a new family of data structures, we can hope better performance of the extracted code.

Indeed, such a structure usually has better performance that its purely functional equivalent. For example, one can implement sets that optimize themselves for the elements that are often needed. It can be very interesting to substitute a persistent data structure for a functional one during extraction. A simple functional structure is written in *Coq* and the extraction substitutes an optimized persistent version for it in the target language. These structures have to be equivalent one with another. This substitution raises some issues:

- How can we define such an equivalence?
- How can we prevent the user from breaking this equivalence?
- How can we prove this equivalence? (If we want to keep the extracted code certified, we have to prove this equivalence)

This paper tackles the first and second issues. The last one is partially answered: we can prove equivalence only if both structures are coded in *Coq*. The proof of an imperative program is far beyond the scope of the paper, though feasible [12].

Persistent arrays are used as a running example throughout the remaining of this paper.

1.3 Persistent Arrays

Some purely functional languages, like *Haskell*, provide *persistent* arrays. These structures can be seen as functional arrays. They provide (at least) the functions:

```
init : (A:Set) Z -> (Z -> A) -> (Array A).
get  : (A:Set) (Array A) -> Z -> A.
set  : (A:Set) (Array A) -> Z -> A -> (Array A).
```

These functions have the usual behavior of arrays. The main difference is the behavior of `set`. Instead of modifying the array `t`, `set A t i a` return a new *persistent array*. The content of `t` is not modified: the functional behavior is kept.

Of course, this is not done by copying the whole array. The array is modified in place. An indirection is added to previous references. It indicates that the array has been modified and stores the index and the old value. The function `get` is consistent with this behavior. If `get` founds an indirection for the index it is looking for, it returns the corresponding value. This method have an advantage: if one uses the array in a *single-threaded* way, both *reads* and *writes* are made in $O(1)$. Figure 1 shows an implementation of persistent arrays written in *Ocaml* and the execution of

```
let t' = set t 1 3
```

This simple implementation is used for this paper, but a final version could be further optimized. For example, one could duplicate the array after several modifications or optimize the “diffs” stack.

This structure is used in the extracted code. In *Coq*, we represent these arrays by function from `Z` to `A`. This equivalent implementation is easy to understand and to work with.

2 Drawbacks of Simple Solutions

Before presenting complex solutions, we present here simple proposals that do not work very well. We explain why these natural proposals are unsatisfying. This allows a better analysis of the difficulties of the problem.

2.1 Axiomatization

Facing the problem of extending the *Coq* system with a new data structure, the most frequently used solution is *axiomatization*. Axiomatization consists in introducing *ex nihilo* types, functions to work with them and properties. For example, we add the new structure by stating its existence in *Coq*:

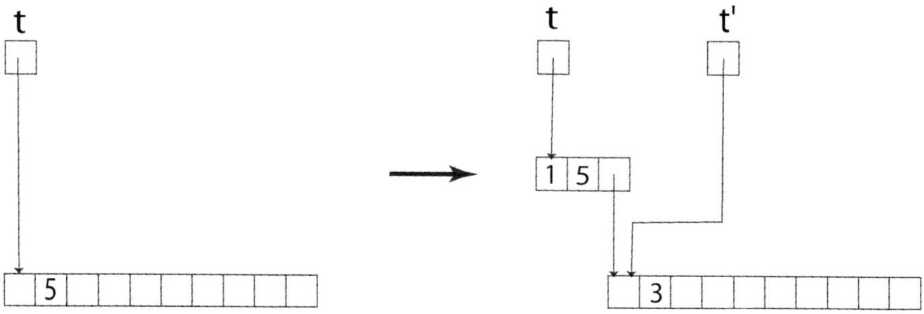


```
module FunArray = struct
  type 'a t = 'a record ref
  and 'a record = Data of 'a array | Diff of int * 'a * 'a t

  let create n v = ref (Data (Array.create n v))

  let rec get arr n = match !arr with
    | Data a -> a.(n)
    | Diff (m,v,arr') -> if n == m then v else get arr' n

  let set arr n v = match !arr with
    | Data a ->
        let old = a.(n) in
        a.(n) <- v;
        let res = ref (Data a) in
        arr := Diff (n,old,res);
        res
    | Diff _ ->
        ref (Diff (n,v,arr))
end
```

Fig. 1. Persistent arrays in *Ocaml*

```
Axiom array : (A : Set)  -> Set.
```

Then, we introduce the functions to manipulate this new type:

```
Axiom init : (A:Set) (size:Z)  (init_val:Z -> A) -> (array A).
Axiom get : (A:Set)  (i:Z) (array A) -> A.
Axiom set : (A:Set)  (i:Z) (v:A) (array A)  -> (array A).
Axiom size : (A:Set) (array A) -> Z.
```

Finally, we declare some axioms giving their specification. The first one could be:

```
Axiom getInit : (A:Set) (size:Z) (f:(Z->A)) (i:Z) `0 <= i < size`
                         -> ((get A  i (init A size f)) = (f i)).
```

This axiom states that if we look inside an array that has just been created we found the initial value. The next ones could be:

```
Axiom getSet1 : (A:Set) (t:(array A)) (i:Z) (j:Z)
 `0 <= j < (size A t)` ->
 `i <> j` -> (v:A) -> ((get A j (set A i v t)) = (get A j t)).

Axiom getSet2 : (A:Set) (r:(array A)) (i:Z)
 `0 <= i < (size A t)` ->
 (v:A) ((get A i (set A i v t)) = v).
```

This set of axioms defines arrays which are not modified when a `set` is called outside of the array bounds.

We can, for the moment, let unknown the value of an array outside his bounds. We do not axiomatize a value for it. So no one can use this value in proofs. The extracted code can send back any value for an "out of bound" `get`. Nevertheless, later in this paper, arrays will be defined in *Coq*. Then, we have to choose a value for "out of bound" `get`. So, we make here some design choices for the remaining of the paper:

- When accessed outside of the array, a default value is given back. We choose to return the value at the first index of the array.
- For this reason, arrays of size strictly less than 1 are just seen as array of size 1. This technical restriction is here to ease the explanation. We plan to remove it in final libraries.

For these reasons, we force the size of arrays to be positive:

```
Axiom sizePos : (A:Set) (t:(array A)) -> `(size A t) > 0 `.
```

Finally, we just add extraction values for the axiomatized functions.

This approach is very simple and natural. It allows to quickly add new data structures. Nevertheless, it has some severe drawbacks:

- First of all, it is always dangerous to add new axioms in a logical system. They can break the system's consistency
- As arrays are not objects of the calculus, we cannot compute with them. For example, it is impossible to use *reflexivity* to prove `(get 0 (set 0 1 (init 100 id))) = 1`. With arrays defined in *Coq*, `Reflexivity` tactic would ask the system to test convertibility and conclude. Here, we have to do rewriting with axioms manually, for each reduction step.
- Moreover, some terms of the calculus cannot be written with axioms. The type system may reduce terms while typing. Axiomatized data structures cannot be reduced—they have no computable content—and some terms cannot be typed. For example, if we have `A` and `B` convertible and `b` of type `B`, we can write `(([x:A] t) b)`. But without convertibility—even with an axiom stating `A=B`—, writing such a term results in a type error.

For these reasons, we drop the idea of axiomatization.

2.2 Arrays as Functions

Since we want arrays to be defined in the system, we try to implement directly arrays as terms. We define, in *Coq*, arrays as functions. Actually, for extraction to detect where functions must be replaced by arrays in the extracted code, we have to create a new type:

```
Inductive array [A:Set] :=
Array : (size:Z) (value:(Z->A)) (array A).
```

This defines array as a tuple of the size and the value of the array.

Then it is easy to define terms representing operations on arrays.

```
Definition init := [A:Set;  size:Z; value: (Z->A)]
       (Array A (max size `1`) (truncate A value size)).
```

where `truncate A value size` returns a function having the same value than `value` on [0...(`size`−1)] and `value(0)` elsewhere.

```
Definition get := [A:Set; i:Z; t:(array A)]
Cases t of
(Array size value) => (value i)
end.

Definition size := [A:Set; t:(array A)]
Cases t of
 (Array size value) => size
end.

Definition set := [A:Set ; i:Z; v:A; t:(array A)]
Cases  t of
(Array size value) =>
       (Cases (interval i size) of
           true => (array A size {...value modified at index i...}
           false => t
       end)
end.
```

where `interval` simply tests whether i is in [0...(`size`−1)].

These terms are computable. This solves the issues described in the previous section. Moreover, it is easy to check that these type definition and functions conform to the axioms—apart `sizePos`— and design choices (value for "out of bound" access) made in previous section.

Axiom `sizePos`, which states that size of an array is positive, is more complex. In fact, it highlights an issue in this design. We cannot enforce the operations we have defined to be the only admissible operations. The user can manually create an array `Array Z `-1` id` which size is not positive. We can solve this by changing the definition of array.

```
Inductive array[A:Set] :=
Array : (size:Z) `size > 1` -> (value:(Z->A)) (array A).
```

It is easy to define—in *Ocaml*—extraction values for the types and operations we have defined.

However, new troubles—coming from the computable nature of these arrays—arise. In the *Coq* system, arrays are seen as functions. But in the extracted code, they are seen as persistent arrays. For example, the user can define his own `get` function. This function is extracted as a function operating on arrays seen as functions. But, in the extracted program, arrays are persistent arrays. The extracted code cannot be typed and is not correct anymore. Conversely, one can create an array by hand—without using `init`. Then the array is extracted as a function and cannot be used as a persistent array.

Moreover, since arrays are functions, one can define a *Coq* array with, for example, a different value for each index in $\mathbb{Z}$. But this function is not really an array and cannot be kept in the extracted code. This is dramatic: some proofs can rely on some values that are lost at extraction. We have lost the correctness of the extracted program.

Some workarounds. It is possible to partially solve these problems with some tricks. For example, one can add, to the type of arrays, the need of a "key" that cannot be extracted. Then, the user cannot extract his own `get` function, because he would have to use the key. It is also possible to add an inductive proof that the array was constructed only by the "authorized" functions. We have studied these solutions. They are inelegant and do not work well in practice. We do not detail them here.

Views. Adding *views* [11] to the extraction solves some of these problems. Views allow to view a type as another type. Conversion from the observed type to the real type is automatic. They can solve the troubles stressed in this section. Nevertheless, we have not used this method, for several reasons:

- We do not want to modify extraction in *Coq*.
- One can do heavy operations (convert from a function to an array) without seeing it. Even some inlining from *Coq* could introduce some conversions.
- It only allows substitution at extraction and not within the *Coq* system.
- It is an ad-hoc solution. Even if we use views, we have to introduce observational equivalence in order to reason about them.

3 Array Objects and Observational Equivalence

3.1 Arrays as Objects

In the previous section, we presented some issues regarding the compatibility between a data structure and functions operating on it. *Coq* world and *Ocaml* world have different incompatible representations of arrays. So, it is difficult to make sure that the representations does not interact. There is a natural and classical solution to this problem: to keep together data structure and function in an *object*. As functions and structure are kept together, we are sure that they are compatible. We generalize the structure over the type used to represent the array. We use an existential on the type of arrays. We obtain, in pseudo-code:

```
exists t,
state :: t,
get :: t -> Z -> A,
set :: t -> Z -> A -> t
```

An array is a bundle of a type, an implementation of arrays on that type and the value of the array.

We first define the type of an implementation, on type `X`, of arrays of size `n`. (The size of the array appears in the type to ease technical details of presentation.)

```
Record ArraImpl [A : Set ; n : Z; X : Set] : Set :=
ArrayCons {
  get_rec : X -> Z -> A;
  set_rec : X -> Z -> A -> X;
  init_rec :  `0<n`-> (Z -> A) -> X}.
```

Then, we create the type of arrays. It is a tuple of a type `X`, an implementation of arrays using that type and the value of the array.

```
Inductive array [A: Set; n : Z]  : Set :=
Array : (X : Set) (implementation : (ArrayImpl A n  X))
           (value : X) (array A n).
```

Then, we write a `generate` function that creates an array—implemented as function—and helper methods. We set extraction of `generate` to an *Ocaml* function that builds a *persistent* array. As functions are bundled with the value, the correctness is ensured. We create in *Coq* helper functions `get`, `set` and `size` over the bundled functions.

This representation uses an existential type, and so it could seem that the system cannot compute with it. In practice, however, the user will always work with a particular implementation whose methods are computable. Then proofs on one implementation can be extended to any implementation. For example, if the user wants to show a property for any array, he can copy the array content in an array with a known implementation. Then, he can use computations in proofs.

The `Inductive` above can look like the unsound strong Sigma–type. In fact, *Coq* allows this kind of declaration, but forbids some eliminations on it. These restrictions make the value of `X` inaccessible in a proof.

3.2 Value Equality and Observational Equivalence

In order to reason about arrays we introduce a way of comparing arrays. In *Coq*, equality is not sufficient for comparing arrays. Indeed, equality is convertibility and some arrays may be the same without being convertible. For example, if arrays are implemented as functions, `set i v (set j v t)` and `set j v (set i v t)` are different. We describe two ways of comparing arrays: one equivalence is *easy* to prove—it only uses the state of the array—and one equivalence is *powerful*—one can deduce that equivalent arrays have the same futures. Then, we try to link them.

Value equality. A first way of comparing arrays is value equality. It is an extensional equality over all the indexes of the arrays.

```
Definition eqValue := [A : Set; n : Z; a, b : (array A n)]
       (i : Z) ((get A n a i ) = (get A n b i)).
```

This equivalence is the easiest to prove and to manipulate. However, it is not sufficient for our arrays: if two arrays have the same value, we cannot prove that after the same modification they still have the same value (we have no information about the `set` method of the object. This method can be different for the arrays and can lead to a different value). We need another equivalence to characterize this fact.

Observational equivalence. We define in *Coq* an *observational equivalence.* This notion is close to equivalence used in *semantics.* One can find some deeper explanation in [13]. This *powerful* equivalence simply expresses that two arrays will have the same behavior in the future. If we apply `get`—with the same arguments—on them, we get back the same value. If we apply `set`—with the same arguments— on them, we get back equivalent arrays.[1]

```
coInductive eqArray [A : Set; n : Z] :
               (array A n) -> (array A n) -> Prop :=
  eq_array : (a,b : (array A n)) (eqValue A n a b) ->
        ((i:Z) (v:A)
            (eqArray A n (set A n a i v) (set A n b i v)) ->
             (eqArray A n a b).
```

This equivalence is powerful: it expresses that two structures have the same future. However it is difficult to use. One has to make infinite proofs to prove this equivalence. For arrays, we expect to be able to prove the equality of indexes—`eqValue`—and to deduce the *observational equivalence.* Indeed, arrays are uniquely defined by values of their cells.

[1] This definition makes use of *Coinductive* types. One can think of them as infinite recursive structures, whose no substructure is undefined. One can construct infinite proofs to work with them. More details on the use of *coinductive* types in *Coq* can be found in [3].

Nevertheless, with the current definition of arrays, this is impossible. The array object can contain any methods for its implementation. In order to assert that these methods implement arrays, we have to thread the property "this object was created by `generate`" everywhere in proofs. Unfortunately, the preservation of this property through `set` is very difficult to prove. To get an infinite proof of `eqArray A (set A i t) (generate n p f')`, we need a coinductive invariant. In the recursive case, both arrays are built using `set`. But, in the invariant, `generate n p f'` is not built using `set`...

This difficulty arises because the model is not restrictive enough. We have to add constraints on the content of the objects. We create objects with axioms.

3.3 Objects with Axioms

In order to restrict the use of object arrays to valid arrays, we have to force the user to provide object with a working implementation of arrays. For this, we add proof obligations of array axioms to the structure of previous section. The objects are now composed of:

- a type `X`
- an implementation `impl` of arrays using `X`
- a proof that `impl` verifies array axioms
- a value of type `X`

We define a property stating that `impl` is an implementation of arrays:

```
Definition arrayAxioms :=
  [A : Set; n : Z; X: Set;  implementation : (ArrayImpl A n X)]
 (getInit_ax A n X implementation)
/\ ...
/\ (getSet2_ax A n X implementation).
```

Then, we add these properties to the definition of array type:

```
Inductive array [A : Set; n : Z] : Set :=
 Array : (X : Set) (impl : (ArrayImpl A n X))
             (proof : (arrayAxioms A n X impl))
                (value : X) -> (array A n).
```

We adjust extraction for this type and make helper properties to use the axioms. With this model we have the best of both worlds: we can use computation and axioms while proving. Moreover, if we have a proof by computation for a given implementation we can get a proof for any implementation. Indeed, we are able to prove:

```
Theorem eqValueArray : (A : Set; n : Z ; a, b : (array A n))
 (eqValue A n a b) -> (eqArray A n a b).
```

This allows to substitute any implementation for any other as soon as they verify properties of arrays. The user only has to check that the values of both arrays—that may have different implementations—are the same. Then, he can deduce that the arrays have the same behavior in the future. One could wonder whether it is possible or not to automatically replace one implementation by another in a proof. There are two ideas that could help to answer this question:

- we could introduce an axiom `eqArray A n a b -> a = b`. A new trouble arises: is this axiom safe?
- we could use tactics to automaticly replace an array by a copy of it in another implementation, and transform the proof with `eqArray`. As this property is an equivalence, it could be useful to use the `Setoid` tactic that does automatic rewriting in a setoid.

As an example, we have created an implementation of array objects with axioms using functions to represent arrays. We have extracted it to persistent arrays in *Ocaml*. As methods and data are stored together in an object, we can be sure of the correctness of the extracted program as soon as the *Ocaml* code for persistent arrays is correct.

We can give a sketch of the proof of this fact. If the data structure used in *Ocaml* is proved to be an array in *Coq*, we can be sure that two arrays with the same value but different implementations are observationally equivalent one to another. As the Ocaml program can only use the methods of these objects—`set`, `get` and `size`—, we know, by observational equivalence—which is proved in *Coq*—,that the arrays have the same behavior: we can safely replace one implementation by another. Proving persistent data structures in *Coq* is beyond the scope of this paper.

Remark. The proof uses meta-theoretical assumptions. *Coq* has to stay consistent with all axioms used in proof. Extraction has to keep correctness of code. This points are beyond the scope of this paper.

4 Practical Use

4.1 Examples and Limitations

Objects with axioms allow to compute with arrays. In this subsection we show how this can be used in practice.

For theorems that use array only with closed values—and not parameters—the computations can be very useful. For example, if one wants to prove that a function sorts a particular array—with known size and values—, the proof will be very easy. If there is a decision function to test whether an array is sorted, the only proof step is to apply this decision function. Computation does the remaining of the proof. This is far better than with axioms: an arbitrary number of rewriting steps would be needed to run the algorithm on the array. Moreover, the size of the proof would be dramatic (computations are not kept in the proof term, while axioms applications are).

Sadly, results are not so good in presence of parameters. This is a common problem with computable terms in the *Coq* system. When trying to compute during a proof, we always end up on a `Cases` expression with a head variable and thus no further reduction. For example, if we try to prove the property `(get (set t i (get t i)) i) = (get t i)` for a particular implementation of arrays, we found after reductions that the result depends of the comparison of i and the size of the array. We have to manually resolve this by using the tactic `Case`.

Nevertheless, this model keeps some interests:

- when the computation leads to a `Cases`, it helps the user to find out the different cases in proof. In the example above, it is easy to found the different cases and finish the proof
- this may lead to a new tactic which recursively eliminates `Cases` expression and tries to conclude
- we do not need to add axioms. This ensures that we do not break the consistency of the system
- having a computable object allows to do simple proofs on closed arrays. This allows to do quick checks on functions and specifications before starting the proof.

4.2 Benchmarks

For benchmarks, we use functional translations of imperative programs. These translations were made by *Why* [12]—a software to generate proof obligations from annotated imperative programs. We have made a compatibility layer from our arrays to the arrays used in this tool, in order to have an important panel of programs to benchmark. One could notice that functional translations of *Why* programs are using arrays linearly. This is fair, as we stressed that persistent arrays are mainly useful for linear—or quasi-linear—algorithms.

We have compared our arrays to arrays implemented as functions—as in *Coq* world. As a good programmer could have used better purely functional implementation—complicating proofs in *Coq*—, we also compare these arrays to arrays implemented with *balanced trees*. We use the `map` structure from *Ocaml* standard library.

We run the tests on *quicksort* with constant arrays—to be in the $O(n^2)$ case—of size 1000 and 10000, *selection-sort* and *heapsort* with randomly generated arrays.

The results were obtained on a Pentium III 933 MHz processor and are expressed in seconds.

test	persistent	functions	map
quicksort 1000	0.64	51.6	1.5
quicksort 10000	133	–	235
selectionsort 1000	0.87	47.1	1.69
selectionsort 10000	112	–	222
heapsort 1000	0.05	3.7	0.11
heapsort 10000	0.97	–	2.14

These results clearly show the necessity of optimized data structures in extracted programs.

5 Conclusion

We have presented a general framework to substitute a data structure to another in *Coq*, or during program extraction: we find a set of properties that completely defines a structure and a *value* that encodes the state of the structure. Then, we can define an *object with axioms* and a value equality for that structure. We can prove the equivalence between *value equality* and *observational equivalence*. This allows substitution of data structures in extracted programs. We have a complete example using persistent arrays, including benchmarking of persistent array versus other implementations.

Future work. There are many ways we could continue this work:

- To keep the correctness of extracted code, we need to prove the code for persistent arrays to be a valid array implementation. This is difficult as it is an imperative implementation. *Why* [12] allows to prove imperative programs but is restricted to programs without memory aliases. Persistence is pertinent only in presence of aliases: what does observationally functional means if nobody is observing?
- By using objects, we break the symmetry of structures. A `concat` function taking two arrays would have to be encoded as a method of the first array taking the second array as parameter, as usual in object oriented programming. Nevertheless, as the method do not know the type of the second array, no optimization can be used in the method. One idea to solve this problem would be to generalize the object paradigm and to use type classes. This has to be investigated.
- This work also raises some logical questions about *observational equivalence*. It seems that two structures observationally equivalent can be swapped without problem. It is obviously false in the *Calculus of Inductive Constructions*, but we think it holds in system $\mathcal{F}$. We need this to prove correctness of substitution during program extraction. This remains to be proved. We also wonder if *Coq* is consistent with an axiom stating that observational equivalence implies equality. A related question is: can we prove more properties with defined data structures than with axiomatized data structures? We have seen that it is often easier to do a proof, but are some properties true

only with defined terms? Given a proof Π of a property P with x and y convertible, can we always make a prove Π' of P' equivalent to P using $x = y$ as hypothesis?

Acknowledgements. Thanks are due to Jean-Christophe Filliâtre and to the referees for their feedback on this paper.

References

1. The Coq proof assistant. `http://coq.inria.fr`.
2. T. Coquand and G. Huet. The Calculus of Constructions. *Information and Computation*, 76(2/3):95–120, 1988.
3. Eduardo Giménez. A tutorial on recursive types in Coq. Technical Report 0221, INRIA, Mai 1998.
4. Haskell, a Purely Functional Language. `http://www.haskell.org`.
5. The Objective Caml language. `http://pauillac.inria.fr/ocaml/`.
6. Chris Okasaki. *Purely Functional Data Structures*. Cambridge University Press, 1998.
7. C. Paulin-Mohring. Inductive Definitions in the System Coq - Rules and Properties. In M. Bezem and J.-F. Groote, editors, *Proceedings of the conference Typed Lambda Calculi and Applications*, volume 664 of *Lecture Notes in Computer Science*. Springer-Verlag, 1993. Also LIP research report 92-49.
8. The Coq Development Team. The Coq proof assistant reference manual–chapter 4. Technical Report 0255, INRIA, Février 2002.
9. The Coq Development Team. The Coq proof assistant reference manual version 7.2. Technical Report 0255, INRIA, Février 2002.
10. The Coq Development Team. The Coq proof assistant tutorial version 7.2. Technical Report 0256, INRIA, Février 2002.
11. Philip Wadler. Views: A way for pattern matching to cohabit with data abstraction. In *14th ACM Symposium on Principles of Programming Languages*, Munich, January 1987.
12. Why: a software certification tool. `http://why.lri.fr`.
13. Glynn Winskel. *The Formal Semantics of Programming Languages — An Introduction*. The MIT Press, 1993.

Permutative Conversions in Intuitionistic Multiary Sequent Calculi with Cuts

José Espírito Santo and Luís Pinto*

Departamento de Matemática, Universidade do Minho
4710-057 Braga, Portugal
{jes,luis}@math.uminho.pt

Abstract. This work presents an extension with cuts of Schwichtenberg's multiary sequent calculus. We identify a set of permutative conversions on it, prove their termination and confluence and establish the permutability theorem. We present our sequent calculus as the typing system of the *generalised multiary λ-calculus* $\lambda\mathbf{J}^{\mathbf{m}}$, a new calculus introduced in this work. $\lambda\mathbf{J}^{\mathbf{m}}$ corresponds to an extension of λ-calculus with a notion of *generalised multiary application*, which may be seen as a function applied to a list of arguments and then explicitly substituted in another term. Proof-theoretically the corresponding typing rule encompasses, in a modular way, generalised eliminations of von Plato and Herbelin's head cuts.

1 Introduction

It is well-known that two intuitionistic sequent calculus derivations determine the same natural deduction proof when they are inter-permutable [9,6]. In [1] this idea is made precise for cut-free sequent calculus by the identification of a basic set of permutations and the definition of a confluent and weakly normalising rewriting system whose normal forms are the normal natural deductions. Schwichtenberg proved in [7] that a variant of this rewriting system is strongly normalising.

In this work we obtain similar results for an extension with cuts of the system in [7]. Specifically: (i) we define the system $\lambda\mathbf{J}^{\mathbf{m}}$ and permutative conversions on it; (ii) we prove confluence and strong normalisation of the corresponding rewriting system; (iii) we introduce an interpretation ϕ of $\lambda\mathbf{J}^{\mathbf{m}}$ into a notational variant of the λ-calculus; (iv) we prove the permutability theorem establishing that two $\lambda\mathbf{J}^{\mathbf{m}}$-terms have the same interpretation iff they are inter-permutable.

Schwichtenberg introduces in [7] the idea of *multiary left rule*, a generalisation of the ordinary binary left rule for implication, by which in a single inference one may introduce $A_1 \supset ... \supset A_k \supset B$, for some $k \geq 1$. Schwichtenberg also introduces the notion *multiary sequent terms*, used in [7] as a convenient tool to represent derivations, in order to obtain termination results.

* Both authors are supported by FCT through the Centro de Matemática da Universidade do Minho, Braga, Portugal

M. Hofmann (Ed.): TLCA 2003, LNCS 2701, pp. 286–300, 2003.

Here we take the view that the notion of multiary sequent terms is of interest on its own, having in mind the computational interpretation of sequent calculus. Indeed, $\lambda\mathbf{J}^{\mathbf{m}}$ may be seen as an extension of the λ-calculus where application is generalised in two directions: (i) "generality" in the sense of von Plato's generalised eliminations [8]; and (ii) "multiarity" here formalised in the style of Herbelin's $\overline{\lambda}$-calculus [4]. We call this new construction *generalised multiary application*. Three particular cases of this construction (*generalised application*, *multiary application* and *simple application*) determine subsystems of $\lambda\mathbf{J}^{\mathbf{m}}$, which turn out to correspond exactly to three previously know calculi: (i) ΛJ [5] (denoted here as $\lambda\mathbf{J}$) —the type-theoretic counterpart to von Plato's natural deduction system with generalised eliminations; (ii) $\lambda\mathcal{P}\mathbf{h}$ [2,3] —the multiary λ-calculus, essentially corresponding to the head-cuts subsystem of Herbelin's $\overline{\lambda}$; (iii) $\lambda\mathcal{G}$ —a sequent calculus isomorphic to natural deduction [2]. Interpretations of $\lambda\mathbf{J}^{\mathbf{m}}$ onto $\lambda\mathbf{J}$ and $\lambda\mathcal{P}\mathbf{h}$, and also from these two systems onto $\lambda\mathcal{G}$, are defined and the commutative diagram in Figure 1 is obtained.

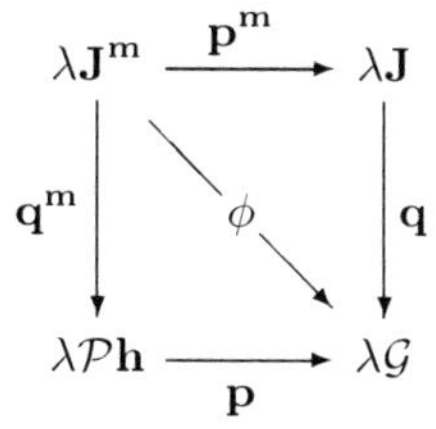

Fig. 1. $\lambda\mathbf{J}^{\mathbf{m}}$ and its subsystems

We consider two kinds of permutative conversions: $\mathbf{p}$-permutations (inspired by [7]) and the $\mathbf{q}$-permutation (specific to this work). Permutations of the former (resp. the latter) kind deal with "generality" (resp. "multiarity") of application and, when combined together, they reduce generalised multiary applications to simple applications. Each of the mappings in Figure 1 produces the normal form w.r.t. the appropriate kind(s) of permutations: mappings $\mathbf{p}$ and $\mathbf{p}^{\mathbf{m}}$ (resp. $\mathbf{q}$ and $\mathbf{q}^{\mathbf{m}}$) produce $\mathbf{p}$- (resp. $\mathbf{q}$-) normal forms, whereas ϕ produces $\mathbf{p}, \mathbf{q}$-normal forms.

2 $\lambda\mathbf{J}^{\mathbf{m}}$: The Generalised Multiary λ-Calculus

2.1 Expressions

The generalised multiary λ-calculus $\lambda\mathbf{J}^{\mathbf{m}}$ is a term calculus for intuitionistic implicational logic, corresponding to an extension with cuts of Schwichtenberg's multiary cut-free sequent calculus presented in [7]. In $\lambda\mathbf{J}^{\mathbf{m}}$, *formulas* (=*types*) A, B, C, ... are built up from *propositional variables* p, q, ... using just $\supset$ (for implication) and *contexts* Γ are finite sets of *variable : formula* pairs, associating at most one formula to each variable. In the following, $\mathbf{V}$ denotes the set of variables and x, y, w, z range over $\mathbf{V}$.

Definition 1 *The terms of* $\lambda\mathbf{J}^{\mathbf{m}}$ *are described in the following grammar:*

$$\begin{array}{ll}(terms\ of\ \lambda\mathbf{J}^{\mathbf{m}}) & t,u,v ::= x \mid \lambda x.t \mid t(u,l,(x)v) \\ (lists\ of\ \lambda\mathbf{J}^{\mathbf{m}}) & l ::= t::l \mid []\end{array}$$

The sets of $\lambda\mathbf{J}^{\mathbf{m}}$*-terms and* $\lambda\mathbf{J}^{\mathbf{m}}$*-lists are denoted by* $\mathbf{\Lambda J}^{\mathbf{m}}$ *and* $\mathcal{L}\mathbf{J}^{\mathbf{m}}$ *respectively. Term constructions of the form* $t(u,l,(x)v)$ *are called* generalised multiary applications. *The list* $[]$ *is called the* empty list *and lists of the form* $t::l$ *are called* cons-lists. *In terms* $\lambda x.v$ *and* $t(u,l,(x)v)$*, occurrences of* x *in* v *are bound.*

Schwichtenberg's multiary sequent terms are obtained from $\mathbf{\Lambda J}^{\mathbf{m}}$ by requiring t to be a variable in every $t(u,l,(x)v)$ and writing $y(u,l,(x)v)$ as $v_x(y,\boldsymbol{L})$, where $\boldsymbol{L}$ is the list u,l. Notice this $\boldsymbol{L}$ is non-empty as required in [7]. In Section 3 we explain in what sense $t(u,l,(x)v)$ is a generalised form of application.

2.2 Typing Rules

Sequents of $\lambda\mathbf{J}^{\mathbf{m}}$ are of one of the following two forms

$$\begin{array}{c}\Gamma;-\vdash t:A \\ \Gamma;B\vdash l:C,\end{array}$$

called *term sequents* and *list sequents* respectively. The distinguished position in the LHS of sequents is called the *stoup* and may either be empty (as in term sequents) or hold a formula (the case of list sequents). Read a list sequent $\Gamma;B\vdash l:C$ as "list l leads the formula B to its instance C in context Γ". C is an *instance* of B if B is of the form $B_1 \supset \ldots \supset B_k \supset C$, for some $k \geq 0$.

Definition 2 *The typing rules of* $\lambda\mathbf{J}^{\mathbf{m}}$ *are as follows:*

$$\frac{}{x:A,\Gamma;-\vdash x:A}\ Axiom \qquad \frac{x:A,\Gamma;-\vdash t:B}{\Gamma;-\vdash \lambda x.t:A\supset B}\ Right$$

$$\frac{\Gamma;-\vdash t:A\supset B \quad \Gamma;-\vdash u:A \quad \Gamma;B\vdash l:C \quad x:C,\Gamma;-\vdash v:D}{\Gamma;-\vdash t(u,l,(x)v):D}\ gm-Elim$$

$$\frac{\Gamma;-\vdash u:A \quad \Gamma;B\vdash l:C}{\Gamma;A\supset B\vdash u::l:C}\ Lft \qquad \frac{}{\Gamma;C\vdash []:C}\ Ax$$

with the proviso that $x:A$ *does not belong to* Γ *in* Right *and the proviso that* $x:C$ *does not belong to* Γ *in* gm-Elim.

An instance of rule $gm-Elim$ is called a *generalised multiary elimination* (or gm-elimination, for short). We explain now why we regard these typing rules as defining a sequent calculus.

Observe that

$$\frac{y:A\supset B,\Gamma;-\vdash u:A \quad y:A\supset B,\Gamma;B\vdash l:C \quad x:C,y:A\supset B,\Gamma;-\vdash v:D}{y:A\supset B,\Gamma;-\vdash y(u,l,(x)v):D}\ m-Left \tag{1}$$

is a derived rule in $\lambda\mathbf{J}^{\mathbf{m}}$, whose instances are called *multiary left* (or m-left, for short) inferences. Rule $m-Left$ may be seen as the particular case of $gm-Elim$ when the leftmost premiss is an instance of $Axiom$. Schwichtenberg's multiary sequent calculus is then obtained by requiring every gm-elimination to be a m-left inference.

In $\lambda\mathbf{J}^{\mathbf{m}}$, "multiarity" is implemented with rules Lft and Ax, a device due to Herbelin [4]. Rule Lft is a special form of a sequent calculus left rule that requires both the active formula of the right premiss and the main formula to be in the stoup. This entails, in a derivation, that if the inference immediately above the right premiss of an instance of Lft is not an instance of Ax, then it is again an instance of Lft. Therefore, the rightmost branch of a derivation $\mathcal{D}$ ending with a list sequent is a sequence of k instances of Lft ($k \geq 0$) topped with an instance of Ax.

Going back to (1), one can now see that the formula $A \supset B$ introduced by $m - Left$ is actually of the form $A \supset B_1 \supset ... \supset B_k \supset C$, for some $k \geq 0$. A particular case of this rule is (2), obtained when the second premiss of $m - Left$ is an instance of Ax and, hence, $k = 0$, $l = []$ and $B = C$.

$$\frac{y:A \supset B, \Gamma; - \vdash u:A \quad x:B, y:A \supset B, \Gamma; - \vdash v:D}{y:A \supset B, \Gamma; - \vdash y(u, [], (x)v):D}\ Left \tag{2}$$

This is the ordinary binary left rule for implication.

We now want to explain what kind of derivations become available when one goes beyond Schwitchenberg's derivations, *i.e.* when one no longer requires every gm-elimination to be a m-left inference. For that purpose, we interpret $\lambda\mathbf{J}^{\mathbf{m}}$ in Herbelin's $\overline{\lambda}$-calculus [4].

2.3 Interpretation into $\overline{\lambda}$-Calculus

In $\overline{\lambda}$ we find the same separation between terms and lists, but there are no variables and no gm-applications. Instead there are the term constructors xl (dereliction), tl (head-cut) and $v\{x := t\}$ (mid-cut), with typing rules

$$\frac{x:A, \Gamma; A \vdash l:B}{x:A, \Gamma; - \vdash xl:B}\ Der$$

$$\frac{\Gamma; - \vdash t:A \quad \Gamma; A \vdash l:B}{\Gamma; - \vdash tl:B}\ h - Cut \qquad \frac{\Gamma; - \vdash t:A \quad x:A, \Gamma; - \vdash v:B}{\Gamma; - \vdash v\{x := t\}:B}\ m - Cut$$

Now, on the one hand, a variable y is interpreted as the dereliction $y[]$ and, on the other hand, a gm-application $t(u, l, (x)v)$ is interpreted as the combination $v\{x := t(u :: l)\}$ of an head-cut and a mid-cut:

$$\frac{\dfrac{\Gamma; - \vdash t:A \supset B \quad \dfrac{\Gamma; - \vdash u:A \quad \Gamma; B \vdash l:C}{\Gamma; A \supset B \vdash u :: l:C}\ Lft}{\Gamma; - \vdash t(u :: l):C}\ h - Cut \qquad x:C, \Gamma; - \vdash v:D}{\Gamma; - \vdash v\{x := t(u :: l)\}:D}\ m - Cut \tag{3}$$

Notice that the computational interpretation of a mid-cut is an explicit substitution [4,2]. The same interpretation was given to $v_x(y, \boldsymbol{L})$ in [7].

When the rightmost premiss of a gm-elimination is an instance of *Axiom*, the gm-elimination is essentially an head-cut $t(u :: l)$, which corresponds to a right-permuted cut. Only head-cuts $t[]$ are missing in $\lambda\mathbf{J^m}$. However, $t[]$ and t are the same, up to a trivial right-permutation of cuts. The particular form $x(u :: l)$ of head-cuts (together with variables x) gives to $\lambda\mathbf{J^m}$ the full power of dereliction xl. In this sense, $\lambda\mathbf{J^m}$ includes the entire cut-free fragment of $\overline{\lambda}$.

According to (3), a gm-elimination is a particular form of mid-cut. In $\lambda\mathbf{J^m}$ we do not have general mid-cuts - and this is what is missing for having a direct simulation of LJ with cuts inside $\lambda\mathbf{J^m}$. Instead, there is the derived rule

$$\frac{\Gamma; - \vdash t : A \quad x : A, \Gamma; - \vdash v : B}{\Gamma; - \vdash \mathbf{s}(t, x, v) : B} \tag{4}$$

where $\mathbf{s}(t, x, v)$ is the *generalised multiary substitution* operation (gm-substitution, for short), to be introduced in Definition 3 below.

2.4 Reduction Rules

In $\lambda\mathbf{J^m}$ we have reduction rules and permutative conversions. Permutative conversions aim to reduce gm-eliminations to a particular form that corresponds to the elimination rule of natural deduction. They are defined in Section 4 and constitute the central topic of our study. Reduction rules are introduced now.

Definition 3 *The* reduction rules *for* $\lambda\mathbf{J^m}$ *are as follows:*

$$\begin{array}{lrl}
(\beta_1) & (\lambda x.t)(u, [], (y)v) & \to \mathbf{s}(\mathbf{s}(u, x, t), y, v) \\
(\beta_2) & (\lambda x.t)(u, v :: l, (y)v) & \to \mathbf{s}(u, x, t)(v, l, (y)v) \\
(\pi) & t(u, l, (x)v)(u', l', (y)v') & \to t(u, l, (x)v(u', l', (y)v')) \\
(\mu) & t(u, l, (x)x(u', l', (y)v)) & \to t(u, \mathbf{append}(l, u', l'), (y)v), \quad x \notin u', l', v
\end{array}$$

where

$$\begin{aligned}
\mathbf{s}(t, x, x) &= t \\
\mathbf{s}(t, x, y) &= y, \; y \neq x \\
\mathbf{s}(t, x, \lambda y.u) &= \lambda y.\mathbf{s}(t, x, u) \\
\mathbf{s}(t, x, u(v, l, (y)v')) &= \mathbf{s}(t, x, u)(\mathbf{s}(t, x, v), \mathbf{s}'(t, x, l), (y)\mathbf{s}(t, x, v'))
\end{aligned}$$

$$\begin{aligned}
\mathbf{s}'(t, x, []) &= [] \\
\mathbf{s}'(t, x, v :: l) &= \mathbf{s}(t, x, v) :: \mathbf{s}'(t, x, l)
\end{aligned}$$

$$\begin{aligned}
\mathbf{append}([], u, l) &= u :: l \\
\mathbf{append}(u' :: l', u, l) &= u :: \mathbf{append}(l', u, l)
\end{aligned}$$

The notations $\to_{\beta,\pi,\mu}$ and $\to^*_{\beta,\pi,\mu}$ stand for the one-step and the zero or more steps reduction relations, respectively, induced by the reduction rules (β_1), (β_2), (π) and (μ), that is $\to_{\beta,\pi,\mu}$ is the compatible closure of these rules and $\to^*_{\beta,\pi,\mu}$ is the reflexive and transitive closure of $\to_{\beta,\pi,\mu}$.

The set of normal forms w.r.t. $\rightarrow_{\beta,\pi,\mu}$, which we write as $\mathbf{nf}(\lambda\mathbf{J^m})$, is given by the following grammar:

$$\begin{aligned} t, u &::= x \mid \lambda x.t \mid x(u, l, (y)v) \\ l &::= u :: l \mid [] \end{aligned}$$

where in the last production for terms, if v is of the form $y(u', l', (y')v')$, y must occur either in u', l' or v'. The normal forms w.r.t. $\rightarrow_{\beta,\pi}$ are as above omitting this last proviso. The latter and the former sets of normal forms correspond to Schwichtenberg's "multiary sequent terms" and their "multiary normal forms", respectively.

Reduction rule (μ) is structural and was already introduced in [7]. Consider the μ-redex in Definition 3 and a typing derivation $\mathcal{D}$ of this redex. When $x \notin u', l', v'$, the name x is in some sense redundant. Its type, of the form $A \supset B$, instead of being introduced in $\mathcal{D}$ by the m-left inference $x(u', l', (y)v')$, could have been introduced in a more "multiary" fashion by a Lft-inference $u' :: l'$. This transformation of $\mathcal{D}$ is the ultimate effect of (μ). Notice that there are in $\mathcal{D}$ subderivations of $\Gamma;D \vdash l : A \supset B$, $\Gamma; - \vdash u' : A$ and $\Gamma;B \vdash l' : C$, for some Γ, C, D, and that

$$\frac{\Gamma;D \vdash l : A \supset B \quad \Gamma; - \vdash u' : A \quad \Gamma;B \vdash l' : C}{\Gamma;D \vdash \mathbf{append}(l, u', l') : C} \tag{5}$$

is a derived rule of $\lambda\mathbf{J^m}$.

Reduction rules (β_1), (β_2) and (π) perform cut-elimination. Consider a gm-elimination $t(u, l, (y)v)$ and let us focus on the head-cut $t(u :: l)$ that exists inside it, in the sense of (3). Such an head-cut is right-permuted, because its right cut-formula $A \supset B$, being in the stoup, is main and linear. Now, if t is a gm-elimination, the head-cut is left-permutable. Rule π permutes the lower gm-elimination, where $t(u :: l)$ lives, over the upper gm-elimination t. If t is a λ-abstraction $\lambda x.t_0$, the head-cut is both left- and right-permuted, and the key-step of cut-elimination applies. A first cut with cut-formula A is generated and immediately eliminated by performing $\mathbf{s}(u, x, t_0)$. Now, if l is empty, we do not have arguments for forming a new gm-elimination, and a another substitution is called - this is rule (β_1). If $l = v_0 :: l_0$, a second cut with cut-formula B is generated, namely the head-cut $(\mathbf{s}(u, x, t_0))(v_0 :: l_0)$ that lives in the new gm-elimination $\mathbf{s}(u, x, t_0)(v_0, l_0, (y)v)$ - this is rule (β_2).

With the help of the fact that typing rules (4) and (5) are derivable, it is routine to prove subject reduction for $\rightarrow_{\beta,\pi,\mu}$.

We end this section by stating two properties of gm-substitution, that follow by routine induction.

Lemma 1 *For all* $t, u \in \mathbf{\Lambda J^m}$, $\mathbf{s}(t, x, u) = u$, *if* $x \notin u$.

Lemma 2 (Substitution Lemma) *For all* $t, u, v \in \mathbf{\Lambda J^m}$ *such that* $y \notin t$, *and* $x \neq y$, $\mathbf{s}(t, x, \mathbf{s}(u, y, v)) = \mathbf{s}(\mathbf{s}(t, x, u), y, \mathbf{s}(t, x, v))$.

3 Various Subsystems of $\lambda \mathbf{J}^{\mathbf{m}}$

In this section we define several subsystems of $\lambda \mathbf{J}^{\mathbf{m}}$ by constraining the construction $t(u, l, (x)v)$, as illustrated in the following diagram:

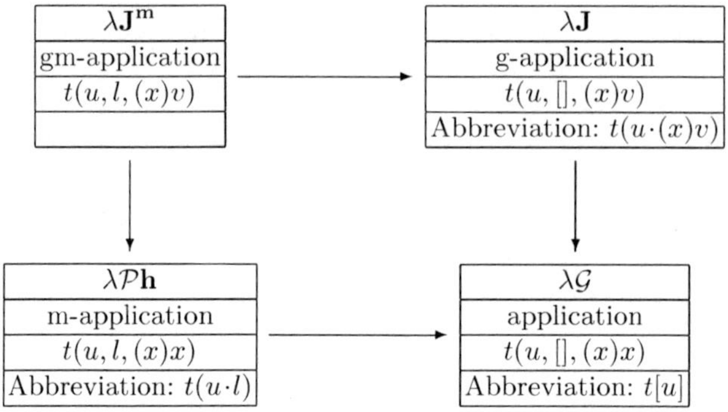

It turns out that, by doing so, we capture a number of previously known systems as subsystems of $\lambda \mathbf{J}^{\mathbf{m}}$, w.r.t. expressions, typing rules and reduction rules allowed.

3.1 λJ: The Generalised λ-Calculus

Definition 4 *The terms of* $\lambda \mathbf{J}$ *are described in the following grammar:*

$$\begin{array}{ll}(\lambda \mathbf{J} - terms)\; t, u, v ::= x \mid \lambda x.t \mid t(u, l, (x)v) \\ (\lambda \mathbf{J} - lists) \qquad l ::= []\end{array}$$

where x ranges over the set $\mathbf{V}$ *of* variables. $\mathbf{\Lambda J}$ *and* $\mathcal{L}\mathbf{J}$ *are used to denote the sets of* $\lambda \mathbf{J}$*-terms and* $\lambda \mathbf{J}$*-lists respectively.*

Hence, the sets $\mathbf{\Lambda J}$ and $\mathcal{L}\mathbf{J}$ are obtained from $\mathbf{\Lambda J}^{\mathbf{m}}$ and $\mathcal{L}\mathbf{J}^{\mathbf{m}}$, respectively, by omitting cons-lists. Since there is only one form of lists in $\lambda \mathbf{J}$, every gm-application in $\lambda \mathbf{J}$ is of the form $t(u, [], (x)v)$, which we call a *generalised application* (or g-application, for short). $\lambda \mathbf{J}$-terms can simply be described as:

$$(\lambda \mathbf{J} - terms)\; t, u, v ::= x \mid \lambda x.t \mid t(u \cdot (x)v) \ ,$$

where $t(u \cdot (x)v)$ is used as an abbreviation to $t(u, [], (x)v)$. This expression can be typed by the derived rule

$$\frac{\Gamma; - \vdash t : A \supset B \quad \Gamma; - \vdash u : A \quad x : B, \Gamma; - \vdash v : C}{\Gamma; - \vdash t(u \cdot (x)v) : C}\; g - Elim \ , \tag{6}$$

with proviso $x : B$ does not belong to Γ, which corresponds to an instance of the rule $gm - Elim$ where the penultimate premiss is an instance of Ax.

Definition 5 *The* reduction rules *for* $\lambda\mathbf{J}$ *are as follows:*

$$(\beta_1) \qquad (\lambda x.t)(u\cdot(y)v) \to \mathbf{s}(\mathbf{s}(u,x,t),y,v)$$
$$(\pi)\ t(u\cdot(x)v)(u'\cdot(y)v') \to t(u\cdot(x)v(u'\cdot(y)v'))$$

where
$$\begin{aligned}
\mathbf{s}(t,x,x) &= x \\
\mathbf{s}(t,x,y) &= y,\ y \neq x \\
\mathbf{s}(t,x,\lambda y.u) &= \lambda y.\mathbf{s}(t,x,u) \\
\mathbf{s}(t,x,u(v\cdot(y)v')) &= \mathbf{s}(t,x,u)(\mathbf{s}(t,x,v)\cdot(y)\mathbf{s}(t,x,v'))
\end{aligned}$$

Comparatively to $\lambda\mathbf{J}^{\mathbf{m}}$, $\lambda\mathbf{J}$ drops all rules and clauses involving *cons*. Since β_2-redexes and μ-*contracta* fall outside $\mathbf{\Lambda J}$ (notice that $\mathbf{append}([],u',l')$ is a cons-list), the rules (β_2) and (μ) are omitted.

The set $\mathbf{nf}(\lambda\mathbf{J})$ of $\lambda\mathbf{J}$ normal forms is given by the following grammar:

$$t,u ::= x \mid \lambda x.t \mid x(u\cdot(y)t)$$

Because of the omission of the (μ)-rule, there are $\lambda\mathbf{J}$ normal forms which are not $\lambda\mathbf{J}^{\mathbf{m}}$ normal forms.

The system thus obtained is the so-called ΛJ-calculus of Joachimski and Matthes [5], which in turn is the Curry-Howard counterpart to a system of natural deduction due to von Plato [8], where the idea of generalised elimination rules originated. These rules allow arbitrary consequences, as in the usual natural deduction rule for disjunction elimination, and the rule for implication (with term annotations) is

$$\frac{t:A\supset B \quad u:A \quad \begin{matrix}[x:B]\\ \vdots \\ v:C\end{matrix}}{t(u\cdot(x)v):C}\supset E\ .$$

The $g-Elim$ rule in (6) is then the "sequent style" formulation of this natural deduction rule.

3.2 $\lambda\mathcal{P}\mathbf{h}$: The Multiary λ-Calculus

Definition 6 *The terms of* $\lambda\mathcal{P}\mathbf{h}$ *are described in the following grammar:*

$$(\lambda\mathcal{P}\mathbf{h}-terms)\ t,u ::= x \mid \lambda x.t \mid t(u,l,(x)x)$$
$$(\lambda\mathcal{P}\mathbf{h}-lists)\quad l ::= t::l \mid []$$

where x *ranges over the set* $\mathbf{V}$ *of* variables. $\mathbf{\Lambda}\mathcal{P}\mathbf{h}$ *and* $\mathcal{L}\mathcal{P}\mathbf{h}$ *are used to denote the sets of* $\lambda\mathcal{P}\mathbf{h}$*-terms and* $\lambda\mathcal{P}\mathbf{h}$*-lists respectively.*

Hence, the sets $\mathbf{\Lambda}\mathcal{P}\mathbf{h}$ and $\mathcal{L}\mathcal{P}\mathbf{h}$ are obtained from $\mathbf{\Lambda J}^{\mathbf{m}}$ and $\mathcal{L}\mathbf{J}^{\mathbf{m}}$, respectively, by constraining v in $t(u,l,(x)v)$ to be x. A gm-application of the form $t(u,l,(x)x)$

is called a *multiary application* (or m-application, for short) and is written as $t(u{\cdot}l)$. It may be typed with the derived rule

$$\frac{\Gamma; - \vdash t : A \supset B \quad \Gamma; - \vdash u : A \quad \Gamma; B \vdash l : C}{\Gamma; - \vdash t(u{\cdot}l) : C}\ m - Elim$$

which corresponds to an instance of the rule $gm - Elim$ where the rightmost premiss is an instance of *Axiom*.

Definition 7 *The* reduction rules *for* $\lambda\mathcal{P}\mathbf{h}$ *are as follows:*

$$\begin{array}{rrl}
(\beta_1) & (\lambda x.t)(u{\cdot}[]) & \rightarrow \mathbf{s}(u, x, t) \\
(\beta_2) & (\lambda x.t)(u{\cdot}v::l) & \rightarrow \mathbf{s}(u, x, t)(v{\cdot}l) \\
(h) & t(u{\cdot}l)(u'{\cdot}l') & \rightarrow t(u{\cdot}\mathbf{append}(l, u', l'))
\end{array}$$

$$\begin{array}{lrl}
\textit{where} & \mathbf{s}(t, x, x) & = x \\
& \mathbf{s}(t, x, y) & = y,\ y \neq x \\
& \mathbf{s}(t, x, \lambda y.u) & = \lambda y.\mathbf{s}(t, x, u) \\
& \mathbf{s}(t, x, u(v{\cdot}l)) & = \mathbf{s}(t, x, u)(\mathbf{s}(t, x, v){\cdot}\mathbf{s}'(t, x, l)) \\
& & \\
& \mathbf{s}'(t, x, []) & = [] \\
& \mathbf{s}'(t, x, v::l) & = \mathbf{s}(t, x, v)::\mathbf{s}'(t, x, l) \\
& & \\
& \mathbf{append}([], u, l) & = u::l \\
& \mathbf{append}(t::l, u', l') & = t::\mathbf{append}(l, u', l')
\end{array}$$

Observe that π-*contracta* and μ-redexes fall outside $\mathbf{\Lambda}\mathcal{P}\mathbf{h}$. The (π) and (μ) rules have been combined into the new rule (h):

$$\begin{array}{rcl}
t(u{\cdot}l)(u'{\cdot}l') & = & t(u, l, (x)x)(u', l', (y)y) \\
& \rightarrow_\pi & t(u, l, (x)x(u', l', (y)y)) \\
& \rightarrow_\mu & t(u, \mathbf{append}(l, u', l'), (y)y) \\
& = & t(u{\cdot}\mathbf{append}(l, u', l'))\ .
\end{array}$$

The set of $\lambda\mathcal{P}\mathbf{h}$ normal forms is the restriction to $\lambda\mathcal{P}\mathbf{h}$ of the set of $\lambda\mathbf{J^m}$ normal forms and can thus be described as follows:

$$\begin{array}{rl}
t, u ::= & x \mid \lambda x.t \mid x(u{\cdot}l) \\
l ::- & u::l \mid []
\end{array}$$

The system thus obtained is identical to the $\lambda\mathcal{P}\mathbf{h}$-calculus defined in [2,3]. Originally this system was seen as a calculus of right-permuted cuts, but it can also be seen as the natural extension of the λ-calculus where functions may be applied to lists of arguments.

3.3 $\lambda\mathcal{G}$: Gentzen-Style λ-Calculus

The calculus $\lambda\mathcal{G}$ is the subsystem obtained from $\lambda\mathbf{J^m}$ by combining the constraints that define $\lambda\mathbf{J}$ and $\lambda\mathcal{P}\mathbf{h}$, that is, by constraining gm-applications to be of the form $t(u, [], (x)x)$, which we call a *(simple) application.* Therefore the set of $\lambda\mathcal{G}$-terms, written as $\mathbf{\Lambda}\mathcal{G}$, can be characterised as follows:

$$
\begin{array}{ll}
(\lambda\mathcal{G} - terms) & t, u ::= x \mid \lambda x.t \mid t(u, l, (x)x) \\
(\lambda\mathcal{G} - lists) & l ::= []
\end{array}
$$

As before an application $t(u, [], (x)x)$ can be abbreviated to $t(u\cdot[])$. Further, we even write this last expression as $t[u]$. The typing rule is the following derived rule:

$$
\frac{\Gamma; - \vdash t : A \supset B \quad \Gamma; - \vdash u : A}{\Gamma; - \vdash t[u] : B}\ Elim
$$

which one can recognize as the ordinary natural deduction elimination rule. This rule is an instance of the rule $gm - Elim$ where the last two premisses are instances of Ax and $Axiom$ respectively. Using the abbreviations above, and since there is only one form of lists in $\lambda\mathcal{G}$, $\lambda\mathcal{G}$-terms can simply be described as follows:

$$
(\lambda\mathcal{G} - terms)\ t, u ::= x \mid \lambda x.t \mid t[u]
$$

Definition 8 *The unique* reduction rule *for $\lambda\mathcal{G}$ is as follows:*

$$
(\beta_1) \qquad (\lambda x.t)[u] \rightarrow \mathbf{s}(u, x, t)
$$

$$
\begin{array}{rrcl}
where & \mathbf{s}(t, x, x) & = & x \\
& \mathbf{s}(t, x, y) & = & y,\ y \neq x \\
& \mathbf{s}(t, x, \lambda y.u) & = & \lambda y.\mathbf{s}(t, x, u) \\
& \mathbf{s}(t, x, u[v]) & = & \mathbf{s}(t, x, u)[\mathbf{s}(t, x, v)]
\end{array}
$$

The system $\lambda\mathcal{G}$ is no more than an isomorphic copy of the λ-calculus. In fact, $\lambda\mathcal{G}$ is the image of Gentzen's mapping $\mathcal{G}$ from natural deduction into sequent calculus [2].

Again, the normal forms of $\lambda\mathcal{G}$ are not necessarily normal forms of its extensions. For example, $t[u][u']$ is not a redex of $\lambda\mathcal{G}$, but it is a redex of its extensions.

4 Permutative Conversions for $\lambda\mathbf{J}^{\mathbf{m}}$

Terms of $\lambda\mathbf{J}^{\mathbf{m}}$ can be interpreted onto $\lambda\mathcal{G}$ via the mapping ϕ introduced below. In this section we identify a set of conversions on $\lambda\mathbf{J}^{\mathbf{m}}$-terms, corresponding to certain oriented permutations on derivations. From the viewpoint of λ-calculi, the goal of permutations is to reduce gm-applications to simple applications. Taking this set of conversions as defining a rewriting system, one obtains a strongly normalising and confluent rewriting system such that the normal form of a $\lambda\mathbf{J}^{\mathbf{m}}$-term is its ϕ-image. Moreover we get a permutability theorem: two $\lambda\mathbf{J}^{\mathbf{m}}$-terms have the same ϕ-image iff they are inter-permutable.

4.1 Mapping ϕ

Definition 9 *$\phi : \mathbf{\Lambda J^m} \longrightarrow \mathbf{\Lambda \mathcal{G}}$ and $\phi' : \mathbf{\Lambda \mathcal{G}} \times \mathbf{\Lambda \mathcal{G}} \times \mathcal{L}\mathbf{J^m} \times \mathbf{V} \times \mathbf{\Lambda \mathcal{G}} \longrightarrow \mathbf{\Lambda \mathcal{G}}$ are defined by simultaneous recursion as follows:*

$$\begin{aligned} \phi(x) &= x \\ \phi(\lambda x.t) &= \lambda x.\phi(t) \\ \phi(t(u,l,(x)v)) &= \phi'(\phi(t),\phi(u),l,x,\phi(v)) \end{aligned}$$

$$\begin{aligned} \phi'(t,u,[],x,v) &= \mathbf{s}(t[u],x,v) \\ \phi'(t,u,v::l,x,v') &= \phi'(t[u],\phi(v),l,x,v') \end{aligned}$$

The following two propositions establish that ϕ commutes with application and that $\lambda\mathcal{G}$-terms are invariant under ϕ, and thus ϕ is onto.

Proposition 1 *For all $t, u \in \mathbf{\Lambda J^m}$, $\phi(t[u]) = \phi(t)[\phi(u)]$.*

Proof: $\phi(t[u]) = \phi'(\phi(t),\phi(u),[],x,\phi(x)) = \mathbf{s}(\phi(t)[\phi(u)],x,x) = \phi(t)[\phi(u)]$. □

Proposition 2 *For all $t \in \mathbf{\Lambda \mathcal{G}}$, $\phi(t) = t$.*

Proof: Routine induction on t. The case of application uses Proposition 1. □

Two other properties relating the mappings ϕ and ϕ' with the notion of substitution are as follows.

Proposition 3 *For all $t, u, v \in \mathbf{\Lambda J^m}$ and for all $l \in \mathcal{L}\mathbf{J^m}$,*

$$\phi'(\phi(t),\phi(u),l,x,\phi(v)) = \mathbf{s}(\phi(t(u \cdot l)),x,\phi(v)) \ .$$

Proof: By induction on l. □

Proposition 4

(*i*) $\phi(\mathbf{s}(t,x,v)) = \mathbf{s}(\phi(t),x,\phi(v))$, *for all $t, v \in \mathbf{\Lambda J^m}$.*
(*ii*) $\phi'(\mathbf{s}(\phi(t),x,\phi(v_1)),\mathbf{s}(\phi(t),x,\phi(v_2)),\mathbf{s}'(t,x,l),y,\mathbf{s}(\phi(t),x,\phi(v)))$
$= \mathbf{s}(\phi(t),x,\phi'(\phi(v_1),\phi(v_2),l,y,\phi(v)))$, *for all $t, v, v_1, v_2 \in \mathbf{\Lambda J^m}$, $l \in \mathcal{L}\mathbf{J^m}$.*

Proof: By simultaneous induction on v and l. □

4.2 Permutative Conversions

Permutative conversions on $\lambda\mathbf{J^m}$-terms, that from here on we call simply permutations, are divided into **p**-*permutations* and **q**-*permutations*. **p**-permutations (resp. **q**-permutations) deal with "generality" (resp. multiarity) of gm-application. The **p**-permutations are:

$$\begin{array}{ll} (p_1) & t(u,l,(x)y) \to y, \quad x \neq y \\ (p_2) & t(u,l,(x)\lambda y.v) \to \lambda y.t(u,l,(x)v) \\ (p_3) & t_1(u_1,l_1,(x)t_2(u_2,l_2,(y)v)) \to \\ & \quad t_1(u_1,l_1,(x)t_2)(t_1(u_1,l_1,(x)u_2),\mathbf{p_3'}(t_1,u_1,l_1,x,l_2),(y)v) \ \text{ if } x \notin v, \end{array}$$

where

$$\begin{aligned} \mathbf{p_3'} &: \mathbf{\Lambda J^m} \times \mathbf{\Lambda J^m} \times \mathcal{L}\mathbf{J^m} \times \mathbf{V} \times \mathcal{L}\mathbf{J^m} \longrightarrow \mathcal{L}\mathbf{J^m} \\ \mathbf{p_3'}(t,u,l,x,[]) &= [] \\ \mathbf{p_3'}(t,u,l,x,u'::l') &= t(u,l,(x)u')::\mathbf{p_3'}(t,u,l,x,l') \ . \end{aligned}$$

Mapping $\mathbf{p}'_3$ is a way of generating a gm-application for each element of a list. $\mathbf{p}$-permutations are essentially as in Schwichtenberg's [7]. In particular, we follow the idea of requiring $x \notin v$ in (p_3), which is crucial in guaranteeing termination. In a sense, our permutations are simpler because we are not obliged to stay in the cut-free fragment and do not need permutations to preserve μ-normal form.

The unique $\mathbf{q}$-permutation is

$$(q)\ \ t(u, v::l, (x)v') \to t[u](v, l, (x)v')\ .$$

This kind of permutation is first pointed out in [2] in the context of the system $\lambda\mathcal{P}\mathbf{h}$. As opposed to [7], a permutation to deal with multiarity is necessary because we are not confined to the cut-free fragment.

We use $\to_p$ and $\to_q$ to denote the one-step reduction relations (compatible closure) induced by $\mathbf{p}$-permutations and by the $\mathbf{q}$-permutation respectively. The notation $\to_{p,q}$ represents the union of $\to_p$ and $\to_q$. As usual, given a one-step reduction relation $\to_r$, $\to_r^*$ and $\leftrightarrow_r^*$ denote respectively its reflexive and transitive closure and its equivalence closure.

Proposition 5 establishes invariance of permutation reducibility under the mapping ϕ. In its proof we use the following lemma, proved by induction on l.

Lemma 3 *$\phi'(t, u, \mathbf{s}'(t_1(u_1 \cdot l_1), x, l), y, v) = \phi'(t, u, \mathbf{p}'_3(t_1, u_1, l_1, x, l), y, v)$, for all $t, u, v, t_1, u_1 \in \mathbf{\Lambda J^m}$, $l, l_1 \in \mathcal{L}\mathbf{J^m}$.*

Proposition 5 *If $t \to_{p,q}^* u$ then $\phi(t) = \phi(u)$, for all $t, u \in \mathbf{\Lambda J^m}$.*

Proof: The proof follows by induction on the relation $\to_{p,q}^*$. □

Proposition 6 asserts that any $\lambda\mathbf{J^m}$-term can be reduced to its ϕ-image using $\mathbf{p}$ and $\mathbf{q}$-permutations. Its proof uses the fact that g-applications can be reduced to substitutions using solely $\mathbf{p}$-permutations, as in the following lemma proved by induction on v.

Lemma 4 *$t(u \cdot (x)v) \to_p^* \mathbf{s}(t[u], x, v)$, for all $t, u, v \in \mathbf{\Lambda}\mathcal{G}$.*

Proposition 6 *(i) $t \to_{p,q}^* \phi(t)$, for all $t \in \mathbf{\Lambda J^m}$.*
(ii) $\phi(t)(\phi(u), l, (x)\phi(v)) \to_{p,q}^* \mathbf{s}(\phi(t(u \cdot l)), x, \phi(v))$,
for all $l \in \mathcal{L}\mathbf{J^m}$ and for all $t, u, v \in \mathbf{\Lambda J^m}$.

Proof: By simultaneous induction on the structure of t and l. □

4.3 Main Results

Propositions 5 and 6, establishing respectively invariance of $\mathbf{p}$ and $\mathbf{q}$-permutations under ϕ and reducibility to the ϕ-image using such permutations, are now used in proving the theorems below.

Theorem 1 (characterization of p, q-normal forms) *For all $t \in \mathbf{\Lambda J^m}$, t is $\mathbf{p}, \mathbf{q}$-normal iff $t \in \mathbf{\Lambda}\mathcal{G}$.*

Proof: On the one hand, terms of $\lambda\mathcal{G}$ have neither **p** or **q**-redexes and so they are **p**, **q**-normal. Consider, on the other hand, that t is **p**, **q**-normal. By Proposition 6, $t \rightarrow^*_{p,q} \phi(t)$ and thus the normality of t implies $t = \phi(t)$. The proof concludes observing that the co-domain of ϕ is $\mathbf{\Lambda}\mathcal{G}$. □

Theorem 2 (confluence) *For all* $t, t_1, t_2 \in \mathbf{\Lambda J^m}$, *if* $t \rightarrow^*_{p,q} t_1$ *and* $t \rightarrow^*_{p,q} t_2$, *then* $t_1 \rightarrow^*_{p,q} t_3$ *and* $t_2 \rightarrow^*_{p,q} t_3$, *for some* $t_3 \in \mathbf{\Lambda J^m}$.

Proof: Assuming $t \rightarrow^*_{p,q} t_1$ and $t \rightarrow^*_{p,q} t_2$, by Proposition 6 follows that $t_1 \rightarrow^*_{p,q} \phi(t_1)$ and $t_2 \rightarrow^*_{p,q} \phi(t_2)$. Yet by Proposition 6 we have $t \rightarrow^*_{p,q} \phi(t)$ and we can now use Proposition 5 to conclude that $\phi(t_1) = \phi(t) = \phi(t_2)$. □

Theorem 3 (permutability) $\phi(t_1) = \phi(t_2)$ *iff* $t_1 \leftrightarrow^*_{p,q} t_2$, *for all* $t_1, t_2 \in \mathbf{\Lambda J^m}$.

Proof: Proposition 5 guarantees $\phi(t_1) = \phi(t_2)$ whenever $t_1 \leftrightarrow^*_{p,q} t_2$. As to the only if part, we use Proposition 6, obtaining $t_1 \rightarrow^*_{p,q} \phi(t_1)$ and $t_2 \rightarrow^*_{p,q} \phi(t_2)$ and thus, as by hypothesis $\phi(t_1) = \phi(t_2)$, t_1 and t_2 are inter-permutable. □

In order to ensure termination of the rewriting system induced by permutations **p** and **q**, we introduce a notion of weight for terms and lists of $\lambda\mathbf{J^m}$.

Definition 10 $\mathbf{w}(t)$ *and* $\mathbf{w}(l)$, *the* weight *of an* $\lambda\mathbf{J^m}$*-term* t *and of an* $\mathcal{L}\mathbf{J^m}$*-list* l *respectively, are defined as follows:*

$$\begin{aligned} \mathbf{w}(x) &= 1 & \mathbf{w}([]) &= 0 \\ \mathbf{w}(\lambda x.t) &= 1 + \mathbf{w}(t) & \mathbf{w}(u::l) &= 2 + \mathbf{w}(u) + \mathbf{w}(l) \\ \mathbf{w}(t(u,l,(x)v)) &= \mathbf{w}(v)(\mathbf{w}(t) + \mathbf{w}(u) + \mathbf{w}(l) + 1) \end{aligned}$$

Theorem 4 (termination) *The rewriting system induced by permutations* **p** *and* **q** *is strongly normalisable.*

Proof: Each permutation can be shown to have a RHS of weight lower than its LHS and thus every sequence of permutations must be finite. For permutation (p_3), we use the inequality below, that can be proved by induction on l_2.

$$\mathbf{w}(\mathbf{p_3'}(t_1, u_1, l_1, x, l_2)) < (\mathbf{w}(l_2) + 1)(\mathbf{w}(t_1) + \mathbf{w}(u_1) + \mathbf{w}(l_1)) + \mathbf{w}(l_2)$$ □

Since $\rightarrow_{p,q}$ is confluent and strongly normalising, each $\lambda\mathbf{J^m}$-term t has a unique normal form that we denote by $\downarrow_{p,q}(t)$.

Theorem 5 (representation of ϕ) $\phi(t) = \downarrow_{p,q}(t)$, *for all* $t \in \mathbf{\Lambda J^m}$.

Proof: By Proposition 6, $t \rightarrow^*_{p,q} \phi(t)$ and $\phi(t)$ is a normal form. □

5 Decomposing the Main Results

This section illustrates two ways of decomposing the mapping ϕ using $\lambda\mathcal{P}\mathbf{h}$ and $\lambda\mathbf{J}$ as intermediate systems, according to Figure 1 in the introductory section. In the spirit of the study in the previous section for mapping ϕ, similar results may be established (i) for the mappings **p** and $\mathbf{p^m}$ relatively to **p**-permutations and (ii) for the mappings **q** and $\mathbf{q^m}$ relatively to **q**-permutations.

Definition 11 *The mappings* $\mathbf{p^m}$, $\mathbf{q^m}$, $\mathbf{p}$ *and* $\mathbf{q}$ *are as follows.*

$$\mathbf{p^m} : \mathbf{\Lambda J^m} \longrightarrow \mathbf{\Lambda \mathcal{P}h}$$

$$\begin{aligned}
\mathbf{p^m}(x) &= x \\
\mathbf{p^m}(\lambda x.t) &= \lambda x.\mathbf{p^m}(t) \\
\mathbf{p^m}(t(u,l,(x)v)) &= \mathbf{s}(\mathbf{p^m}(t)(\mathbf{p^m}(u)\cdot\mathbf{p^m}'(l)), x, \mathbf{p^m}(v)) \\
\mathbf{p^m}'([]) &= [] \\
\mathbf{p^m}'(u::l) &= \mathbf{p^m}(u)::\mathbf{p^m}'(l)
\end{aligned}$$

$$\mathbf{q^m} : \mathbf{\Lambda J^m} \longrightarrow \mathbf{\Lambda J}$$

$$\begin{aligned}
\mathbf{q^m}(x) &= x \\
\mathbf{q^m}(\lambda x.t) &= \lambda x.\mathbf{q^m}(t) \\
\mathbf{q^m}(t(u,l,(x)v)) &= \mathbf{q^m}'(\mathbf{q^m}(t), \mathbf{q^m}(u), l, x, \mathbf{q^m}(v)) \\
\mathbf{q^m}'(t,u,v::l,x,v') &= \mathbf{q^m}'(t[u], \mathbf{q^m}(v), l, x, v') \\
\mathbf{q^m}'(t,u,[],x,v) &= t(u\cdot(x)v)
\end{aligned}$$

$$\mathbf{p} : \mathbf{\Lambda J} \longrightarrow \mathbf{\Lambda \mathcal{G}}$$

$$\begin{aligned}
\mathbf{p}(x) &= x \\
\mathbf{p}(\lambda x.t) &= \lambda x.\mathbf{p}(t) \\
\mathbf{p}(t(u\cdot(x)v)) &= \mathbf{s}(\mathbf{p}(t)[\mathbf{p}(u)], x, \mathbf{p}(v))
\end{aligned}$$

$$\mathbf{q} : \mathbf{\Lambda \mathcal{P}h} \longrightarrow \mathbf{\Lambda \mathcal{G}}$$

$$\begin{aligned}
\mathbf{q}(x) &= x \\
\mathbf{q}(\lambda x.t) &= \lambda x.\mathbf{q}(t) \\
\mathbf{q}(t(u\cdot l)) &= \mathbf{q}'(\mathbf{q}(t), \mathbf{q}(u), l)) \\
\mathbf{q}'(t,u,v::l)) &= \mathbf{q}'(t[u], \mathbf{q}(v), l)) \\
\mathbf{q}'(t,u,[])) &= t[u]
\end{aligned}$$

A mapping corresponding to $\mathbf{p}$ from $\lambda\mathbf{J}$ into the λ-calculus is given in [5]. In [2] an interpretation $\mathcal{Q}$ of $\lambda\mathcal{P}\mathbf{h}$ into the λ-calculus corresponding to $\mathbf{q}$ is studied.

Let us now consider $\mathbf{p}$ and $\mathbf{q}$-permutations in the context of the subsystems $\lambda\mathcal{P}\mathbf{h}$ and $\lambda\mathbf{J}$. In $\lambda\mathcal{P}\mathbf{h}$ we have that no $\mathbf{p}$-permutation applies to its terms and that the $\mathbf{q}$-permutation can be simplified as follows:

$$(q)\ t(u\cdot v::l) \to t[u](v\cdot l)\ .$$

Similarly, $\lambda\mathbf{J}$ is irreducible for $\mathbf{q}$-permutations and $\mathbf{p}$-permutations assume the following special forms.

$$\begin{array}{ll}
(p_1) & t(u\cdot(x)y) \to y,\ \ x \neq y \\
(p_2) & t(u\cdot(x)\lambda y.v) \to \lambda y.t(u\cdot(x)v) \\
(p_3) & t_1(u_1\cdot(x)t_2(u_2\cdot(y)v)) \to t_1(u_1\cdot(x)t_2)(t_1(u_1\cdot(x)u_2)\cdot(y)v)\ \text{ if } x \notin v
\end{array}$$

The sequence of theorems in the previous section, leading to the representation of mapping ϕ via $\mathbf{p},\mathbf{q}$-normal forms, can also be established by analogous arguments for mappings $\mathbf{p}$, $\mathbf{q}$, $\mathbf{p^m}$ and $\mathbf{q^m}$. In particular, we have permutability theorems for all these mappings and we have confluence and strong normalisation of the rewriting systems induced by $\to_q$ on $\mathbf{\Lambda\mathcal{P}h}$, $\to_p$ on $\mathbf{\Lambda J}$, $\to_p$ on $\mathbf{\Lambda J^m}$

and $\to_q$ on $\mathbf{\Lambda J^m}$. The unique normal forms of a $\lambda\mathbf{J^m}$-term t w.r.t. each of these systems is written as $\downarrow_q^{\lambda\mathcal{P}\mathbf{h}}(t)$, $\downarrow_p^{\lambda\mathbf{J}}(t)$, $\downarrow_p(t)$ and $\downarrow_q(t)$, respectively.

Theorem 6 (representation of p, q, $\mathbf{p^m}$ and $\mathbf{q^m}$) *2mm*

(i) $\mathbf{q}(t) = \downarrow_q^{\lambda\mathcal{P}\mathbf{h}}(t)$, *for all* $t \in \mathbf{\Lambda\mathcal{P}h}$. *(ii)* $\mathbf{p}(t) = \downarrow_p^{\lambda\mathbf{J}}(t)$, *for all* $t \in \mathbf{\Lambda J}$.
(iii) $\mathbf{p^m}(t) = \downarrow_p(t)$, *for all* $t \in \mathbf{\Lambda J^m}$. *(iv)* $\mathbf{q^m}(t) = \downarrow_q(t)$, *for all* $t \in \mathbf{\Lambda J^m}$.

6 Conclusion

This paper is a direct successor of [1,7] in revisiting with the help of new λ-calculi —specifically, variants of Herbelin's calculus— Zucker and Pottinger classical results [9,6]. The novelty here is that, at the same time, we pay attention to how the λ-calculus we arrive at, the system $\lambda\mathbf{J^m}$, more than being a term-annotation or book-keeping device [6], gives computational interpretations to fragments of sequent calculus. We believe it does so because: on the one hand $\lambda\mathbf{J^m}$ may be seen as an extension of the λ-calculus with a notion of generalised multiary application; on the other hand it captures as subsystems, in a modular fashion, a system isomorphic to the λ-calculus, the system ΛJ of Joachimski and Matthes and the system $\lambda\mathcal{P}\mathbf{h}$ in [3]. So, we consider worthwhile to investigate more into its properties. In particular, we intend to study confluence and normalisation of the reduction relation $\to_{\beta,\pi}$, introduced in Section 2 to perform cut-elimination. Also of interest is the study of properties of the rewriting system obtained by adding permutations **p** and **q** on top of the reduction relation $\to_{\beta,\pi}$.

References

1. R. Dyckhoff and L. Pinto, *Permutability of inferences in intuitionistic sequent calculi*, *Theoretical Computer Science*, **212** (1999) 141–155.
2. J. Espírito Santo, *Conservative extensions of the λ-calculus for the computational interpretation of sequent calculus*, *PhD thesis*, University of Edinburgh, 2002.
3. J. Espírito Santo, *An isomorphism between a fragment of sequent calculus and an extension of natural deduction, in: M. Baaz and A. Voronkov (Eds.), Proc. of LPAR'02*, 2002, Springer-Verlag, Lecture Notes in Artificial Intelligence, vol. **2514**, 354–366.
4. H. Herbelin, *A λ-calculus structure isomorphic to a Gentzen-style sequent calculus structure, in: L. Pacholski and J. Tiuryn (Eds.), Proceedings of CSL'94*, 1995, Springer-Verlag, Lecture Notes in Computer Science, vol. **933**, 61–75.
5. F. Joachimski and R. Matthes, *Short proofs of normalisation for the simply typed λ-calculus, permutative conversions and Gödel's* **T**, (accepted for publication in *Archive for Mathematical Logic*).
6. G. Pottinger, *Normalization as a homomorphic image of cut-elimination, Annals of Mathematical Logic* **12** (1977) 323–357.
7. H. Schwichtenberg, *Termination of permutative conversions in intuitionistic Gentzen calculi*, *Theoretical Computer Science*, **212** (1999) 247–260.
8. J. von Plato, *Natural deduction with general elimination rules, Annals of Mathematical Logic*, **40** (2001) 541–567.
9. J. Zucker, *The correspondence between cut-elimination and normalization, Annals of Mathematical Logic* **7** (1974) 1–112.

A Universal Embedding for the Higher Order Structure of Computational Effects

John Power*

Laboratory for the Foundations of Computer Science,
University of Edinburgh,
King's Buildings,
Edinburgh EH9 3JZ,
Scotland, UK
Fax +44 131 667 7209
ajp@dcs.ed.ac.uk

Abstract. We give a universal embedding of the semantics for the first order fragment of the computational λ-calculus into a semantics for the whole calculus. In category theoretic terms, which are the terms of the paper, this means we give a universal embedding of every small *Freyd*-category into a closed *Freyd*-category. The universal property characterises the embedding as the free completion of the *Freyd*-category as a $[\rightarrow, Set]$-enriched category under conical colimits. This embedding extends the usual Yoneda embedding of a small category with finite products into its free cocompletion, i.e., the usual category theoretic embedding of a model of the first order fragment of the simply typed λ-calculus into a model for the whole calculus, and similarly for the linear λ-calculus. It agrees with an embedding previously given in an ad hoc way without a universal property, so it shows the definitiveness of that construction.

1 Introduction

A fundamental construction in the semantics of programming lanuages based on the simply typed λ-calculus is the Yoneda embedding [4,6,12,13,19,20]. A major part of its significance, for both the usual simply typed λ-calclus [4,6,12,13] and the linear simply typed λ-calculus [13,19,20], is that it embeds a model of the first order fragment of the λ-calculus into a model of the whole calculus, while respecting the structure that already exists. In category-theoretic terms, for the simply typed λ-calculus, this says the Yoneda embedding $Y : C \longrightarrow [C^{op}, Set]$ respects finite products and any existing exponentials in C while embedding C into a cartesian closed category; and for the linear λ-calculus, it amounts to the same statement but with finite product structure replaced systematically by symmetric monoidal structure. The Yoneda embedding has a subtle universal property: it is the free colimit completion of C that respects the cartesian or

* This work is supported by AIST Japan, and by EPSRC grant GR/M56333: The structure of programming languages: syntax and semantics.

M. Hofmann (Ed.): TLCA 2003, LNCS 2701, pp. 301–315, 2003.

symmetric monoidal structure of C [3,5,7]. That fact is crucial to the development of all of the above-cited papers and is analysed conceptually in [2]. In this paper, we adapt the construction to the computational λ-calulus, i.e., to the archetypal call-by-value simply typed λ-calculus for reasoning about computational effects [9,10,11,17,18,16,19]. Specifically, we give a universal property for a Yoneda-like embedding of the semantics for the first-order fragment of the computational λ-calculus into a semantics for the whole calculus.

The above summary of the content of this paper begs some important questions. Exactly what does one mean by the first order fragment of the computational λ-calculus? What are the models of the first order fragment? Exactly what is a "Yoneda-like" embedding? And exactly what universal property is desirable for such an embedding and why?

There has been considerable work on giving a definitive notion of the first order fragment of the computational λ-calculus and its semantics [8,9,10,11,17, 18,19,21,22]. The reason the definition is not obvious is because, for many years, the usual description of the computational λ-calculus included a T-type constructor, i.e. for every type X, one would have a type TX, but the rules of the calculus implied that TX must agree, up to isomorphism, with $1 \rightarrow X$. So it was unclear whether one should regard T as a first order type constructor or as a higher order type constructor. More recent work, such as that on data refinement in [8], that on continuations in [21,22], and the more general work on computational effects in [9,18,21,22], have strongly indicated that T is better regarded as a higher order construct. Indeed, it is becoming more common nowadays not to have a type constructor T in the standard definition of the computational λ-calculus at all as it is redundant. Nowadays, the computational λ-calculus is often described as having exactly the same type and term structure as the simply typed λ-calculus, the only difference with the simply typed λ-calculus being the presence of an additional unary predicate for definedness, satisfying natural rules [14,16,18] equivalent to Moggi's original ones [10,11] (see Section 2). So the first order fragment of the computational λ-calculus may be defined by exactly the same types and terms as is the first order fragment of the simply typed λ-calculus (including a *let* expression of course), but subject to the two predicates of the computational λ-calculus restricted to those terms. A definitive sound and complete class of models is given by the class of $Freyd$-categories [9,18]. A sound and complete class of models for the whole calculus is given by the class of *closed* $Freyd$-categories [9,18], which is (non-trivially) equivalent to the class of categories with finite products, a strong monad, and Kleisli exponentials (the heart of the proof is in [19]). So here, we seek a Yoneda-like embedding of a small $Freyd$-category into a closed $Freyd$-category, with the embedding satisfying a universal property agreeing with those for the simply typed λ-calculus and the linear λ-calculus.

A $Freyd$-category consists of a category C_0 with finite products, a symmetric *premonoidal* category C_1, and an identity-on-objects strict symmetric premonoidal functor $J : C_0 \longrightarrow C_1$ (see Section 2). So it is not given by a single category with additional structure on that category or by a single category satisfying a property, but rather by a pair of categories and a functor, together with additional structure. So the notion of an embedding, in particular a "Yoneda-like embedding", in this setting, requires care. That may be resolved by using the theory of enriched categories [7]. Given a cartesian closed (more generally a symmetric monoidal closed) category V, one can define a notion of V-category. Under reasonable size and completeness conditions, V is itself a V-category, and for every small V-category C one has a presheaf V-category $[C^{op}, V]$ and a fully faithful V-functor $Y : C \longrightarrow [C^{op}, V]$ that provides a definitive notion of a V-enriched Yoneda embedding. Letting V be the cartesian closed category $[\rightarrow, Set]$, a V-category consists exactly of a pair of categories and an identity-on-objects functor $J : C_0 \longrightarrow C_1$, which is our basic data for a $Freyd$-category. Enriched category theory provides a notion of an embedding of one such structure into another such structure, and reference to the enriched Yoneda embedding largely resolves our search for a "Yoneda-like" embedding for a $Freyd$-category.

We need one further extension of our notion of "Yoneda-like" embedding. If C is a small category with finite products, the ordinary Yoneda embedding $Y : C \longrightarrow [C^{op}, Set]$ preserves them. And if C is a small symmetric monoidal category, $[C^{op}, Set]$ acquires a symmetric monoidal structure that makes the ordinary Yoneda embedding into a strong symmetric monoidal functor. But, using the enrichment we have just outlined, if $J : C_0 \longrightarrow C_1$ is a small $Freyd$-category, it is not the case that $[J^{op}, V]$ has a non-trivial $Freyd$-structure. So we must make a mild generalisation of the notion of "Yoneda-like" embedding, following [7], so that for any factorisation

$$C \xrightarrow{\ Y'\ } D \xrightarrow{\ I\ } [C^{op}, V]$$

of the enriched Yoneda embedding, with I an inclusion of a full sub-V-category (Y' is necessarily fully faithful), we call $Y' : C \longrightarrow D$ a "Yoneda-like" embedding. This is a very mild extension: the behaviour of the embedding is not allowed to change.

One can characterise Yoneda-like embeddings. For every small V-category C, every Yoneda-like embedding of C is the free completion of C under a class of colimits and conversely [7]. But in order to prove that enriched result, one needs a surprisingly delicate notion of enriched colimit called a *weighted* colimit, sometimes also called an *indexed* colimit. The *conical* colmits, which amount to the first obvious guess for a notion of enriched colimit, are among the enriched colimits but are not all of them. Nevertheless, they are a natural class of colimits, and for a small V-category C, the free completion of C under conical colimits duly yields a Yoneda-like embedding. That completion under conical colimits is the universal property we need here in the special case of $V = [\rightarrow, Set]$.

With all of these concepts, we can now make the universal statement we seek: for any small $Freyd$-category, its free completion as a $[\rightarrow, Set]$-enriched category under conical colimits has a canonical closed $Freyd$-structure, and the Yoneda-like embedding of the $Freyd$-category into its free conical colimit completion preserves the $Freyd$-structure and yields its free conical colimit completion as a $Freyd$-category. This universal construction agrees with the ad hoc embedding of a premonoidal $[\to, Set]$-category into a closed premonoidal $[\to, Set]$-category given in [17], agrees with the above-cited embeddings for cartesian and symmetric monoidal categories, i.e., for simply typed λ-calculus and for linear λ-calculus, and agrees with the universal properties of those constructions.

The paper is organised as follows. In Section 2, we give a version of the computational λ-calculus, and recall the notions of *Freyd*-category, which supplies a sound and complete class of models for the first order fragment of the computational λ-calculus, and closed *Freyd*-category, which gives a sound and complete class of models for the whole calculus. In Section 3, we recall the relevant basic definitions and constructions of enriched category theory and show how they apply to this setting. We characterise those $[\to, Set]$-categories that have conical colimits. In Section 4, we study the enriched Yoneda embedding in the setting of enrichment in $[\to, Set]$. In Section 5, we characterise the free conical colimit completion of a small $[\to, Set]$-enriched category. And in Section 6, we complete the proof that the free conical cocompletion of a small *Freyd*-category has a closed *Freyd*-structure, that the embedding preserves it, and that it yields the free conical colimit completion of the *Freyd*-category as a *Freyd*-category.

2 The Computational λ-Calculus, *Freyd*-Categories and Closed *Freyd*-Categories

In this section, we describe a version of the computational λ-calculus, or λ_c-calculus, and we describe sound and complete classes of models for both the whole calculus and its first order fragment. This is background material for this paper (see for instance [18]), but it is fundamental to understanding the point of the paper, hence our including it.

Our version of the λ_c-calculus has type constructors

$$X ::= B \mid X_1 \times X_2 \mid 1 \mid X \Rightarrow Y$$

where B is a base type and $X \Rightarrow Y$ denotes the exponential. We do not assert the existence of a type constructor TX: this formulation is equivalent to the original one because TX may be defined to be $1 \Rightarrow X$.

The terms of the λ_c-calculus are

$$e ::= x \mid b \mid e'e \mid \lambda x.e \mid * \mid (e, e') \mid \pi_i(e)$$

where x is a variable, b is a base term of arbitrary type, $*$ is of type 1, with π_i existing for $i = 1$ or 2, all subject to the evident typing. We do not explicitly have a *let* constructor or constructions $[e]$ or $\mu(e)$ in the full calculus, but we tacitly assume its first order fragment contains *let*: in the full calculus, we consider *let* $x = e$ *in* e' as syntactic sugar for $(\lambda x.e')e$, and $[e]$ as syntactic sugar for $(\lambda x).e$ where x is of type 1, and $\mu(e)$ as syntactic sugar for $e(*)$.

The λ_c-calculus has two predicates, existence, denoted by $\downarrow$, and equivalence, denoted by $\equiv$. The $\downarrow$ rules may be expressed as saying $* \downarrow$, $x \downarrow$, $\lambda x.e \downarrow$ for all e, if $e \downarrow$ then $\pi_i(e) \downarrow$, and similarly for (e, e'). A value is a term e such that $e \downarrow$.

There are two classes of rules for $\equiv$. The first class say that $\equiv$ is a congruence, with variables allowed to range over values. And the second class are rules for the basic constructions and for unit, product and functional types. It follows from the rules for both predicates that types together with equivalence classes of terms form a category, with a subcategory determined by values.

It is straightforward, using the original formulation of the λ_c-calculus in [10], to spell out the inference rules required to make this formulation agree with the original one: one just bears in mind that the models are the same, and we use syntactic sugar as detailed above.

The λ_c-calculus represents a fragment of a call by value programming language. In particular, it was designed to model fragments of ML, but is also a fragment of other languages such as the idealised language FPC (see for instance [4]). For category theoretic models, the key feature is that there are two entities, expressions and values. So the most direct way to model the language as we have formulated it is in terms of a pair of categories C_0 and C_1, together with an identity-on-objects inclusion functor $J : C_0 \longrightarrow C_1$. This train of thought leads directly to the notion of closed $Freyd$-category, which we now describe. But the first sound and complete class of models for the λ_c-calculus was given by Moggi in [11].

In order to define the notions of $Freyd$-category and closed $Freyd$-category, we must recall the definitions of premonoidal category and strict premonoidal functor, and symmetries for them, as introduced in [19] and further studied in [17]. A premonoidal category is a generalisation of the concept of monoidal category: it is essentially a monoidal category except that the tensor need only be a functor of two variables and not necessarily be bifunctorial, i.e., given maps $f : X \longrightarrow Y$ and $f' : X' \longrightarrow Y'$, the evident two maps from $X \otimes X'$ to $Y \otimes Y'$ may differ.

Example 1. Given a symmetric monoidal category C together with a specified object S, define the category K to have the same objects as C, with $K(X, Y) = C(S \otimes X, S \otimes Y)$, and with composition in K determined by that of C. For any object X of C, one has functors $X \otimes - : K \longrightarrow K$ and $- \otimes X : K \longrightarrow K$, but they do not satisfy the bifunctoriality condition above, hence do not yield a monoidal structure on K. They do yield a premonoidal structure, as we define below.

In order to make precise the notion of a premonoidal category, we need some auxiliary definitions.

Definition 1. *A* binoidal category *is a category K together with, for each object X of K, functors $h_X : K \longrightarrow K$ and $k_X : K \longrightarrow K$ such that for each pair (X, Y) of objects of K, $h_X Y = k_Y X$. The joint value is denoted $X \otimes Y$.*

Definition 2. *An arrow $f : X \longrightarrow X'$ in a binoidal category is* central *if for every arrow $g : Y \longrightarrow Y'$, the two composites from $X \otimes Y$ to $X' \otimes Y'$ agree. Moreover, given a binoidal category K, a natural transformation $\alpha : G \Longrightarrow H : C \longrightarrow K$ is called* central *if every component of α is central.*

Definition 3. *A* premonoidal category *is a binoidal category K together with an object I of K, and central natural isomorphisms a with components $(X \otimes Y) \otimes Z \longrightarrow X \otimes (Y \otimes Z)$, l with components $X \longrightarrow X \otimes I$, and r with components $X \longrightarrow I \otimes X$, subject to two equations: the pentagon expressing coherence of a, and the triangle expressing coherence of l and r with respect to a.*

Now we have the definition of a premonoidal category, it is routine to verify that Example 1 is an example of one. There is a general construction that yields premonoidal categories too: given a strong monad T on a symmetric monoidal category C, the Kleisli category $Kl(T)$ for T is always a premoidal category, with the functor from C to $Kl(T)$ preserving premonoidal structure strictly: of course, a monoidal category such as C is trivially a premonoidal category. That construction is fundamental, albeit implicit, in Eugenio Moggi's work on monads as notions of computation [10], as explained in [19].

Definition 4. *Given a premonoidal category K, define the* centre *of K, denoted $\mathsf{Z}(K)$, to be the subcategory of K consisting of all the objects of K and the central morphisms.*

For an example of the centre of a premonoidal category, consider Example 1 for the case of C being the category Set of small sets, with symmetric monoidal structure given by finite products. Suppose S has at least two elements. Then the centre of K is precisely Set. In general, given a strong monad on a symmetric monoidal category, the base category C need not be the centre of $Kl(T)$, but, modulo a faithfulness condition sometimes called the mono requirement [10,19], must be a subcategory of the centre.

The functors h_X and k_X preserve central maps. So we have

Proposition 1. *The centre of a premonoidal category is a monoidal category.*

This proposition allows us to prove a coherence result for premonoidal categories, directly generalising the usual coherence result for monoidal categories. Details appear in [19].

Definition 5. *A* symmetry *for a premonoidal category is a central natural isomorphism with components $c : X \otimes Y \longrightarrow Y \otimes X$, satisfying the two conditions $c^2 = 1$ and equality of the evident two maps from $(X \otimes Y) \otimes Z$ to $Z \otimes (X \otimes Y)$. A* symmetric *premonoidal category is a premonoidal category together with a symmetry.*

All of the examples of premonoidal categories we have discussed so far are symmetric.

Definition 6. *A* strict premonoidal functor *is a functor that preserves all the structure and sends central maps to central maps.*

One may similarly generalise the definition of strict symmetric monoidal functor to strict symmetric premonoidal functor. This all allows us to define the notion of a *Freyd*-category.

Definition 7. *A Freyd-category is a category C_0 with finite products, a symmetric premonoidal category C_1, and an identity-on-objects strict symmetric premonoidal functor $J : C_0 \longrightarrow C_1$.*

It follows from the definition of *Freyd*-category that every map in C_0 must be sent by J to a map in the centre $Z(C_1)$ of C_1. So it is generally safe to think of C_0 as an identity-on-objects subcategory of central maps of C_1.

Definition 8. *A Freyd-category $J : C_0 \longrightarrow C_1$ is* closed *if for every object X of C_0 (equivalently of C_1), the functor*

$$J(- \times X) : C_0 \longrightarrow C_1$$

has a right adjoint $X \to -$.

It is proved but only stated implicitly in [19] and it is stated explicitly in [9,18] that we have

Theorem 1. *To give a category C with finite products and a strong monad on it, such that Kleisli exponentials exist, is equivalent to giving a closed Freyd-category $J : C \longrightarrow C'$.*

This all means that the class of closed *Freyd*-categories provides a sound a complete class of models for the computational λ-calculus, and, using the evident notion of its first order fragment (including *let* of course), the class of *Freyd*-categories is a sound and complete class of models for its first order fragment.

3 Enrichment in $[\to, \boldsymbol{Set}]$

In this section, we describe enriched categories, in particular with respect to enrichment in $[\to, Set]$, we outline some of the main definitions and results that are relevant to us here, and we show how those definitions correspond, in the particular case at hand, to natural concepts that we will need later. The standard reference for enriched categories is [7].

For simplicity of exposition, we shall restrict our attention to enrichment in a complete and cocomplete cartesian (rather than just monoidal or symmetric monoidal) closed category V.

Definition 9. *A V-category C consists of*

- *a set ObC of* objects
- *for every pair (X, Y) of objects of C, an object $C(X, Y)$ of V*
- *for every object X of C, a map $\iota : 1 \longrightarrow C(X, X)$*
- *for every triple (X, Y, Z), a map*

$$\cdot : C(Y, Z) \times C(X, Y) \longrightarrow C(X, Z)$$

subject to an associativity axiom for $\cdot$ and an axiom making ι a left and right unit for $\cdot$.

The leading example has $V = Set$, in which case the notion of V-category agrees exactly with the usual notion of (locally small) category. Other standard examples involve $V = Poset$, yielding (locally small) locally ordered categories, and $V = Cat$, yielding (locally small) 2-categories. The example of primary interest to us has $V = [\rightarrow, Set]$: an object of this category consists of a pair of sets (X_0, X_1) and a function from one to the other, $f : X_0 \longrightarrow X_1$, and an arrow amounts to a commutative square in Set. Products in this category are given pointwise; the closed structure is more complicated.

Proposition 2. *To give an $[\rightarrow, Set]$-category is equivalent to giving a pair of categories and an identity-on-objects functor $J : C_0 \longrightarrow C_1$.*

Proof. Given an $[\rightarrow, Set]$-category C, put $ObC_0 = ObC_1 = ObC$. For any pair (X, Y) of objects of C, the data for an $[\rightarrow, Set]$-category gives us an object $C(X, Y)$ of $[\rightarrow, Set]$, i.e., a pair of sets and a function $f : A \longrightarrow B$. So *define* $C_0(X, Y) = A$ and $C_1(X, Y) = B$, and *define* the behaviour of the putative functor $J : C_0 \longrightarrow C_1$ on the homset $C_0(X, Y)$ to be $f : C_0(X, Y) \longrightarrow C_1(X, Y)$. The rest of the data and the axioms for an $[\rightarrow, Set]$-category provide the rest of the data and axioms to make C_0 and C_1 into categories and to make J functorial. The converse follows by similarly routine calculation.

Based on this result, we henceforth identity the notion of $[\rightarrow, Set]$-category with a pair of categories and an identity-on-objects functor $J : C_0 \longrightarrow C_1$.

In general, every V-category C has an underlying ordinary category $U(C)$ defined by $Ob(U(C)) = ObC$ and with the homset $(UC)(X, Y)$ defined to be the set of maps in V from the terminal object 1 to $C(X, Y)$. The composition of the V-category C routinely induces a composition for $U(C)$, and similarly for the identity maps.

Proposition 3. *The underlying ordinary category of an $[\rightarrow, Set]$-category $J : C_0 \longrightarrow C_1$ is the category C_0.*

This result follows from routine checking.

There is a general notion of what it means for an enriched category to have conical colimits. There is a more general notion of weighted colimit, which is the definitive notion of colimit for an enriched category, but it is complicated and we do not need it for this paper, So we just explain the notion of conical colimit here. The following is a little ad hoc, but it is equivalent to the general elegant tretment in [7].

Given a V-category C and a small ordinary category L, one can construct a V-category $[L, C]$. An object of $[L, C]$ is a functor from L to $U(C)$. Given functors $H, K : L \longrightarrow U(C)$, one defines the homobject $[L, C](H, K)$ of V to be an equaliser in V of two maps of the form

$$\Pi_{ObL} C(HX, KX) \longrightarrow \Pi_{ArrL} C(HX, KY)$$

one determined by postcomposition with Kf, the other given by precomposition with Hf, for each map f in L, thus internalising the notion of natural transformation. When $V = Set$, this construction agrees with the usual definition of the functor category.

Definition 10. *For an arbitrary V-category C and a small ordinary category L, given a functor $H : L \longrightarrow U(C)$, a* conical colimit *of H is a* cocone over H *with vertex defined to be $colimH$, such that composition with the* cocone *yields, for every object X of C, an isomorphism in V of the form*

$$C(colimH, X) \cong [L, C](H, \Delta X)$$

A V-category C is said to have all conical colimits if, for every small category L, every functor $H : L \longrightarrow U(C)$ has a conical colimit. If $V = Set$, this definition agrees with the usual notion of cocompletenss of a locally small category. But we can also characterise this definition when $V = [\to, Set]$.

Proposition 4. *An $[\to, Set]$-category $J : C_0 \longrightarrow C_1$ has all conical colimits if and only if C_0 has all colimits and J preserves all colimits.*

Proof. Suppose the $[\to, Set]$-category $J : C_0 \longrightarrow C_1$ has all conical colimits. In general, if an arbitrary V-category has all conical colimits, it follows that its underlying ordinary category has all colimits. So C_0 has all colimits. Now, by direct use of the definition of conical colimits in the case of $V = [\to, Set]$, it follows that J must preserve them. The converse holds by direct calculation.

Summarising, in this section we have shown that if $V = [\to, Set]$, a V-category is exactly a pair of categories (C_0, C_1) and an identity-on-objects functor $J : C_0 \longrightarrow C_1$, that the underlying ordinary category of J is C_0, and that J has conical colimits as an $[\to, Set]$-category if and only if C_0 is cocomplete and J preserves colimits.

4 The Yoneda Embedding

In this section, we investigate the enriched Yoneda embedding $Y : C \longrightarrow [C^{op}, V]$ in the setting of $V = [\to, Set]$. To do that, we first observe that V itself has the structure of a V-category, with homobject $V(X, Y)$ given by the exponential Y^X of V. This yields the following result in the case of $V = [\to, Set]$.

Proposition 5. *The cartesian closed category $[\to, Set]$ extends canonically to the $[\to, Set]$-category*

$$inc : [\to, Set] \longrightarrow [\to, Set]_1$$

where $[\to, Set]_1(f : X \longrightarrow Y, f' : X' \longrightarrow Y')$ is defined to be the set of functions from Y to Y'. The behaviour of the functor inc is evident.

For any small V-category C, one has a functor V-category $[C, V]$: an object is a V-functor from C to V and the homobject $[C, V](H, K)$ is given by an equaliser extending that used to define $[L, C]$ above. Details appear in [7]. We spell out the situation in the case of $V = [\to, Set]$ here.

Proposition 6. *Given a $[\to, Set]$-category $J : C_0 \longrightarrow C_1$, the functor $[\to, Set]$-category $[J, [\to, Set]]$ is defined as follows: an object consists of*

- *a functor* $H_0 : C_0 \longrightarrow Set$
- *a functor* $H_1 : C_1 \longrightarrow Set$
- *a natural transformation* $\phi : H_0 \Rightarrow H_1 J$

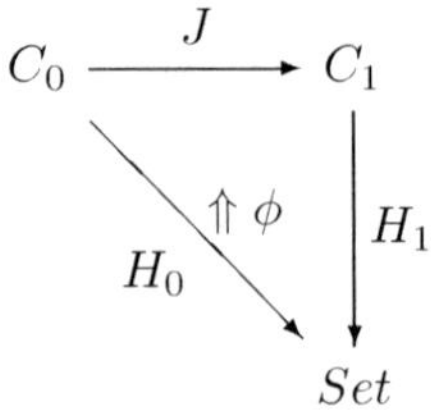

An arrow in $[J, [\rightarrow, Set]]_0$ *from* (H_0, K_0, ϕ_0) *to* (H_1, K_1, ϕ_1) *consists of a pair of natural transformations* $(H_0 \Rightarrow K_0, H_1 \Rightarrow K_1)$ *making the evident diagram involving the* ϕ*'s commute. An arrow in* $[J, [\rightarrow, Set]]_1$ *between the same objects consists of a natural transformation* $H_1 \Rightarrow K_1$*. Composition and the behaviour of the identity-on-objects functor are evident.*

Proof. This follows quite directly from consideration of the definition of the functor V-category $[C, V]$ where $V = [\rightarrow, Set]$. An object of $[C, V]$ in this setting consists of an $[\rightarrow, Set]$-functor from J to $[\rightarrow, Set]$ regarded as a $[\rightarrow, Set]$-category using the construction of Proposition 5. Such a functor assigns, to each object X of C_0, equivalently each object X of C_1, an arrow in Set, giving precisely the data for the object parts of H_0 and H_1 and the natural transformation ϕ. The behaviour of the $[\rightarrow, Set]$-functor on homs is equivalent to the behaviour of H_0 and H_1 on arrows. And the various axioms for a $[\rightarrow, Set]$-functor are equivalent to functoriality of H_0 and H_1 and naturality of ϕ. Similarly routine calculations yield the characterisations of the two sorts of arrow in the functor $[\rightarrow, Set]$-category.

We can further characterise this $[\rightarrow, Set]$-category by means of a lax colimit in the 2-category Cat [1].

Definition 11. *Given a functor* $J : C_o \longrightarrow C_1$*, denote by* $l(J)$ *the category determined by being universal of the form*

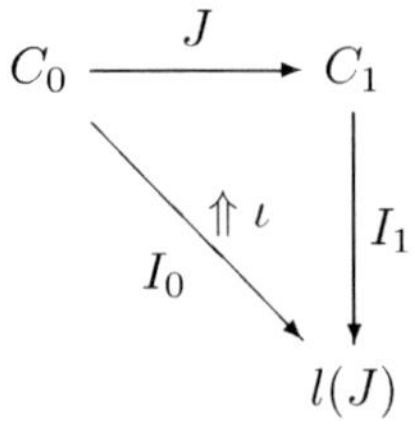

I.e., for every such diagram with an arbitrary vertex D*, there is a unique functor from* $l(J)$ *to* D *making corresponding functors and natural transformations agree.*

One can provably extend the condition of the definition uniquely to yield an isomorphism of categories between a category with such lax cocones with vertex

D as objects and the functor category from $l(J)$ to D. It is easy to construct $l(J)$: it is freely generated by having C_0 and C_1 as full subcategories, together with, for each object X of C_0, an arrow from $I_0(X)$ to $I_1(X)$, subject to the collection of such arrows being made natural in C_0. Note that the coprojections I_0 and I_1 are fully faithful. The universal property tells us that to give an object of $[J, [\to, Set]]$ is equivalent to giving a functor from $l(J)$ to Set, allowing us to deduce the following proposition.

Proposition 7. *The functor $[\to, Set]$-category $[J, [\to, Set]]$ is given by the identity-on-objects/fully faithful factorisation of*

$$[I_1, Set] : [l(J), Set] \longrightarrow [C_1, Set]$$

i.e., $[J, [\to, Set]]_0$ is isomorphic to $[l(J), Set]$, and $[J, [\to, Set]]_1$ and the identity-on-objects functor are given by the identity-on-objects/fully faithful factorisation of $[I_1, Set]$.

Finally, we investigate the enriched Yoneda embedding $Y : C \longrightarrow [C^{op}, V]$ when $V = [\to, Set]$. Given an $[\to, Set]$-category $J : C_0 \longrightarrow C_1$, the Yoneda embedding is an $[\to, Set]$-functor from J to $[J^{op}, [\to, Set]]$, i.e., a pair of functors

$$(Y_0 : C_0 \longrightarrow [J^{op}, [\to, Set]]_0, Y_1 : C_1 \longrightarrow [J^{op}, [\to, Set]]_1)$$

subject to the evident commutativity, where we can characterise the codomain categories using Proposition 7.

Proposition 8. *The functor $Y_0 : C_0 \longrightarrow [J^{op}, [\to, Set]]_0$ is the composite of the ordinary Yoneda embedding $Y : C_0 \longrightarrow [C_0^{op}, Set]$ with the (fully faithful) functor $Lan_{I_0} : [C_0^{op}, Set] \longrightarrow [l(J^{op}), Set] = [J^{op}, [\to, Set]]_0$. And the functor $Y_1 : C_1 \longrightarrow [J^{op}, [\to, Set]]_1$ is given by the defining property of a factorisation system applied to the square*

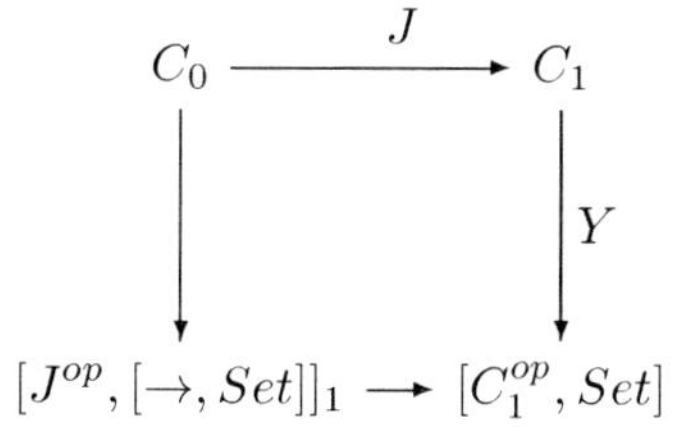

where the bottom (fully faithful) functor is given by (applying $(-)^{op}$ to) Proposition 7 and the left-hand functor is given by the composite of Y_0 with the identity-on-objects functor determined by (applying $(-)^{op}$ to) Proposition 7. The definition of a factorisation system (this can be proved directly) yields a unique functor from C_1 to $[J^{op}, [\to, Set]]_1$ making both triangles commute.

Proof. The enriched Yoneda embedding takes an object X of C_0, equivalently of C_1, to the $[\to, Set]$-functor $J(-, X) : J^{op} \longrightarrow [\to, Set]$, which may be described as the pair of functors

$$(C_0(-, X) : C^{op} \longrightarrow Set, C_1(-, X) : C_1^{op} \longrightarrow Set)$$

together with the natural transformation from the first to the second determined by J. But it is routine to verify that $C_1(-,X) : C_1^{op} \longrightarrow Set$ is the left Kan extension of $C_0(-,X)$ along J^{op}. And by composition of left Kan extensions, its left Kan extension along I_1 agrees with $Lan_{I_0}C_0(-,X)$. And, since I_0 and I_1 are fully faithful, it follows that $Lan_{I_0}C_0(-,X)$ commutes with both $C_0(-,X)$ and $C_1(-,X)$, respecting ι. This proves the characterisation we claim for Y_0, and that for Y_1 follows because, by its definition, it makes the two triangles commute.

The behaviour of the two functors on maps follows routinely if we can see that Lan_{I_0} and Lan_{I_1} are fully faithful. The proof is the same for both, so let us just consider I_0. Since I_0 is fully faithful, it follows (see for instance [7]) as used above that for any functor $H : C_0^{op} \longrightarrow Set$, we have that H_0 is coherently isomorphic to the composite $(Lan_{I_0}H_0)I_0$. But $Lan_{I_0} : [C_0^{op}, Set] \longrightarrow [l(J^{op}), Set]$ has a right adjoint given by sending a functor to its composite with I_0, and the above-mentioned isomorphism tells us that the unit of the adjunction is an isomorphism, and hence that the adjunction is a coreflection, and hence that Lan_{I_0} is fully faithful.

5 The Free Conical Colimit Completion of a Small $[\rightarrow, Set]$-Category

In this section, we characterise the conical colimit completion of a small V-category in the setting where $V = [\rightarrow, Set]$. If V were Set, the small conical colimit completion of a small V-category C would be exactly $[C^{op}, Set]$, but that is not true for general V, and in particular, it is not true for $V = [\rightarrow, Set]$, as we shall see. But why should we consider this at all in the context of this paper?

In general, for a small V-category C, the functor V-category $[C^{op}, V]$ has finite products, indeed all limits and colimits. We have already characterised conical colimits in the setting of $V = [\rightarrow, Set]$ in Proposition 4. A dual result applies to conical limits. And finite products are conical limits. So, for any small $[\rightarrow, Set]$-category $J : C_0 \longrightarrow C_1$, it follows that in the presheaf $[\rightarrow, Set]$-category

$$[J^{op}, [\rightarrow, Set]]_0 \longrightarrow [J^{op}, [\rightarrow, Set]]_1$$

the category $[J^{op}, [\rightarrow, Set]]_0$ has and the functor preserves finite products. So it does not have non-trivial $Freyd$-structure. So, in particular, it does not provide a closed $Freyd$-category into which J, equipped with a non-trivial $Freyd$-structure, can embed as a $Freyd$-category. So we cannot generalise the embedding of a category with finite products into a cartesian closed category, or that of a symmetric monoidal category into a symmetric monoidal closed category, as used for simply typed λ-calculus and linear λ-calculus respectively, with references as in the introduction, simply by immediate reference to enrichment of the Yoneda embedding to the setting of $V = [\rightarrow, Set]$ in order to embed a $Freyd$-category, i.e., a model of the first order fragment of the computational

λ-calculus, into a closed *Freyd*-category, i.e., into a model of the whole of the computational λ-calculus. We provably (by the above) need to add some measure of subtlety. That subtlety is given by restricting the Yoneda embedding to the conical colimit completion. An instance of one of the central theorems of [7] is the following.

Theorem 2. *The free conical colimit completion of a small V-category C is given by the closure of C in $[C^{op}, V]$ with respect to the Yoneda embedding $Y : C \longrightarrow [C^{op}, V]$ under conical colimits.*

We use this result to characterise the conical colimit completion of a small $[\rightarrow, Set]$-category $J : C_0 \longrightarrow C_1$.

Theorem 3. *The free conical colimit completion of a small $[\rightarrow, Set]$-category $J : C_0 \longrightarrow C_1$ is given by*

- *the category $[C_0^{op}, Set]$*
- *the identity-on-objects/fully faithful factorisation of*

$$Lan_{J^{op}} : [C_0^{op}, Set] \longrightarrow [C_1^{op}, Set]$$

Proof. First observe, using Proposition 4, that this $[\rightarrow, Set]$-category has conical colimits: $[C_0^{op}, Set]$ is cocomplete and $Lan_{J^{op}}$ has a right adjoint, and factoring a colimit preserving functor into an identity-on-objects functor followed by a fully faithful functor makes the former also preserve all colimits. Now observe that the canonical $[\rightarrow, Set]$-functor into $[J^{op}, [\rightarrow, Set]]$ preserves colimits: the canonical $[\rightarrow, Set]$-functor is given by Lan_{I_0}, which has a right adjoint, so preserves colimits, together with the functor determined by the universal property of a factorisation system applied to the commutative square

$$\begin{array}{ccccc}
[C_0^{op}, Set] & \longrightarrow & C' & \longrightarrow & [C_1^{op}, Set] \\
{\scriptstyle Lan_{I_0}} \downarrow & & & & \downarrow {\scriptstyle id} \\
[J^{op}, [\rightarrow, Set]]_0 & \longrightarrow & [J^{op}, [\rightarrow, Set]]_1 & \longrightarrow & [C_1^{op}, Set]
\end{array}$$

where the top and bottow rows are given by identity-on-objects/fully faithful factorisations, and where the diagram commutes by calculations with left Kan extensions and using fully faithfulness of I_1. The canonical $[\rightarrow, Set]$-functor is fully faithful: we established fully faithfulness of Lan_{I_0} in the previous section, and the intermediary functor as above is fully faithful as composite with the bottom right-hand functor in the diagram is fully faithful. Next observe, by Proposition 8, that the Yoneda embedding factors through the canonical $[\rightarrow, Set]$-functor. Finally, observe that every object of this full sub-$[\rightarrow, Set]$-category is generated by a conical colimit of representables: that is routine as every functor $H : C_0^{op} \longrightarrow Set$ is a conical colimit of representables (see for instance [7]). Combining all these observations yields the result.

6 The Universal Property

The central technical work of the paper has now been done, culminating in Theorem 2. In [17], we gave an ad hoc description of an embedding of something very similar to a small *Freyd*-category into something very similar to a closed *Freyd*-category; that construction agrees with the construction of Theorem 2, and now it is little more than an observation that the proof in [17] extends to yield a precise statement of a universal characterisation of the construction. The point of this paper is that the ad hoc embedding of [17] is a definitively natural category theoretic construct.

There is just one delicate point here: exactly what do we mean by the "free conical colimit-complete closed *Freyd*-category on a *Freyd*-category? The point to note is that, by a map of closed *Freyd*-categories, we mean a map that preserves the *Freyd*-structure but need not preserve the closed structure. This should not come as a great surprise: the maps of primary interest between cartesian closed categories are functors that preserve finite products but need not preserve the closed structure; most forgetful functors to *Set* are examples. It is also consistent with the work of [8] on data refinement.

Theorem 4. *The free conical colimit completion of a small Freyd-category* $J : C_0 \longrightarrow C_1$ *is the free conical colimit-complete closed Freyd-category on* J*, i.e., the* $[\to, Set]$*-category of Theorem 2 with a natural Freyd-structure.*

Proof. Theorem 2 characterises the free conical colimit completion of any $[\to, Set]$-category $J : C_0 \longrightarrow C_1$. We need only show that that construction acts as we wish with respect to *Freyd*-structure. But $[C_0^{op}, Set]$ is cartesian closed, with $Y : C_0 \longrightarrow [C^{op}, Set]$ preserving finite products, and $Lan_{J^{op}}$ has a right adjoint. So the only remaining non-trivial point is to construct, for functors $F, H, K : C_0^{op} \longrightarrow Set$ and for every natural transformation $\alpha : Lan_{J^{op}} H \Rightarrow Lan_{J^{op}} K$, a natural transformation $Lan_{J^{op}}(F \times H) \Rightarrow Lan_{J^{op}}(F \times K)$; and that must be done coherently. But to do that, we just make two uses of the fact that $[C^{op}, Set]$ is the free colimit completion of C. It follows from this free cocompleteness that, for any object X of C_0, equivalently of C_1, the functor $X \otimes - : C_1 \longrightarrow C_1$ extends to $[C_1^{op}, Set]$. This yields $F \otimes \alpha$ for any representable $F = C_0(-, X)$. For an arbitrary F, one deduces the construction by use of symmetry and by centrality of the maps in the canonical colimiting cocone of F.

References

1. R. Blackwell, G.M. Kelly, and A.J. Power. Two-dimensional monad theory. *J. Pure Appl. Algebra*, 59:1–41, 1989.
2. E. Cheng, M. Hyland, and A.J. Power. Pseudo-distributive laws. Submitted.
3. B.J. Day. On closed categories of functors. . In *Midwest Category Seminar Reports IV*, volume 137 of *Lectures Notes in Mathematics*, pages 1–38, 1970.

4. M. Fiore, G.D. Plotkin, and D. Turi. Abstract syntax and variable binding. In *Proceedings of 14th Symposium on Ligic in Computer Science*, pages 214–224, IEEE Computer Society Press, 1999.
5. Guen Bin Im and G.M. Kelly. A universal property of the convolution monoidal structure. *J. Pure Appl. Algebra*, 43:75–88, 1986.
6. A. Joyal, M. Nielsen, and G. Winskel. Bisimulation from open maps. *Information and Computation*, 127: 164–185, 1996.
7. G.M. Kelly. *Basic concepts of enriched category theory*, volume 64 of *London Mathematical Society Lecture Notes Series*, Cambridge University Press, 1982.
8. Y. Kinoshita and A.J. Power. Data-refinement in call-by-value languages. In *Proceedings of Computer Science Logic*, in volume 1683 of *Lecture Notes in Computer Science*, pages 562–576, 1999.
9. P.B. Levy, A.J. Power, and H. Thielecke. Modelling environments in call-by-value programming languages. *Theoretical Computer Science*, to appear.
10. E. Moggi. Computational lambda-calculus and monads. In *Proceedings of 4th Symposium on Logic in Computer Science*, pages 14–32, IEEE Computer Society Press, 1989.
11. E. Moggi. Notions of computation and monads. *Information and Computation*, 93:55–92, 1991.
12. M. Nygaard and G. Winskel. Linearity in Process Languages. In *Proceedings of 17th Symposium on Logic in Computer Science*, pages 433–444, IEEE Computer Society Press, 2002.
13. P.W. O'Hearn and D.J. Pym. The logic of bunched implications. *Bull. Symbolic Logic* 5:215–403, 1999.
14. G.D. Plotkin and A.J. Power. Adequacy for algebraic effects. In *Proceedings of Foundations of Software Science and Compuation Structures*, in volume 2030 of *Lecture Notes in Computer Science*, pages 1–24, 2001.
15. G.D. Plotkin and A.J. Power. Semantics for algebraic operations. In *Proceedings of Mathematical Foundations of Programming Semantics*, in volume 45 of *Electronic Notes in Theoretical Computer Science*, 2001.
16. G.D. Plotkin and A.J. Power. Algebraic operations and generic effects. *Applied Categorical Structures*, to appear.
17. A.J. Power. Premonoidal categories as categories with algebraic structure. *Theoretical Computer Science*, 278:303–321, 2002.
18. A.J. Power. Models for the computational λ-calculus. In *Proceedings of 1st Irish Conferenence on Mathematical Foundations of Computer Science and Information Technology*, in volume 40 of *Electronic Notes in Theoretical Computer Science*, 40, 2001.
19. A.J. Power and E.P. Robinson. Premonoidal categories and notions of computation. In: *Proc. LDPL 96, Math Structures in Computer Science* 7:453–468, 1997.
20. Miki Tanaka. Abstract syntax and variable binding for linear binders. In *Proceedings of 25th Mathematical Foundations of Computer Science*, in volume 1893 of *Lecture Notes in Computer Science*, pages 670–679, 2000.
21. Hayo Thielecke. Continuation passing style and self-adjointness. In *Proceedings 2nd ACM SIGPLAN Workshop on Continuations*, number NS-96-13 in BRICS Notes Series, December 1996.
22. Hayo Thielecke. Continuations semantics and self-adjointness. In *Proceedings of 13th Conference on Mathematical Foundations of Programming Semantics*, in volume 6 of *Electronic Notes in Theoretical Computer Science*.

Author Index

Lecture Notes in Computer Science

For information about Vols. 1–2592

please contact your bookseller or Springer-Verlag

Vol. 2593: A.B. Chaudhri, M. Jeckle, E. Rahm, R. Unland (Eds.), Web, Web-Services, and Database Systems. Proceedings, 2002. XI, 311 pages. 2003.

Vol. 2594: A. Asperti, B. Buchberger, J.H. Davenport (Eds.), Mathematical Knowledge Management. Proceedings, 2003. X, 225 pages. 2003.

Vol. 2595: K. Nyberg, H. Heys (Eds.), Selected Areas in Cryptography. Proceedings, 2002. XI, 405 pages. 2003.

Vol. 2596: A. Coen-Porisini, A. van der Hoek (Eds.), Software Engineering and Middleware. Proceedings, 2002. XII, 239 pages. 2003.

Vol. 2597: G. Păun, G. Rozenberg, A. Salomaa, C. Zandron (Eds.), Membrane Computing. Proceedings, 2002. VIII, 423 pages. 2003.

Vol. 2598: R. Klein, H.-W. Six, L. Wegner (Eds.), Computer Science in Perspective. X, 357 pages. 2003.

Vol. 2599: E. Sherratt (Ed.), Telecommunications and beyond: The Broader Applicability of SDL and MSC. Proceedings, 2002. X, 253 pages. 2003.

Vol. 2600: S. Mendelson, A.J. Smola, Advanced Lectures on Machine Learning. Proceedings, 2002. IX, 259 pages. 2003. (Subseries LNAI).

Vol. 2601: M. Ajmone Marsan, G. Corazza, M. Listanti, A. Roveri (Eds.) Quality of Service in Multiservice IP Networks. Proceedings, 2003. XV, 759 pages. 2003.

Vol. 2602: C. Priami (Ed.), Computational Methods in Systems Biology. Proceedings, 2003. IX, 214 pages. 2003.

Vol. 2603: A. Garcia, C. Lucena, F. Zambonelli, A. Omicini, J. Castro (Eds.), Software Engineering for Large-Scale Multi-Agent Systems. XIV, 285 pages. 2003.

Vol. 2604: N. Guelfi, E. Astesiano, G. Reggio (Eds.), Scientific Engineering for Distributed Java Applications. Proceedings, 2002. X, 205 pages. 2003.

Vol. 2606: A.M. Tyrrell, P.C. Haddow, J. Torresen (Eds.), Evolvable Systems: From Biology to Hardware. Proceedings, 2003. XIV, 468 pages. 2003.

Vol. 2607: H. Alt, M. Habib (Eds.), STACS 2003. Proceedings, 2003. XVII, 700 pages. 2003.

Vol. 2609: M. Okada, B. Pierce, A. Scedrov, H. Tokuda, A. Yonezawa (Eds.), Software Security – Theories and Systems. Proceedings, 2002. XI, 471 pages. 2003.

Vol. 2610: C. Ryan, T. Soule, M. Keijzer, E. Tsang, R. Poli, E. Costa (Eds.), Genetic Programming. Proceedings, 2003. XII, 486 pages. 2003.

Vol. 2611: S. Cagnoni, J.J. Romero Cardalda, D.W. Corne, J. Gottlieb, A. Guillot, E. Hart, C.G. Johnson, E. Marchiori, J.-A. Meyer, M. Middendorf, G.R. Raidl (Eds.), Applications of Evolutionary Computing. Proceedings, 2003. XXI, 708 pages. 2003.

Vol. 2612: M. Joye (Ed.), Topics in Cryptology – CT-RSA 2003. Proceedings, 2003. XI, 417 pages. 2003.

Vol. 2613: F.A.P. Petitcolas, H.J. Kim (Eds.), Digital Watermarking. Proceedings, 2002. XI, 265 pages. 2003.

Vol. 2614: R. Laddaga, P. Robertson, H. Shrobe (Eds.), Self-Adaptive Software: Applications. Proceedings, 2001. VIII, 291 pages. 2003.

Vol. 2615: N. Carbonell, C. Stephanidis (Eds.), Universal Access. Proceedings, 2002. XIV, 534 pages. 2003.

Vol. 2616: T. Asano, R. Klette, C. Ronse (Eds.), Geometry, Morphology, and Computational Imaging. Proceedings, 2002. X, 437 pages. 2003.

Vol. 2617: H.A. Reijers (Eds.), Design and Control of Workflow Processes. Proceedings, 2002. XV, 624 pages. 2003.

Vol. 2618: P. Degano (Ed.), Programming Languages and Systems. Proceedings, 2003. XV, 415 pages. 2003.

Vol. 2619: H. Garavel, J. Hatcliff (Eds.), Tools and Algorithms for the Construction and Analysis of Systems. Proceedings, 2003. XVI, 604 pages. 2003.

Vol. 2620: A.D. Gordon (Ed.), Foundations of Software Science and Computation Structures. Proceedings, 2003. XII, 441 pages. 2003.

Vol. 2621: M. Pezzè (Ed.), Fundamental Approaches to Software Engineering. Proceedings, 2003. XIV, 403 pages. 2003.

Vol. 2622: G. Hedin (Ed.), Compiler Construction. Proceedings, 2003. XII, 335 pages. 2003.

Vol. 2623: O. Maler, A. Pnueli (Eds.), Hybrid Systems: Computation and Control. Proceedings, 2003. XII, 558 pages. 2003.

Vol. 2624: H.G. Dietz (Ed.), Languages and Compilers for Parallel Computing. Proceedings, 2001. XI, 444 pages. 2003.

Vol. 2625: U. Meyer, P. Sanders, J. Sibeyn (Eds.), Algorithms for Memory Hierarchies. Proceedings, 2003. XVIII, 428 pages. 2003.

Vol. 2626: J.L. Crowley, J.H. Piater, M. Vincze, L. Paletta (Eds.), Computer Vision Systems. Proceedings, 2003. XIII, 546 pages. 2003.

Vol. 2627: B. O'Sullivan (Ed.), Recent Advances in Constraints. Proceedings, 2002. X, 201 pages. 2003. (Subseries LNAI).

Vol. 2628: T. Fahringer, B. Scholz, Advanced Symbolic Analysis for Compilers. XII, 129 pages. 2003.

Vol. 2631: R. Falcone, S. Barber, L. Korba, M. Singh (Eds.), Trust, Reputation, and Security: Theories and Practice. Proceedings, 2002. X, 235 pages. 2003. (Subseries LNAI).

Vol. 2632: C.M. Fonseca, P.J. Fleming, E. Zitzler, K. Deb, L. Thiele (Eds.), Evolutionary Multi-Criterion Optimization. Proceedings, 2003. XV, 812 pages. 2003.

Vol. 2633: F. Sebastiani (Ed.), Advances in Information Retrieval. Proceedings, 2003. XIII, 546 pages. 2003.

Vol. 2634: F. Zhao, L. Guibas (Eds.), Information Processing in Sensor Networks. Proceedings, 2003. XII, 692 pages. 2003.

Vol. 2636: E. Alonso, D, Kudenko, D. Kazakov (Eds.), Adaptive Agents and Multi-Agent Systems. XIV, 323 pages. 2003. (Subseries LNAI).

Vol. 2637: K.-Y. Whang, J. Jeon, K. Shim, J. Srivastava (Eds.), Advances in Knowledge Discovery and Data Mining. Proceedings, 2003. XVIII, 610 pages. 2003. (Subseries LNAI).

Vol. 2638: J. Jeuring, S. Peyton Jones (Eds.), Advanced Functional Programming. Proceedings, 2002. VII, 213 pages. 2003.

Vol. 2639: G. Wang, Q. Liu, Y. Yao, A. Skowron (Eds.), Rough Sets, Fuzzy Sets, Data Mining, and Granular Computing. Proceedings, 2003. XVII, 741 pages. 2003. (Subseries LNAI).

Vol. 2641: P.J. Nürnberg (Ed.), Metainformatics. Proceedings, 2002. VIII, 187 pages. 2003.

Vol. 2642: X. Zhou, Y. Zhang, M.E. Orlowska (Eds.), Web Technologies and Applications. Proceedings, 2003. XIII, 608 pages. 2003.

Vol. 2643: M. Fossorier, T. Høholdt, A. Poli (Eds.), Applied Algebra, Algebraic Algorithms and Error-Correcting Codes. Proceedings, 2003. X, 256 pages. 2003.

Vol. 2644: D. Hogrefe, A. Wiles (Eds.), Testing of Communicating Systems. Proceedings, 2003. XII, 311 pages. 2003.

Vol. 2645: M.A. Wimmer (Ed.), Knowledge Management in Electronic Government. Proceedings, 2003. XI, 320 pages. 2003. (Subseries LNAI).

Vol. 2646: H. Geuvers, F, Wiedijk (Eds.), Types for Proofs and Programs. Proceedings, 2002. VIII, 331 pages. 2003.

Vol. 2647: K.Jansen, M. Margraf, M. Mastrolli, J.D.P. Rolim (Eds.), Experimental and Efficient Algorithms. Proceedings, 2003. VIII, 267 pages. 2003.

Vol. 2648: T. Ball, S.K. Rajamani (Eds.), Model Checking Software. Proceedings, 2003. VIII, 241 pages. 2003.

Vol. 2649: B. Westfechtel, A. van der Hoek (Eds.), Software Configuration Management. Proceedings, 2003. VIII, 241 pages. 2003.

Vol. 2651: D. Bert, J.P. Bowen, S. King, M, Waldén (Eds.), ZB 2003: Formal Specification and Development in Z and B. Proceedings, 2003. XIII, 547 pages. 2003.

Vol. 2652: F.J. Perales, A.J.C. Campilho, N. Pérez de la Blanca, A. Sanfeliu (Eds.), Pattern Recognition and Image Analysis. Proceedings, 2003. XIX, 1142 pages. 2003.

Vol. 2653: R. Petreschi, Giuseppe Persiano, R. Silvestri (Eds.), Algorithms and Complexity. Proceedings, 2003. XI, 289 pages. 2003.

Vol. 2656: E. Biham (Ed.), Advances in Cryptology – EUROCRPYT 2003. Proceedings, 2003. XIV, 649 pages. 2003.

Vol. 2657: P.M.A. Sloot, D. Abramson, A.V. Bogdanov, J.J. Dongarra, A.Y. Zomaya, Y.E. Gorbachev (Eds.), Computational Science – ICCS 2003. Proceedings, Part I. 2003. LV, 1095 pages. 2003.

Vol. 2658: P.M.A. Sloot, D. Abramson, A.V. Bogdanov, J.J. Dongarra, A.Y. Zomaya, Y.E. Gorbachev (Eds.), Computational Science – ICCS 2003. Proceedings, Part II. 2003. LV, 1129 pages. 2003.

Vol. 2659: P.M.A. Sloot, D. Abramson, A.V. Bogdanov, J.J. Dongarra, A.Y. Zomaya, Y.E. Gorbachev (Eds.), Computational Science – ICCS 2003. Proceedings, Part III. 2003. LV, 1165 pages. 2003.

Vol. 2660: P.M.A. Sloot, D. Abramson, A.V. Bogdanov, J.J. Dongarra, A.Y. Zomaya, Y.E. Gorbachev (Eds.), Computational Science – ICCS 2003. Proceedings, Part IV. 2003. LVI, 1161 pages. 2003.

Vol. 2663: E. Menasalvas, J. Segovia, P.S. Szczepaniak (Eds.), Advances in Web Intelligence. Proceedings, 2003. XII, 350 pages. 2003. (Subseries LNAI).

Vol. 2665: H. Chen, R. Miranda, D.D. Zeng, C. Demchak, J. Schroeder, T. Madhusudan (Eds.), Intelligence and Security Informatics. Proceedings, 2003. XIV, 392 pages. 2003.

Vol. 2667: V. Kumar, M.L. Gavrilova, C.J.K. Tan, P. L'Ecuyer (Eds.), Computational Science and Its Applications – ICCSA 2003. Proceedings, Part I. 2003. XXXIV, 1060 pages. 2003.

Vol. 2668: V. Kumar, M.L. Gavrilova, C.J.K. Tan, P. L'Ecuyer (Eds.), Computational Science and Its Applications – ICCSA 2003. Proceedings, Part II. 2003. XXXIV, 942 pages. 2003.

Vol. 2669: V. Kumar, M.L. Gavrilova, C.J.K. Tan, P. L'Ecuyer (Eds.), Computational Science and Its Applications – ICCSA 2003. Proceedings, Part III. 2003. XXXIV, 948 pages. 2003.

Vol. 2670: R. Peña, T. Arts (Eds.), Implementation of Functional Languages. Proceedings, 2002. X, 249 pages. 2003.

Vol. 2674: I.E. Magnin, J. Montagnat, P. Clarysse, J. Nenonen, T. Katila (Eds.), Functional Imaging and Modeling of the Heart. Proceedings, 2003. XI, 308 pages. 2003.

Vol. 2675: M. Marchesi, G. Succi (Eds.), Extreme Programming and Agile Processes in Software Engineering. Proceedings, 2003. XV, 464 pages. 2003.

Vol. 2676: R. Baeza-Yates, E. Chávez, M. Crochemore (Eds.), Combinatorial Pattern Matching. Proceedings, 2003. XI, 403 pages. 2003.

Vol. 2678: W. van der Aalst, A. ter Hofstede, M. Weske (Eds.), Business Process Management. Proceedings, 2003. XI, 391 pages. 2003.

Vol. 2679: W. van der Aalst, E. Best (Eds.), Applications and Theory of Petri Nets 2003. Proceedings, 2003. XI, 508 pages. 2003.

Vol. 2686: J. Mira, J.R. Álvarez (Eds.), Computational Methods in Neural Modeling. Proceedings, Part I. 2003. XXVII, 764 pages. 2003.

Vol. 2687: J. Mira, J.R. Álvarez (Eds.), Artificial Neural Nets Problem Solving Methods. Proceedings, Part II. 2003. XXVII, 820 pages. 2003.

Vol. 2692: P. Nixon, S. Terzis (Eds.), Trust Management. Proceedings, 2003. X, 349 pages. 2003.

Vol. 2694: R. Cousot (Ed.), Static Analysis. Proceedings, 2003. XIV, 505 pages. 2003.

Vol. 2701: M. Hofmann (Ed.), Typed Lambda Calculi and Applications. Proceedings, 2003. VIII, 317 pages. 2003.

Vol. 2706: R. Nieuwenhuis (Ed.), Rewriting Techniques and Applications. Proceedings, 2003. XI, 515 pages. 2003.

Vol. 2707: K. Jeffay, I. Stoica, K. Wehrle (Eds.), Quality of Service – IWQoS 2003. Proceedings, 2003. XI, 517 pages. 2003.